The Balance of Nature and Human Impact

It is clear that nature is undergoing rapid changes as a result of human activities such as industry, agriculture, travel, fisheries and urbanization. What effects do these activities have? Are they disturbing equilibria in ecological populations and communities, thus upsetting the balance of nature, or are they enhancing naturally occurring disequilibria, perhaps with even worse consequences?

It is often argued that large-scale fluctuations in climate and sea levels have occurred over and over again in the geological past, long before human activities could possibly have had any impact, and that human effects are very small compared to those that occur naturally. Should we conclude that human activity cannot significantly affect the environment, or are these naturally occurring fluctuations actually being dangerously enhanced by humans?

This book examines these questions, first by providing evidence for equilibrium and nonequilibrium conditions in relatively undisturbed ecosystems, and second by examining human-induced effects.

Klaus Rohde is Professor Emeritus at the University of New England, Armidale, Australia. He is well known for his work on the ecology, biogeography and ultrastructure of parasites, particularly marine parasites, and on latitudinal gradients in biodiversity. He has published extensively on parasite ecology, non-equilibrium ecology and marine parasitology.

The Balance of Nature and Human Impact

Edited by

KLAUS ROHDE

University of New England
Armidale, Australia

CAMBRIDGE
UNIVERSITY PRESS

CAMBRIDGE UNIVERSITY PRESS
Cambridge, New York, Melbourne, Madrid, Cape Town,
Singapore, São Paulo, Delhi, Mexico City

Cambridge University Press
The Edinburgh Building, Cambridge CB2 8RU, UK

Published in the United States of America by Cambridge University Press, New York

www.cambridge.org
Information on this title: www.cambridge.org/9781107019614

© Cambridge University Press 2013

This publication is in copyright. Subject to statutory exception
and to the provisions of relevant collective licensing agreements,
no reproduction of any part may take place without the written
permission of Cambridge University Press.

First published 2013

Printed and bound in the United Kingdom by the MPG Books Group

A catalogue record for this publication is available from the British Library

Library of Congress Cataloging-in-Publication Data
The balance of nature and human impact / edited by Klaus Rohde,
University of New England, Armidale, Australia.
pages   cm
Includes index.
ISBN 978-1-107-01961-4
1. Population biology.   2. Biotic communities.   3. Nature – Effect of human beings on.
I. Rohde, Klaus, 1932–
QH352.B34   2013
576.8–dc23

2012028838

ISBN 978-1-107-01961-4 Hardback

Cambridge University Press has no responsibility for the persistence or
accuracy of URLs for external or third-party internet websites referred to
in this publication, and does not guarantee that any content on such
websites is, or will remain, accurate or appropriate.

Contents

Color plates are to be found between pp. 48 and 49.

Foreword

As a somewhat obsessive, but far from expert, sea kayaker, I attend to issues of stability a good deal. Kayakers distinguish two kinds. Primary or initial stability is a measure of how much a kayak rocks in the water when it is displaced from the level. Secondary or final stability is a measure of how readily a kayak capsizes. Beginners instinctively confuse the two. They are concerned that the rocking associated with the primary stability characteristics of a boat reflects its secondary stability, and in overreacting to the former they can overwhelm the latter, and get very wet as a consequence. With practice, one learns to improve the primary stability that is experienced under an increasingly wide variety of conditions of wind and wave, to reduce the likelihood of unintentionally exceeding a kayak's secondary stability, and then to be able to use boats with inherently lower stability of both kinds. In a world in which the abundances and distributions of the majority of organisms are heavily influenced, and often continually buffeted, directly or indirectly by anthropogenic activities, we need to become the population management equivalent of more expert kayakers. This book provides one further step along that course.

Historically, ecologists have spent a lot of time debating whether populations have the equivalent of primary stability (i.e. whether they have their dynamics in some way bounded), how it varies intraspecifically and interspecifically, and what that might mean in turn for the structure and composition of assemblages and communities. Indeed, a plethora of measures have been developed to assess that stability, far in advance of what any typical group of kayakers might think of, but perhaps not dissimilar to the technicalities of boat and ship design. This book reveals that the debate is not over. However, in broad strokes, it is clear that (i) there is an enormous range of population behavior, and it is difficult to categorize in simple terms; (ii) that behavior can be very context specific, with regard to species, space, and time; and (iii) in consequence it can be difficult a priori to predict for any given case without knowing a good deal about those specifics. What is then key to understand is how natural population behavior is reshaped by anthropogenic pressures, just as the wake of a passing vessel can dominate the primary stability experienced by a kayaker, by setting up movements that are commonly at odds with those that were otherwise being felt. Put crudely, do those pressures act in a similar way to those more naturally experienced by a population or are they substantially different in character? This book offers some answers to that question.

More and more, however, the concerns of ecologists are becoming focused on issues of secondary stability. Are there "tipping" points that rapidly throw populations and communities from one state to another (the equivalent of the kayak being the correct or the wrong way up), how can those points be recognized, and under what circumstances do they occur? The challenge here is that the answers may well once again be very context specific, begging the question of whether there are practically (as opposed perhaps to theoretically) useful generalizations, or whether every case has to be considered in its own terms. Again, the chapters of this book have something to say on the issue. For the kayaker, a shift in balance, a paddle stroke and a brace, each exacted at the right time, can prevent the loss of secondary stability or indeed bring that loss about. For the population manager, much the same is likely commonly true for the tools they have at their disposal to influence movements, births, and deaths. For both kayaker and population manager, what is needed is sufficient experience to gauge what actions are most appropriate and when. It has been argued that it takes a skillful kayaker to complete an Eskimo roll when they have lost secondary stability, but far more skill not to lose that stability in the first place.

Of course, for the population manager, as for the kayaker, ultimately what is required is a synthetic understanding of both primary and secondary stability, and how they interact. Only then will it be possible to ensure some degree of understanding or control over the futures of species, assemblages, and communities. This book helps in bringing both sets of issues within the same covers.

Kevin J. Gaston
Environment & Sustainability Institute
University of Exeter

List of contributors

Professor Inger G. Alsos

Tromsø University Museum, University of Tromsø, NO-9037, Tromsø, Norway
inger.g.alsos@uit.no

Associate Professor Nigel R. Andrew

Zoology, University of New England, Armidale 2351, Australia nigel.andrew@une.edu.au

Associate Professor Michael Box

School of Physics, University of New South Wales, Sydney NSW 2052, Australia
M.Box@unsw.edu.au

Professor Christian Brochmann

National Centre for Biosystematics, Natural History Museum, University of Oslo,
P.O. Box 1172 Blindern NO-0318 Oslo, Norway christian.brochmann@nhm.uio.no

Professor Daniel Brooks

Department of Ecology & Evolutionary Biology, University of Toronto, 25 Harbord St.
Toronto, ON Canada, M5S 3G5. Postal address: 1821 Greenbriar Lane, Lincoln, NE
68506, USA dan.brooks@utoronto.ca

Associate Professor Peter J. Clarke

Botany, University of New England, Armidale NSW 2351, Australia
pclarke1@une.edu.au

Professor Yihong Du

Mathematics, School of Science and Technology, University of New England, Armidale
NSW 2351, Australia ydu@turing.une.edu.au

Professor Mary E. Edwards

Geography and Environment, University of Southampton, Highfield, Southampton,
SO17 1BJ, United Kingdom M.E.Edwards@soton.ac.uk

Professor Hugh A. Ford

Zoology, University of New England, Armidale 2351, Australia hford@une.edu.au

Dr. Graham Forrester

Dept. of Natural Resources Science, University of Rhode Island, Kingston, RI 02881, USA forrester.graham@gmail.com

Associate Professor Len Gillman

School of Applied Science, Auckland University of Technology, Auckland, Private Bag 92006, New Zealand len.gillman@aut.ac.nz

Professor Harold Heatwole

Zoology, University of North Carolina, NC, Raleigh, NC 27695–7617, USA (postal address), and Zoology, University of New England, Armidale, Australia halfh@ncsu.edu

Dr. Eric Hoberg

Chief Curator, US National Parasite Collection, ARS, USDA Animal Parasitic Diseases Laboratory BARC East 1180, 10300 Baltimore Avenue, Beltsville, MD 20705 Eric.Hoberg@ARS.USDA.GOV

Dr. Aneta Kostadinova

Institute of Parasitology, Biology Centre of the Academy of Sciences of the Czech Republic, Branišovská 31, 370 05 České Budějovice, Czech Republic aneta.kostadinova@uv.es

Professor Boris R. Krasnov

Mitrani Department of Desert Ecology, Jacob Blaustein Institutes for Desert Research, Ben-Gurion University of the Negev, Sede Boqer Campus, 84990 Midreshet Ben-Gurion, Israel krasnov@bgu.ac.il

Professor Michael J. Lawes

Research Institute for the Environment and Livelihoods, Charles Darwin University, Darwin, NT 0909, Australia michael.lawes@cdu.edu.au

Professor Harvey B. Lillywhite

Department of Biology and Director, Seahorse Key Marine Laboratory, University of Florida, Gainesville, Florida 32611–8525, USA hblill@ufl.edu

Associate Professor Brian McGill

School Biology and Ecology & Sustainable Solutions Initiative, University of Maine, Deering Hall 303 Orono, ME 04469, USA mail@brianmcgill.org

Professor Camilo Mora

Department of Geography, University of Hawaii Manoa, USA cmora@hawaii.edu

Professor Serge Morand

Institut des Sciences de l'Evolution – CNRS, CC065, Université Montpellier 2, Montpellier Cedex, France serge.morand@univ-montp2.fr

Dr. Lloyd W. Morrison

Department of Biology, Missouri State University, 901 S. National Ave., Springfield MO 65897, USA LloydMorrison@MissouriState.edu

Dr. Ana Pérez-del-Olmo

Departament de Biologia Animal, de Biologia Vegetal i d'Ecologia, Universitat Autònoma de Barcelona, 08193 Cerdanyola del Vallès, Barcelona, Spain ana.perez@uab.es

Dr. Annapaola Rizzoli

Centro Ricerca ed Innovazione, Dipartimento di Biodiversita ed Ecologia Molecolare, Fondazione Edmund Mach, 38010 San Michele all'Adige (TN), Italy annapaola.rizzoli@mach.it

Professor Klaus Rohde

Zoology, University of New England, Armidale NSW 2351, Australia krohde@une.edu.au

Professor Peter F. Sale

Assistant Director, United Nations University, Institute for Water, Environment and Health, Hamilton ON, and Professor Emeritus, University of Windsor, Canada sale@uwindsor.ca

Dr. Andrea Šimková

Department of Botany and Zoology, Faculty of Science, Masaryk University, Kotlářská 2, 61137 Brno, Czech Republic simkova@sci.muni.cz

Dr. Mark Steele

Department of Biology, California State University, Northridge, CA 91330–8303 mark.steele@csun.edu

Associate Professor G. H. Walter

School of Biological Sciences, The University of Queensland, Brisbane Qld 4072 Australia g.walter@uq.edu.au

Dr. Shane Wright

School of Biological Sciences, University of Auckland, Auckland, Private Bag 92019, New Zealand sd.wright@auckland.ac.nz

Professor Fernando A. Zapata

Department of Biology, Universidad del Valle, Apartado Aéreo 25360, Cali, Colombia fernando.zapata@correounivalle.edu.co

Acknowledgments

The authors and editor wish to thank the following for reviewing chapters in this book.

Associate Professor Nigel Andrew, Zoology, University of New England, Armidale, Australia

Professor Stuart Barker, Animal Science, University of New England, Armidale, Australia

Professor William Bond, Department of Botany, University of Cape Town, Rondebosch, South Africa

Associate Professor Michael Box, Physics, University of New South Wales, Sydney, Australia

Associate Professor Peter Clarke, Botany, University of New England, Armidale, Australia

Associate Professor Indraneil Das, Institute of Biodiversity and Environmental Conservation, Universiti Malaysia Sarawak, Kota Samarahan, Sarawak

Professor Fordyce Davidson, Mathematics, University of Dundee, Dundee, UK

Professor Christopher Dickman, Biological Sciences, University of Sydney, Australia

Dr. Andrew Glikson, Earth and Marine Science, Australian National University, Canberra, Australia

Professor Harold Heatwole, Zoology, University of North Carolina, Raleigh NC, USA, and University of New England, Armidale, Australia

Professor Ingibjörg Svala Jónsdóttir, University of Iceland, Reykjavik, Iceland

Professor David Karoly, Earth Sciences, University of Melbourne, Australia

Professor Boris Krasnov, Desert Research, Ben Gurion University, Mizpe-Ramon, Israel

Dr. Lesley T. Lancaster, National Center for Ecological Analysis and Synthesis, University of California, Santa Barbara, CA, USA

Dr. Tommy Leung, Zoology, University of New England, Armidale, Australia

Professor Serge Morand, Institut des sciences de l'évolution, Université Montpellier 2, France

Dr. Chris Pavey, CSIRO Ecosystem Sciences, Business & Innovation Centre, Desert Knowledge Precinct, Alice Springs, Australia

Professor Robert Poulin, Zoology, University of Otago, Dunedin, New Zealand

Professor Harry F. Recher, AM, FRZS, POB 154, Brooklyn, NSW, Australia

Professor Thomas W. Schoener, Evolution and Ecology, University of California at Davis, CA, USA

Dr. John Terblanche, Conservation Ecology and Entomology, Stellenbosch University, Matieland, South Africa

Dr. John Veron, Australian Institute of Marine Science, Townsville, Australia

Ass. Professor J. Wilson White, Biology and Marine Biology, University of North Carolina, Wilmington, NC, USA

Associate Professor Ian Whittington, Earth and Environmental Science, University of Adelaide, Australia

Dr. John W. Wilkinson, Amphibian & Reptile Conservation, Boscombe, Bournemouth, Dorset, UK

Professor Yihong Du, Mathematics, University of New England, Armidale, Australia

We thank **Cesar Luis Barrio-Amorós**, Instituto de Biodivesidad Tropical, San José, Pérez Zeledón, San Isidro, Costa Rica for the cover photo.

Professor David Jablonski, Geophysical Sciences, University of Chicago, kindly provided advice and references for Chapter 11.

E.P. Hoberg and D.R. Brooks are grateful to **Alycia Stigall** at the OHIO Center for Ecology and Evolutionary Studies, Ohio University, for sharing preprints of several papers and for discussions about invasion and faunal structure.

Introduction

Klaus Rohde

It is obvious that nature is undergoing rapid changes as a result of human activities such as industry, agriculture, travel, fisheries, urbanization, etc. What effects do these activities have? Are they disturbing equilibria in ecological populations and communities, i.e., are they upsetting the balance of nature, or are they enhancing naturally occurring disequilibria, perhaps with even worse consequences? This book examines these questions, first by providing evidence for equilibrium and nonequilibrium (= disequilibrium) conditions in natural systems, and second by examining human-induced effects, among them those due to climate change, habitat destruction and introduction of alien species. One often hears the argument, not only from non-scientists but also from some scientists, that large-scale fluctuations in climate, sea levels etc. have occurred over and over again in the geological past, long before human activities could possibly have had any impact, that human effects are very small compared to those naturally occurring anyway, and that they cannot significantly affect the environment. Is this indeed so? Or is it possible that naturally occurring fluctuations are being dangerously enhanced by humans?

(1) The concept of ecological equilibrium is used differently by different authors. A detailed historical discussion of the meaning of the concept with many examples was given in Rohde (2005). According to the most widely accepted usage, those ecological systems are in equilibrium which fluctuate around some stable point, i.e., return to it after disturbance due to "self-correcting mechanisms" (Hutchinson, 1948). Competition is usually considered to be the most important of these mechanisms. How common are equilibrium and nonequilibrium states in ecological systems, under what conditions are populations, communities and ecosystems likely to be in or approaching equilibrium? In particular, how important are they in evolution? Richard Dawkins (1976) has claimed that the development of the concept of evolutionarily stable strategies (ESSs) may be the most important contribution to evolutionary theory since Darwin. The concept of ESS relies on equilibrium assumptions, and so does that of evolutionarily stable states (a concept closely related to ESS). Maynard Smith defines an evolutionarily stable state of a population as a genetic composition that is restored by selection after a disturbance, provided the disturbance is not too large (see Maynard Smith & Price, 1973, and Maynard Smith, 1982). However, large disturbances are frequent in many ecosystems and for many

The Balance of Nature and Human Impact, ed. Klaus Rohde. Published by Cambridge University Press.

populations, and establishment of evolutionarily stable states will therefore often be difficult or impossible. Communities – as shown in theoretical and experimental studies (e.g., of plankton dynamics) – may never reach equilibrium even in homogeneous and relatively constant environments, because multi-species interactions may lead to oscillations and chaos. The fate of ecological systems, because of such factors, is often unpredictable, and interspecific competition, often thought to be the major factor in "regulating" community structure, may in fact bring about the opposite: disequilibrium. In view of this, we can re-phrase the major aim of this book as an evaluation of how common evolutionarily stable strategies (or states) in populations and establishment of equilibria in communities due to competition are.

(2) Almost all of the world's ecosystems have now been modified by human activity, through habitat loss, fragmentation and degradation, as well as by pollution and invasive species. Furthermore, they are now being subjected to climate change, largely the result of human activity. A second major aim of the book is to examine the extent to which equilibrium in ecological communities (where it existed) has been upset by human-induced changes, and to what extent disequilibria have been enhanced. There is abundant evidence that species' distributions and population sizes are changing, in many cases detrimentally. Communities are losing species and gaining new ones, so that interspecific interactions, such as competition, predation and mutualism, are being reframed.

The book examines the various topics, beginning with a discussion of examples of nonequilibrium and equilibrium in populations and metapopulations, in communities, and in ecosystems at geographical scales. Latitudinal gradients in biodiversity, i.e., the most pervasive and best-documented trends in the geographical distribution of animals and plants, are discussed, as well as some important hypotheses explaining them, including the hypothesis of effective evolutionary time, which does not rely on equilibrium assumptions, and molecular evidence for it.

The part on human-induced effects includes discussions of invading species and climate change. An environmental physicist discusses the physics of climate change; other chapters deal with current and predicted effects on changes in species diversity and species ranges due to climate change, species invasions and habitat losses of amphibians, reptiles, birds and insects, as well as discussions of the expected fates of the most diverse marine ecosystems on Earth, i.e., coral reefs. This part also contains a chapter by a mathematician on the features of invading species that determine successful invasions, using an equation-based model, and another chapter on the effects of invasions in evolutionary history and what a study of such effects can teach us about present events. Also discussed is the potential for the development of new infectious diseases due to changing climatic conditions.

Emphasis is laid on the importance of autecological studies, that is of the importance of long-term studies of single species. There is an account of such long-term studies of some flatworms which suggests that not competitive interactions between species but evolutionary and ecological contingencies are largely responsible for ecological adaptations and niche selection. It also shows that at least some species have such intricate morphological and ecological adaptations that they cannot easily be replaced if habitats are destroyed, and that each species will react differently to environmental disturbance and climate change.

In the final part, "An overall view", some important conclusions based on the book chapters and literature accounts, and some important consequences of climate change and ecological disturbance, are discussed. We return to the questions asked in the first paragraph: how common are evolutionarily stable states in populations and how important is interspecific competition for determining community structure, and what are some of the consequences of the enhanced environmental fluctuations in a changing world? The conclusion is: evolutionarily stable states are not as common and interspecific competition is not as important as assumed by many, and consequences of climate change are often unpredictable in detail, but overall may be catastrophic. Patterns of disease transmission will be changed, and the enhanced disequilibrium poses important challenges for managing biodiversity and conservation. In a joint section, several authors who have contributed chapters to the book examine the question on what steps should be taken to secure future biodiversity.

This book is aimed at scientists, students and interested lay people. Many scientists at the forefront of research contributed chapters, which provide a review of what is known in their fields and stimulate ideas for future research. The book can be used as a source for university courses, and it is hoped to provide useful information to lay people who wish to inform themselves on current topics of ecological research, particularly that related to human impact on the environment, including climate change. Examples have been selected from a large range of animal and plant groups on the basis of how well they have been examined with regard to equilibrium and nonequilibrium strategies and states, or the extent to which human impact on them has been documented. The book provides a basis for public discussions of human impact on the environment and what to do about it, with easy access to a large body of (mainly recent) literature and discussions of aspects insufficiently discussed in standard texts on ecology.

References

Dawkins, R. (1976). *The Selfish Gene*. Oxford: Oxford University Press.

Hutchinson, G. E. (1948). Circular causal systems in ecology. *Annals of the New York Academy of Sciences*, **50**, 221–246.

Maynard Smith, J., & Price, G. R. (1973). The logic of animal conflict. *Nature*, **246**, 15–18.

Maynard Smith, J. (1982). *Evolution and the Theory of Games*. Cambridge: Cambridge University Press.

Rohde, K. (2005). *Nonequilibrium Ecology*. Cambridge: Cambridge University Press.

Part I

Nonequilibrium and Equilibrium in Populations and Metapopulations

1 Reef fishes: density dependence and equilibrium in populations?

Graham E. Forrester and Mark A. Steele

Summary

Small colorful coral reef fishes have been excellent subjects for ecological field experiments (Sale, 2002). They have provided substantial insight about the strength of density-dependent interactions and their underlying biological causes. In this chapter we first summarize what we know about the biological mechanisms of density-dependent regulation in coral reef fishes, highlighting one pervasive mechanism – competition for structural refuges used to avoid predators. We then summarize the evidence for ongoing coral declines and the progressive loss of architectural complexity on reefs. We argue that as reefs become architecturally simpler they provide fewer refuges from predation, and so the carrying capacity for many fish populations is declining. As a result of ongoing competition for gradually diminishing supply of refuges, we hypothesize that some species of reef fishes will continue to experience density-dependent mortality even as their populations decline globally. This hypothesis contradicts conventional views on the regulation of marine populations, which hold that density-dependent interactions should diminish in importance as populations decline (Figures 1.1 and 1.2).

Background: the development of ideas about density dependence in reef fishes

A long-standing controversy in ecology has revolved around the relative effects of density-dependent and density-independent processes on population dynamics (Murdoch, 1994; Cappuccino & Price, 1995). Although it is necessary for long-term persistence, recognizing that density dependence need not have a strong effect at all times on the dynamics of a population was a key step in resolving a major historical controversy over density dependence and density independence as mutually exclusive alternatives (Andrewartha & Birch, 1954; Nicholson, 1957). In fact, both sorts of processes affect most populations (Turchin et al., 1995). Ideas about populations of reef fishes and other marine species have followed a similar path, following early controversy over which processes control dynamics (Doherty, 1991; Jones, 1991).

The Balance of Nature and Human Impact, ed. Klaus Rohde. Published by Cambridge University Press.
© Cambridge University Press 2013.

Figure 1.1.   The bridled goby (*Coryphopterus glaucofraenum*) suffers density-dependent mortality when crevices used to hide from predators are limited. See plate section for color version.

Figure 1.2.   Most coral reef fishes depend upon shelter provided by the physical structure of living corals. As coral reefs continue to degrade, shelter may become more limited and even if fish populations decline in density they may still experience density-dependent mortality. See plate section for color version.

Like many marine organisms, coral reef fishes produce pelagic larvae that can disperse over wide areas. Density dependence could be occurring during the pelagic larval phase (Sandin & Pacala, 2005), but because of the difficulty of estimating mortality of pelagic individuals there is no empirical evidence for or against this possibility. Instead, early debates revolved around whether the dynamics of benthic populations are controlled by the influx of pelagic larvae (called settlement) or by density-dependent interactions among juveniles and adults in the benthic habitat. These explanations for population dynamics were initially presented as dichotomous alternatives. The hypothesis that abundance was limited by the rate of larval input (called recruitment limitation) predicts that mortality rates in the benthic habitat are density-independent, so that benthic populations fluctuate in correspondence with the rate of replenishment via larval influx. This hypothesis was presented as an alternative to the hypothesis that abundance was limited by density-dependent interactions (competition and predation) affecting individuals within the benthic habitat (Doherty, 1983; Victor, 1983; Connell, 1985).

The recruit-adult hypothesis

It has been acknowledged more recently that processes influencing both the rate of input to populations via settlement and output via mortality will influence the dynamics of most benthic marine populations (Gaines & Roughgarden, 1985; Menge & Sutherland, 1987; Caley *et al.*, 1996). Considerable progress has been made in developing models that can evaluate the relative effects of density-dependent and density-independent processes. A simple idea that captures the essence of most models is the "recruit-adult hypothesis" (Menge, 2000). This hypothesis was developed for marine invertebrates that occupy primary space on the substratum. These benthic populations are ultimately resource limited because of the obvious limit to two-dimensional space for species that require attachment sites on the substratum. The recruit-adult hypothesis posits that variation in oceanographic processes controls the influx of larvae to populations and determines the extent to which populations in different locations reach their carrying capacity: sites with high settlement are chronically space limited, whereas sites with low settlement rarely reach their carrying capacity and adult densities are sensitive to larval influx (Connolly & Roughgarden, 1998; Menge, 2000; Connolly *et al.*, 2001; Muko *et al.*, 2001) (Figure 1.3).

The recruit-adult hypothesis is also applicable when density-dependent mortality arises from causes other than limitation of two-dimensional space, and analyses of reef fish populations have confirmed two key predictions of this hypothesis. Firstly, the sensitivity of adult densities to larval input is greater where settlement is low (Schmitt *et al.*, 1999). Second, mortality after settlement was shown to be consistently density-dependent in dozens of manipulations in small patches of habitat, as long as experimental densities encompassed the high end of the natural range (Osenberg *et al.*, 2002).

Competition for refuges is a widespread cause of density dependence

Moving beyond simple manipulations of population density, several researchers studying reef fishes have manipulated both local density and putative causes of density dependence (food, predators, shelter) in cross-factored experiments (Forrester, 1990; Hixon & Jones, 2005). For animals like reef fishes, for whom causes of death are difficult to

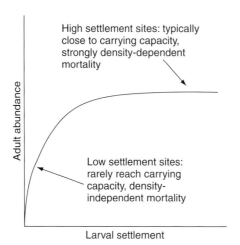

Figure 1.3. The recruit-adult hypothesis: a simple conceptual model for benthic marine populations. Oceanographic processes controlling larval influx determine whether populations are typically recruitment-limited or resource-limited.

establish by observation, this is a powerful method to identify the biological interactions responsible for density-dependent mortality.

Our experiments with two small reef fishes, the bridled goby (*Coryphopterus glaucofraenum*) and the goldspot goby (*Gnatholepis thompsoni*), provide a good example of this approach, and yielded results that appear to be representative of many species (White *et al.*, 2010). Establishing a gradient of goby density in the presence and absence of larger predatory fishes showed that predation was the agent of most mortality, and the proximate cause of density dependence (Forrester & Steele, 2000). Like many reef fishes, gobies use structural habitat features as refuges from predation. By manipulating both goby density and the abundance of refuges within small plots, we showed that the ultimate cause of density dependence was competition for a limited supply of refuges (Forrester & Steele, 2004). Gobies rapidly retreat to shared unguarded refuges when a predator approaches or attacks, and only when refuges are relatively abundant do all prey individuals actually escape. Competition for refuges thus resembles the childhood game of musical chairs, and this process is well described by simple mortality functions in which the per capita prey mortality rate depends on the ratio of prey (gobies) to refuges (Samhouri *et al.*, 2009). These manipulations, like most ecological experiments, were performed at a small spatial scale (using replicate reef patches a few meters in extent). Interactions between density-dependent processes and local heterogeneity in density and environmental factors often cause population dynamics at large scales to differ from predictions based on small-scale measurements (Rastetter *et al.*, 1992; Gardner *et al.*, 2001). In some cases, the so-called "scale transition" is strong enough to make extrapolating from small-scale experiments wildly inaccurate (Chesson, 1998). Gobies, however, displayed similar relationships between density and mortality on both small and large reefs (Steele & Forrester, 2005; Forrester *et al.*, 2008a) and a simulation model tailored to bridled goby demography suggests that the scale-transition should be modest (Vance *et al.*, 2010).

The recruit-adult hypothesis expanded: spatial variation in carrying capacity based on availability of refuges

The recruit-adult hypothesis assumes that differences among populations are due only to oceanographic processes altering the influx of larvae (Connolly & Roughgarden, 1998; Menge, 2000; Connolly *et al.*, 2001; Muk, *et al.*, 2001). The carrying capacity of the benthic habitat and the post-settlement interactions within it are assumed to be consistent among sites (Figure 1.3). Based on our understanding of the interactive effect of predation and competition on local-scale mortality of gobies, we present an expanded version of the recruit-adult hypothesis (Figure 1.4). We hypothesize that large reefs differ in refuge abundance, which in turn creates differences in carrying capacity among sites. Consequently, differences among sites in population dynamics are a function of both the influx of larvae and the carrying capacity of the benthic habitat (Figure 1.4).

A key prediction of this hypothesis is that reefs providing more shelter from predation have a higher carrying capacity. We recently tested this prediction experimentally using five large reefs in the Bahamas, each comprising several thousand square meters of goby habitat (Steele & Forrester, 2005). We increased the availability of shelter for gobies on half of each reef by adding dead coral and rubble (Forrester *et al.*, 2008b), and the large size of the reefs meant each half was populated by an effectively separate population of gobies. Refuge densities varied prior to the experiment. Each year for 3 years, we

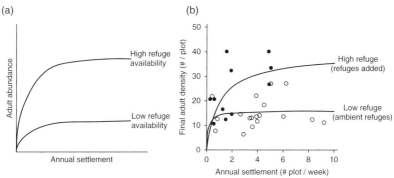

Figure 1.4.   An expanded version of the recruit-adult hypothesis with spatial variation in carrying capacity. We hypothesized that, in addition to varying in larval settlement, populations at different sites also vary in carrying capacity. (a) We use a set of hypothetical reefs to illustrate a key prediction of this hypothesis: asymptotic adult abundance will be higher on reefs with more shelter. (b) Results of an experimental test of this prediction using bridled gobies occupying large reefs in the Bahamas (see text for details). Each data point is a different year class (= one generation) of fish on one of 10 large reefs (1800–7800 m^2). Black symbols indicate reefs to which we added rocks to increase refuge densities. White symbols represent treatment reefs prior to the manipulation, and control reefs to which rocks were never added. Data for each group were fitted to the function $A = aS/[1+(a/b)S]$, where A = adult density, S = annual settlement, and a and b are fitted parameters that represent density-independent mortality and maximum adult density respectively (Schmitt *et al.*, 1999). Maximum adult density (b) differs significantly between treatments and the model explains 39% of variation in the data.

monitored larval settlement over the entire summer reproductive season and the density of adult gobies in October. Because these gobies are annuals, each year represents a separate generation and the adults present in October represent the survivors of the year's settlement. A preliminary analysis of the results provides some support for this expanded version of the recruit-adult hypothesis (Figure 1.4).

Reef fish populations in decline

Since we found some evidence for spatial difference in carrying capacity between reefs, we now consider how competition for refuges and the effects of density dependence might change over time given the progressive changes to coral reefs that have occurred over the past half century, and which appear likely to continue on a similar trajectory in future (Sale, 2011).

Although their amenability to experimentation has allowed us to identify the biological mechanisms underlying density dependence in reef fishes, a major challenge when trying to assess the effects of these interactions on population dynamics is the shortage of long-term data on abundance. Parenthetically, the types of data available for small reef-associated fishes are exactly the opposite of those available for larger, commercially harvested fishes. The relative wealth of good time-series for large, harvested species means we have been able to document their long-term dynamics in great detail, but the dearth of experimental data means identifying the underlying biological mechanisms has been difficult (Rose *et al.*, 2001; Myers *et al.*, 2002).

Although time-series data for reef fishes are sparse prior to the mid-1980s, a comprehensive meta-analysis of records from the Caribbean reveals a general decline in abundance starting in the mid-1990s (Paddack *et al.*, 2009). Importantly, declines are apparent for most small- and medium-sized fishes, across all major trophic and taxonomic groups. Although a corresponding region-wide meta-analysis for the Indo-Pacific is lacking, there are examples of declines in this region (Graham *et al.*, 2007). There are also, of course, exceptions to this general pattern; for example at some sites herbivores have increased in abundance since the 1990s (Pratchett *et al.*, 2008) whereas cryptobenthic fishes have remained stable (Bellwood *et al.*, 2006), but the majority of reef fish species appear to be in decline. Large-bodied apex predators were severely depleted much earlier than the 1990s, and their loss is attributed to overfishing (Munro, 1983; Jackson, 1997). Overfishing cannot, however, explain the more recent decline of the majority of smaller fishes, which are not harvested for food.

Possible links between reef degradation and fish declines

An alternative explanation for widespread recent declines in the abundance of small reef fishes is that they occurred in response to die-offs in the benthic community that provides habitat for most fish species (Paddack *et al.*, 2009). The ongoing degradation of benthic coral reef communities has been well documented. Coral populations have been in progressive decline worldwide for at least the last 30 years (Gardner *et al.*, 2003; Schutte *et al.*, 2010) and, consequently, reefs are steadily becoming flatter and less architecturally complex (Alvarez-Filip *et al.*, 2009). As corals decline in cover, they are sometimes replaced by macroalgae, whereas in other places sea fans or sponges are increasing, apparently in response (Norstrom *et al.*, 2009).

Two pieces of circumstantial evidence are consistent with the hypothesis that habitat-related factors are somehow linked to fish declines (Graham *et al.*, 2007; Paddack *et al.*, 2009). First, numerous spatial comparisons show more fish occupying reefs that have higher coral cover (Ault & Johnson, 1998; Holbrook *et al.*, 2000) or that are more architecturally complex (Luckhurst & Luckhurst, 1978; Caley & St John, 1996). A simple space-for-time substitution suggests a similar association between reef fish and reef architecture might occur over time. Second, widespread fish declines have lagged behind the degradation of reef habitats by several years, which is consistent with increased fish mortality being triggered by loss of habitat (Graham *et al.*, 2007; Paddack *et al.*, 2009). The lag between loss of coral and reductions in fish density appears to be shorter (5–10 years) in the Indo-Pacific than in the Caribbean (10–20 years), perhaps because fewer Caribbean fishes depend on live corals for habitat than their Indo-Pacific counterparts (Paddack *et al.*, 2009). Indeed the most compelling examples of rapid reductions in fish abundance after episodes of mass coral mortality come from the Pacific (Jones *et al.*, 2004), and are most striking for fish that are either specialist coral dwellers (Munday, 2004; Feary *et al.*, 2007) or rely on live corals for food (Pratchett *et al.*, 2006, 2008).

The recruit-adult hypothesis expanded: declining carrying capacity for reef fish populations based on loss of refuges

For the gobies we studied, and a few other reef fishes (Holbrook & Schmitt, 2002; Hixon & Jones, 2005), experimental evidence has established causal links between the architectural complexity of reefs, refuge availability for small fishes, competition for refuges and mortality from predation. For most species, however, a mechanistic under-standing of the putative association between habitat loss and fish declines is lacking. There is, however, evidence suggesting that measures of architectural complexity may provide an index of the availability of shelter from predators for many small-medium fishes, so these causal links may apply to a quite large set of species (Sano *et al.*, 1984; Caley & St John, 1996; Lewis, 1997; Syms & Jones, 2000; Gratwicke & Speight, 2005). Just as spatial variation in refuge availability causes differences across sites in carrying capacity, losses over time of architectural complexity may be causing a progressive reduction in the number of refuges from predation, and so may be gradually reducing the carrying capacity of the habitat (Figure 1.5). We therefore predict that the likelihood of density-dependent regulation is not necessarily reduced as populations decline, if that decline is due to a deterioration in carrying capacity of the environment. This prediction is counter to the original recruit-adult hypothesis, which posits that low adult densities are the result of low input of settling larvae followed by density-independent mortality.

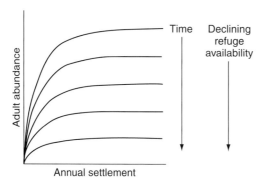

Figure 1.5. A new expansion of the recruit-adult hypothesis with a temporal decline in carrying capacity. We hypothesize that ongoing declines in coral cover and/or topographic relief are reducing the carrying capacity of reefs. For many small reef fishes we suggest that the carrying capacity at a site is based on the abundance of shelters from predation. We illustrate a key prediction of this hypothesis: asymptotic adult abundance will decline over time as the abundance of refuges decreases. Larval settlement is assumed to vary independently among years independent of shelter availability.

Progressive diminution of the carrying capacity of their stream habitat may explain why juvenile salmon in the Pacific Northwest continue to suffer density-dependent mortality despite long-term declines in their abundance (Achord *et al.*, 2003). In these streams, the key limited resource for juvenile salmon appears to be marine-derived nutrients from adult salmon carcasses, which provide food for juveniles directly and also fertilize stream food webs to increase the production of invertebrate prey (Larkin & Slaney, 1997). Historical declines in the number of adult salmon returning to rivers appear to have steadily reduced the carrying capacity of juvenile habitats and so, even though juvenile densities have decreased by over 90% since the 1960s, they continue to experience the effects of resource limitation (Gresh *et al.*, 2000). Although the limiting resource differs (food rather than shelter), this example clearly illustrates the potential for populations to experience density-dependent regulation throughout a protracted and severe decline in abundance.

If mesopredator release is occurring, its likely effect is to intensify density-dependent mortality in small prey fishes

For most small reef fishes, although the ultimate cause of density-dependent mortality is competition for refuges, the proximate agent of mortality is predation. We therefore now turn to consider the possible effects of changes in predator abundance. The predators of small reef fishes are a diverse group of medium-sized fishes that includes small groupers, snappers, lizardfish and trumpetfish, collectively known as "mesopredators" (Hixon & Carr, 1997; Forrester & Steele, 2000; Hixon *et al.*, 2002; Holbrook & Schmitt, 2002; Johnson, 2006; Overholtzer-McLeod, 2006; White, 2007). Mesopredators are themselves prey of very large piscivorous fishes such as sharks, barracudas, large groupers and jacks. The very large piscivores that are the apex predators on coral reefs have been in decline from overharvesting for at least a century (Munro, 1983; Jackson, 1997). The "mesopredator release hypothesis" (Prugh *et al.*, 2009) posits that the decline of apex predators has led to a corresponding increase in either the numerical abundance or activity of their mesopredator prey (Stallings, 2008). The mesopredator release has not yet been widely tested on coral reefs, but increased apex predator biomass inside one Bahamian reserve was associated with reduced mesopredator biomass and increased biomass of some (but not all) small fishes (Lamb & Johnson, 2010), which is consistent with a cascading effect of mesopredator release on small reef fishes. Our aim here is simply to point out that, should mesopredator release occur, it is likely to act synergistically with ongoing losses of architectural complexity to steadily intensify density-dependent predation on small fishes.

Cryptic density dependence and responses of settling larvae to habitat degradation

The original recruit-adult hypothesis, and our expanded version that allows for spatial and temporal variation in carrying capacity, both assume that input to benthic populations by the arrival of larvae is independent of the "quality" of the benthic habitat. In many species, however, pelagic larvae colonizing benthic habitats larvae choose sites according to a variety of physical attributes (Sale *et al.*, 1984), and sometimes based on the presence of conspecifics (Sweatman, 1983) or other species (Sale *et al.*, 1984). Larvae use a variety of sensory cues to make these finely resolved choices about where to settle (Kingsford *et al.*, 2002). Sites selected are expected to be of higher quality relative to others. Operationally, high-quality sites possess features that result in lower per-capita mortality at any given level of settlement and so have a higher carrying capacity (Shima & Osenberg, 2003).

When there is spatial covariation between settlement and site quality, newly settled fishes should be distributed among sites differing in quality in an approximately ideal-free manner (Fretwell & Lucas, 1969). As a result, density-dependent mortality is "cryptic" and difficult to detect in observational studies (Shima & Osenberg, 2003). For example, two coral-dwelling gobies (*Elacatinus evelynae* and *E. prochilos*) settled preferentially among coral heads, and preferred corals were assumed to be of higher "quality". Mortality after settlement was low on preferred corals, with the result that mortality was not correlated with settler density. However, a subsequent manipulation in which density treatments were randomly assigned to corals eliminated the correlation between settler density and coral "quality" and revealed that mortality was density-dependent (Wilson & Osenberg, 2002).

In reality, site "quality" is likely to be multifaceted and challenging to measure operationally. For example, our work with gobies allowed us to identify a simple component of site quality (the density of suitable refuges) that we could measure and manipulate, and so were able to demonstrate its influence on adult mortality. We expected that settling goby larvae would select reefs with higher refuge densities because (all else being equal) adult mortality is lower where refuges are abundant. Surprisingly, goby larvae displayed the opposite pattern and settled preferentially to reefs with low refuge densities (Forrester *et al.*,

2008b). We cannot yet explain this apparently paradoxical larval preference, but its consequence is that density dependence is not cryptic in bridled gobies.

We bring up the concept of cryptic density dependence here because, in principle, this phenomenon might occur in time, as well as in space, if site "quality" progressively declines (Shima & Osenberg, 2003). For example, some fish selectively colonize particular species of live coral at the time of settlement, and appear to select species that provide shelter from predators late in life (Gutierrez, 1998). If preferred species of coral decline at a site, site quality for these fishes should decline and fewer larvae may settle. For these habitat specialists, declines in adult abundance may thus be primarily due to diminishing larval settlement and density-dependent mortality will be "cryptic", making it difficult to test our hypothesis that carrying capacity is declining using observational time-series data. For fishes whose habitat requirements are broader and more flexible, settling individuals may switch to settle in marginal habitats as their preferred habitats decline (Wilson *et al.*, 2008) and relationships between density and mortality should be easier to detect.

Conclusion

Populations that have declined to become rare and are at risk of extinction are often assumed to have vital rates that are density-independent and be on a "random walk". For marine populations with planktonic larvae, the recruit-adult hypothesis predicts that when benthic populations have become sparse they will fluctuate in response to the rate of larval delivery. The possibility that populations experiencing severe declines are resource limited is often not considered by conservation biologists, and has not been considered in the design of management strategies for coral reefs, simply because at-risk populations are typically small relative to past levels. Our aim in this chapter is to present an alternative possibility, which might apply to many small reef fishes. If there is a causal link between fish declines and the loss of architectural complexity on reefs, as we hypothesize, many species of reef fishes may continue to compete for a gradually disappearing supply of refuges, even when their populations become small relative to historical levels. The fact populations at low abundances may continue to experience density-dependent mortality will affect estimates of extinction risk and, perhaps more importantly, means that the probability of recovery from decline may be lower than otherwise expected.

References

Achord, S., Levin, P. S., & Zabel, R. W. (2003). Density-dependent mortality in Pacific salmon: the ghost of impacts past? *Ecology Letters*, **6**, 335–342.

Alvarez-Filip, L., Dulvy, N. K., Gill, J. A., Cote, I. M., & Watkinson, A. R. (2009). Flattening of Caribbean coral reefs: region-wide declines in architectural complexity. *Proceedings of the Royal Society of London B*, **276**, 3019–3025.

Andrewartha, H. G., & Birch, L. C. (1954). *The Distribution And Abundance of Animals*. Chicago, IL: University of Chicago Press.

Ault, T. R., & Johnson, C. R. (1998). Spatially and temporally predictable fish communities on coral reefs. *Ecological Monographs*, **68**, 25–50.

Bellwood, D. R., Hoey, A. S., Ackerman, J. L., & Depczynski, M. (2006). Coral bleaching, reef fish community phase shifts and the resilience of coral reefs. *Global Change Biology*, **12**, 1587–1594.

Caley, M. J., Carr, M. H., Hixon, M. A., *et al.* (1996). Recruitment and the local dynamics of open marine populations. *Annual Review of Ecology and Systematics*, **27**, 477–500.

Caley, M. J., & St John, J. (1996). Refuge availability structures assemblages of tropical reef fishes. *Journal of Animal Ecology*, **65**, 414–428.

Cappuccino, N., & Price, P. W. (1995). *Population Dynamics: New Approaches and Synthesis*. San Diego, CA: Academic Press.

Chesson, P. (1998). Spatial scales in the study of reef fishes: a theoretical perspective. *Australian Journal of Ecology*, **23**, 209–215.

Connell, J. H. (1985). The consequences of variation in initial settlement vs postsettlement mortality in rocky intertidal communities. *Journal of Experimental Marine Biology and Ecology*, **93**, 11–45.

Connolly, S. R., Menge, B. A., & Roughgarden, J. (2001). A latitudinal gradient in recruitment of intertidal invertebrates in the northeast Pacific Ocean. *Ecology*, **82**, 1799–1813.

Connolly, S. R., & Roughgarden, J. (1998). A latitudinal gradient in Northeast Pacific intertidal community structure: evidence for an oceanographically based synthesis of marine community theory. *American Naturalist*, **151**, 311–326.

Doherty, P. J. (1983). Tropical territorial damselfishes: is density limited by aggression or recruitment? *Ecology*, **64**, 176–190.

Doherty, P. J. (1991). Spatial and temporal patterns of recruitment. In P. F. Sale (Ed.), *The Ecology of Fishes on Coral Reefs* (pp. 261–293). San Diego, CA: Academic Press.

Feary, D. A., Almany, G. R., Jones, G. P., & McCormick, M. I. (2007). Coral degradation and the structure of tropical reef fish communities. *Marine Ecology-Progress Series*, **333**, 243–248.

Forrester, G. E. (1990). Factors influencing the juvenile demography of a coral reef fish. *Ecology*, **71**, 1666–1681.

Forrester, G. E., & Steele, M. A. (2000). Variation in the presence and cause of density-dependent mortality in three species of reef fishes. *Ecology*, **81**, 2416–2427.

Forrester, G. E., & Steele, M. A. (2004). Predators, prey refuges, and the spatial scaling of density-dependent prey mortality. *Ecology*, **85**, 1332–1342.

Forrester, G. E., Steele, M. A., Samhouri, J. F., Evans, B., & Vance, R. R. (2008a). Spatial density dependence scales up but does not produce temporal density dependence in a coral reef fish. *Ecology*, **89**, 2980–2985.

Forrester, G. E., Steele, M. A., Samhouri, J. F., & Vance, R. R. (2008b). Settling larvae of a small coral-reef fish discriminate reef features at large, but not small, spatial scales. *Limnology and Oceanography*, **53**, 1956–1962.

Fretwell, S. D., & Lucas, H. L. (1969). On territorial behavior and other factors influencing habitat distribution in birds. *Acta Biotheoretica*, **19**, 16–36.

Gaines, S. D., & Roughgarden, J. (1985). Larval settlement rate: a leading determinant of structure in an ecological community of the marine intertidal zone. *Proceedings of the National Academy of Sciences of the USA*, **82**, 3707–3711.

Gardner, R. H., Kemp, M. W., Kennedy, V. S., & Petersen, J. E. (2001). *Scaling Relations in Experimental Ecology*. New York: Columbia University Press.

Gardner, T. A., Cote, I. M., Gill, J. A., Grant, A., & Watkinson, A. R. (2003). Long-term region-wide declines in Caribbean corals. *Science*, **301**, 958–960.

Graham, N. A. J., Wilson, S. K., Jennings, S., *et al.* (2007). Lag effects in the impacts of mass coral bleaching on coral reef fish, fisheries, and ecosystems. *Conservation Biology*, **21**, 1291–1300.

Gratwicke, B., & Speight, M. R. (2005). Effects of habitat complexity on Caribbean marine fish assemblages. *Marine Ecology-Progress Series*, **292**, 301–310.

Gresh, T., Lichatowich, J., & Schoonmaker, P. (2000). An estimation of historic and current levels of salmon production in the Northeast Pacific ecosystem: evidence of a nutrient deficit in the freshwater systems of the Pacific Northwest. *Fisheries*, **25**, 15–21.

Gutierrez, L. (1998). Habitat selection by recruits establishes local patterns of adult distribution in two species of damselfishes: *Stegastes dorsopunicans* and *S. planifrons*. *Oecologia*, **115**, 268–277.

Hixon, M. A., & Carr, M. H. (1997). Synergistic predation, density dependence, and population regualtion in marine fish. *Science*, **277**, 946–949.

Hixon, M. A., & Jones, G. P. (2005). Competition, predation, and density-dependent mortality in demersal marine fishes. *Ecology*, **86**, 2847–2859.

Hixon, M. A., Webster, M. S., & Sale, P. F. (2002). Density dependence in marine fishes: coral-reef populations as model systems. In P. F. Sale (Ed.), *Coral Reef Fishes: Dynamics and Diversity in a Complex Ecosystem* (pp. 303–325). San Diego, CA: Academic Press.

Holbrook, S. J., Forrester, G. E., & Schmitt, R. J. (2000). Spatial patterns in abundance of a damselfish reflect availability of suitable habitat. *Oecologia*, **122**, 109–120.

Holbrook, S. J., & Schmitt, R. J. (2002). Competition for shelter space causes density-dependent predation mortality in damselfishes. *Ecology*, **83**, 2855–2868.

Jackson, J. B. C. (1997). Reefs since Columbus. *Coral Reefs*, **16**, S23–S32.

Johnson, D. W. (2006). Predation, habitat complexity, and variation in density-dependent mortality of temperate reef fishes. *Ecology*, **87**, 1179–1188.

Jones, G. P. (1991). Postrecruitment processes in the ecology of coral reef fish populations: a multifactorial perspective. In P. F. Sale (Ed.), *The Ecology of Fishes on Coral Reefs* (pp. 293–328). San Diego, CA: Academic Press.

Jones, G. P., McCormick, M. I., Srinivasan, M., & Eagle, J. V. (2004). Coral decline threatens fish biodiversity in marine reserves. *Proceedings of the National Academy of Sciences of the USA*, **101**, 8251–8253.

Kingsford, M. J., Leis, J. M., Shanks, A., *et al.* (2002). Sensory environments, larval abilities and local self-recruitment. *Bulletin of Marine Science*, **70**, 309–340.

Lamb, R. W., & Johnson, D. W. (2010). Trophic restructuring of coral reef fish communities in a large marine reserve. *Marine Ecology-Progress Series*, **408**, 169–180.

Larkin, G. A., & Slaney, P. A. (1997). Implications of trends in marine-derived nutrient influx to south coastal British Columbia salmonid production. *Fisheries*, **22**, 16–24.

Lewis, A. R. (1997). Effects of experimental coral disturbance on the structure of fish communities on large patch reefs. *Marine Ecology-Progress Series*, **161**, 37–50.

Luckhurst, B. E., & Luckhurst, K. (1978). Analysis of the influence of substrate variables on coral reef fish communities. *Marine Biology*, **49**, 317–323.

Menge, B. A. (2000). Recruitment vs. postrecruitment processes as determinants of barnacle population abundance. *Ecological Monographs*, **70**, 265–288.

Menge, B. A., & Sutherland, J. P. (1987). Community regulation: variation in disturbance, competition, and predation in relation to environmental stress and recruitment. *The American Naturalist*, **130**, 730–757.

Muko, S., Sakai, K., & Iwasa, Y. (2001). Dynamics of marine sessile organisms with space-limited growth and recruitment: application to corals. *Journal of Theoretical Biology*, **210**, 67–80.

Munday, P. L. (2004). Habitat loss, resource specialization, and extinction on coral reefs. *Global Change Biology*, **10**, 1642–1647.

Munro, J. L. (1983). *Caribbean Coral Reef Fishery Resources* (Vol. 7). Manila, Philippines: International Center for Living Aquatic Resources Management.

Murdoch, W. W. (1994). Population regulation in theory and practice. *Ecology*, **75**, 271–287.

Myers, R. A., Hart, P. J. B., & Reynolds, J. D. (2002). Recruitment: understanding density-dependence in fish populations. In P. J. B. Hart & J. D Reynolds (Eds.), *Handbook of Fish Biology and Fisheries Vol. 1: Fish Biology* (pp. 123–148). Oxford: Blackwell Scientific.

Nicholson, A. J. (1957). The self adjustment of populations to change. *Cold Spring Harbor Symposium in Quantitative Biology*, **22**, 153–172.

Norstrom, A. V., Nystrom, M., Lokrantz, J., & Folke, C. (2009). Alternative states on coral reefs: beyond coral-macroalgal phase shifts. *Marine Ecology-Progress Series*, **376**, 295–306.

Osenberg, C. W., St Mary, C. M., Schmitt, R. J., *et al.* (2002). Rethinking ecological inference: density dependence in reef fishes. *Ecology Letters*, **5**, 715–721.

Overholtzer-McLeod, K. L. (2006). Consequences of patch reef spacing for density-dependent mortality of coral-reef fishes. *Ecology*, **87**, 1017–1026.

Paddack, M. J., Reynolds, J. D., Aguilar, C., *et al.* (2009). Recent region-wide declines in Caribbean reef fish abundance. *Current Biology*, **19**, 590–595.

Pratchett, M. S., Munday, P. L., Wilson, S. K., *et al.* (2008). Effects of climate-induced coral bleaching on coral-reef fishes – ecological and economic consequences. In R. N. Gibson, R. J. A. Atkinson & J. D. M. Gordon (Eds.), *Oceanography and Marine Biology: An Annual Review, Vol. 46* (pp. 251–296). Boca Raton, FL: CRC Press.

Pratchett, M. S., Wilson, S. K., & Baird, A. H. (2006). Long-term monitoring of the Great Barrier Reef. *Journal of Fish Biology*, **69**, 1269–1280.

Prugh, L. R., Stoner, C. J., Epps, C. W., *et al.* (2009). The rise of the mesopredator. *Bioscience*, **59**, 779–791.

Rastetter, E. B., King, A. W., Cosby, B. J., *et al.* (1992). Aggregating fine-scale ecological knowledge to model coarser-scale attributes of ecosystems. *Ecological Applications*, **2**, 55–70.

Rose, K. A., Cowan, J. H., Winemiller, K. O., Myers, R. A., & Hilborn, R. (2001). Compensatory density dependence in fish populations: importance, controversy, understanding and prognosis. *Fish and Fisheries*, **2**, 293–327.

Sale, P. F. (2002). *Coral Reef Fishes: Dynamics and Diversity in a Complex Ecosystem*. Amsterdam: Elsevier.

Sale, P. F. (2011). *Our Dying Planet: An Ecologist's View of the Crisis we Face*. Berkeley, CA: University of California Press.

Sale, P. F., Douglas, W. A., & Doherty, P. J. (1984). Choice of microhabitats by coral-reef fishes at settlement. *Coral Reefs*, **3**, 91–99.

Samhouri, J. F., Vance, R. R., Forrester, G. E., & Steele, M. A. (2009). Musical chairs mortality functions: density-dependent deaths caused by competition for unguarded refuges. *Oecologia*, **160**, 257–265.

Sandin, S. A., & Pacala, S. W. (2005). Demographic theory of coral reef fish populations with stochastic recruitment: comparing sources of population regulation. *American Naturalist*, **165**, 107–119.

Sano, M., Shimizu, M., & Nose, Y. (1984). Changes in structure of coral reef fish communities by destruction of hermatypic corals: observational and experimental views. *Pacific Science*, **38**, 51–79.

Schmitt, R. J., Holbrook, S. J., & Osenberg, C. W. (1999). Quantifying the effects of multiple processes on local abundance: a cohort approach for open populations. *Ecology Letters*, **V2**, 294–303.

Schutte, V. G. W., Selig, E. R., & Bruno, J. F. (2010). Regional spatio-temporal trends in Caribbean coral reef benthic communities. *Marine Ecology-Progress Series*, **402**, 115–122.

Shima, J. S., & Osenberg, C. W. (2003). Cryptic density dependence: effects of covariation between density and site quality in reef fish. *Ecology*, **84**, 46–52.

Stallings, C. D. (2008). Indirect effects of an exploited predator on recruitment of coral-reef fishes. *Ecology*, **89**, 2090–2095.

Steele, M. A., & Forrester, G. E. (2005). Small-scale field experiments accurately scale up to predict density dependence in reef fish populations at large-scales. *Proceedings of the National Academy Of Sciences of the USA*, **102**, 13513–13516.

Sweatman, H. P. A. (1983). Influence of conspecifics on choice of settlement sites by larvae of two pomacentrid fishes (*Dascyllus aruanus* and *D. reticulatus*) on coral reefs. *Marine Biology*, **75**, 225–229.

Syms, C., & Jones, G. P. (2000). Disturbance, habitat structure, and the dynamics of a coral-reef fish community. *Ecology*, **81**, 2714–2729.

Turchin, P., Cappuccino, N., & Price, P. W. (1995). Population regulation: old arguments and a new synthesis. In N. Cappuccino & P. W. Price (Eds.), *Population Dynamics: New Approaches and Synthesis* (pp. 19–40). San Diego, CA: Academic Press.

Vance, R. R., Steele, M. A., & Forrester, G. E. (2010). Using an individual-based model to quantify scale transition in demographic rate functions: deaths in a coral reef fish. *Ecological Modelling*, **221**, 1907–1921.

Victor, B. C. (1983). Recruitment and population dynamics of a coral reef fish. *Science*, **219**, 419–420.

White, J. W. (2007). Spatially correlated recruitment of a marine predator and its prey shapes the large-scale pattern of density-dependent prey mortality. *Ecology Letters*, **10**, 1054–1065.

White, J. W., Samhouri, J. F., Stier, A. C., *et al.* (2010). Synthesizing mechanisms of density dependence in reef fishes: behavior, habitat configuration, and observational scale. *Ecology*, **91**, 1949–1961.

Wilson, J., & Osenberg, C. W. (2002). Experimental and observational patterns of density-dependent settlement and survival in the marine fish *Gobiosoma*. *Oecologia*, **130**, 205–215.

Wilson, S. K., Burgess, S. C., Cheal, A. J., *et al.* (2008). Habitat utilization by coral reef fish: implications for specialists vs. generalists in a changing environment. *Journal of Animal Ecology*, **77**, 220–228.

2 Population dynamics of ectoparasites of terrestrial hosts

Boris R. Krasnov and Annapaola Rizzoli

Introduction

The main unit of ecological interest is not an individual organism but rather an assemblage of individuals belonging to the same species and coexisting in time and space. Contrary to most free-living species, spatial distribution of parasites is not continuous but consists of a set of more or less uniform inhabited patches represented by the host organisms, while the environment between these patches is decidedly unfavorable and strongly affects the probability of those parasites with free-living stages completing their life cycle and thus persisting. Thus, spatial distribution of an ensemble of conspecific ectoparasites is heterogeneous and fragmented among (a) host individuals, (b) host species within a location, and (c) locations. In this chapter, we will consider the lowest hierarchical level of this fragmentation, namely ectoparasite infrapopulations, i.e., assemblages of conspecific parasites infesting an individual host. We will focus on several common taxa of arthropod ectoparasites of mammalian hosts. We will start with variation in patterns of parasite abundance among parasite species as well as among host species, gender and age cohorts. Then, we will discuss relationships between abundance and distribution of ectoparasites. Finally, we will focus on host-related and environment-related factors affecting ectoparasite abundance and distribution. We will demonstrate that ectoparasite populations are affected by intrinsic and extrinsic factors whose actions promote equilibrium and nonequilibrium conditions, respectively.

Measurements of abundance and distribution

The fragmented distribution pattern of a parasite among host individuals prevents us from characterizing the abundance of this parasite by a single value. This pattern mainly stems from the fact that the distribution of a parasite population across a host population is usually aggregated. In other words, most parasite individuals occur in a few host individuals, while most host individuals have only a few, if any, parasites (Anderson & May, 1978; Poulin, 1993; Shaw & Dobson, 1995; Shaw *et al.*, 1998; Wilson *et al.*, 2001). The aggregated distribution of parasite individuals among hosts is caused by a variety of

The Balance of Nature and Human Impact, ed. Klaus Rohde. Published by Cambridge University Press.
© Cambridge University Press 2013.

factors (Poulin, 2007) and can have important consequences for different aspects of the evolutionary ecology of parasites (e.g., Morand *et al.*, 1993). As a result, the fraction of uninfested hosts should also be taken into account when abundance of a parasite is considered. In other words, given the aggregated distribution of a parasite across hosts, a parasite's abundance should be considered in conjunction with its distribution.

Common measures of parasite abundance and distribution are mean abundance, intensity of infestation and prevalence. Mean abundance is simply the mean number of parasites per host individual and is calculated by summing both infested and uninfested hosts. Intensity of infestation (sometimes called parasite burden or parasite load) is the mean abundance of parasites per infested host individual, whereas prevalence is the proportion of infested hosts. Obviously, intensity of infestation is a product of mean abundance and prevalence. These measures are straightforward, simple to understand and are easily calculated.

Is abundance a parasite species character?

It is commonly accepted that the density (abundance per unit area) of a species in a location results from the interplay between the intrinsic properties of that species (Hughes *et al.*, 2000; Blackburn & Gaston, 2001; Lopez-Sepulcre & Kokko, 2005) and the extrinsic properties of the local habitat, both biotic and abiotic (Rosenzweig, 1981; Newton, 1998). As a result, the predictability of the density level of any given species of parasite is often low (Fieberg & Ellner, 2000).

High intraspecific variation in the population parameters of parasites, such as their intensity of infestation, prevalence and abundance, is well documented. For example, the abundance of parasites is strongly dependent on the abundance of their hosts (Anderson & May 1978; see below), which, in turn, is spatially variable. Dependence of survival and, consequently, abundance of parasites on spatially variable abiotic factors (e.g., climate) has been reported for both endo- (Galaktionov, 1996) and ectoparasites (Metzger & Rust, 1997). However, in spite of the strong dependence of parasite population parameters on extrinsic factors and, therefore, the expected spatial variation of these parameters, species-specific features of parasites such as body size and egg production could constrain this variation (Poulin, 1999). Indeed, Arneberg *et al.* (1997), studying nematodes parasitic in mammals, demonstrated that abundance was less variable within than among nematode species. The conclusion from their study is, therefore, that the level of abundance is a "true" attribute of a nematode species, i.e., that the biological features of parasite species can potentially override local environmental conditions in driving parasite population dynamics.

Ectoparasites are much more strongly influenced by their off-host environment than endoparasites studied by Arneberg *et al.* (1997). This suggests that the reported patterns may not be valid for them, although low variation of within-species density on a temporal or a spatial scale has been reported for some ectoparasites (e.g., Launay, 1989). To compare within-species and among-species variation in abundance, Krasnov *et al.* (2006a) and Korallo-Vinarskaya *et al.* (2009) used data on fleas and gamasid mites parasitic on small mammals, respectively. First, a strong positive correlation was found between the lowest observed abundances and all other observed abundance values across flea and mite species. In other words, different parasites demonstrated a relatively narrow range of abundances when exploiting the same host species in different regions. Second, the results of the repeatability analysis (Arneberg *et al.*, 1997) showed that abundances of the same parasite species on the same host species but in different regions were more similar to each other than expected by chance, and varied significantly among parasite species, with about 50% of the variation among samples accounted for by differences between parasite species. Similar results were obtained for abundance and prevalence of larvae and nymphs of two tick species parasitic on small mammalian hosts in central Europe (Krasnov *et al.*, 2007). Abundance and prevalence of each tick species/stage within each host and across all hosts appeared to have characteristic limits of variation.

The above analyses demonstrated that patterns found for mammalian endoparasites (Arneberg *et al.*, 1997) and parasites of fish (Poulin, 2006) are also valid for ectoparasites despite their greater sensitivity to external factors. This implies that some flea, mite or tick species-specific life history traits determine the limits of abundance. Lower limits of abundance can be affected by species-specific mating systems, relationship between mating and blood feeding, and time necessary for a blood meal, whereas upper limits of abundance can be determined by species-specific reproductive outputs, generation times, preferences for blood-sucking on a specific body part of a host, mortality rate and/ or the ability of both imagos and pre-imagos to withstand crowding.

Relationship between abundance and prevalence

A positive relationship between local abundance and occupancy is a well-known pattern (Gaston, 2003). In the application of this relationship to host-parasite systems, a positive correlation between the mean abundance of parasites and their prevalence (i.e., host occupancy) has been supported in many studies (Shaw & Dobson, 1995; Morand & Guégan, 2000; Krasnov *et al.*, 2002; Simkova *et al.*, 2002). The positive abundance-occupancy relationship has been explained by a variety of mechanisms. In fact, Gaston *et al.* (1997) and Gaston (2003) listed nine different hypotheses aimed at explaining it. Morand and Guégan (2000) tested several of these hypotheses using nematodes parasitic on mammals. In particular, they found that prevalence of nematodes could be success-fully predicted using an epidemiological model with a minimal number of parameters such as mean abundance of a parasite, its variance and an indicator of aggregation. The latter parameter, in turn, can be calculated from the positive relationship between mean abundance and its variance (Taylor's power law; see Morand & Krasnov, 2008 for details). Consequently, Morand and Guégan (2000) concluded that the abundance-distribution relationship in parasites could be explained by demographic and stochastic mechanisms revealed by simple epidemiological models without invoking more com-plex explanations such as, for example, the niche breadth hypothesis (Brown, 1984).

A positive relationship between abundance and distribution of ectoparasites within and across host species has been confirmed for fleas (e.g., Krasnov *et al.*, 2005a), ixodid ticks (Stanko *et al.*, 2007), gamasid mites (Krasnov *et al.*, 2010) and lice (Matthee & Krasnov, 2009) on various hosts and in various geographic regions. In general, the relationship between the prevalence of a parasite and its mean abundance fitted the logistic curve. Prevalences were low at low parasite abundances and rose rapidly to high asymptotes with an increase in abundances, although the rate of the rise varied between parasites, even those exploiting the same host species in the same locations and at the same time (Matthee & Krasnov, 2009).

Whatever the mechanism behind the positive abundance-prevalence relationship may be, the most important findings in these studies are that (a) the pattern of the relationship between abundance and prevalence is surprisingly similar in parasites that differ substantially in their origin, physiology, behavior and ecology and (2) a large proportion of the variance in parasite prevalence can be explained solely by their mean abundance. The latter finding suggests that prevalence of ectoparasites can be reliably predicted from the data on abundance. Indeed, using a simple epidemiological model that takes into account parasite mean abundance and its variance (Taylor *et al.*, 1979; Anderson & May, 1985; Morand and Guégan, 2000), observed prevalence has been successfully predicted in different ectoparasites (Krasnov *et al.*, 2005a, 2005b; Stanko *et al.*, 2007; Matthee & Krasnov, 2009; Krasnov *et al.*, 2010). These results supported the demographic hypothesis of parasite abundance and distribution (Anderson *et al.*, 1982), which suggested that the observed distributions of parasites across host individuals are generated by two opposing forces, namely those leading to over-dispersion (aggregation) and those leading to under-dispersion (regularity). Stochastic variability in demographic parameters may generate both over- (pure birth process) and under-dispersion (pure death process), whereas stochasticity in environmental processes creates over-dispersion.

Biases in ectoparasite infestation

Any field zoologist knows that some host species are characterized by higher parasite abundance than other species. Consequently, a level of parasite abundance can be determined not only by parasite, but also by host identity. In other words, one may ask whether parasite abundance can be seen also as a property of the host. For example, Arneberg *et al.* (1997) investigated abundance of nematodes not only among nematode species, but also among host species, and demonstrated that independently of nematode species, some mammal species have many nematodes per individual, whereas other mammals have only a few nematodes per individual. Along the same lines, repeatability of ectoparasite abundance within as opposed to among host species has been shown for fleas (Krasnov *et al.*, 2006a), ixodid ticks (Krasnov *et al.*, 2007) and gamasid mites (Krasnov *et al.*, 2008), although the proportion of the variance accounted for by differences among host species as opposed to within a host species was lower than that accounted for by differences among parasite species. The repeatability of parasite abundance within host species suggests that some host properties also constrain to some extent the number of parasites harbored by an individual. For fleas, these constraints can be related to processes on the host body that affect imago and/or to processes within a host burrow or nest that affect pre-imago. Furthermore, this suggests that some host species represent better habitats for ectoparasites than other host species due to differences in the amount of resource provided by a host (e.g., body size; Morand & Guégan, 2000) and/or pattern of acquisition of these resources (e.g., defense abilities; Klein & Nelson, 1998; Mooring et al., 2000).

Biases in parasite infestation related to host gender are known for various host-parasite systems (see review in Zuk & McKean, 1996) both endo- (e.g., Poulin, 1996) and ectoparasites (Ulmanen & Myllimäki, 1971; Botelho & Linardi, 1996; Bursten et al., 1997; Anderson & Kok, 2003; Morand et al., 2004). In most cases, males of higher vertebrates (birds and mammals) are infested by more parasites than females. Male-biased parasitism is often related to gender differences in body size, mobility and immunocompetence. Indeed, males in higher vertebrates are usually larger and more mobile than females. For example, Perkins et al. (2003) found that sexually mature male woodmice represented the functional transmission cohort which supported the majority of co-feeding ticks. The male-biased spatial organization of the woodmice at high population density, and in particular the tendency of males to aggregate more than females, leading also to a high contact rate between individuals, may be an explanation for the greater parasite burden of males (Stradiotto et al., 2009). In addition, individuals of a smaller sex (females) with a larger body surface:body mass ratio can afford fewer parasites per unit surface area than individuals of a larger sex and, thus, should be more defensive (Gallivan & Horak, 1997). The difference in mobility increases the chances of males to be exposed to a larger variety and number of parasites (Randolph, 1977; Lang, 1996). Differences in the immunocompetence between males and females may occur because of the immunosuppressive effect of androgens (Zuk & McKean, 1996, Lee et al., 2001). It was suggested that if ectoparasites increase their fitness through dispersal, then they would benefit from a tighter association with male hosts (Bursten et al., 1997; Smith et al., 2005). Examples of female-biased parasitism also exist. For example, females of some bat species hosted higher densities of ectoparasites than did males (Dick et al., 2003; Zahn & Rupp, 2004; Kanuch et al., 2005). Sometimes female-biased parasitism was found even in those host taxa for which male-biased parasitism is usually the case (e.g., rodents; Haitlinger, 1973; Krasnov et al., 2005c). Moreover, gender differences in the infestation by ectoparasites can vary seasonally (Krasnov et al., 2005c; Perkins et al., 2003) as well as in dependence on ectoparasite taxon and environmental conditions (Matthee et al., 2010).

Parasite abundance and patterns of distribution often vary between younger and older hosts of different taxa (e.g., Goater & Ward, 1992). Data from a number of host-parasite systems demonstrated that relationships between host age and parasite abundance and distribution may be either linearly positive, asymptotic or convex (see review in Hudson & Dobson, 1995). Each of these patterns can be generated by various mechanisms such as parasite load-dependent mortality (Rousset et al., 1996) and age-dependent development of host defense mechanisms (Hudson & Dobson, 1995). Convex or asymptotic relationships between host age and prevalence can be expected if the level of acquired resistance against parasites is low either in younger (by definition) and older (because of immunosenescence; see Møller & de Lope, 1999) or in younger-only host cohorts, respectively. Age-dependent behavioral defense can also generate asymptotic age-intensity curves, especially in the case of hematophagous ectoparasites (e.g., Schofield & Torr, 2002). Age-bias in parasitism by ectoparasites has been reported in various parasite-host associations, but no common trend could be gleaned from these reports (see Marshall, 1981; Krasnov, 2008). Krasnov *et al.* (2006b) found two different patterns of the change in flea aggregation and prevalence with host age in seven species of rodents from Slovakia. These results suggested that age-dependent patterns of flea parasitism pattern could be generated by various processes and strongly affected by natural history parameters of a host species such as dispersal pattern, spatial distribution, and structure of shelters. In addition, the relationship between ectoparasite density and host age could be parasite density-dependent and also mediated by other factors such as ambient temperature and presence of ectoparasites belonging to another taxon (Hawlena *et al.*, 2005).

Factors affecting ectoparasite abundance and distribution

It is obvious that some kind of relationship exists between abundance of a consumer and abundance of a resource. Classic models describing the dynamics of host and parasite populations were presented by Anderson and May (1978), May and Anderson (1978), Grenfell (1992) and Arneberg *et al.* (1998). In general, these models predict that the abundance of a parasite should increase in a curvilinear fashion to a plateau with increasing host density because the greater the host density, the greater the probability that each parasite's individual or respective transmission stage will contact a host (Haukisalmi & Henttonen, 1990). In particular, this pattern has been reported for fleas (e.g., Launay, 1989; Krasnov *et al.*, 2002). Krasnov *et al.* (2002) studied how the abundance of a rodent host in Middle Eastern desert affected the abundance and distribution of two fleas. The abundance of a flea reproducing during the reproductive period of its host increased with an increase in host density to a plateau. This plateau may be due to the limited carrying capacity of host individuals that can harbor only a limited number of parasites and/or to flea-induced host mortality. In contrast, the relationship between abundance of a seasonal flea, which reproduces in the periods when its host does not reproduce, and host density appeared to be linear. This result suggests that the explanation of a plateau caused by host carrying capacity and/or mortality when flea intensity exceeds a specific threshold is unsatisfactory. An alternative explanation implies the heterogeneity of hosts in relation to burrows as habitats for successful flea breeding, as well as the carrying capacity of host habitat. The increase in density of small mammal populations, particularly in solitary species, often results in a surplus of individuals ("transients") that have no individual home ranges (Gliwicz, 1992). In theory, these "homeless" individuals should not be putative hosts for fleas, because they do not possess burrows that are necessary for flea reproduction and development of pre-imaginal stages, although they can take part in flea transmission. In turn, the density of resident hosts is determined by the carrying capacity of a given habitat. As overall host density increases, the number of residents attains a particular level determined by the carrying capacity of host habitat. Further increases in host density result in an increase in the number of transients, while the number of residents remains stable. Because transient hosts do not participate in the siphonapteran life cycle, the overall increase in host density after saturation of a habitat by residents has no effect on the flea burden. This may also explain the absence of a plateau in a seasonal flea with an increase of host abundance because this flea reproduces in periods when most host individuals in a given habitat are residents, whereas transient individuals are absent.

Stanko *et al.* (2006) studied the effect of host abundance on flea abundance and distribution using data on different flea-host associations from central Europe. Surprisingly and in contrast to theoretical predictions, relationships between flea abundance or prevalence and host abundance were found to be either negative or absent. Negative relationships between flea and host abundance were always accompanied by negative relationships between flea prevalence and host density. Thus, the link between host and flea abundances differed among host species exploited by the same flea.

Several studies have been carried out to understand the relationships among abundances of hosts and ixodid ticks (e.g., deer and *Ixodes ricinus*). The greater the abundance of deer in certain locations, the higher the abundance of questing ticks on vegetation within a suitable habitat was found to be (Tagliapietra *et al.*, 2011). Deer increase in habitats suitable for *I. ricinus* could also be used as a predictor of tick-borne disease risk (Rizzoli *et al.*, 2009). Deer removal was found to be followed by decline of questing ticks, although this could lead to higher infestation level on alternative hosts with an unexpected increase in TBD disease risk (Perkins *et al.*, 2006)

A strong link between host and parasite population dynamics suggests close interactions between host and parasite demographic parameters (Anderson & Gordon, 1982). For example, temporal variation of the rodent abundance was found to affect abundance of larval and imaginal ticks (Rosà *et al.*, 2007). The lack of this link in a parasite-host system hints that demographies of a given parasite and a given host are unrelated. In other words, a parasite is not equally dependent on all its host species, but rather the parasite's population dynamics is determined by populations of some but not other hosts.

Patterns of relationships between parasite distribution and host abundance contradicting the predictions of epidemiological models have been reported in other studies too (e.g., Sorci *et al.*, 1997). A negative relationship or a lack of relationship between parasite abundance with an increase in host abundance can arise for a number of reasons. One of these may be the lower rate of parasite reproduction and transmission compared to the rate of reproduction and dispersal of hosts. In other words, the rate of establishment of new patches (newly born or dispersing young mammals) is faster than the rate of their infestation. Consequently, under high host density, a fraction of host individuals may remain "under-used" by the parasites merely because they cannot keep pace with host reproduction and dispersal. The contradictions between theoretical models and real data demonstrate that nature is much more complicated than any human-constructed model. This contradiction was found not only in fleas but in other ectoparasites as well (e.g., ticks; Daniels *et al.* 1993 versus Krasnov *et al.*, 2007). The diversity of patterns of the relationship between host abundance and flea abundance and distribution mentioned above may be a result of different regulating mechanisms governing different flea-host associations (parasite-induced host mortality, host-induced parasite mortality and density-dependent reductions in parasite fecundity and survival; Grenfell & Dobson, 1995). Furthermore, different regulating mechanisms may act simultaneously within the same flea-host association.

Contradictions between real data and theoretical predictions may be explained by factors that are not usually taken into account in simple epidemiological models. One of these factors is density-dependent changes in the host's spatial behavior. As mentioned above, "homeless" host individuals are characteristic of any population of small mammals at high density. Krasnov *et al.* (2002) corrected the pre-existing models of Anderson and May (1978), substituting the overall host density by the density of resident host individuals. Although both basic and "corrected" models describing the relationships between ectoparasite burden and host density fitted the observational data well, simulations of the fraction of resident hosts demonstrated that this parameter influenced the relationship between host density and flea abundance only when residents constituted no more than 50% of all host individuals. In other words, when the percentage of non-residents is relatively low, they do not contribute heavily toward flea transmission, so their influence on flea dynamics and distribution is negligible. However, when the fraction in the host population is relatively high, their contribution to flea transmission becomes significant. In such a case, a model that takes into account this non-resident component should describe the observational data better than one which considers the overall host density only.

Changes of host community structure can strongly affect populations of a parasite that exploits all members of a community (Norman *et al.*, 1999; Ostfeld & Keesing, 2000; Holt *et al.*, 2003). For example, a resident parasite may respond to the introduction of a new host species into a community in two different ways. Its abundance and/or prevalence could either increase because of an increase in the amount of available resources (i.e., total number of hosts) or decrease in a resident host because a parasite will be "diluted" among local and introduced hosts (Ostfeld & Keesing, 2000; Telfer *et al.*, 2005). However, a "dilution" effect may be manifested differently in different parasites. For example, Krasnov *et al.* (2007) found that abundance and/or prevalence of a tick *I. ricinus* decreased with an increase in the number of host species in a community, while this was not the case for *Ixodes trianguliceps*. This difference could stem from the difference between tick species in the level of ecological specialization. Larvae, nymphs and imago of *I. trianguliceps* inhabit mainly burrows and underground nests of the hosts, whereas *I. ricinus* quest for their hosts outside their shelters. Therefore, an increase in the number of host species in a location would not affect *I. trianguliceps* because the growth of a host community would not generally increase the number and diversity of small mammalian inhabitants per burrow (e.g., Gliwicz, 1992). Furthermore, the "dilution" effect seems to be only one of the varieties of mechanisms by which diversity of host community may alter parasite transmission and dynamics (Johnson & Thieltges, 2010).

Numerous studies confirmed the importance of the effect of environmental factors on ectoparasite population dynamics (see Marshall, 1981, and references therein). In particular, landscape fragmentation has been shown to be a factor that affects ectoparasite population dynamics as strongly as factors associated with host abundance and diversity (Estrada-Peña, 2009). In repeatability analyses of flea and mite abundances mentioned above, Krasnov *et al.* (2006a) and Korallo-Vinarskaya *et al.* (2009) tested the repeatability of this parameter not only within parasite and/or host species, but also within regions. In both ectoparasite taxa, the proportion of the variance accounted for by differences among regions, as opposed to within-region differences, was rather low, but significant, suggesting that some locations are characterized by higher ectoparasite abundance than other locations. However, it should be remembered that the effect of environmental factors on ectoparasite density dynamics may differ among ectoparasite species and higher taxa due not only to difference in natural history among taxa but also to difference in sensitivity within a taxon (Harper *et al.*, 1992; Lang, 1996).

Concluding remarks

Although abundance of an ectoparasite species appears to be a true species character, it is determined to some extent and within species-specific limits by host species identity, density and spatial behavior as well as by local biotic and abiotic conditions. On the one hand, this supports the ideas of Arneberg *et al.* (1997) and Poulin (2006) that the biological attributes of parasite species are primary determinants of parasite dynamics. On the other hand, expression of ectoparasite dynamics depends on a variety of factors. Although distribution of ectoparasites may be described by relatively simple models with a limited set of parameters, the relationships between these parameters are complicated and variable, so they may be manifested differently among ectoparasite species and higher taxa, host species, seasons and geographic locations. This suggests that despite determination of population size of an ectoparasite by some intrinsic causes, the strong effect of host-related and environmental factors as well as factors associated with co-occurring ectoparasites results in nonequilibrium conditions.

References

Anderson, P. C., & Kok, O. B. (2003). Ectoparasites of springhares in the Northern Cape Province, South Africa. *South African Journal of Wildlife Research*, **33**, 23–32.

Anderson, R. M., & Gordon, D. M. (1982). Processes influencing the distribution of parasite numbers within host populations with special emphasis on parasite-induced host mortality. *Parasitology*, **85**, 373–398.

Anderson, R. M., & May, R. M. (1978). Regulation and stability of host-parasite population interactions. I. Regulatory processes. *Journal of Animal Ecology*, **47**, 219–247.

Anderson, R. M., & May, R. M. (1985). Helminth infection of humans: mathematical models, population dynamics and control. *Advances in Parasitology*, **24**, 1–101.

Anderson, R. M., Gordon, D. M., Crawley, M. J., & Hassell, M. P. (1982). Variability in the abundance of animal and plant species. *Nature*, **296**, 245–248.

Arneberg, P., Skorping, A., & Read, A. F. (1997). Is population density a species character? Comparative analyses of the nematode parasites of mammals. *Oikos*, **80**, 289–300.

Arneberg, P., Skorping, A., Grenfell, B., & Read, A. F. (1998). Host densities as determinants of abundance in parasite communities. *Proceedings of the Royal Society of London B*, **265**, 1283–1289.

Blackburn, T. M., & Gaston, K. J. (2001). Linking patterns in macroecology. *Journal of Animal Ecology*, **70**, 338–352.

Botelho, J. R., & Linardi, P. M. (1996). Interrelações entre ectoparasitos e roedores em ambientes silvestre e urbano de Belo Horizonte, Minas Gerais, Brasil. *Revista Brasileira de Entomologia*, **40**, 425–430.

Brown, J. H. (1984). On the relationship between abundance and distribution of species. *The American Naturalist*, **124**, 255–279.

Bursten, S. N., Kimsey, R. B., & Owings, D. H. (1997). Ranging of male *Oropsylla montana* fleas via male California ground squirrel (*Spermophilus beecheyi*) juveniles. *Journal of Parasitology*, **83**, 804–809.

Daniels, T. J., Fish, D., & Schwartz, I. (1993). Reduced abundance of *Ixodes scapularis* (Acari, Ixodidae) and Lyme disease risk by deer exclusion. *Journal of Medical Entomology*, **30**, 1043–1049

Dick, C. W., Gannon, M. R., Little, W. E., & Patrick, M. J. (2003). Ectoparasite associations of bats from central Pennsylvania. *Journal of Medical Entomology*, **40**, 813–819.

Estrada-Peña, A. (2009). Diluting the dilution effect: a spatial Lyme model provides evidence for the importance of habitat fragmentation with regard to the risk of infection. *Geospatial Health*, **3**, 143–155.

Fieberg, J., & Ellner, S. P. (2000). When is it meaningful to estimate an extinction probability? *Ecology*, **81**, 2040–2047.

Galaktionov, K. V. (1996). Life cycles and distribution of seabird helminths in Arctic and subArctic regions. *Bulletin of the Scandinavian Society for Parasitology*, **6**, 31–49.

Gallivan, G. J., & Horak, I. G. (1997). Body size and habitat as determinants of tick infestations of wild ungulates in South Africa. *South African Journal of Wildlife Research*, **27**, 63–70.

Gaston, K. J. (2003). *The Structure and Dynamics of Geographic Ranges*. Oxford: Oxford University Press.

Gaston, K. J., Blackburn, T. M., & Lawton, J. H. (1997). Interspecific abundance-range size relationships: an appraisal of mechanisms. *Journal of Animal Ecology*, **66**, 579–601.

Gliwicz, J. (1992). Patterns of dispersal in non-cyclic populations of small rodents. In N. C. Stenseth & W. Z. Lidicker (Eds.), *Animal Dispersal: Small Mammals as a Model* (pp. 147–159). London: Chapman and Hall.

Goater, C. P., & Ward, P. I. (1992). Negative effects of *Rhabdias bufonis* (Nematoda) on the growth and survival of toads (*Bufo bufo*). *Oecologia*, **89**, 161–165.

Grenfell, B. T. (1992). Parasitism and the dynamics of ungulate grazing systems. *The American Naturalist*, **139**, 907–929.

Grenfell, B. T., & Dobson, A. P. (Eds.) (1995). *Ecology of Infectious Diseases in Natural Populations*. Cambridge: Cambridge University Press.

Haitlinger, R. (1973). The parasitological investigation of small mammals of the Góry Sowie (Middle Sudetes). I. Siphonaptera (Insecta). *Polskie Pismo Entomologiczne*, **43**, 499–519.

Harper, G. H., Marchant, A., & Boddington, D. G. (1992). The ecology of the hen flea *Ceratophyllus gallinae* and the moorhen flea *Dasypsyllus gallinulae* in nestboxes. *Journal of Animal Ecology*, **61**, 317–327.

Haukisalmi, V., & Henttonen, H. (1990). The impact of climatic factors and host density on the long-term population dynamics of vole helminths. *Oecologia*, **83**, 309–315.

Hawlena, H., Abramsky, Z., & Krasnov, B. R. (2005). Age-biased parasitism and density-dependent distribution of fleas (Siphonaptera) on a desert rodent. *Oecologia*, **146**, 200–208.

Holt, R. D., Dobson, A. P., Begon, M., Bowers, R. G., & Schauber, E. M. (2003). Parasite establishment in host communities. *Ecology Letters*, **6**, 837–842.

Hudson, P. J., & Dobson, A. P. (1995). Macroparasites: observed patterns. In B. T. Grenfell & A. P. Dobson (Eds.), *Ecology of Infectious Diseases in Natural Populations* (pp. 144–176). Cambridge: Cambridge University Press.

Hughes, T. P., Baird, A. H., Dinsdale, E. A., *et al.* (2000). Supply-side ecology works both ways: the link between benthic adults, fecundity, and larval recruits. *Ecology*, **81**, 2241–2249.

Johnson, P. T. J., & Thieltges, D. W. (2010). Diversity, decoys and the dilution effect: how ecological communities affect disease risk. *Journal of Experimental Biology*, **213**, 961–970.

Kanuch, P., Kristin, A., & Kristofik, J. (2005). Phenology, diet, and ectoparasites of Leisler's bat (*Nyctalus leisleri*) in the western Carpathians (Slovakia). *Acta Chiropterologica*, **7**, 249–257.

Klein, S. L., & Nelson, R. J. (1998). Adaptive immune responses are linked to the mating system of arvicoline rodents. *The American Naturalist*, **151**, 59–67.

Korallo-Vinarskaya, N. P., Krasnov, B. R., Vinarski, M. V., *et al.* (2009). Stability in abundance and niche breadth of gamasid mites across environmental conditions, parasite identity and host pools. *Evolutionary Ecology*, **23**, 329–345.

Krasnov, B. R. (2008). *Functional and Evolutionary Ecology of Fleas. A Model for Ecological Parasitology*. Cambridge: Cambridge University Press.

Krasnov, B. R., Khokhlova, I. S., & Shenbrot, G. I. (2002). The effect of host density on ectoparasite distribution: an example with a desert rodent parasitized by fleas. *Ecology*, **83**, 164–175.

Krasnov, B. R., Stanko, M., Miklisova, D., & Morand, S. (2005a). Distribution of fleas (Siphonaptera) among small mammals: mean abundance predicts prevalence via simple epidemiological model. *International Journal for Parasitology*, **35**, 1097–1101.

Krasnov, B. R., Morand, S., Khokhlova, I. S., Shenbrot, G. I., & Hawlena, H. (2005b). Abundance and distribution of fleas on desert rodents: linking Taylor's power law to ecological specialization and epidemiology. *Parasitology*, **131**, 825–837.

Krasnov, B. R., Morand, S., Hawlena, H., Khokhlova, I. S., & Shenbrot, G. I. (2005c). Sex-biased parasitism, seasonality and sexual size dimorphism in desert rodents. *Oecologia*, **146**, 209–217.

Krasnov, B. R., Shenbrot, G. I. Khokhlova, I. S., & Poulin, R. (2006a). Is abundance a species attribute of haematophagous ectoparasites? *Oecologia*, **150**, 132–140.

Krasnov, B. R., Stanko, M., & Morand, S. (2006b). Age-dependent flea (Siphonaptera) parasitism in rodents: a host's life history matters. *Journal of Parasitology*, **92**, 242–248.

Krasnov, B. R., Stanko, M., & Morand, S. (2007). Host community structure and infestation by ixodid ticks: repeatability, dilution effect and ecological specialization. *Oecologia*, **154**, 185–194.

Krasnov, B. R., Korallo-Vinarskaya, N. P., Vinarski, M. V., *et al.* (2008). Searching for general patterns in parasite ecology: host identity vs. environmental influence on gamasid mite assemblages in small mammals. *Parasitology*, **135**, 229–242.

Krasnov, B. R., Korallo-Vinarskaya, N. P., Vinarski, M. V., & Lareschi, M. (2010). Prediction of prevalence from mean abundance via a simple epidemiological model in mesostigmate mites from two geographic regions. *Parasitology*, **137**, 1227–1237.

Lang, J. D. (1996). Factors affecting the seasonal abundance of ground squirrel and wood rat fleas (Siphonaptera) in San Diego County, California. *Journal of Medical Entomology*, **33**, 790–804.

Launay, H. (1989). Ecological factors acting on the distribution and the population dynamics of *Xenopsylla cunicularis* Smit, 1957 (Insecta: Siphonaptera) a flea parasitic on the European rabbit, *Oryctolagus cuniculus* (L.). *Vie et Milieu*, **39**, 111–120.

Lee, C. Y., Alexander, P. S., Yang, V. V. C., & Yu, J. Y. L. (2001). Seasonal reproductive activity of male formosan wood mice (*Apodemus semotus*): relationships to androgen levels. *Journal of Mammalogy*, **82**, 700–708.

Lopez-Sepulcre, A., & Kokko, H. (2005). Territorial defense, territory size, and population regulation. *The American Naturalist*, **166**, 317–329.

Marshall, A. G. (1981). *The Ecology of Ectoparasitic Insects*. London: Academic Press.

Matthee, S., & Krasnov, B. R. (2009). Searching for mechanisms of generality in the patterns of parasite abundance and distribution: ectoparasites of a South African rodent, *Rhabdomys pumilio*. *International Journal for Parasitology*, **39**, 781–788.

Matthee, S., McGeoch, M. A., & Krasnov, B. R. (2010). Gender-biased ectoparasite infections: species-specific variation and the extent of male-biased parasitism. *Parasitology*, **137**, 651–660.

May, R. M., & Anderson, R. M. (1978). Regulation and stability of host-parasite population interactions. II. Destabilizing processes. *Journal of Animal Ecology*, **47**, 455–461.

Metzger, M. E., & Rust, M. K. (1997). Effect of temperature on cat flea (Siphonaptera: Pulicidae) development and overwintering. *Journal of Medical Entomology*, **34**, 173–178.

Møller, A. P., & de Lope, F. (1999). Senescence in a short-lived migratory bird: age-dependent morphology, migration, reproduction and parasitism. *Journal of Animal Ecology*, **68**, 163–171.

Mooring, M. S., Benjamin, J. E., Harte, C. R., & Herzog, N. B. (2000). Testing the interspecific body size principle in ungulates: the smaller they come, the harder they groom. *Animal Behaviour*, **60**, 35–45.

Morand, S., Göuy De Bellocq, J., Stanko, M., & Miklisova, D. (2004). Is sex-biased ectoparasitism related to sexual size dimorphism in small mammals of Central Europe? *Parasitology*, **129**, 505–510.

Morand, S., & Guégan, J.-F. (2000). Distribution and abundance of parasite nematodes: ecological specialization, phylogenetic constraints or simply epidemiology? *Oikos*, **88**, 563–573.

Morand, S., & Krasnov, B. R. (2008). Why apply ecological laws to epidemiology? *Trends in Parasitology*, **24**, 304–309.

Morand, S., Pointier, J.-P., Borel, G., & Theron, A. (1993). Pairing probability of schistosomes related to their distribution among the host population. *Ecology*, **74**, 2444–2449.

Newton, I. (1998). *Population Limitation in Birds*. London: Academic Press.

Norman, R., Bowers, R. G., Begon, M., & Hudson, P. J. (1999). Persistence of tick-borne virus in the presence of multiple host species: tick reservoirs and parasite mediated competition. *Journal of Theoretical Biology*, **200**, 111–118.

Ostfeld, R., & Keesing, F. (2000). The function of biodiversity in the ecology of vector-borne zoonotic diseases. *Canadian Journal of Zoology*, **78**, 2061–2078.

Perkins, S. E., Cattadori, I. M., Tagliapietra, V., Rizzoli, A., & Hudson P. J. (2003). Evidence for functional groups in disease transmission, *International Journal for Parasitology*, **33**, 909–917.

Perkins, S. E., Cattadori, A., Tagliapietra, V., Rizzoli. A., & Hudson, P. J. (2006). Localized deer absence leads to loss of the dilution effect and tick amplification. *Ecology*, **87**, 1981–1986.

Poulin, R. (1993). The disparity between observed and uniform distibutions: a new look at parasite aggregation. *International Journal for Parasitology*, **23**, 937–944.

Poulin, R. (1996). Sexual inequalities in helminth infections: a cost of being male? *The American Naturalist*, **147**, 289–295.

Poulin, R. (1999). Body size vs abundance among parasite species: positive relationships? *Ecography*, **22**, 246–250.

Poulin, R. (2006). Variation in infection parameters among populations within parasite species: intrinsic properties versus local factors. *International Journal for Parasitology*, **36**, 877–885.

Poulin, R. (2007). *Evolutionary Ecology of Parasites: From Individuals to Communities* (2nd edn). Princeton, NJ: Princeton University Press.

Randolph, S. E. (1977). Changing spatial relationships in a population of *Apodemus sylvaticus* with the onset of breeding. *Journal of Animal Ecology*, **46**, 653–676.

Rizzoli, A., Hauffe, H. C., Tagliapietra, V., Neteler, M., & Rosà R. (2009). Forest structure and roe deer abundance predict tick-borne encephalitis risk in Italy. *PLoS ONE*, **4**, E4336.

Rosà, R., Pugliese, A., Ghosh, M., Perkins, S. E., & Rizzoli, A. (2007). Temporal variation of *Ixodes ricinus* intensity on the rodent host *Apodemus flavicollis* in relation to local climate and host dynamics. *Vector-Borne and Zoonotic Diseases*, **7**, 285–295

Rosenzweig, M. L. (1981). A theory of habitat selection. *Ecology*, **62**, 327–335.

Rousset, F., Thomas, F., de Meeûs, T., & Renaud, F. (1996). Inference of parasite-induced host mortality from distribution of parasite loads. *Ecology*, **77**, 2203–2211.

Schofield, S., & Torr, S. J. (2002). A comparison of feeding behaviour of tsetse and stable flies. *Medical and Veterinary Entomology*, **16**, 177–185.

Shaw, D. J., & Dobson, A. P. (1995). Patterns of macroparasite abundance and aggregation in wildlife populations: a quantitative review. *Parasitology*, **111**, S111–S127.

Shaw, D. J., Grenfell, B. T., & Dobson, A. P. (1998). Patterns of macroparasite aggregation in wildlife host populations. *Parasitology*, **117**, 597–610.

Simkova, A., Kadlec, D., Gelnar, M., & Morand, S. (2002). Abundance-prevalence relationship of gill congeneric ectoparasites: testing the core satellite hypothesis and ecological specialization. *Parasitology Research*, **88**, 682–686.

Smith, A., Telfer, S., Burthe, S., Bennett, M., & Begon, M. (2005). Trypanosomes, fleas and field voles: ecological dynamics of a host-vector-parasite interaction. *Parasitology*, **131**, 355–365.

Sorci, G., Defraipont, M., & Clobert, J. (1997). Host density and ectoparasite avoidance in the common lizard (*Lacerta vivipara*). *Oecologia*, **11**, 183–188.

Stanko, M., Krasnov, B. R., & Morand, S. (2006). Relationship between host density and parasite distribution: inferring regulating mechanisms from census data. *Journal of Animal Ecology*, **75**, 575–583.

Stanko, M., Krasnov, B. R., Miklisova, D., & Morand, S. (2007). Simple epidemiological model predicts the relationships between prevalence and abundance in ixodid ticks. *Parasitology*, **134**, 59–68.

Stradiotto, A., Cagnacci, F., Delahay, R., Tioli, S., Nieder, L., & Rizzoli, A. (2009). Spatial organization of the yellow-necked mouse: effects of density and resource availability. *Journal of Mammalogy*, **90**, 704–714.

Tagliapietra, V., Rosà, R., Arnoldi, D., *et al.* (2011). Saturation deficit and deer density affect questing activity and local abundance of *Ixodes ricinus* (Acari, Ixodidae) in Italy. *Veterinary Parasitology*, **183**, 114–124.

Taylor, L. R., Woiwod, I. P., & Perry, J. N. (1979). The negative binomial as a dynamic ecological model and density-dependence of *k*. *Journal of Animal Ecology*, **48**, 289–304.

Telfer, S., Bown, K. J., Sekules, R., *et al.* (2005). Disruption of a host-parasite system following the introduction of an exotic host species. *Parasitology*, **130**, 661–668.

Ulmanen, I., & Myllymäki, A. (1971). Species composition and numbers of fleas (Siphonaptera) in a local population of the field vole, *Microtus agrestis* (L.). *Annales Zoologici Fennici*, **8**, 374–384.

Wilson, K., Bjørnstad, O. N., Dobson, A. P., *et al.* (2001). Heterogeneities in macroparasite infections: patterns and processes. In P. J. Hudson, A. Rizzoli, B. T. Grenfell, H. Heesterbeek & A. P. Dobson (Eds.), *The Ecology of Wildlife Diseases* (pp. 6–44). Oxford: Oxford University Press.

Zahn, A., & Rupp, D. (2004). Ectoparasite load in European vespertilionid bats. *Journal of Zoology*, **262**, 383–391.

Zuk, M., & McKean, K. A. (1996). Sex differences in parasite infections: patterns and processes. *International Journal for Parasitology*, **26**, 1009–1024.

3   Metapopulation dynamics in marine parasites

Ana Pérez-del-Olmo, Aneta Kostadinova and Serge Morand

Introduction

The metapopulation framework stemming from Levins's (1969, 1970) seminal concept and which evolved into a modern ecological theory (Hanski & Gilpin, 1997; Hanski, 1999a, 1999b; Hanski & Gaggiotti, 2004; Kritzer & Sale, 2006) is based on the development of ideas from, and applications to, terrestrial systems. However, key environmental differences exist between marine and terrestrial ecosystems, such as the larger scale of chemical, material and organism transport resulting in the greater "openness" of local marine environments (Carr et al., 2003; Sale et al., 2006) and higher marine population connectivity. There are relatively few barriers that might delineate dispersal and migration in the ocean compared with those in terrestrial or freshwater environments that are physically fragmented into discrete patches of habitat supporting discrete local populations (Waples, 1998). Further terrestrial-marine differences with relevance for the application of metapopulation theory in marine systems are the high per capita fecundity and dispersal potential of many marine species, leading to a more open spatial structure of the populations (via decoupling of local offspring production from recruitment to a parental population; see, e.g., Roughgarden et al., 1988; Carr et al., 2003; Kinlan & Gaines, 2003; Sale et al., 2006).

The overall greater dispersal potential in the ocean as evidenced by the different population genetic structures of marine animals (i.e., large effective population sizes, higher gene flow and higher genetic homogeneity maintained by pelagic dispersal; e.g., Waples, 1998; Carr et al., 2003, and references therein) may have direct effects on the structure and dynamics of the populations of their parasites. Among important adaptations to the dilute environment and longer food chains in the ocean are the generalist ability of parasites to infect both new intermediate and definitive hosts and to exploit new food-web pathways through the use of paratenic hosts that ensure wider dispersal in time and space (Marcogliese, 2002, 2007). What are the implications of these characteristics of marine parasite metapopulations for recognizing repeatable patterns suggestive of equilibrium or nonequilibrium conditions, and how can a disturbance of such patterns by human activities (e.g., overfishing) be recognized?

The Balance of Nature and Human Impact, ed. Klaus Rohde. Published by Cambridge University Press.
© Cambridge University Press 2013.

Metapopulations within metapopulations

The generally more open spatial population structures of marine species have led to an emphasis on the demographic influence of dispersal of individuals among a system of discrete local populations (versus local extinctions) thereby stressing the importance of the coupling of spatial scales for metapopulations in the ocean (Kritzer & Sale, 2004; Sale *et al.*, 2006). These ideas are incorporated in the most recent "marine" definition of a metapopulation as a system of populations of a species in which (i) local populations inhabit discrete habitat patches and (ii) interpatch dispersal is neither so low as to negate significant demographic connectivity nor so high as to eliminate any independence of local population dynamics, including a degree of asynchrony with other local populations (Sale *et al.*, 2006).

Parasites exist as demographically coupled systems, in the way their free-living hosts (as well as non-host organisms) do. Hereafter, we will use the term "parasite" broadly, focusing for simplicity on macroparasites that cannot increase their populations without leaving the host (e.g., helminths, crustaceans) and our examples are largely derived from ectoparasitic and endoparasitic helminths that do not reproduce directly within their hosts. The view that parasite populations constitute metapopulations is well established in parasitological literature (e.g., Esch & Fernández, 1993; Bush *et al.*, 1997; Margolis *et al.*, 1981; Esch *et al.*, 2002; Dobson, 2003; Šimková & Morand, 2005). Parasites live in hosts which represent small, well-delimited patches of suitable habitat (host individuals) of larger habitat fragments (host populations) that are further nested within spatially structured metapopulations. Because parasite populations are at least partially embedded within host populations, the mechanisms defining population structure of the hosts would inevitably affect parasite population structuring. Some structural differences due to the fragmented structure of host populations and the complexity of host-parasite relationships resulting in compositional heterogeneity (stage-structuring depending on life-cycle patterns; see Bush *et al.*, 1997) exist, so that specific terminology has been established to address the nested hierarchical nature of parasite populations. Three hierarchical levels are currently recognised: (i) infrapopulation, defined as all individuals of a parasite species inhabiting an individual host; (ii) component population, including all individuals of a specified life history phase of a parasite species, in an ecosystem; and (iii) suprapopulation, comprising all individuals of a parasite species, in all stages of development, in an ecosystem.

Bush *et al.* (1997) stated that "both component populations and suprapopulations are metapopulations because they are found in spatially discontinuous habitats (often hosts)". Establishment of terminological frameworks, however, does not solve all problems in defining the nature and extent of parasite populations. Thus the question of whether all individuals of a single parasite species within a single host or whether all life-cycle stages of a single parasite species within a given ecosystem constitute a population (Esch *et al.*, 2002) remains unsettled (Granovitch, 1999; Zelmer & Seed, 2004). This calls for a functional rather than spatial delineation of parasite populations that would advance our understanding of the mechanisms governing parasite dynamics within a metapopulation framework (Zelmer & Seed, 2004). One such fundamental mechanism of metapopulation dynamics is population connectivity (i.e., the exchange of individuals between discrete local populations) acting in marine systems through different degrees of coupling of local reproduction and recruitment (supply of offspring) (Kritzer & Sale, 2004; Sale *et al.*, 2006).

Adult macroparasites do not generally migrate between individual host patches (directly transmitted parasites being an exception especially at high host densities; e.g., Ritchie, 1997) so that infrapopulations can increase in size through recruitment only. We here illustrate the demographic coupling of the spatial scales for marine parasites by considering local parasite populations as ensembles of adult individuals of the same species inhabiting sympatric populations of one or more definitive host species, that are coupled by the production and supply of offspring at the local scale and by host movement (active dispersal or drifting/rafting) at the regional scale. Adult infra- and component populations are linked by contribution to a common larval pool and thus may represent replicates (and exhibit demic structure) of the local populations depending on the degree of pre-recruitment offspring mixing (Nadler, 1995; Criscione & Blouin, 2005; Criscione *et al.*, 2011). Figure 3.1 illustrates a continuum of demographic connectivity (low to high) among local parasite populations considering the degree of host specificity and parasite dispersal abilities mediated by the mobility of the intermediate, paratenic and

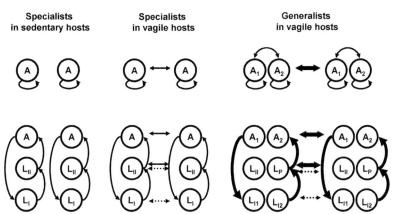

Figure 3.1.   A simplified continuum of parasite population structures relative to life cycle (direct, first row, and complex, second row) and linkages between two local populations (suprapopulations) (columns: demographically "closed" to "open" systems). Populations of parasitic larval stages for parasites with complex life cycles are represented as two layers corresponding to the two-stage classes for larval habitats (hosts in the life cycle): first intermediate (bottom layer) and second (third)/paratenic (middle layer) hosts. Top layer indicates adult populations. Single-tipped arrows indicate local unidirectional larval dispersal paths: propagule production (left) and supply of offspring (right). Double-tipped arrows indicate large-scale bi-directional dispersal paths: dispersal via host drifting (invertebrate hosts, dashed arrows) and dispersal via active host movement (vertebrate hosts, solid arrows); arrow width is proportional to the strength of the linkages between populations. A, populations of adults (numbers indicate populations in different host species); L_I, populations of larval stages in the fisrt intermediate hosts; L_{II}, populations of larval stages in the second intermediate hosts; LP, populations of larval stages in the paratenic hosts.

definitive hosts. At one extreme are the narrow host specialists in sedentary hosts that would tend to exist in populations with closed recruitment, whereas a relaxed specificity to both definitive and intermediate hosts and high host mobility would lead to high (and demographically important) connectivity at the other extreme (populations with open recruitment); in the latter case local populations within a metapopulation are connected by both host-mediated adult and larval dispersal. Parasites with complex life cycles possess spatially segregated juvenile and adult habitats (second row in Figure 3.1) and thus should exhibit a strong reliance on local environmental conditions (eggs, free-living dispersive larval stages) or the spatial distribution of the intermediate hosts (a successful transmission requires overlap of the distributions of all hosts).

Understanding the degree of independence of local populations would profit from empirical data on lifespan duration and dispersal abilities of parasites (e.g., measurements of the rates of larval release, transport and settlement), in association with knowledge of larval behaviors that unfortunately is critically lacking (but interest is emerging; see Fingerut *et al.*, 2003a, 2003b; Zimmer *et al.*, 2009). However, the focal nature of parasite transmission (Esch *et al.*, 2002; Esch & Fernández, 1993) indicates that recruitment to and demographic exchange of individuals among (if at all) host patches and fragments occur on very small spatial scales. Thus the larval free-living helminth dispersive stages are lecithotrophic (non-feeding), short-lived and travel short distances (Esch & Fernández, 1993). For example Zimmer *et al.* (2009), in a series of field measurements and mechanistic experiments, revealed that actual dispersal of *Himasthla rhigedana* cercariae emitted by their snail host *Cerithidea californica* at Carpinteria Salt Marsh Reserve was about 130 times less than passive transport (1.95 m vs. 259 m); both measures indicate a rather limited dispersal. On the other hand, large-scale dispersal distances (hundreds and thousands of kilometres) in the ocean are achieved through highly vagile populations of the vertebrate definitive hosts (birds, mammals and fish) and the intermediate/paratenic fish hosts, and possibly via passive drifting/rafting of invertebrate intermediate hosts (e.g., migration across the Pacific Ocean, from New Zealand to Chile, of adult trochid gastropods has been inferred recently; see Donald *et al.*, 2005).

Spatial patterns of populations and epidemiology

The third-order scaling of habitat fragmentation for parasites outlined above provides an excellent system to study the distribution of species abundance reflecting species-environment relationships in both space and time. This scaling allows comparative analyses at different scales using a large number of replicates (e.g., infra- and component populations of a species) and its hierarchical nature provides a setting for examination of the effect of larger-scale processes on local population dynamics, thus linking the approaches based on metapopulation theory and landscape ecology. The clear patch delimitation in the naturally fragmented parasite populations and the aggregated nature of host populations bear strong analogies to metapopulation modeling frameworks (Grenfell & Harwood, 1997; Nee *et al.*, 1997) and this has resulted in the burst of epidemiological models that include a much larger amount of detail in their structure than metapopulation models (Dobson, 2003, and references therein), thus providing insights into metapopulation processes at several spatial and temporal scales.

Two general patterns that capture essential fundamentals of the structuring of the distributions of free-living terrestrial species have received thorough attention in the contexts of metapopulation dynamics during the last decade (Gaston, 2003; Gaston *et al.*, 2006). The first is the positive relationship between the mean local abundance (μ) of a species (e.g., across patches) and its variance (σ^2). The second is the positive relationship between the mean local abundance (μ) of a species and its probability of occurrence (*p*; e.g., the proportion of the available patches occupied) (Hanski, 1982; Brown, 1984; Gaston, 1996; Gaston *et al.*, 2000, 2006). These two patterns are most widely documented for a wide range of populations at different spatial scales, and a number of models and hypotheses have been suggested to describe and explain them (e.g., Holt *et al.*, 2002; He & Gaston, 2003; Gaston *et al.*, 2006, and references therein).

The variance-abundance relationship depends on the underlying spatial distribution of individuals among sampling units (e.g., host patches): the variance (σ^2), for example, (i) equals the mean (i.e., $\sigma^2 = \mu$) of a random (Poisson) distribution; (ii) increases more rapidly with the mean of a negative binomial distribution (NBD) but less than proportional to the squared mean (i.e., $\sigma^2 = \mu + \mu^2/k$; Anderson & Gordon, 1982); (iii) is proportional to the squared mean of gamma or log-normal distribution (i.e., $\sigma^2 = c\mu^2$; Dennis & Patil, 1984; Engen & Lande, 1996); or (iv) rises even faster with the mean; i.e., is proportional to a fractional power of the mean (Taylor's power law; Taylor, 1961; Taylor & Taylor, 1977; Eqn (1)):

$$\sigma^2 = a\,\mu^b \qquad\qquad (1)$$

where *a* and *b* are population parameters; *a* depends upon the size of the sampling units and *b* is a measure of spatial heterogeneity. The parameters *a* and *b* are estimated by linear regression of log-transformed variance (σ^2) against log-transformed mean (μ).

NBD and Taylor's power law appear to be the most flexible models for aggregated distributions, the latter covering a wider range of distributions, and perhaps the most intensively applied in parasite epidemiology and evolutionary ecology of parasites (reviewed by Morand & Krasnov, 2008; see also Wilson *et al.*, 2002; Pérez-del-Olmo *et al.*, 2011, and references therein). Aggregated distribution of parasites among host patches appears to be an almost universal pattern among populations of metazoan parasites and therefore a characteristic feature of parasitism (Crofton, 1971; Shaw & Dobson, 1995). Parasite aggregation, best described by the NBD (Anderson & May, 1978; May & Anderson, 1978; Shaw & Dobson, 1995; Shaw *et al.*, 1998), is a central feature of epidemiological models due to its important role in the population dynamics of both the parasite and its host (Anderson & May, 1978). An aggregated distribution is characterized statistically by a variance-to-mean (or σ^2/μ) ratio significantly greater than unity and by agreement with the NBD; a number of measures of aggregation are detailed in Elliott (1977) and Wilson *et al.* (2002), for example. The spatial aggregation parameter *k* of the NBD ($0 < k <\sim 20$, typically $k < 1$; see Shaw & Dobson, 1995) can be estimated from the variance and the mean using simple moment estimates, for large and small samples (number of sampling units < 50), respectively (Elliott, 1977), or from Taylor's power law as:

$$1/k = a\mu^{b-2} - 1/\mu \qquad (2)$$

The abundance-occupancy relationship (abundance-prevalence in parasitological literature) derived from basic epidemiological models predicted by the NBD takes the form

$$p = 1 - (1 + \mu/k)^{-k} \qquad (3)$$

Substituting and recognizing that σ^2 is defined by Eqn (1) gives a general model linking mean abundance, its spatial variance and occupancy (He & Gaston, 2003):

$$p = 1 - \left(1 + \mu/\sigma^2\right)^{\mu^2/\sigma^2 - \mu} \qquad (4)$$

Notably, examination of the two distributional patterns in association using terrestrial mammalian host-parasite systems has revealed their interconnection in defining species spatial distribution, i.e., that a positive abundance-occupancy relationship in parasites can be explained by demographic parameters defined by epidemiological models (gastrointestinal nematodes, see Morand & Guégan, 2000; ectoparasites, see Krasnov *et al.*, 2005; Matthee & Krasnov, 2009), a connection also depicted by a general model linking the mean abundance, the spatial variance in abundance, and the occupancy in free-living species populations (Eqn (4); see He & Gaston, 2003; Gaston *et al.*, 2006).

Parasites of fish may represent a useful system in the search for general patterns of parasitism in the ocean since they are perhaps the best-studied group of marine parasites with respect to patterns in both taxonomic and ecological diversity. However, ecological studies have been focused on the community rather than population level and then at fairly local scales (e.g., Lo *et al.*, 1999; Osset *et al.*, 2005; Vignon & Sasal, 2010). One might expect that the wealth of community studies, some carried out at larger spatial scales, may serve as sources of comparative data to test hypotheses at the population level. However, surprisingly few provide empirical data useful to look at the abundance-occupancy or abundance-variance relationships for individual parasites.

Here we focus on the findings based on three large datasets of marine parasites that may inspire further search of patterns and possible mechanisms of parasite spatial structuring in the marine environment. The first comprises data for fish head and gill ectoparasites originating from the most comprehensive marine parasite survey by Rohde *et al.* (1995) (Morand *et al.*, 1999, 2002). The second comprises data on population samples for 20 species (predominantly endoparasites) of the sparid *Boops boops* from the North-East Atlantic and the Mediterranean (Pérez-del-Olmo *et al.*, 2011). The third represents a compilation of previously published data for 164 parasite populations in crustaceans (49 host-parasite combinations) and 338 populations in bivalves (36 host-parasite combinations) (Thieltges *et al.*, 2009).

Morand *et al.* (1999) investigated the patterns of spatial distribution of parasites in marine fish populations using data for 171 population samples of 36 species-rich ectoparasite communities from a large-scale survey by Rohde *et al.* (1995). They revealed a strong positive abundance-variance relationship interspecifically and demonstrated stronger intraspecific than interspecific aggregation, thus supporting the "aggregation model of coexistence" (Shorrocks, 1996). In a further analysis of this dataset, Morand *et al.* (2002) depicted prerequisites of the "core-satellite" hypothesis, i.e., an overall bimodal frequency distribution of ectoparasite prevalence and a positive abundance-prevalence relationship.

Recently we examined a large taxonomically consistent dataset of parasites of *B. boops*, along a coastal positional gradient from northern North-East Atlantic to northern Mediterranean coasts of Spain, which allowed a number of key questions regarding the structure of parasite populations and communities in this marine system to be addressed at different spatial scales. Parasites of *B. boops* exhibited a highly significant positive interspecific abundance-occupancy pattern at three spatial scales. The species formed a continuum at all scales but a large group at the lower abundance-distribution extreme (14 rare species) and a smaller group of species present in a substantial proportion of the component populations in all localities at the upper extreme were distinguishable (Pérez-del-Olmo *et al.*, 2009). These "core" species were also found to exhibit a wide geographical distribution (Pérez-del-Olmo *et al.*, 2007) and to infect host populations earlier than rare species (Pérez-del-Olmo *et al.*, 2008).

Examination of a subset of *B. boops* data consisting of 294 population samples for 20 metazoan parasite species within a metapopulation framework resulted in five key findings concerning the variation in infection parameters and patterns of aggregation of parasites among individual hosts and the effect of taxonomy and ecology of transmission on the abundance-occupancy and abundance-variance relationships (Pérez-del-Olmo *et al.*, 2011). Along with published results, we here present a wider treatment of these data. We found a strong positive abundance-occupancy relationship both inter- and intraspecifically, with clearly bimodal distribution of parasite prevalence (Figure 3.2A) and core-satellite switching, and therefore agreement with three premises of the "core-satellite" model (Hanski, 1982). We also revealed that prevalence and abundance are true species characteristics, i.e., repeatable among populations of the same species across its regional distribution. This finding extends the validity of the patterns revealed for mammalian endo- and ectoparasites (Arneberg *et al.*, 1997; Krasnov *et al.*, 2006), parasites in freshwater fish (Poulin, 2006) and larval trematodes in their invertebrate second intermediate hosts (Thieltges *et al.*, 2009; see below) to marine fish parasites.

Parasites of *B. boops* exhibited a strong abundance-variance relationship with a fit to a power function, relating spatial variance to mean abundance (i.e., Taylor's power law, Eqn (1)) both interspecifically and intraspecifically. The slope of the linear regression (Figure 3.2B) is significantly greater than unity ($b = 1.49 \pm 0.02$, $r^2 = 0.94$, $P < 0.0001$, $n = 294$) and comparable with those reported for a wide range of macroparasites ($b = 1.55 \pm 0.04$, $n = 269$; Shaw & Dobson, 1995) and free-living animal populations ($b = 1.45 \pm 0.39$; Taylor & Taylor, 1977) albeit somewhat lower than those recorded in marine fish

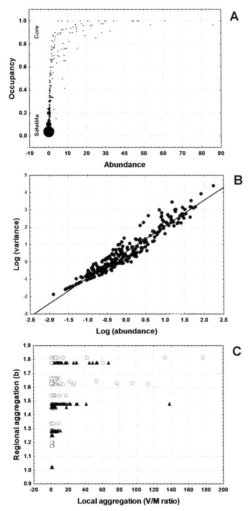

Figure 3.2. Abundance-variance-occupancy and aggregation relationships for parasites in the dataset of Pérez-del-Olmo *et al.* (2011). (A) Abundance-occupancy (prevalence). "Bubbles" indicate the relative frequencies of the number of points represented by a single plot position. (B) Abundance-variance (log-transformed data). (C) Local-regional relationship for parasite spatial heterogeneity. Aggregation levels assessed by the slope (*b*) of Taylor's power law (regional) and by variance-to-mean ratio (V/M) (local). Solid triangles indicate populations of the core species. Top two rows of points represent populations of specific parasites of *B. boops*: *Magnibursatus bartolii*, *Wardula bartolii* (open circles) and *Bacciger israelensis* (solid triangles).

ectoparasite ($b = 1.71 \pm 0.04$, $n = 177$; Morand *et al.*, 2002) and mammalian nematode ($b = 1.80 \pm 0.04$; $n = 45$; Morand & Guegan, 2000) populations. Tests of the effect of ecological determinants provided strong evidence that parasite prevalence and abundance and the abundance-variance relationship are dependent on host specificity and regional distribution patterns of the parasites, but not on their mode of transmission or region of origin. Ecological specialization resulted in higher infection parameters and an increase in the levels of aggregation, corresponding to a pattern observed in free-living organisms, i.e., high spatial variability associated with a small range size (Morand & Krasnov, 2008). Core parasites of *B. boops* had distinctly higher abundance, prevalence and lower spatial variation than rare species, thus supporting Taylor and Taylor's (1977) concept of dispersal as a process counteracting aggregation. These were predominantly generalist parasites utilizing multiple species of final and intermediate hosts, and alternative routes of transmission (including paratenic hosts) (Pérez-del-Olmo *et al.*, 2009). Therefore, low specificity decreasing the stochasticity of host-parasite encounters resulted in homogenization of parasite population densities across localities, a process responsible for the spatial synchrony patterns of parasite communities (Pérez-del-Olmo *et al.*, 2009).

Figure 3.2C illustrates a statistically significant (Spearman's rho 0.45, $P < 0.001$) hierarchical relationship of the estimates of parasite aggregation within a local population (measured by σ^2/μ ratio of abundance in individual fish) and among populations (*b* estimated from Taylor's power law relationship), indicating that aggregative processes at the two scales are linked, i.e., parasites which are aggregated within individual host patches also tend to exhibit more heterogeneous distribution among local populations. The triangular spread of the data shows that species within the 0–20 range of the σ^2/μ ratio (typically >1) may exhibit varying degrees of spatial clustering at the higher spatial scale and that high aggregation levels at the lower scale effectively result in high spatial heterogeneity at the regional scale (or the reverse).

Using the abundance-variance relationship we predicted both intraspecifically and replicating across species the occupancy (prevalence) in the epidemiological model developed for macroparasites by estimating the parameter *k* of the negative binomial distribution from the parameters *a* and *b* of the Taylor power law. The strong correlation between the observed and predicted prevalences suggested a strong linkage between the three measures of parasitism and their relationships and that demographic and stochastic processes explain parsimoniously three of the central premises of Hanski's (1982) core-satellite hypothesis (Pérez-del-Olmo *et al.*, 2011).

Thieltges *et al.* (2009) examined patterns of geographical variation of trematode parasitism in crustacean and bivalve intermediate hosts and found a significant positive correlation between abundance and prevalence interspecifically in both host groups. They reported strong repeatability patterns for trematode intensity and abundance in crustacean hosts, with a large proportion of the variation associated with parasite–host pair variation (81% and 78%, respectively) compared to much lower repeatability in bivalves (37.4% and 40.2%, respectively), and suggested space limitations (i.e., host size) and density-dependent mortalities as possible mechanisms for these differences. Thieltges *et al.* (2009) also tested the effect of the spatial scale (geographical distances spanning from <10 km to >1000 km) on trematode prevalence, intensity and abundance, and revealed a considerable geographical consistency in parasite load in crustacean hosts and higher variation in infection levels in bivalves at larger scales (>100 km), suggesting a higher importance of environmental factors at larger scales for trematode parasitism in the latter host group.

Concluding remarks

In spite of the rapidly growing integration of metapopulation ecology into both terrestrial and marine ecology, the use of metapopulation concepts in addressing marine parasite dynamics remains a neglected field of research. Likewise, spatial ecology is still in its infancy in spite of the powerful insights being rapidly gained from spatial modeling of parasite distribution patterns in the terrestrial and freshwater systems. It is apparent that further advancement in our understanding of the parasite population dynamics in the ocean would largely benefit from comparative studies at the level of local populations. Assessment of various spatial and temporal scales might be influential for distinguishing equilibrium vs. nonequilibrium behavior of the marine host-parasite systems especially in relation to predicting disturbance effects. We would like to suggest the following avenues of research, which would advance our knowledge of metapopulation structure and dynamics of both marine parasites and their hosts.

- Assessing the spatial scales of parasite dispersal: parasites have long been used as natural tags for assessment of spatial partitioning of fish populations. The application of advanced individual assignment techniques, especially genetic methodologies, and a holistic approach would provide powerful tools for inferring connectivity among fish and parasite populations.
- Addressing disturbance in the marine environment: fishing practices affect parasite population dynamics via degradation of benthic habitats with major effects on larval parasites in intermediate hosts, and reduction of fish populations with major effects on trophic cascades, and consequently on adult and larval parasites in fish. Examination of these effects in a metapopulation framework would help advance our predictions for habitat and parasite population recovery as a result of the development of marine reserves and from natural disasters.

References

Anderson, R. M., & Gordon, D. M. (1982). Processes influencing the distribution of parasite numbers within host populations with special emphasis on parasite-induced host mortalities. *Parasitology*, **85**, 373–398.

Anderson, R. M., & May, R. M. (1978). Regulation and stability of host-parasite population interactions. I. Regulatory processes. *Journal of Animal Ecology*, **47**, 219–247.

Arneberg, P., Skorping, A., & Read, A. F. (1997). Is population density a species character? Comparative analyses of the nematode parasites of mammals. *Oikos*, **80**, 289–300.

Brown, J. H. (1984). On the relationship between abundance and distribution of species. *American Naturalist*, **124**, 255–279.

Bush, A. O., Lafferty, K. D., Lotz, J. M., & Shostak, A. W. (1997). Parasitology meets ecology on its own terms: Margolis *et al.* revisited. *Journal of Parasitology*, **84**, 575–583.

Carr, M. H., Neigel, J. E., Estes, J. A., Andelman, S., Warner, R. R., & Largier, J. (2003). Comparing marine and terrestrial ecosystems: implications for the design of coastal marine reserves. *Ecological Applications*, **13** (Suppl.), S90–S107.

Criscione, C. D., & Blouin, M. S. (2005). Effective sizes of macroparasite populations: a conceptual model. *Trends in Parasitology*, **21**, 212–217.

Criscione, C. D., Vilas, R. N., Paniagua, E., & Blouin, M. (2011). More than meets the eye: detecting cryptic microgeographic population structure in a parasite with a complex life cycle. *Molecular Ecology*, **20**, 2510–2524.

Crofton, H. D. (1971). A quantitative approach to parasitism. *Parasitology*, **62**, 179–193.

Dennis, B., & Patil, G. P. (1984). The Gamma-distribution and weighted multimodal Gamma-distribution as model of population abundance. *Mathematical Biosciences*, **68**, 187–212.

Dobson, A. (2003). Metalife! *Science*, **301**, 1488–1490.

Donald, K. M., Kennedy, M., & Spenser, H. (2005). Cladogenesis as the result of long-distance rafting events in South Pacific topshells (Gastropoda, Trochidae). *Evolution*, **59**, 1701–1717.

Elliott, J. M. (1977). *Some Methods for the Statistical Analysis of Samples of Benthic Invertebrates* (2nd edn). Ambleside: Freshwater Biological Association.

Engen, S., & Lande, R. (1996). Population dynamic models generating the lognormal species abundance distribution. *Mathematical Biosciences*, **132**, 169–183.

Esch, G. W., & Fernández, J. G. (1993). *A Functional Biology of Parasitism: Ecological and Evolutionary Implications*. London: Chapman and Hall.

Esch, G. W., Barger, M. A., & Fellis, K. J. (2002). The transmission of digenetic trematodes: style, elegance, complexity. *Integrative and Comparative Biology*, **42**, 304–312.

Fingerut, J. T., Zimmer, C. A., & Zimmer, R. K. (2003a). Patterns and processes of larval emergence in an estuarine parasite system. *Biological Bulletin*, **205**, 110–120.

Fingerut, J. T., Zimmer, C. A., & Zimmer, R. K. (2003b). Larval swimming overpowers turbulent mixing and facilitates transmission of a marine parasite. *Ecology*, **84**, 2502–2515.

Gaston, K. J. (1996). The multiple forms of the interspecific abundance-distribution relationship. *Oikos*, **75**, 211–220.

Gaston, K. J. (2003). *The Structure and Dynamics of Geographic Ranges*. Oxford: Oxford University Press.

Gaston, K. J., Blackburn, T. M., Greenwood, J. J. D., *et al.* (2000). Abundance-occupancy relationships. *Journal of Applied Ecology*, **37** (Suppl. 1), 39–59.

Gaston, K. J., Borges, P. A. V, He, F., & Gaspar, C. (2006). Abundance, spatial variance and occupancy: arthropod species distribution in the Azores. *Journal of Animal Ecology*, **75**, 646–656.

Granovitch, A. I. (1999). Parasitic systems and the structure of parasite populations. *Helgoland Marine Research*, **53**, 9–18.

Grenfell, B., & Harwood, J. (1997). (Meta)population dynamics of infectious diseases. *Trends in Ecology and Evolution*, **12**, 395–399.

Hanski, I. (1982). Dynamics of regional distribution: the core and satellite species hypothesis. *Oikos*, **38**, 210–221.

Hanski, I. (1999a). Habitat connectivity, habitat continuity, and metapopulations in dynamic landscapes. *Oikos*, **87**, 209–219

Hanski, I. (1999b). *Metapopulation Ecology*. Oxford: Oxford University Press.

Hanski, I., & Gaggiotti, O. E. (2004). *Ecology, Genetics and Evolution of Metapopulations*. London: Elsevier Academic Press.

Hanski, I., & Gilpin, M. E. (1997). *Metapopulation Biology: Ecology, Genetics and Evolution*. New York: Academic Press.

He, F., & Gaston, K. J. (2003). Occupancy, spatial variance, and the abundance of species. *The American Naturalist*, **162**, 366–375.

Holt, A. R., Gaston, K. J., & He, F. (2002). Occupancy-abundance relationships and spatial distribution. *Basic and Applied Ecology*, **3**, 1–13.

Kinlan, B. P., & Gaines, S. D. (2003). Propagule dispersal in marine and terrestrial environments: a community perspective. *Ecology*, **84**, 2007–2020.

Krasnov, B. R., Shenbrot, G. I., Khokhlova, I. S., & Poulin, R. (2006). Is abundance a species attribute? An example with haematophagous ectoparasites. *Oecologia*, **150**, 132–140.

Krasnov, B. R., Stanko, M., Miklisova, D., & Morand, S. (2005). Distribution of fleas (Siphonaptera) among small mammals: mean abundance predicts prevalence via simple epidemiological model. *International Journal for Parasitology*, **35**, 1097–1101.

Kritzer, J. P., & Sale, P. F. (2004). Metapopulation ecology in the sea: from Levins' model to marine ecology and fisheries science. *Fish and Fisheries*, **5**, 131–140.

Kritzer, J. P., & Sale, P. F. (2006). *Marine Metapopulations*. Amsterdam: Elsevier Academic Press.

Levins, R. (1969). Some demographic and genetic consequences of environmental heterogeneity for biological control. *Bulletin of the Entomological Society of America*, **15**, 237–240.

Levins, R. (1970). Extinction. Some mathematical problems in biology. In M. Gesternhaber (Ed.), *Some Mathematical Problems in Biology* (pp. 77–107). Providence, RI: American Mathematical Society.

Lo, C. M., Morand, S., & Calzin, R. (1999). Le parasitisme des poissons coralliens. Reflet de l'habitat? *Comptes Rendus de l'Académie des Sciences Paris, Sciences de la vie*, **322**, 281–287.

Marcogliese, D. J. (2002). Food webs and the transmission of parasites to marine fish. *Parasitology*, **124** (Suppl.), S83–S99.

Marcogliese, D. J. (2007). Evolution of parasitic life in the ocean: paratenic hosts enhance lateral incorporation. *Trends in Parasitology*, **23**, 519–521.

Margolis, L. G., Esch, G. W., Holmes, J. C., Kuris, A. M., & Schad, G. A. (1981). The use of ecological terms in parasitology (Report of an *ad hoc* committee of the American Society of Parasitologists). *Journal of Parasitology*, **68**, 131–133.

Matthee, S., & Krasnov, B. R. (2009). Searching for generality in the patterns of parasite abundance and distribution: ectoparasites of a South African rodent, *Rhabdomys pumilio*. *International Journal for Parasitology*, **39**, 781–788.

May, R. M., & Anderson, R. M. (1978). Regulation and stability of host-parasite population interactions. II. Destabilising processes. *Journal of Animal Ecology*, **47**, 249–267.

Morand, S., & Guégan, J.-F. (2000). Distribution and abundance of parasite nematodes: ecological specialization, phylogenetic constraints or simply epidemiology? *Oikos*, **88**, 563–573.

Morand, S., & Krasnov, B. R. (2008). Why apply ecological laws to epidemiology? *Trends in Parasitology*, **24**, 304–309.

Morand, S., Poulin, R., Rohde, K., & Hayward, C. (1999). Aggregation and species coexistence of ectoparasites of marine fishes. *International Journal for Parasitology*, **29**, 663–672.

Morand, S., Rohde, K., & Hayward, C. (2002). Order in ectoparasite communities of marine fish is explained by epidemiological processes. *Parasitology*, **124** (Suppl.), S57–S63.

Nadler, S. A. (1995). Microevolution and the genetic structure of parasite populations. *Journal of Parasitology*, **81**, 395–403.

Nee, S., May, R. M., & Hassell, P. (1997). Two-species metapopulation models. In I. Hanski & M. E. Gilpin (Eds.), *Metapopulation Biology: Ecology, Genetics and Evolution* (pp. 123–147). San Diego, CA: Academic Press.

Osset, E. A., Fernandez, M., Raga, J. A., & Kostadinova, A. (2005). Mediterranean Diplodus annularis (Teleostei: Sparidae) and its brain parasite: unforeseen outcome. *Parasitology International*, **54**, 201–206.

Pérez-del-Olmo, A., Fernández, M., Gibson, D. I., Raga, J. A., & Kostadinova, A. (2007). Descriptions of some unusual digeneans from *Boops boops* L. (Sparidae) and a complete checklist of its metazoan parasites. *Systematic Parasitology*, **66**, 137–158.

Pérez-del-Olmo, A., Fernández, M., Raga, J. A., Kostadinova, A., & Poulin, R. (2008). Halfway up the trophic chain: development of parasite communities in the sparid fish *Boops boops. Parasitology*, **135**, 257–268.

Pérez-del-Olmo, A., Fernández, M., Raga, J. A., Kostadinova, A., & Morand, S. (2009). Not everything is everywhere: similarity-decay relationship in a marine host-parasite system. *Journal of Biogeography*, **36**, 200–209.

Pérez-del-Olmo, A., Morand, S., Raga, J. A., & Kostadinova, A. (2011). Abundance–variance and abundance–occupancy relationships in a marine host–parasite system: the importance of taxonomy and ecology of transmission. *International Journal for Parasitology*, **41**, 1361–1370.

Poulin, R. (2006). Variation in infection parameters among populations within parasite species: intrinsic properties versus local factors. *International Journal for Parasitology*, **36**, 877–885.

Ritchie, G. (1997). The host transfer ability of *Lepeophtheirus salmonis* (Copepoda: Caligidae) from farmed Atlantic salmon, *Salmo salar* L. *Journal of Fish Diseases*, **20**, 153–157.

Rohde, K., Hayward, C., & Heap, M. (1995). Aspects of the ecology of metazoan ectoparasites of marine fishes. *International Journal for Parasitology*, **25**, 945–970.

Roughgarden, J., Gaines, S. D., & Possingham, H. P. (1988). Recruitment dynamics in complex life cycles. *Science*, **241**, 1460–1466.

Sale, P. F., Hanski, I., & Kritzer, J. P. (2006). The merging of metapopulation theory and marine ecology: establishing the historical context. In J. P. Kritzer & P. F. Sale (Eds.), *Marine Metapopulations* (pp. 3–28). Amsterdam: Elsevier Academic Press.

Shaw, D. J., Grenfell, B. T., & Dobson, A. P. (1998). Patterns of macroparasite aggregation in wildlife host populations. *Parasitology*, **117**, 597–610.

Shaw, D. J., & Dobson, A. P. (1995). Patterns of macroparasite abundance and aggregation in wildlife populations: a quantitative review. *Parasitology*, **111** (Suppl.), S111–S133.

Shorrocks, B. (1996). Local diversity: a problem with too many solutions. In M. Hochberg, J. Clobert & R. Barbault (Eds.), *The Genesis and Maintenance of Biological Diversity* (pp. 104–122). Oxford: Oxford University Press.

Šimková, A., & Morand, S. (2005). Metapopulation biology of marine parasites. In K. Rohde (Ed.), *Marine Parasitology* (pp. 302–309). Melbourne and Wallingford: CSIRO and CAB International.

Taylor, L. R. (1961). Aggregation, variance and the mean. *Nature*, **189**, 732–735.

Taylor, L. R., & Taylor, R. A. J. (1977). Aggregation, migration and population dynamics. *Nature*, **265**, 415–421.

Thieltges, D. W., Fredensborg, B. L., & Poulin, R. (2009). Geographical variation in metacercarial infection levels in marine invertebrate hosts: parasite species character versus local factors. *Marine Biology*, **156**, 983–990.

Vignon, M., & Sasal, P. (2010). Multiscale determinants of parasite abundance: a quantitative hierarchical approach for coral reef fishes. *International Journal for Parasitology*, **40**, 443–451.

Waples, R. S. (1998). Separating the wheat from the chaff: patterns of genetic differentiation in high gene flow species. *Journal of Heredity*, **89**, 438–450.

Wilson, K., Bjørnstad, O. N., Dobson, A. P., *et al.* (2002). Heterogeneities in macroparasite infections: patterns and processes. In P. J. Hudson, A. Rizzoli, B. T. Grenfell, H. Heesterbeek & A. P. Dobson (Eds.), *The Ecology of Wildlife Diseases* (pp. 6–44). Oxford: Oxford University Press.

Zelmer, D. A., & Seed, J. R. (2004). A path hath smaller patches: delineating ecological neighbourhoods for parasites. *Comparative Parasitology*, **71**, 93–103.

Zimmer, R. K., Figerut, J. T., & Zimmer, C. A. (2009). Dispersal pathways, seed rains, and the dynamics of larval behavior. *Ecology*, **90**, 1933–1947.

Figure 1.1.   The bridled goby (*Coryphopterus glaucofraenum*) suffers density-dependent mortality when crevices used to hide from predators are limited.

Figure 1.2.   Most coral reef fishes depend upon shelter provided by the physical structure of living corals. As coral reefs continue to degrade, shelter may become more limited and even if fish populations decline in density they may still experience density-dependent mortality.

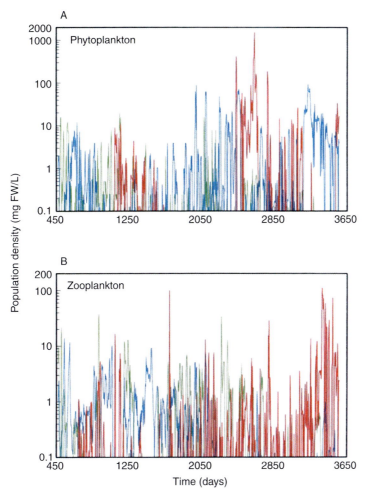

Figure 4.1. Nonequilibrium dynamics in an experimental community of plankton that consisted of more than 20 species and developed in a long-term laboratory experiment under constant external conditions. Data show the observed time course of (A) phyto- and (B) zooplankton. (A: green = green flagellates, blue = prokaryotic pico-phytoplankton, red = the diatom Melosira. B: green = the rotifer Brachionus, blue = the copepod *Eurytemora*, red = protozoans). After Scheffer *et al.* (2003). With permission of the authors and Kluwer Academic Publishers (Springer).

Figure 5.2.   (A) Intense crown fire that has burned tall *Eucalyptus* forest and adjacent rainforest. Such fires were thought to trigger instability leading to fire feedbacks, but both these Australian forests show remarkable floristic stability after disturbance. (B) Grassland fire adjacent to forest thicket in the high-altitude grassland biome of South Africa. Repeated fires are predicted to decrease the less flammable thickets and convert these patches to grassland as fire regimes change. Increased atmospheric CO_2 may, however, favor forest expansion. Photos Peter Clarke.

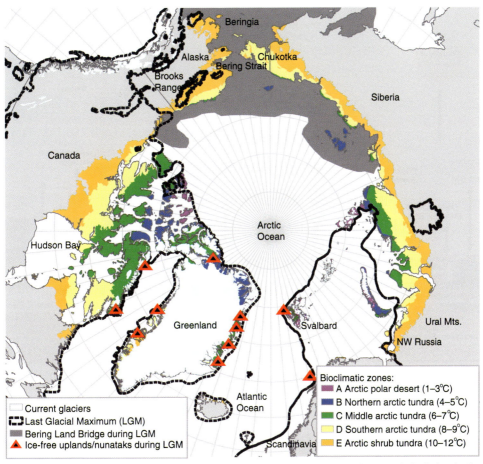

Figure 10.1.   Bioclimatic zones (http://www.arcticatlas.org/maps/catalog/index.shtml) and glaciation in the Arctic. Glacial limits are according to Ehlers and Gibbard (2004) for Europe and Dyke *et al.* (2003) for North America. Ice-free uplands and nunataks are according to Brochmann *et al.* (2003).

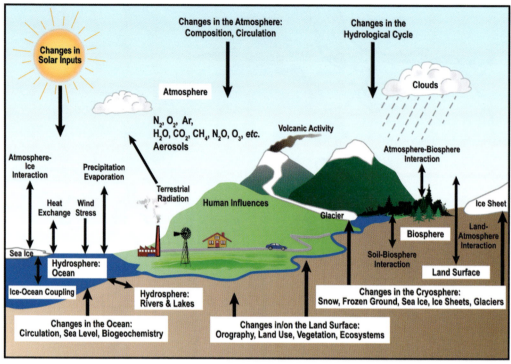

Figure 13.1. Important interactions in the dynamic climate system. [IPCC AR4 WG1, faq-1-2-fig-1]

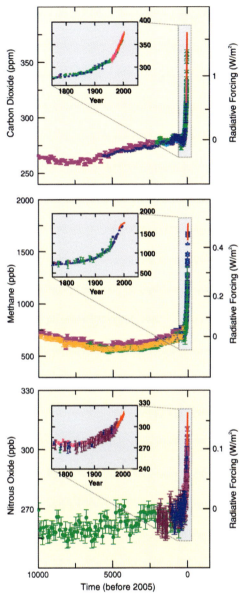

Figure 13.2. Concentrations of key greenhouse gases over the past 10 000 years. [IPCC AR4 WG1, fig2–3]

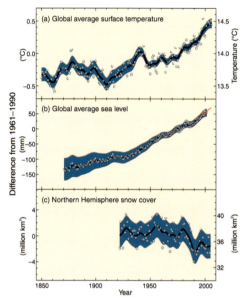

Figure 13.3. Changes in temperature, sea level and snow cover since 1850. [IPCC AR4 WG1, fig1–1]

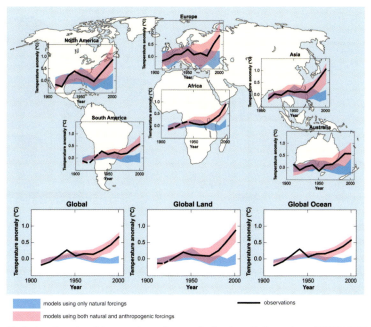

Figure 13.4. Global and regional temperature changes in the twentieth century. [IPCC AR4 WG1, fig2–5]

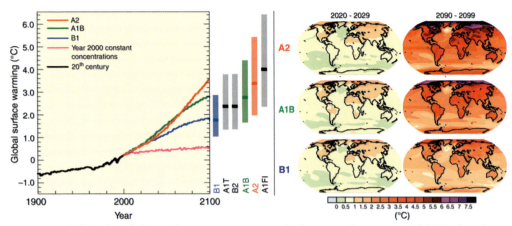

Figure 13.5. (left) Projected changes in average temperature in the twenty-first century. (right) Projected warming patterns for 2020–2029 and 2090–2099. [IPCC AR4 WG1, fig3–2]

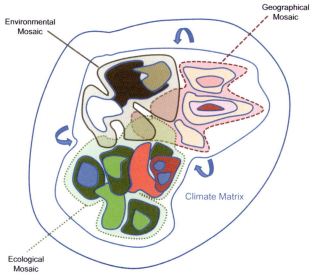

Figure 14.1. Mosaic structure in space and time. Faunas are structured relative to mosaics resulting from an
intricate interaction among ecological (= species richness, population structure, diversity, faunal
associations and processes), environmental (= habitat) and geographical (= structural homogeneity
and heterogeneity, barriers, invasion corridors) drivers unfolding within a climatological matrix.
Autonomous, overlapping and interdependent mosaics emerge and interact within regimes of
climate that contribute to periods or episodes of stasis, perturbation and transition.

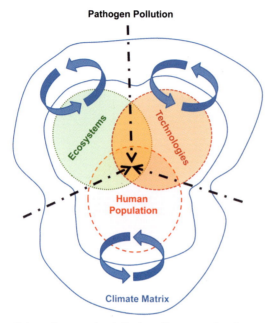

Figure 15.1. Drivers for emerging infectious diseases and pathogen pollution. A relationship for pathogen pollution and processes for emerging infectious diseases arising from interactions among human populations, ecosystem structure (and biodiversity), and the development of technologies which influence globalization. Feedback loops within a scale-dependent matrix for climate and accelerated warming influence patterns of invasion and emergence for pathogens.

Part II

Nonequilibrium and Equilibrium in Communities

4   The paradox of the plankton

Klaus Rohde

General background

Freshwater streams and lakes are habitats for complex ecosystems, of which plankton is an important component. Even more extensive are the oceans, which cover about 70% of the Earth's surface. Marine ecosystems including their plankton have very great ecological and economic significance. Although our knowledge of biodiversity patterns in marine phyto- and zooplankton (compared to terrestrial systems) is still very limited (Irigoien *et al.*, 2004), much work, some of it theoretical, some experimental, has led to important insights.

The study of plankton has played a crucial historical role in our understanding of ecological processes. The famous "paradox of the plankton" formulated by Hutchinson (1961) drew attention to the fact that many more species coexist in the supposedly homogeneous habitat than permitted under the competitive exclusion principle of Gause. Hutchinson suggested that nonequilibrium conditions might lead to the greater than expected diversity, a suggestion shown to be correct by many subsequent studies. Hutchinson himself thought that seasons and weather-induced fluctuations were responsible. But, in addition, as reviewed by Scheffer *et al.* (2003), homogeneity due to mixing hardly exists, and even in the open ocean meso-scale vortices and fronts result in spatial heterogeneity. Moreover, modeling of plankton communities has shown that even in homogeneous and constant environments plankton may never reach equilibrium, because multi-species interactions may lead to oscillations and chaos. This is supported by laboratory experiments, which have shown highly irregular and unpredictable long-term fluctuations at the species level (Figure 4.1), although total algal biomass and other indicators at higher aggregation levels may show regular patterns.

In the following I discuss some of the more important studies on which these conclusions are based. Some deal with communities at the species level, others with higher levels of aggregation over large spatial scales.

Competition for resources, competitive exclusion, coexistence and chaos

An early monographic treatment of plankton ecology is by Harris (1986), who showed that environmental disturbances are so frequent that competitive exclusion in phytoplankton

The Balance of Nature and Human Impact, ed. Klaus Rohde. Published by Cambridge University Press.
© Cambridge University Press 2013.

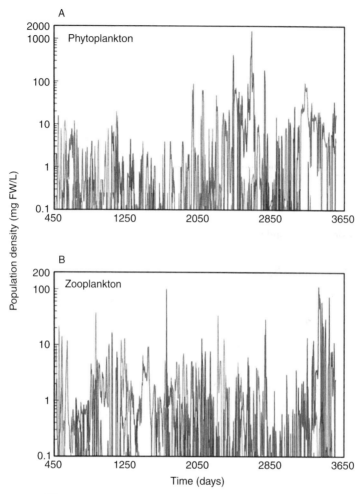

Figure 4.1. Nonequilibrium dynamics in an experimental community of plankton that consisted of more than 20 species and developed in a long-term laboratory experiment under constant external conditions. Data show the observed time course of (A) phyto- and (B) zooplankton. (A: green = green flagellates, blue = prokaryotic pico-phytoplankton, red = the diatom Melosira. B: green = the rotifer Brachionus, blue = the copepod *Eurytemora*, red = protozoans). After Scheffer *et al.* (2003). With permission of the authors and Kluwer Academic Publishers (Springer). See plate section for color version.

species does not occur, leading to nonequilibrium and explaining the paradox of the plankton, although strong environmental pressures such as seasonality may make patterns more predictable (see Rohde, 2005, p. 34). Extensive and thorough theoretical and experimental studies by a group around Jef Huisman over many years have led to significant insights, although many problems remain unresolved.

Huisman and Weissing (1999) have shown that a resource competition model (based on a standard model for phytoplankton competition) can generate oscillations and chaos when species compete for three or more resources, which may allow the coexistence of many species competing for a few resources. They conclude that "competitive systems may display highly dynamical phenomena, with continuous shifts and changes in species composition. . . . Competitive interactions that generate oscillations and chaos may allow the persistence of a great diversity of competitors on only a few limiting resources".

Using the same resource competition model, Huisman and Weissing (2001a) subsequently showed that when several species compete for three resources there can be several alternative outcomes. The winner depends on a fractal, so detailed that it is impossible to tell in advance which species will win. Here, the authors distinguish the "time course" and the "outcome" of competition. In chaotic systems, the time course depends on minute differences of initial conditions, and it is therefore not surprising that it is unpredictable. However, the outcome may still be predictable if there is a single chaotic attractor. Huisman and Weissing were able to show more than just unpredictability of the course in long-term dynamics, they were able to show that the dynamics leading to alternative outcomes may show transient chaos, and that, importantly, the basins of attraction of these outcomes may have an intermingled fractal geometry, which makes predictions about who wins impossible. "Only predictions in terms of probabilities make sense."

How common are such chaotic systems in nature? Several papers deal with this problem. Thus, Huisman *et al.* (2001) simulated competition between many species using different physiological scenarios (such as random species parameters, physiological trade-offs) and concluded that physiological and life-history patterns are important in determining whether species interactions generate nonequilibrium with resulting increase in diversity. A very high biodiversity, with sometimes more than 100 species competing for three resources, occurred in simulations with a cyclic relation between competitive abilities and resource contents.

In view of these intriguing findings, Huisman and Weissing (2001b) decided to analyze this standard resource competition model in full detail. The model is concerned with competition for abiotic essential resources; that is, resources required for growth. The model is based on a number of assumptions, which are: (1) population dynamics depend on resource availabilities which in turn depend on resource supply rates and the amounts of resources consumed by the organisms; (2) specific growth rates of species are determined by the most limiting resource (Von Liebig's "Law of the Minimum") and are governed by the Monod equation; (3) there is a continuous supply of resources. The authors point out that this competition model has been tested extensively in experiments with phytoplankton species (e.g., Tilman, 1977, 1981; Sommer, 1985, 1986). Indeed, because phytoplankton species, as well as terrestrial vegetation, use "quite a number of essential resources", the model is particularly relevant for such species. If species consume most of those resources (i.e., if species have high contents of those resources) for which they have high requirements (i.e., for which they have a high R^*), the model predicts stable coexistence of species; the model predicts oscillations and chaos if species consume most of those resources for which they have intermediate requirements; it predicts competitive exclusion if species consume most of those resources for which they have low requirements, with the winner determined by initial conditions. Concerning the general applicability of the model, the authors point out that the results should, at least qualitatively, also apply if mathematical expressions other than the

Monod equation are used for specific growth rates. Likewise, the authors conjectured that the predictions should also be valid if Liebig's Law of the Minimum is not followed, i.e., in cases where essential resources show interactive effects. Indeed, in a later publication, Huisman and Weissing (2002) demonstrated that competition for interactively essential resources leads to dynamics as complex as competition for non-interactive resources following the Law of the Minimum. However, the model did not consider resource storage, and storage-based models often make more reliable predictions in fluctuating environments (resource storage was considered by Revilla and Weissing (2008), who found, using an extension of the Droop model, currently considered to be the standard quota model for resource competition, that oscillations occur (almost) as readily in a model with nutrient storage as in a model without storage).

Beninca *et al.* (2008) provided experimental evidence for chaos in long-term (over 2300 days) experiments. The system they used contained bacteria, a number of phytoplankton species, herbivorous as well as predatory zooplankton species, and detritivores. External conditions were kept constant. Very large fluctuations with different periodicities shown to be chaotic were found. The authors concluded that predictions of abundance of particular species are in principle impossible and that complex food webs can persist over long periods even if conditions are not stable.

Huisman and Weissing (2002) further consider the necessity to address the following empirical questions: (1) how many resources are limiting in phytoplankton communities? (2) What are the trade-offs in resource requirements? (3) How are resource requirements and resource consumption related? The first question is important, because the model predicts that competition may generate nonequilibrium dynamics if species compete for more than two resources. The second question is important because complex dynamics are predicted only if there are trade-offs in resource requirements. The third question is important because the dynamics depend on the order of species ranked according to their resource requirements versus species ranked according to resource consumption. Even for three species there are $3! \times 3! \times 3!$ (= 216) possible ways to order their resource requirements for three resources. Hence, the model predictions discussed above cover only a minute fraction of all possible scenarios, i.e., the model predictions can only be considered as "rules of thumb".

Studies of phytoplankton (references in Huisman & Weissing 2001b) have shown that there are other limiting resources in addition to the traditionally recognized phosphorus, nitrogen, silica and light, such as various trace metals and, during dense phytoplankton blooms, inorganic carbon. Also, limitation by two resources appears to be more common than limitation by a single one, and little is known about trade-offs between some of the essential resources. The authors conclude that "on the basis of the available data, we can neither confirm nor reject the hypothesis of complex dynamics in phytoplankton communities", and "More data, for more species and especially for more limiting resources, would be most welcome". Also and importantly, the simulations did not consider the effect of predators, such as zooplankton and fish, which will be discussed below.

Huisman *et al.* (1999a) investigated the effects of critical light intensities (the lowest light intensity at which a species can just survive) on the competitive abilities of two species of green algae and two cyanobacteria in continuous cultures with limited light.

Monoculture experiments were used to estimate the critical light intensity of each species. According to theory, the species with the lowest critical light intensity should be the superior competitor for light. Competition experiments showed that this was indeed the case for almost all species combinations.

Passarge *et al*. (2006) extended this combined theoretical and experimental approach, and tested three alternative hypotheses for phytoplankton species, that competition for nutrients and light leads to stable coexistence of species, to alternative stable states, or to competitive exclusion. In nutrient-poor waters with good light penetration, phytoplankton can be expected to compete for limiting nutrients, a prediction well supported by experiments (references in Passarge *et al*.), whereas in nutrient-rich waters, species can be expected to compete for light. More specifically, in nutrient-rich waters that are well mixed and constant, the species with the lowest critical light intensity (corresponding to the light intensity at the bottom where the species can just survive) will be the superior competitor, a prediction also supported by experimental evidence (Huisman *et al*., 1999a; Litchman, 2003; Kardinaal *et al*., 2007). Passarge *et al*. (2006) examined the question of how competition affects species composition in waters of intermediate productivity. The model predictions are (1) if there is no trade-off, i.e., if a strong competitor for light is also a strong competitor for nutrients, the strong competitor should exclude all other species; (2) if there is a trade-off and the superior nutrient competitor uses relatively more light and the superior light competitor uses relatively more nutrients, there should be stable coexistence at intermediate nutrient supply rates; (3) if there is a trade-off but the superior nutrient competitor uses up relatively more nutrients and the superior light competitor uses relatively more light, the winner depends on the initial conditions, resulting in alternative stable states. Experiments were conducted in phosphorus-limited and light-limited chemostats. Phosphorus levels ranged from oligotrophic to eutrophic conditions. The monoculture experiments revealed that species that were strong competitors for phosphorus were strong competitors for light as well, i.e., there were no trade-offs. The population dynamics in the competition experiments were as predicted by the model: all competition experiments led to competitive exclusion. However, physiological characteristics of the species indicated that, if trade-offs in competitive abilities were found, competition for phosphorus and light would lead to alternative stable states rather than to stable coexistence. Therefore, stable coexistence resulting from competition for phosphorus and light is rare, and an unlikely explanation for the high biodiversity commonly found in phytoplankton communities. The authors suggest that strong competitors contain significantly more phosphorus than weak ones and may therefore be a preferred source of food for predators, i.e., they constitute an example for "keystone predation" (Leibold, 1996). Such keystone predation might lead to an increased diversity.

Again using a combined modeling and experimental approach, Agawin *et al*. (2007) have shown that the outcome of competition between nitrogen-fixing cyanobacteria and other phytoplankton species can be accurately predicted at different nitrate and light levels. Importantly, there is an intricate interplay between competition and facilitation: at low nitrate concentrations, nitrogen released by the nitrogen-fixing species facilitated an increase in abundance of the non-nitrogen-fixing species, whereas at high nitrate concentrations, the species with the lowest critical light intensity excluded the other.

But light intensities are far from being the only factor that determines the outcome of competition for light. As shown by Engelmann (1882, 1883a, 1883b, cit. Stomp *et al.*, 2007a, 2007b) and many others after him, different species of phytoplankton possess different photosynthetic pigments and can utilize different wavelengths of light. Stomp *et al.* (2004) were able to show theoretically and experimentally that phytoplankton species can differ in spectral characteristics allowing niche differentiation and coexistence of species in the light spectrum. They point out that earlier studies which had concluded that competition for light must lead to competitive exclusion did not consider spectral properties of the light. Stomp *et al.* (2004) used two species of picocyanobacteria isolated from the Baltic Sea for their laboratory experiments. One species is red due to the pigment phycoerythrin, the other is green-blue due to the pigment phycocyanin. When light was red, only the green species survived, while when light was green only the red species survived. Under white light both species coexisted for at least 60 days. They shared the light spectrum. A third species, a filamentous cyanobacterium, contained both the red and the green-blue pigment but could adapt its pigment composition (a phenomenon known as complementary chromatic adaptation), thus enabling it to use the light color not used by its competitors. Under white light, this chameleon among the cyanobacteria survived in the presence of the green species by turning red, while it survived in the presence of the red species by turning green.

In a follow-up study, Stomp *et al.* (2008) examined the timescale of phenotypic plasticity and found that the "adaptable" species (with phenotypic plasticity allowing it to change color in 7 days) always excluded the red and green species, whatever the frequency of switching from one light regime to the other, and even before it had time to completely adjust its color.

Stomp *et al.* (2007a) examined the question of whether this niche differentiation in the light spectrum, which they had demonstrated in the laboratory, can be demonstrated in natural waters as well. They used a competition model for predicting the outcome of competition between red and green phytoplankton species and then tested the results by sampling picocyanobacteria from 70 water bodies as diverse as clear blue oceans to turbid brown peat lakes and found that, under certain conditions, species can indeed coexist. The model did not show nonequilibrium dynamics or multiple stable states. Simulations were run until equilibrium was reached. The final state was always independent of the initial abundances of the species. The model predicted that red picocyanobacteria should win when turbidity is low, that green picocyanobacteria should win when turbidity is high, and that both species should coexist at intermediate turbidity. Coexistence also depends on the depth of the surface mixed layer. If it is shallow, the green and red bacteria can coexist using the white light spectrum near the surface. These model predictions were consistent with their sampling data from 70 water bodies, which indeed showed that red picocyanobacteria dominated in clear waters, and green bacteria dominated in turbid waters. Coexistence was found in waters of intermediate turbidity. Brilliant follow-up studies by Stomp *et al.* (2007b) examined the question of which spectral niches are available for photosynthetic micro-organisms. They were able to demonstrate that water molecules absorb light at specific wavebands that match the energy required for their stretching and bending vibrations. Light absorption at these specific wavelengths appears only as weak shoulders in the absorption spectrum of pure water; however, these "shoulders create large gaps in the underwater light spectrum due to the exponential nature of light attenuation", defining distinct niches by wavebands between these gaps. These wavebands match the light absorption spectra of the major photosynthetic pigments.

Much recent progress has been made on the effects of mixing in water columns on plankton dynamics. Both numerical simulation and rigorous mathematical investigation on the reaction-diffusion-advection models proposed by Huisman *et al.* suggest that in completely mixed water columns (population distribution does not depend on spatial location) competitive exclusion dominates the phytoplankton dynamics (Huisman & Weissing, 1994, 1995; Weissing & Huisman, 1994), while in incompletely mixed water columns (population distribution is space-dependent and subject to diffusion) coexistence of multiple phytoplankton species is possible (Huisman *et al.*, 1999b; Du & Hsu, 2010; Mei & Zhang, 2012). Moreover, in stratified water columns (where there is a well-mixed surface layer and a poorly mixed deep water layer), even for a single phytoplankton species whose growth is limited by both light and nutrient, multiple stable steady states may exist (Yoshiyama & Nakajima, 2002; Ryabov *et al.*, 2010). These mathematical models suggest that the complexity of mixing of the water column enhances the chance of coexistence and the variety of ways of existence of phytoplankton species.

Huisman *et al.* (2004) used a modeling and experimental approach to examine the question of how changes in turbulent mixing affect competition for light between buoyant and sinking phytoplankton species in eutrophic waters. Their model made certain predictions, which were all confirmed by experiments in which the turbulence of an entire lake was manipulated: the balance between buoyant and sinking species changed in a predictable fashion. Turbulence changes are indeed important factors determining plankton composition under natural conditions, mixing depending on climatic factors such as heat exchange and wind action. Sinking species become dominant as the result of strong mixing as induced by storms, buoyant species become more dominant during warm weather, when wind mixing is low. For instance, it is known that, in eutrophic lakes, buoyant cyanobacteria dominate when mixing is low, and diatoms dominate when mixing is intense. Similarly, in many coastal ecosystems, dinoflagellates dominate when mixing is low, while diatoms dominate when mixing is intense. Therefore, weather-induced changes in turbulence, for instance by periodic storms, may be important contributors to maintaining high biodiversity.

Huisman *et al.* (2006) extended these turbulence models, and showed that reduced vertical mixing in water columns may induce oscillations and chaos in the biomass of phytoplankton and its species composition in oligotrophic waters. They proceeded from the observation that in oligotrophic waters with a mixed, nutrient-poor surface layer, there are often deeper layers with chlorophyll maxima caused by greater phytoplankton biomass. Such deep chlorophyll maxima (DCMs) are permanently found in tropical and subtropical oceans, and seasonally at higher latitudes. In the model, light intensity decreases exponentially with depth and nutrients near the surface are gradually depleted by the phytoplankton; consequently the nutricline and with it the phytoplankton move slowly downwards. When turbulent mixing is high (turbulent diffusivity is ~0.5 $cm^2 s^{-1}$), the phytoplankton settles at a stable equilibrium where the downward flux of consumed nutrients equals the upward flux of nutrients from below. When turbulent mixing is less intense the plankton densities in the DCM oscillate and may even be chaotic. The reason is that, when turbulent mixing is very weak, the downward flux of phytoplankton is fast relative to the upward flux of nutrients, and the phytoplankton, because of lack of light, cannot use up the rising nutrients, which now can rise further towards the surface to where light conditions facilitate plankton growth and the next peak of DCM. Complexity of the dynamics is further increased by introducing seasonality into the model. In nature, detailed time series show that seasonal changes in light do indeed have significant effects on DCM dynamics. When replacing a single entity "phytoplankton" by multiple phytoplankton species in the model, each with a different growth rate, as well as with different light and nutrient requirements and different sinking velocities, again in a seasonal environment, oscillations and chaos were again apparent, and all species could coexist. The authors conclude that, although a simple model does not consider all the real-world phenomena, studies of DCMs show that many features of the real world are indeed adequately reproduced by the model.

Passarge *et al.* (2006), without experimental evidence, considered predators such as zooplankton and fish, or parasites such as viruses. Viruses are very abundant in the oceans and infect many hosts, including bacteria and eukaryotic primary producers (Hilker *et al.*, 2006, further references therein). Although they may affect mortality and diversity of phytoplankton, termination of algal blooms, etc., we still know little about their role in phytoplankton populations. However, it is known that viruses infecting phytoplankton have two replication cycles, i.e., there are lytic (or virulent) infections with destruction of but without reproduction in the host, and lysogenic (or temperate) infections in which the genome of the virus is integrated in that of the host, and the virus multiplies with the host, until the lytic cycle begins. The latter may occur spontaneously or may be induced by environmental triggers (radiation, pollution, temperature changes, nutrient depletion). A number of studies have modeled virally infected plankton populations. In such a model based on reaction-diffusion equations, Malchow *et al.* (2004, 2005) observed oscillations and waves in food chains with frequency-dependent disease transmission. Hilker *et al.* (2006), proceeding from this model, investigated the transition from the lysogenic to the lytic phase and its effects on the dynamics of interacting phyto- and zooplankton in a biomass-based stochastic model of phytoplankton-zooplankton dynamics. The assumption was that part of the phytoplankton population is infected by a virus which switches from lysogeny to lysis. The model generated a complex spatio-temporal structure, as indeed found in natural plankton. In the theoretical study by Medvinsky *et al.* (2002), the dynamics of fish populations are incorporated as well. They demonstrated that the interplay between turbulent mixing and matter fluxes may lead to transient and irregular spatial structure in plankton distribution. Pattern formation is characterized by an intrinsic length of about 1 km, which corresponds to what is found in nature. The dynamics corresponding to pattern formation is chaotic. Planktivorous fish schools may induce irregular spatial patterns as well. Field data showed that relatively stable heterogeneities in temperature, salinity and biogen (substances derived from living organisms) concentration of intermediate size may affect the dynamics of aquatic communities. The kind of dynamics depends on patch size and distance between patches.

There is a continuous competition between pattern creation by biological mechanisms and pattern destruction by turbulences. However, models incorporating more detailed turbulence motion are needed. The authors conclude that chaotic regimes may play a vital role in "organizing" aquatic ecosystems. Beninca *et al.* (2009) examined predator-prey oscillations in long-term experiments of marine plankton. Phytoplankton species competed strongly, zooplankton relatively little. Species coexisted in two predator-prey cycles that shifted from one to the other chaotically.

Conclusions

In summary, the studies discussed here and many others have revealed an amazing complexity of the supposedly homogeneous aquatic habitats and the plankton communities within them. Many problems need much further work, including the effects of predators and disease agents such as viruses. Nevertheless, we can conclude that the paradox of the plankton resolves itself as follows (Scheffer *et al.* 2003): (1) homogeneity due to mixing hardly exists, and even in the open ocean meso-scale vortices and fronts result in spatial heterogeneity; (2) aquatic habitats provide many more niches for niche differentiation than originally thought (different wave lengths of white light; additional essential resources); (3) modeling of plankton communities and experimental studies have shown that even in homogeneous and constant environments plankton may never reach equilibrium, because multi-species competition may lead to oscillations and chaos, contributing to the maintenance of a great biodiversity.

Many of the predictions derived from modeling are supported by field data. Importantly, the detailed plankton studies have provided convincing evidence that, in contrast to many communities in which nonequilibrium conditions occur in largely non-saturated niche space with little interspecific competition, nonequilibrium and chaos in plankton may be caused by such competition.

Acknowledgments

This chapter is an updated version of my earlier discussion, published as an appendix on the website of my book *Nonequilibrium Ecology* (2005). Used here with the permission of Cambridge University Press. I wish to thank Jef Huisman for sending me many of his papers and for critical comments on the earlier version, and Yihong Du for the introductory paragraph on the effects of water mixing and references for it.

References

Agawin, N. S. R., Rabouille, S., Veldhuis, M. J. W., *et al.* (2007). Competition and facilitation between unicellular nitrogen-fixing cyanobacteria and non-nitrogen-fixing phytoplankton species. *Limnology and Oceanography*, **52**, 2233–2248.

Beninca, E., Huisman, J., Heerkloss, R., *et al.* (2008). Chaos in a long-term experiment with a plankton community. *Nature*, **451**, 822–825.

Beninca, E., Johnk, K. D., Heerkloss, R., & Huisman, J. (2009). Coupled predator-prey oscillations in a chaotic food web. *Ecology Letters*, **12**, 1367–1378.

Du, Y., & Hsu, S.-B, (2010). On a nonlocal reaction-diffusion problem arising from the modeling of phytoplankton growth. *SIAM Journal of Mathematical Analysis*, **42**, 1305–1333.

Engelmann, T. W. (1882). Über Sauerstoffausscheidung von Pflanzenzellen im Mikrospectrum. *Botanische Zeitschrift*, **40**, 419–426.

Engelmann, T. W. (1883a). Bacterium photometricum: ein Beitrag zur vergleichenden Physiologie des Licht- und Farbensinnes. *Archiv für Physiologie*, **30**, 95–124.

Engelmann, T. W. (1883b). Farbe und Assimilation. *Botanische Zeitschrift*, **41**, 1–13.

Harris, G. P. (1986). *Phytoplankton Ecology. Structure, Function and Fluctuation*. London: Chapman and Hill.

Hilker, F. M., Malchow, H., Langlais, M., & Petrovskii, S. V. 2006). Oscillations and waves in a virally infected plankton system. Part II: transition from lysogeny to lysis. *Ecological Complexity*, **3**, 200–208.

Huisman, J., & Weissing, F. J. (1994). Light-limited growth and competition for light in well-mixed aquatic environments: an elementary model. *Ecology*, **75**, 507–520.

Huisman, J., & Weissing, F. J. (1995). Competition for nutrients and light in a mixed water column – a theoretical analysis. *The American Naturalist*, **146**, 536–564.

Huisman, J., & Weissing, F. J. (1999). Biodiversity of plankton by species oscillations and chaos. *Nature*, **402**, 407–410.

Huisman, J., & Weissing, F. J. (2001a). Fundamental unpredictability in multispecies competition. *The American Naturalist*, **157**, 488–494.

Huisman, J., & Weissing, F. J. (2001b). Biological conditions for oscillations and chaos generated by multispecies competition. *Ecology*, **82**, 2682–2695.

Huisman, J., & Weissing, F. J. (2002). Oscillations and chaos generated by competition for interactively essential resources. *Ecological Research*, **17**, 175–181.

Huisman, J., van Oostveen, P., & Weissing, F. J. (1999a). Species dynamics in phytoplankton-blooms: incomplete mixing and competition for light. *The American Naturalist*, **154**, 46–67.

Huisman J., Jonker, R. R., Zonneveld, C., & Weissing, F. J. (1999b). Competition for light between phytoplankton species: experimental tests of mechanistic theory. *Ecology*, **80**, 211–222.

Huisman, J., Johansson, A. M., Folmer, E. O., & Weissing, F. J. (2001). Towards a solution of the plankton paradox: the importance of physiology and life history. *Ecology Letters*, **4**, 408–411.

Huisman, J., Sharples, J., Stroom, J. M., *et al.* (2004). Changes in turbulent mixing shift competition for light between phytoplankton species. *Ecology*, **85**, 2960–2970.

Huisman, J., Pham Thi, N. N., Karl, D. M., & Sommeijer, B. (2006). Reduced mixing generates oscillations and chaos in the oceanic deep chlorophyll maximum. *Nature*, **439**, 322–325.

Hutchinson, G. E., (1961). The paradox of the plankton. *The American Naturalist*, **95**, 137–145.

Irigoien, X., Huisman, J., & Harris, R. P. (2004). Global biodiversity patterns of marine phytoplankton and zooplankton. *Nature*, **429**, 863–867.

Kardinaal, W. E. A., Tonk, L., Janse, I., *et al.* (2007). Competition for light between toxic and nontoxic strains of the harmful cyanobacterium Microcystis. *Applied and Environmental Microbiology*, **73**, 2939–2946.

Leibold, M. A. (1996). A graphical model of keystone predators in food webs: trophic regulation of abundance, incidence, and diversity patterns in communities. *The American Naturalist*, **147**, 784–812.

Litchman, E. (2003). Competition and coexistence of phytoplankton under fluctuating light: experiments with two cyanobacteria. *Aquatic Microbial Ecology*, **31**, 241–248.

Malchow, H., Hilker, F. M., Petrovskii, S. V., & Brauer, K. (2004). Oscillations and waves in a virally infected plankton system: Part I: The lysogenic stage. *Ecological Complexity*, **1**, 211–223.

Malchow, H., Hilker, F. M., Sarkar, R. R., & Brauer, K. (2005). Spatiotemporal patterns in an excitable plankton system with lysogenic viral infection. *Mathematical and Computer Modelling*, **42**, 1035–1048.

Medvinsky, A. B., Petrovskii, S. V., Tikhonova, I. A., Malchow, H., & Li, B.-L. (2002). Spatiotemporal complexity of plankton and fish dynamics. *SIAM Review*, **44**, 311–370.

Mei, L., & Zhang, X. (2012). Existence and nonexistence of positive steady states in multi-species phytoplankton dynamics. *Journal of Differential Equations*, **253**, 2025–2063.

Passarge, J., Hol, S., Escher, M., & Huisman, J. (2006). Competition for nutrients and light: stable coexistence, alternative stable states, or competitive exclusion? *Ecological Monographs*, **76**, 57–72.

Revilla, T., & Weissing, F. J. (2008). Nonequilibrium coexistence in a competition model with nutrient storage. *Ecology*, **89**, 865–877.

Rohde, K. (2005). *Nonequilibrium Ecology*. Cambridge: Cambridge University Press.

Ryabov, A. B., Rudolf, L., & Blasius, B. (2010). Vertical distribution and composition of phytoplankton under the influence of an upper mixed layer. *Journal of Theoretical Biology*, **263**, 120–133.

Scheffer, M., Rinaldi, S., Huisman, J., & Weissing, F. J. (2003). Why plankton communities have no equilibrium: solutions to the paradox. *Hydrobiologia*, **491**, 9–18.

Sommer, U. (1985). Comparison between steady state and non-steady state competition: experiments with natural phytoplankton. *Limnology and Oceanography*, **30**, 335–346.

Sommer, U. (1986). Nitrate- and silicate-competition among Antarctic phytoplankton. *Marine Biology*, **91**, 345–351.

Stomp, M., Huisman, J., de Jongh, F., *et al.* (2004). Adaptive divergence in pigment composition promotes phytoplankton diversity. *Nature*, **432**, 104–107.

Stomp, M., Huisman, J., Vörös, L., *et al.* (2007a). Colourful coexistence of red and green picocyanobacteria in lakes and seas. *Ecology Letters*, **10**, 290–298.

Stomp, M., Huisman, J., Stal, J. L., & Matthijs, H. C. (2007b). Colourful niches of phototrophic microorganisms shaped by vibrations of the water molecule. *The ISME Journal*, **1**, 271–282.

Stomp, M., van Dijk, M. A., van Overzee, H. M. J., *et al.* (2008). The timescale of phenotypic plasticity and its impact on competition in fluctuating environments. *The American Naturalist*, **172**, E169–E185.

Tilman, D. (1977). Resource competition between planktonic algae: an experimental and theoretical approach. *Ecology*, **58**, 338–348.

Tilman, D. (1981). Tests of resource competition theory using four species of Lake Michigan algae. *Ecology*, **62**, 802–815.

Weissing, F. J., & Huisman, J. (1994). Growth and competition in a light gradient. *Journal of Theoretical Biology*, **168**, 323–336.

Yoshiyama, K. and Nakajima, H. (2002). Catastrophic transition in vertical distributions of phytoplankton: alternative equilibria in a water column. *Journal of Theoretical Biology*, **216**, 397–408.

5   A burning issue: community stability and alternative stable states in relation to fire

Peter J. Clarke and Michael J. Lawes

Introduction

Fire regimes have long been thought to drive plant community change in biomes by altering feedbacks that maintain "stable" community assemblages (see Jackson, 1968; Mutch, 1970). The occurrence of contrasting vegetation types in otherwise comparable environments, where flammable communities are juxtaposed with those that rarely burn, is often explained by alternative stable state (ASS) theory, where shifts in equilibrium are triggered by catastrophic fire (Petraitis & Latham, 1999; Scheffer & Carpenter, 2003). When high-intensity fires burn into less flammable communities, compositional change is thought to occur because, firstly, an ecological threshold is reached beyond which the disturbance is large enough to remove species that perpetuate the exclusion of fire, secondly, space is then opened up for colonization by more flammable species and finally self-reinforcing pyrogenic dominance is achieved (Figure 5.1). Alternative stable states are often invoked in fire-prone regions and climates where there are sharp boundaries between communities (Figure 5.2). One of the most cited examples of alternative stable states is the fire-triggered transformation of rainforest to more flammable assemblages (Jackson, 1968; Webb, 1968; Bowman, 2000; Beckage *et al.*, 2009; Hoffmann *et al.*, 2009; Warman & Moles, 2009). Other examples of ASS include the conversion of boreal forest to deciduous forest (Johnstone *et al.*, 2010), tropical savanna to grassland (Hoffmann & Jackson, 2000) and the replacement of arid shrublands by more flammable grassland (Nicholas *et al.*, 2011). Fluctuations in community composition are the norm for most plant communities and in this sense they are "meta-stable" at Holocene timescales even under the effects of severe disturbance. In contrast, transformation in structure and complete floristic turnover at decadal timescales are those associated with ASS.

Globally, climate warming is predicted to change fire regimes and increase the severity and frequency of fires that could result in major biome shifts (Bowman *et al.*, 2009). Here we provide a critical overview of the feedback mechanisms that both maintain and disrupt community stability. Recent global analyses of tree cover now reveal how important such feedbacks are in maintaining bistable states because vast areas of the globe have strong

The Balance of Nature and Human Impact, ed. Klaus Rohde. Published by Cambridge University Press.
© Cambridge University Press 2013.

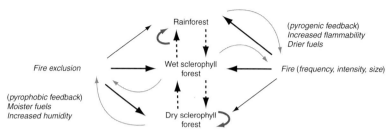

Figure 5.1.   Conceptual model of alternative stable states in forest assemblages with fire regime as a switch that drives rainforest (closed forest) to wet sclerophyll forest (open forest) and then to dry sclerophyll forest (low open forest) through a positive fire feedback where fire promotes flammable species. Size of solid black arrows indicates the proportional strength of driver of ASS. Stability in rainforest through disturbances is maintained through resilience traits such as resprouting and soil seedbanks. Fire maintains floristic stability in the sclerophyll forests by ensuring fire cycle recruitment and regeneration of species with fire-adaptive traits (shown as a right-hand gray feedback). Conversely, fire exclusion maintains the stability of rainforest, and could cause instability in sclerophyll forests by causing recruitment failure (left-hand gray feedbacks).

bimodal tree cover distributions (Staver *et al.*, 2011a, 2011b). These remarkable findings prompt the question – what community and species properties confer resistance to fire? In particular, we examine the extent to which evolutionarily stable traits promote resilience to fire and how this can rapidly change with the invasion of neophytes.

Fire begets fire – the stability paradox

A requirement for change to an alternative state is that disturbance is extreme enough to remove or alter the density of community dominants (Petraitis & Latham, 1999). Fire is a pervasive disturbance because it can kill the buds (apical and axillary meristems) that otherwise allow community dominants to recover from less severe disturbances such as herbivory, wind, drought and flood. Nevertheless, in fire-prone ecosystems, species dominance can be maintained through severe fire events by adaptive traits that protect buds (e.g., thick bark and embedded meristem bud strands) (Crisp *et al.*, 2011) and/or promote rapid growth to re-establish dominance (e.g., fire-cued seedbanks, lignotuber storage organs, low leaf mass areas) (Keeley *et al.*, 2011). Such species not only have life-history traits that are resilient to variation in fire regime components, but are thought to have characteristics that promote repeated fire (Mutch, 1970; Bond & Midgley, 1995; Schwilk & Kerr, 2002). This is particularly so in the ecosystems of Mediterranean climates where the germination of many species is fire (smoke and heat) cued, seed is held in canopy-held seedbanks and where the dominant species have bark, leaf and morphological character-istics that make them highly combustible (Keeley *et al.*, 2011). While fire introduces structural instability through fire cycles, the ability to resprout and store seed banks ensure floristic continuity through the positive neighbor effect (Frelich & Reich, 1999). A pivotal issue in the debate over community stability is the extent to which species response traits to fire regimes predict stability; an issue to which we return to later in this review.

Figure 5.2.   (A) Intense crown fire that has burned tall *Eucalyptus* forest and adjacent rainforest. Such fires were thought to trigger instability leading to fire feedbacks, but both these Australian forests show remarkable floristic stability after disturbance. (B) Grassland fire adjacent to forest thicket in the high-altitude grassland biome of South Africa. Repeated fires are predicted to decrease the less flammable thickets and convert these patches to grassland as fire regimes change. Increased atmospheric CO_2 may, however, favor forest expansion. Photos Peter Clarke. See plate section for color version.

The distinction between processes that initiate the switches in community assemblages and those associated with the feedback that maintains those states is often blurred (Petraitis & Latham, 1999). This is because events that initiate switches between alternative states need not be the same as the feedback mechanisms that maintain these states (Beckage & Ellingwood, 2008). For example, it may take a single severe fire event to initiate change, but fires of less intensity and frequency are required for a flammability feedback to maintain a vegetation state (Cochrane *et al.*, 2003; Nepstad *et al.*, 2001). Despite increased reports of large fires burning across community boundaries (e.g., Flannigan *et al.*, 2009), and the importance of large fire events in initiating community change, it is notable how few studies have reported the demographic consequences of extreme fires burning "non-flammable" vegetation. Often these fire events are rare in space and time, such as in the burning of mangrove forests, alpine heaths, freshwater wetlands and cloud forests, but are more common in systems such as vine thickets and closed forests that are juxtaposed with more flammable grasslands, savannas and heathlands (see Figure 5.2).

Evidence for transformative fire feedbacks

Flammability or pyrogenic feedbacks have been demonstrated in forest assemblages where severe fire events increase the probability that they will burn again (Cochrane et al., 2003; Cochrane & Schulz, 1999; Nepstad et al., 2001; Cochrane, 2003; Odion et al., 2010). This is because severe fires consume canopies, resulting in more rapid drying of ground fuels (Nepstad et al., 2001). Some stable systems, however, show resilience to fire feedbacks even after large "catastrophic" events. This is highlighted in temperate Australian rainforests where severe antecedent fire (20 years before a recent fire) decreased subsequent burn severity (Knox & Clarke, 2012) (Figure 5.2). Long-term reinforcement of fire was, however, prominent in adjacent sclerophyll forests where the same antecedent fires increased burn severity, thus sharpening the contrast between the two ecosystems (Knox & Clarke, 2012) (Figure 5.2). Such examples showing complex feedback effects from antecedent fires highlight the need for more comprehensive assessment of fire feedbacks using historic remote sensing, which is now readily available.

On-ground measures of fire feedbacks through measures of ground temperature, litter, standing fuels and soil variables are more difficult to obtain, especially through time. Hence these proposed feedback effects have relied more on theory from fires burning in quasi-stable fire-prone communities rather than fires burning into communities where fire is likely to destabilize them (e.g., Shakesbury & Doerr, 2006). Such studies report fires of high intensity consuming and altering soil organic matter that changes the moisture-holding capacity and nutrient status of soils (e.g., Hatten & Zabowski, 2009), thus providing the basis for community change. Yet some studies contrasting the effects of fire severity on soil properties detect rapid recovery of soil nutrients and organic matter (Knox & Clarke, 2012). In contrast, space for time comparisons across sharp community boundaries often show nutrient differences independent of lithology. Weak inference has been used to suggest that repeated fires are the mechanism for nutrient depletion (Bowman et al., 2008; Wood & Bowman, 2011) despite the low volatility of critical nutrients such as phosphorus. Given the importance of soil in mediating plant growth and flammability, more long-term controlled experiments on the effects of fire regime and herbivory soil properties are needed to understand soil-fire interactions as a driver of community disequilibrium (e.g., van Langevelde et al., 2003).

Evidence of feedbacks from fire exclusion

Feedbacks associated with fire exclusion are theoretically well supported in forest, woodland and savanna assemblages where accumulation of nutrients, soil organic matter and canopy closure occur in the long-term absence of fire (Beckage et al., 2009; Coetsee *et al.*, 2010). These changes are predicted to inhibit the recruitment of pyrogenic species into the understorey due to shading and lack of favorable microsites for germination (Barrett & Ash, 1992; Close *et al.*, 2009; Woinarski *et al.* 2004). Plant species that colonize assemblages in the intervals between fire have characteristics that are thought to inhibit the ignition and spread of fires (Jackson, 1968; Ash, 1988; Russell-Smith *et al.*, 2004; McGlone *et al.*, 2005; Banfai & Bowman, 2006; Harvest *et al.*, 2008; Close *et al.*, 2009; Fairfax *et al.*, 2009). Nevertheless, evidence is lacking that shows colonization of flammable assemblages (e.g., heaths and sclerophyll forests) by adjacent less flammable assemblages (e.g., rainforest) significantly influences the overall flammability of a site (Bowman and Wilson, 1988; Knox & Clarke, 2012), at least in the short term.

Global evidence of alternative stable states

Regardless of our imprecise knowledge of fire feedbacks, the development of dynamic global vegetation models (DGVMs) has shown large discrepancies between the predicted and actual distributions of forests based on climate (Bond *et al.*, 2005). This is highlighted in high rainfall savannas (> 650 mm MAP) where grass and trees coexist but where DGVMs predict forest cover (Bond *et al.*, 2005; Bond, 2008). In such productive systems, high fire frequencies can be maintained and there is strong empirical evidence of fire-regulated demography (e.g., Lehmann *et al.*, 2009; Werner & Franklin, 2010). Tree densities, however, may be in equilibrium with core fire regimes as they have been shown to be unresponsive to fire frequency and season (Russell-Smith *et al.*, 2003; Higgins *et al.*, 2007; Prior *et al.*, 2009; Lawes *et al.*, 2011). Instead, individuals survive fire, but repeated stem kill creates "sapling bottlenecks" (Bond, 2008; Prior *et al.*, 2010), thereby preventing woody cover from reaching its climate potential.

Global analyses show a strong bimodal distribution of tree cover at intermediate rainfall (1000–2500 mm MAR) in Africa and South America where both forest and savanna co-occur (Staver *et al.*, 2011a) but Australia stands out for its relative lack of closed forest. These bimodal responses in tree cover suggest that savanna is an alternative stable state, because, in the absence of fire, tree cover is high in Africa and South America (Staver *et al.*, 2011a).

Rainforest stability

Repeated fires following large-scale disturbance by wind, logging and/or a drought event that induces catastrophic fire are widely cited as the driver that converts closed forests into savanna or some other vegetation state such as sclerophyll forest (Cochrane & Schulze, 1999; Bowman, 2000; Banfai & Bowman, 2005). In particular, tropical rainforests are highly susceptible to fires reducing tree canopy cover and stem density as well as reducing overall species richness (Uhl & Kauffman, 1990; Cochrane & Schulze, 1999; Cochrane, 2003; Fensham *et al.*, 2003; Laurance *et al.*, 2005; Van Nieuwstadt *et al.*, 2005; Malhi *et al.*, 2009; Slik *et al.*, 2010). Remarkably, longitudinal studies of the change in structure and composition of tropical evergreen forests following severe fire are lacking despite the apparent instability of these biomes and the increased ignitions associated with human activity and clearing (Cochrane, 2003; Staver *et al.*, 2011a, 2011b). The exclusion of fires from savannas adjacent to rainforests has, however, shown the slow colonization of savannas by rainforest species (e.g., Westfall *et al.*, 1983; Duncan & Duncan, 2000; Woinarski *et al.*, 2004; Russell-Smith *et al.*, 2004). Curiously, rainforest boundaries in the tropical savannas of northern Australia – where there is no logging – have been expanding despite ongoing fire activity (Bowman *et al.*, 2010). Similar patterns in expanding thickets have also been measured in South African savannas and are attributed to increasing rainfall and atmospheric CO_2 concentrations (Wigley *et al.*, 2010) (Figure 5.2B). Globally, however, vast areas of equatorial African and South American forests are predicted to be bistable, where climate and fire can interact to convert forest to savanna (Staver *et al.*, 2011a).

By contrast to Africa and South America, Australia is dominated by sclerophyllous savanna and open forest with infrequent closed forest in the 1000–2500 rainfall zone (Staver *et al.*, 2011a). The lack of widespread rainforests and thickets in Australia is not related to a paucity of rainforest lineages because there is abundant evidence from the Tertiary that both rainforests and sclerophyllous forests widely coexisted (Hill, 1994). Instead, it is likely that climate instablity coupled with the development of crown fires that accompanied the expansion of woodlands, dominated by eucalypts in the mesic zone and the sclerophyllous C_4 grass *Triodia* in the arid zone, transformed the continent so that closed forest exist as "Islands of green in a land of fire" (Bowman, 2000). Why rainforest persists and remains in a relatively stable state in Australia is perplexing. Clues to the resilence of rainforests to fire come from a mid-latitude study in eastern Australia comprising mixed lineages out of both Gondwana and Asia (Knox & Clarke, 2012). Severe fires are a relatively common occurrence in adjacent sclerophyll forests and these burn into the rainforest under extreme fire weather (Figure 5.2a). Rainforest canopy closure after crown scorch is rapid and forests previously burned have less severe crown scorch than those where there is no history of fire (Knox & Clarke, 2012). Ground measures comparing the floristic composition of burnt and unburnt rainforest, some 6 years after fire, also revealed surprising community stability despite the presence of top-killed trees (Knox & Clarke, 2012). What traits allow these rainforests to resist the conversion to sclerophyll forest? Unlike tropical forests, nearly all woody growth forms in Australia resprout after fire, and while many tree species exhibit top kill, several dominants resprout from stems where bark thickness protects buds. Trait analyses of understory shrubs, paired with species from more flammable habitats, also reveal selection for enhanced post-fire regrowth (Knox & Clarke, 2012). Ultimately, soil fertility is the selective force influencing plant growth traits where water is not limiting (Poorter *et al.*, 2009); hence, it is not surprising that the boundaries between rainforests and sclerophyll forests correspond with soil fertility differences (Knox & Clarke, 2012). The key to the current resilience and stability of these rainforests to crown fire appears to be "edaphic compensation", which maintains a strong neighborhood effect and lower flammability of live and dead fuels. The edaphic (soil fertility) compensation effect occurs because recovery from fire is promoted on nutrient-rich soils so that height growth and bark thickness are quickly recovered.

Aridity, fire and stability

As rainfall decreases, savannas (< 650 mm MAP) are predicted to be "stable" because tree density is limited by water availability (Sankaran *et al*., 2005; Staver *et al*., 2011a). Correspondingly, there is also strong climatic control over fire frequency as grass cover and other fuel levels decrease (Bradstock, 2010). Nevertheless, a conclusion that fire has no role in semi-arid systems is premature because shrubland-grassland mosaics are common in semi-arid systems and abrupt boundaries are a prominent feature in some biomes (Nano & Clarke, 2010; Nicholas *et al*., 2011). Additionally, fire regime effects are important in shrublands where short fire intervals are known to remove canopy shrub dominants (Nano & Clarke, 2010). Hence, the potential for fire to influence community composition and alternative stable states extends into arid biomes.

The presumed rarity of fire in regions with rainfall less than 650 mm MAR was once thought to indicate that fires would be extremely disruptive to communities since stabilizing selection would be driven by drought rather than fire (Sankaran *et al*., 2005). Long-term satellite imagery has revealed that fire is common in the *Triodia* grasslands of Australia, with some 30 000 fires detected in an 8-year period, burning over 50% of the landscape (*c*. 500 000 km^2) (Nano *et al*., 2012). These grasslands comprise both herbaceous and woody components, developing into savannas as rainfall increases. Like tropical savannas, diversity is highest early in the fire cycle and they are floristically resilient to frequent fires (Wright & Clarke, 2007). Fire does not strongly affect shrub and tree densities because most species basally resprout and seedling recruitment is tied to rainfall rather than fire. Community stability is therefore maintained by seedbanks for ephemeral species and bud banks for the dominant species. In contrast to these grasslands, adjacent *Acacia* shrublands are responsive to fire because it kills the canopy dominants and canopy recovery is extremely slow due to climate limits on recruitment and growth, hence functional and structural diversity increases with time-since-fire (Nano *et al*., 2012). While fire is clearly a destabilizing force in the shrublands, there has been little boundary change in these communities because *Triodia* is unable to colonize the heavy textured soils on which the shrublands grow. Fire therefore appears to sharpen community boundaries, removing fire-killed shrubs from flammable environment, rather than driving alternative states per se. This "meta-stability" is, however, changing with the invasion of exotic grasses, as the next overview highlights.

Community stability and invasive species

Invasive species are known to change fire regimes through flammability feedbacks that promote population establishment and spread (Brooks *et al*., 2004). This occurs because invasive species are often more ignitable and have different fuel bed properties from native species (Berry *et al*., 2011; Setterfield *et al*., 2010), resulting in more frequent and sometime more intense fires of greater residence time (Keeley & Brennan, 2012). Such feedbacks not only promote fires in systems where fire rarely burn, such as rainforests

(D'Antonio & Vitousek, 1992), but also in systems where fire frequency and intensity is increased such as Mediterranean shrublands (Keeley & Brennan, 2012), tropical savanna (Rossiter *et al.*, 2003; Setterfield *et al.*, 2010) and arid woodland (Clarke *et al.*, 2005).

Invasive grasses have characteristics that, in particular, drive fire feedbacks because they change the fuel loads and fire intensity, resulting in the so-called "grass-fire" cycle (D'Antonio & Vitousek, 1992). For example, native grass fuel loads in the tropical savannas of northern Australia are typically 2–4 tonnes/ha, whereas areas invaded by gamba grass (*Andropogon gayanus*) sustain 11–15 tonnes/ha (Setterfield *et al.*, 2010). These high fuel loads promote earlier and more intense fires that reduce the density of normally fire-tolerant trees by 50% and deplete the nitrogen status of savanna soils (Rossiter *et al.*, 2003; Rossiter-Rachor *et al.*, 2008, 2009). Similar fire feedbacks have also been reported in semi-arid grasslands and savannas where the invasion of buffel grass (*Cenchris ciliaris*) is associated with increased fire severity (Miller *et al.*, 2010). In one of the few long-term studies of the grass-fire invasion cycle Clarke *et al.* (2005) showed the complete floristic transformation of a semi-arid grassland over 26 years. Such changes are not restricted to perennial grass transformers because annual grasses have been shown to rapidly transform Mediterranean shrublands into annual grasslands (Keeley & Brennan, 2012).

Conclusion

There has been a revolution in fire science that now acknowledges fire as a critical determining factor in the evolution of plant life and the shaping of biomes. Fire both stabilizes communities, where lineages have been exposed to a long history of fire regimes, but also destabilizes those where adaptive traits are lacking. Our ability to predict the effects of changing climate and fire regimes on community stability is limited by the scarcity of empirical and experimental evidence about the interactions of fire, plant traits and other factors such as rainfall and soil fertility in driving feedbacks. The potential instability of ecosystems that sustain much of the Earth's biodiversity in the face of novel disturbance regimes triggered by invasive species is of profound concern.

References

Ash, J. (1988). The location and stability of rainforest boundaries in North-Eastern Queensland, Australia. *Journal of Biogeography*, **15**, 619–630.

Banfai, D. S., & Bowman, D. M. J. S. (2005). Dynamics of a savanna-forest mosaic in the Australian monsoon tropics inferred from stand structures and historical aerial photography. *Australian Journal of Botany*, **53**, 185–194

Banfai, D. S., & Bowman, D. M. J. S. (2006). Forty years of lowland monsoon rainforest expansion in Kakadu National Park, Northern Australia. *Biological Conservation*, **131**, 553–565.

Barrett, D. J., & Ash, J. E. (1992). Growth and carbon partitioning in rainforest and eucalypt forest species of south coastal New South Wales, Australia. *Australian Journal of Botany*, **40**, 13–25.

Beckage, B., & Ellingwood, C. (2008). Fire feedbacks with vegetation and alternative stable states. *Complex Systems*, **18**, 159–173.

Beckage, B., Platt, W. J., & Gross, L. J. (2009). Vegetation, fire, and feedbacks: a disturbance-mediated model of savannas. *The American Naturalist*, **174**, 805–818.

Berry, Z. C., Wevill, K., & Curran, T. J. (2011). The invasive weed *Lantana camara* increases fire risk in dry rainforest by altering fuel beds. *Weed Research*, **51**, 525–533.

Bond, W. J. (2008). What limits trees in C4 grasslands and savanna? *Annual Review of Ecology, Evolution and Systematics*, **39**, 641–659.

Bond, W. J., & Midgley, J. J. (1995). Kill thy neighbour: an individualistic argument for the evolution of flammability. *Oikos*, **73**, 79–85.

Bond, W. J., Woodward, F. I., & Midgley, G. F. (2005). The global distribution of ecosystems in a world without fire. *New Phytologist*, **165**, 525–538.

Bowman, D. M. J. S. (2000). *Australian Rainforests: Islands of Green in a Land of Fire.* Cambridge: Cambridge University Press.

Bowman, D. M. J. S., & Wilson, B. A. (1988). Fuel characteristics of coastal monsoon forests, Northern Territory, Australia. *Journal of Biogeography*, **15**, 807–817.

Bowman, D. M. J. S., Boggs, G. S., & Prior, L. D. (2008). Fire maintains an *Acacia aneura* shrub-land and *Triodia* hummock grassland mosaic in central Australia. *Journal of Arid Environments*, **72**, 34–47.

Bowman, D. M. J. S., Balch, J. K., Artaxo, P., *et al.* (2009). Fire in the earth system. *Science*, **324**, 481–484.

Bowman, D. M. J. S., Murphy, B. P., & Banfai, D. S. (2010). Has global environmental change caused monsoon rainforest to expand in the Australian monsoon tropics? *Landscape Ecology*, **25**, 1247–1260.

Bradstock, R. A. (2010). A biogeographic model of fire regimes in Australia: current and future implications. *Global Ecology and Biogeography*, **19**, 145–158.

Brooks, M. L., D'Antonio, C. M., Richardson, D. M., *et al.* (2004). Effects of invasive alien plants on fire regimes. *BioScience*, **54**, 677–688.

Clarke, P. J., Latz, P. K., & Albrecht, D. E. (2005). Long-term changes in semi-arid vegetation: invasion of an exotic perennial grass has larger effects than rainfall variability. *Journal of Vegetation Science*, **16**, 237–248.

Close, D. C., Davidson, N. J., Johnson, D. W., *et al.* (2009). Premature decline of *Eucalyptus* and altered ecosystem processes in the absence of fire in some Australian forests. *Botanical Review*, **75**, 191–202.

Cochrane, M. A. (2003). Fire science for rainforests. *Science*, **421**, 913–918.

Cochrane, M. A., & Schulze, M. D. (1999). Fire as a recurrent event in tropical forests of eastern Amazon: effects of forest structure, biomass and species composition. *Biotropica*, **31**, 2–16.

Cochrane, M. A., Alencar, A., Schulze, M. D., *et al.* (2003). Positive feedbacks in the fire dynamic of closed canopy tropical forests. *Science*, **284**, 1832–1835.

Coetsee, C., Bond, W. J., & February, E. C. (2010). Frequent fire affects soil nitrogen and carbon in an African savanna by changing woody cover. *Oecologia*, **162**, 1027–1034.

Crisp, M. D., Burrows, G. E., Cook, L. G., Thornhill, A. H., & Bowman, D. M. J. S. (2011). Flammable biomes dominated by eucalypts originated at the Cretaceous-Palaeogene boundary. *Nature Communications*, **2**, 193.

D'Antonio, C. M., & Vitousek, P. M. (1992). Biological invasions by exotic grasses, the grass/fire cycle, and global change. *Annual Review of Ecology and Systematics*, **23**, 63–87.

Duncan, R. S., & Duncan, V. E. (2000). Forest succession and distance from forest edge in an Afro-tropical grassland. *Biotropica*, **32**, 33–41.

Fairfax, R., Fensham, R., Butler, D., *et al.* (2009). Effects of multiple fires on tree invasion in montane grasslands. *Landscape Ecology*, **24**, 1363–1373.

Fensham, R. J., Fairfax, R. J., Butler, D. W., & Bowmann, D. M. J. S. (2003). Effects of fire and drought in a tropical eucalypt savanna colonized by rain forest. *Journal of Biogeography*, **30**, 1405–1414.

Flannigan, M. D., Krawchuk, M. A., de Groot, W. J., Wotton, B. M., & Gowman, L. M. (2009). Implications of changing climate for global wildland fire. *International Journal of Wildland Fire*, **18**, 483–507.

Frelich, L. E., & Reich, P. B. (1999). Neighbourhood effects, disturbance severity and community stability in forests. *Ecosystems*, **2**, 151–166.

Harvest, T., Davidson, N. J., & Close, D. C. (2008). Is decline in high altitude eucalypt forests related to rainforest understorey development and altered soil bacteria following the long absence of fire? *Austral Ecology*, **33**, 880–890.

Hatten, J. A., & Zabowski, D. (2009). Changes in soil organic matter pools and carbon mineralization as influenced by fire severity. *Soil Science Society of America Journal*, **73**, 262–273.

Higgins, S. I., Bond, W. J., February, E. C., *et al.* (2007). Effects of four decades of fire manipulation on woody vegetation structure in savanna. *Ecology*, **88**, 1119–1125.

Hill, R. S. (1994). *History of Australian Vegetation*. Melbourne: Cambridge University Press.

Hoffmann, W. A., & Jackson, R. B. (2000). Vegetation-climate feedbacks in the conversion of tropical savanna to grassland. *Journal of Climate*, **13**, 1593–1602.

Hoffmann, W. A., Adasme, R., Haridasan, M., *et al.* (2009). Tree topkill, not mortality, governs the dynamics of savanna-forest boundaries under frequent fire in central Brazil. *Ecology*, **90**, 1326–1337.

Jackson, W. D. (1968). Fire, air, water and earth – an elemental ecology of Tasmania. *Proceedings of the Ecological Society of Australia*, **3**, 9–16.

Johnstone, J. F., Hollingsworth, T. N., Chapin, F. S. III, & Mack, M. C. (2010). Change in fire regime break the legacy lock on successional trajectories in Alaskan boreal forest. *Global Change Biology*, **16**, 1281–1295.

Keeley, J. E., & Brennan, T. J. (2012). Fire-driven alien invasion in a fire-adapted ecosystem. *Oecologia* **69**, 1043–1052. doi: 10.1007/s0.

Keeley, J. E., Pausas, J. G., Rundel, P. W., Bond, W. J., & Bradstock, R. A. (2011). Fire as an evolutionary pressure shaping plant traits. *Trends in Plant Science*, **16**, 406–411.

Knox, K. J. E., & Clarke, P. J. (2012). Fire severity, feedback effects and resilience to alternative community states in forest assemblages. *Forest Ecology and Management*, **265**, 47–54.

Laurance, W. F., Oliveira, A. A., Laurance, S. G., *et al.* (2005). Altered tree communities in undisturbed Amazonian forests: a consequence of global change? *Biotropica*, **37**, 160–162.

Lawes, M. J., Murphy, B. P., Midgley, J. J., & Russell-Smith, J. (2011). Are the eucalypt and non-eucalypt components of Australian tropical savannas independent? *Oecologia*, **166**, 229–239.

Lehmann, C. E. R., Prior, L. D., & Bowman, D. M. J. S. (2009). Fire controls population size structure in four dominant tree species in Australian mesic eucalyptus savanna. *Oecologia*, **161**, 505–515.

Malhi, Y., Aragao, L., Galbraith, D., *et al.* (2009). Exploring the likelihood and mechanism of a climate-change-induced dieback of the Amazon rainforest. *Proceedings of the National Academy of Sciences of the USA*, **106**, 20610–20615.

McGlone, M. S., Wilmshurst, J. M., & Leach, H. M. (2005). An ecological and historical review of bracken (Pteridium esculentum) in New Zealand, and its cultural significance. *New Zealand Journal of Ecology*, **29**, 165–184.

Miller, G., Friedel, M., Adam, P., & Chewings, V. (2010). Ecological impacts of buffel grass (*Cenchrus ciliaris* L.) invasion in central Australia – does field evidence support a fire-invasion feedback? *The Rangeland Journal*, **32**, 352–265.

Mutch, R. W. (1970). Wildland fires and ecosystems-a hypothesis. *Ecology*, **51**, 1046–1051.

Nano, C. E. M., & Clarke, P. J. (2010). Woody-grass ratios in a grassy arid system are limited by multi-causal interactions of abiotic constraint, competition and fire. *Oecologia*, **162**, 719–732.

Nano, C. E. M., Clarke, P. J., & Pavey, C. R. (2012). Fire regimes in arid hummock grasslands and Acacia shrublands. In R. A. Bradstock, M. A. Gill & R. J. Williams (Eds.), *Flammable Australia: Fire Regimes, Biodiversity and Ecosystems in a Changing World*. Melbourne: CSIRO.

Nepstad, D., Carvalho, G., Barros, A. C., *et al.* (2001). Road paving, fire regime feedbacks, and the future of Amazon forests. *Forest Ecology and Management*, **154**, 395–407.

Nicholas, A. M. M., Franklin, D. C., & Bowman, D. M. J. S. (2011). Florsitic uniformity across abrupt boundaries between *Triodia* hummock grassland and *Acacia* shrubland on an Australian desert sandplain. *Journal of Arid Environments*, **75**, 1090–1096.

Odion, D. C., Moritz, M. A., & DellaSala, D. A. (2010). Alternative community states maintained by fire in the Klamath Mountains, USA. *Journal of Ecology*, **98**, 96–105.

Petraitis, P. S., & Latham, R. E. (1999). The importance of scale in testing the origins of alternative community states. *Ecology*, **80**, 429–442.

Poorter, H., Niinemets, U., Poorter, L., Wright, I. J., & Villar, R. (2009). Cause and consequence of variation in leaf mass per unit area (LMA): a meta-analysis. *New Phytologist*, **182**, 565–588.

Prior, L. D., Murphy, B. P., & Russell-Smith, J. (2009). Environmental and demographic correlates of tree recruitment and mortality in north Australian savannas. *Forest Ecology and Management*, **257**, 66–74.

Prior, L. D., Williams, R. J., & Bowman, D. M. J. S. (2010). Experimental evidence that fire causes a tree recruitment bottleneck in an Australian tropical savanna. *Journal of Tropical Ecology*, **26**, 595–603.

Rossiter, N. A., Setterfield, S. A., Douglas, M. M., & Hutley, L. M. (2003). Testing the grass fire cycle: exotic grass invasion in the tropical savannas of northern Australia. *Diversity and Distributions*, **9**, 169–176.

Rossiter-Rachor, N. A., Setterfield, S. A., Douglas, M. M., Hutley, L. B., & Cook, G. D. (2008). *Andropogon gayanus* (Gamba grass) invasion increases fire-mediated nitrogen losses in the tropical savannas of northern Australia. *Ecosystems*, **11**, 77–88.

Rossiter-Rachor, N. A., Setterfield, S. A., Douglas, M. M., *et al.* (2009). Invasive *Andropogon gayanus* (Gamba grass) is an ecosystem transformer of nitrogen relations in Australia's tropical savanna. *Ecological Applications*, **19**, 1546–1560.

Russell-Smith, J., Whitehead, P. J., Cook, G. D., & Hoare, J. L. (2003). Response of Eucalyptus-dominated savanna to frequent fires: lessons from Munmarlary, 1973–1996. *Ecological Monographs*, **73**, 349–375.

Russell-Smith, J., Stanton, P. J., Edwards, A. C., & Whitehead, P. J. (2004). Rain forest invasion of eucalypt-dominated woodland savanna, Iron Range, north-eastern Australia: II. Rates of landscape change. *Journal of Biogeography*, **31**, 1305–1316.

Sankaran, M., Hanan, N. P., Scholes, R. J., *et al.* (2005). Determinants of woody cover in African savannas. *Nature*, **438**, 846–849.

Scheffer, M., & Carpenter, S. R. (2003). Catastrophic regime shifts in ecosystems: linking theory to observation. *Trends in Ecology and Evolution*, **18**, 648–656.

Schwilk, D. W., & Kerr, B. (2002). Genetic niche-hiking: an alternative explanation for the evolution of flammability. *Oikos*, **99**, 431–442.

Setterfield, S. A., Rossiter-Rachor, N. A., Hutley, L. B., Douglas, M. M., & Williams, R. J., (2010). Turning up the heat: the impacts of *Andropogon gayanus* (gamba grass) invasion on fire behaviour in northern Australian savannas. *Diversity and Distributions*, **16**, 854–861.

Shakesbury, R. A., & Doerr, S. H. (2006). Wildfire as a hydrological and geomorphological agent. *Earth Science Review*, **74**, 269–307.

Slik, J., Breman, F., Bernard, C., *et al*. (2010). Fire as a selective force in a Bornean tropical everwet forest. *Oecologia*, **164**, 841–849.

Staver, A. C., Archibald, S., & Levin, S. (2011a). Tree cover in sub-Saharan Africa: rainfall and fire constrain forest and savanna as alternative stable states. *Ecology*, **92**, 1063–1072.

Staver, A. C., Archibald, S., & Levin, S. (2011b). The global extent of determinants of savanna and forest as alternative biome states. *Science*, **334**, 230–232.

Uhl, C., & Kauffman, J. B. (1990). Deforestation, fire susceptibility, and potential tree responses to fire in the eastern Amazon. *Ecology*, **71**, 437–449.

van Langevelde, F., van de Vijver, C., Kumar, L., *et al*. (2003). Effects of fire and herbivory on the stability of savanna ecosystems. *Ecology*, **84**, 337–350.

Van Nieuwstadt, M. G. L., & Sheil, D. (2005). Drought, fire and tree survival in a Borneo rain forest, East Kalimantan, Indonesia. *Journal of Ecology*, **93**, 191–201.

Warman, L., & Moles, A. T. (2009). Alternative stable states in Australia's Wet Tropics: a theoretical framework for the field data and a field-case for the theory. *Landscape Ecology*, **24**, 1–13.

Webb, L. J. (1968). Environmental relationships of the structural types of Australian rain forest vegetation. *Ecology*, **49**, 296–311.

Werner, P. A., & Franklin, D. C. (2010). Resprouting and mortality of juvenile eucalypts in an Australian savanna: impacts of fire season and annual sorghum. *Australian Journal of Botany*, **58**, 619–628.

Westfall, R. H., Everson, C. S., & Everson, T. M. (1983). The vegetation of the protected plots at Thabamhlope research station. *South African Journal of Botany*, **2**, 15–25.

Wigley, B. J., Bond, W. J., & Hoffman, M. T. (2010). Thicket expansion in a South African savanna under divergent land use: local vs. global drivers? *Global Change Biology*, **16**, 964–976.

Woinarski, J. C. Z., Risler, J., & Kean, L. (2004). Response of vegetation and vertebrate fauna to 23 years of fire exclusion in a tropical *Eucalyptus* open forest, Northern Territory, Australia. *Austral Ecology*, **29**, 156–176.

Wood, S. W., & Bowman, D. M. J. S. (2011). Alternative stable states and the role of fire-vegetation-soil feedbacks in the temperate wilderness of southwest Tasmania. *Landscape Ecology*, **27**, 13–28.

Wright, B. R., & Clarke, P. J. (2007). Fire regime (recency, interval and season) changes the composition of spinifex (*Triodia* spp.)-dominated desert dunes. *Australian Journal of Botany*, **55**, 709–724.

6 Community stability and instability in ectoparasites of marine and freshwater fish

Andrea Šimková and Klaus Rohde

Introduction

Marine and freshwater fish are hosts to a rich fauna of ectoparasites, living on their gills and skin feeding on blood, mucus and epithelial cells. Fish can easily be obtained and examined in large numbers. Fish ectoparasites represent a highly diverse group including monogeneans, crustaceans, isopods, mollusks and hirudineans. This makes them almost ideal objects for ecological studies. Such studies have been conducted by several researchers, using a range of host species and ecological techniques, with the aim of identifying patterns and processes in parasite communities. Studies have concentrated on different levels of community organization, i.e., those of infra-, component and compound communities, and examined questions of saturation vs. non-saturation of communities, degree of aggregation, temporal and spatial variability of organization, limiting similarity and niche segregation, host specificity, nestedness, and degree of structuring in communities as revealed by null model analyses. All these aspects are of significance in an evaluation of how common equilibrium and nonequilibrium conditions are in ecological communities, the main topic of this book. In this chapter, we provide an up-to-date account of relevant studies.

Parasite communities

Parasite communities have been commonly studied at different levels, i.e., those of infracommunity, component community and compound community (Holmes & Price, 1986). A parasite infracommunity consists of all the infrapopulations (populations of all species) within a host individual. Infracommunities are incapable of self-perpetuation (because most parasites disperse their propagules into the free environment where they usually develop before reaching a host). A parasite component community consists of all infracommunities within a host population. The boundaries for the component community depend on spatial scale (Aho & Bush, 1993). For example, one can consider a component community as (1) all parasites in all individuals of a given host species from

The Balance of Nature and Human Impact, ed. Klaus Rohde. Published by Cambridge University Press.
© Cambridge University Press 2013.

a specific collection site of a water body, or (2) all parasites of all individuals throughout the host's geographical distribution or of a range of host distributions for which the data were obtained. The majority of the studies investigating the structure of parasite communities in fish have been performed at this level. A compound community consists of all parasite communities in an ecosystem. At this level, several important features determining the structure of parasite communities can be investigated (e.g., host specificity).

Non-saturated communities of fish ectoparasites

Different parasite communities may be differently structured. Interactions among species are considered as a major determinant of community structure. Current theories in parasite ecology distinguish two types of parasite communities: isolationist (i.e., non-interactive) communities, in which niche space is not saturated with parasite individuals, and species can coexist without major or no interference from other species; and interactive communities, in which niche space is saturated, and interspecific interactions "regulate" species co-occurrence. Many studies have presented parasite communities as a continuum from interactive to non-interactive, depending on available niche space, as proposed by Cornell and Lawton (1992). Rohde (2005) concluded that non-saturation is strong evidence for nonequilibrium in many communities, and communities of fish ectoparasites can be considered to live under nonequilibrium conditions (Morand *et al.*, 2002a). Such communities are not structured by interspecific competition, they are not saturated with parasite species and many potential niches for ectoparasites are vacant. Gills of some fish were found to harbor thousands of parasites of up to 30 species, whilst most fish species even of similar body size and living in similar habitats were found to harbor only a few specimens of one or two parasite species or were not infected at all (Rohde, 2005). No evidence for interactions has been found in many ectoparasite communities of fish even in cases of high parasite abundance (Rohde, 1979).

To determine whether the community is saturated or unsaturated, many studies focused on relationships between local and regional parasite species richness. An increase of local species richness with regional species richness is expected in unsaturated communities, whilst an asymptotic pattern suggests (but does not prove) community saturation (see Rohde, 2005, pp. 73–76: different transmission rates and intrinsic life spans can cause asymptotic patterns even without interspecific competition). Morand *et al.* (1999) found a positive relationship between infracommunity species richness and total parasite species richness, which supports the assumption that ectoparasite communities of marine fishes are not saturated.

Aggregation of fish parasites promotes the stability of fish ectoparasite communities

Parasites are not uniformly distributed among host individuals, i.e., some hosts harbor more parasites than others. Typically, most host individuals harbor few or no parasites and few hosts harbor many parasites. Also, distribution of parasites is not static through time; different host individuals are exposed to parasites at different times. In addition, parasite aggregation is influenced by variability in host immune response, size and age, physiological and behavioral factors (e.g., Shaw & Dobson, 1995; Zelmer & Arai, 1998; Poulin, 2007). Aggregation may be one of the most powerful factors leading to species coexistence in communities (Ives, 1988, 1991; Jaenike & James, 1991): the aggregation model of coexistence suggests facilitation of species coexistence by aggregation (Shorrocks & Rosewell, 1986; Jaenike & James, 1991). Applied to parasites, species coexistence is facilitated by reducing the overall intensity of competition via aggregated utilization of fragmented host resources. This model postulates that if parasite species are distributed in a way that interspecific aggregation is reduced relative to intraspecific aggregation, species coexistence is facilitated and interspecific competition is reduced. This was indeed shown for ectoparasite communities of marine and freshwater fish (Morand *et al.*, 1999; Šimková *et al.*, 2000). High levels of intraspecific aggregation and interspecific aggregation were found for nine congeneric gill monogenean parasites of *Dactylogyrus* parasitizing roach (*Rutilus rutilus*), at the level of individual hosts or their microhabitats (i.e., gill arches) (Šimková *et al.*, 2000, 2001a).

Temporal and spatial variability affect the structure of ectoparasite communities in freshwater fish

In the case of ectoparasites of freshwater fish, environmental and particularly seasonal variability is an important factor influencing the composition of parasite communities and especially promoting species coexistence (Koskivaara *et al.*, 1992; Koskivaara & Valtonen, 1992). For instance, *Dactylogyrus* (Monogenea) communities living on the gills of roach during late spring and summer are composed of dominant species, whose abundance depends on high temperature. On the other hand, in autumn, when the abundance of such "summer-dominant" species decreased, rare *Dactylogyrus* species, which were absent in summer, were also present in gill ectoparasite communities at low abundances (Šimková *et al.*, 2001b). It seems that seasonality determines interspecific interactions among congeneric ectoparasite species especially in the periods when species diversity and abundance are high. Koskivaara and Valtonen (1992) proposed that ectoparasite communities of freshwater fish could be considered as non-interactive during the period of low parasite abundance and interactive during the period of high parasite abundance. This was also suggested by other studies of gill monogeneans of freshwater fish. Kadlec *et al.* (2003) showed seasonal variation in abundance of *Dactylogyrus carpathicus* and *D. malleus* parasitizing freshwater barbel (*Barbus barbus*) and demonstrated that the increase of *D. malleus* abundance is associated with strong microhabitat selection and reduced niche overlap between two species. A similar finding was made in a study of gill ectoparasite communities of European eel (*Anguilla anguilla*) composed of two congeneric monogenean species, *Pseudodactylogyrus bini* and *P. anguillae*, which show temporal variation in their abundance (Matějusová *et al.*, 2003).

Limiting similarity and niche restriction in ectoparasite species of fish: factors stabilizing the structure of their communities

Hutchinson (1959) suggested that the morphology of organs directly associated with exploitation of niches and especially with food intake becomes less similar in related species such as congeners (which supposedly utilize similar resources) when such species co-occur in the same habitat. In other words, there is a limiting similarity below which species cannot coexist. Several hypotheses about limiting similarity in parasite communities have been proposed; here they are summarized for congeneric ectoparasites of marine and freshwater fish. According to Rohde (1991), niche selection is linked to morphological differences of the attachment apparatus (haptor) of fish ectoparasites. This may permit host exploitation avoiding interspecies competition by "specialization" for certain niches (Rohde, 1979). However, niches of parasite species in fish are restricted even in the absence of potentially competing species (Rohde, 1977), and since fish ectoparasites have the tendency to be aggregated in their niches (either at the level of hosts or at the level of microhabitats), and since gill ectoparasites have restricted niches even in the case of low population densities, Rohde (1977) postulated that niches are restricted in order to facilitate mating encounters ("mating hypothesis of niche restriction").

Furthermore, congeneric parasite species living in the same habitat (i.e., within the same fish host and in overlapping or very close microhabitats on the gills) differ in the shape and size of their copulatory organ, which prevents interspecific hybridization (Rohde & Hobbs, 1986). Marine examples are *Lamellodiscus acanthopagri*, *L. squamosus* and *L. major*, which live in the same microhabitats on *Acanthopagrus australis* on the coast of southeastern Australia, and various monogeneans on the gills of *Lethrinus miniatus* on the Great Barrier Reef; the sclerotized parts of male and female copulatory organs in species living in the same or widely overlapping niches are markedly different (Figure 6.1). On the other hand, congeneric ectoparasites with entirely or largely segregated niches have more or less identical copulatory organs (e.g., *Kuhnia scombri* living on the gills and *K. sprostonae* living on the pseudobranches of *Scomber scombrus* in the North Sea) (two species H in Figure 6.1). According to the hypothesis of reproductive character displacement (Butlin, 1989) such morphological differences in copulatory organs are the result of interactions between species leading to the divergence in mate recognition and to avoidance of interspecific hybridization. Concerning gill parasites, interspecific differences in the morphology of copulatory organs are likely to reinforce reproductive segregation among congeneric species (Rohde & Hobbs, 1986; Rohde 1989). This is important, because ectoparasite communities of fish are often composed of several or many congeneric species, especially of gill and skin monogeneans. Freshwater cyprinid fish, for example, harbor many species of the monogenean *Dactylogyrus*, which are highly host specific, many of them exhibiting narrow host specificity, i.e., a given *Dactylogyrus* species often

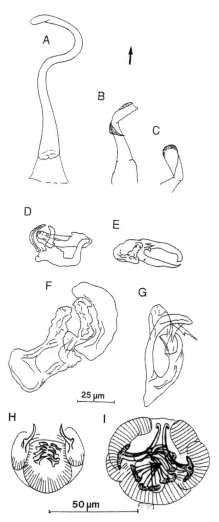

Figure 6.1.   Monogeneans infecting the gills of marine fish can use identical microhabitats on the gills when they differ in the shape (and/or size) of their copulatory organs; they are spatially segregated when they have identical copulatory organs, suggesting that not interspecific competition but reproductive segregation is the primary reason for species segregation. The following examples illustrate this; they show copulatory sclerites of monopisthocotylean monogeneans infecting the gills of *Lethrinus miniatus*: A, *Haliotrema lethrini*; B, *H. fleti*; C, *H. chrysostomi*; D, *Calydiscoides gussevi*; E, *Protolamellodiscus* sp.; F, *Calydiscoides difficilis*; G, *C. australis*, and of polyopisthocotylean monogeneans on the gills of *Scomber* spp.: four species with copulatory organ H, and one with copulatory organ I. All species on *L. miniatus* except for E inhabit identical or overlapping microhabitats, the four species H on *Scomber* spp. are segregated in different microhabitats or hosts, species I overlaps with some of the others. Arrow points to anterior end. Reprinted from Rohde, K., Hayward, C., Heap, M., & Gosper, D. A. (1994), with permission of Elsevier.

parasitizes only one fish species, or alternatively parasitizes a few closely related cyprinid fish species. *Dactylogyrus* diversity on some cyprinid host species is impressive (for instance among European cyprinid fish species, chub, *Squalius cephalus*, harbors 14 *Dactylogyrus* species, and roach, *Rutilus rutilus*, harbors 13 within the central European range of distribution of these fish species). It was shown that in *Dactylogyrus* communities on one host species, *Dactylogyrus* species with similar morphology of attachment apparatus tend to be more positively aggregated among gill arches than *Dactylogyrus* species with different attachment apparatus (Šimková *et al.*, 2001a). Considering niche position on fish gills, *Dactylogyrus* species with similar haptor morphology (i.e., low morphometrical distances of attachment apparatus between species) occur in overlapping niches. On the other hand, *Dactylogyrus* species living in closely located niches have different morphology and size of copulatory organs (Figures 6.2, 6.3) (i.e., they have high interspecies morphometrical distances of the copulatory organ) (Šimková *et al.*, 2002). Using *Dactylogyrus* communities coexisting on roach or chub, two European cyprinid fish species with the highest *Dactylogyrus* species richness, negative relationships between morphometric distance matrices for haptor and copulatory organs were found (Morand *et al.*, 2002b), which is in line with the hypothesis of reinforcement of reproductive barriers in congeneric ectoparasites living on the same host. Nevertheless, fish phylogeny, their life traits like body size or longevity as well as their ecology impose strong constraints on haptor morphology in congeneric parasites. It was shown that *Dactylogyrus* with similar haptor morphology tend to parasitize phylogenetically related fish host species. Even when they occupy closely related niches, the niche positions of such morphologically similar congeneric parasites within a host differ at least in one niche parameter (for instance they occupy the same gill segment within the same gill arch, but their preferred niche differs in gill area position) (Šimková *et al.*, 2006).

Host specificity in structuring parasite communities

Jeffries and Lawton (1984) hypothesized that species specialize in enemy-free space, i.e., species specialize when competitors are absent. Following this hypothesis, specialists should occur in species-poor communities. However, in the case of monogeneans of freshwater fish (at the level of host family and at the level of host species) it was shown that species-rich communities are composed of a higher proportion of specialists whilst species-poor communities are formed mainly by generalists (Poulin, 1997; Šimková *et al.*, 2001c). *Dactylogyrus* communities are a good example to illustrate the importance of host specificity for structuring parasite communities of congeneric monogeneans in freshwater fish (Šimková *et al.*, 2002). Strictly host-specific congeneric ectoparasite species, i.e., specialists, occupy niches located close together (for instance on the same gill arch or the same segment or area within a given gill arch). On the other hand, generalists that are able to colonize more than one host species occupy more distant niches within hosts (for instance they often occupy different gill arches or different segments and also different areas within the same gill arches).

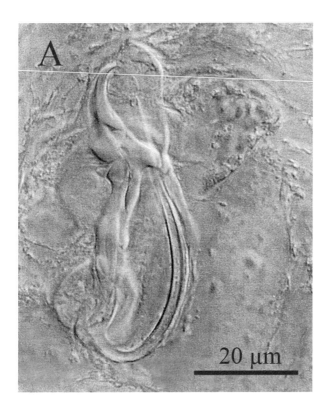

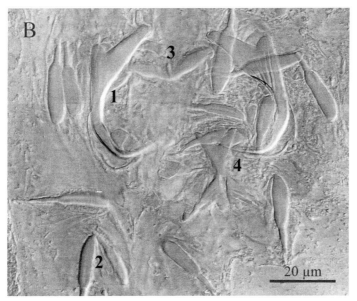

Figure 6.2.   *Dactylogyrus caballeroi* from the roach, *Rutilus rutilus*. (A) Sclerotized part of copulatory organ. (B) Sclerotized parts of attachment organ (haptor): 1, central hooks (anchors); 2, marginal hooks; 3, dorsal connective bar; 4, ventral connective bar. Photos taken by A. Simkova with a light microscope equipped with differential interference contrast.

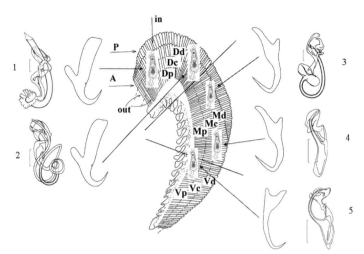

Figure 6.3.   Niche segregation in gill parasites. Each gill arch is separated into three segments (D, dorsal; M, medial; V, ventral), three areas (d, distal; c, central; p, proximal), two surfaces (in, inner; out, outer) and two hemibranches (A, anterior; P, posterior). The positions of five congeneric *Dactylogyrus* species are shown. *Dactylogyrus* with the same morphology of attachment apparatus occupy similar niches; their copulatory organs are of different morphology and size (i.e., species 1 and 2 on the left of the diagram, species 3 and 4 on the right of the diagram). *Dactylogyrus* with different morphology of attachment apparatus occupy different niches (e.g., species 1, 4 and 5) and some of them may possess similar copulatory organs (species 4 and 5).

Nested structure of fish ectoparasite communities

As mentioned above, species assemblages living in fragmented habitats are not distributed randomly across their habitats. Thus, parasite communities in hosts may be structured and predictably so. Therefore, one of the central questions in parasite ecology is whether infracommunities comprise all parasite species living in a component community, or whether infracommunities are predictable subsets formed by certain species of component communities that can coexist. Considering the fact that species richness of the component community is influenced by local availability of parasite species and their probability of colonization, infracommunities of parasites might be subsets of the parasite species occurring in component communities.

A simple pattern of non-random species distribution was described as a nested subset pattern by Patterson and Atmar (1986). In such a case, depauperate assemblages are non-random subsets of richer species assemblages. Nested patterns are usually investigated in a biogeographical context, i.e., this pattern is supposed to be generated by extinction and colonization events on archipelagos or isolated habitats. Species in assemblages are never completely nested. Therefore, the objective of many studies is to investigate the degree of nestedness i.e., how a given assemblage departs from a perfectly nested pattern towards a more random pattern. Whereas ectoparasite infracommunities of freshwater tropical fish showed a nested structure (Guégan & Huguény, 1994; Huguény & Guégan, 1997), it seems that parasite communities of marine fish form random and unstructured assemblages (Worthen & Rohde, 1996; Rohde *et al.*, 1998). Nested patterns in parasite communities may result from different processes. In the case of metazoan parasites of fish, different colonization probabilities are the most likely cause (Worthen & Rohde, 1996; Rohde *et al.*, 1998).

Culter (1998) pointed out that nested patterns emerge only at the appropriate spatial and temporal scales, i.e., if the compared sites have similar ecological characteristics and the species are shared by all sites. Šimková *et al.* (2001d) investigated nested subset patterns in *Dactylogyrus* communities parasitizing roach at three different scales, i.e., (1) among sites (i.e., among parasite component communities), (2) at the sites over seasons and (3) among parasite infracommunities of a local host population over one season. A nested subset pattern was found in *Dactylogyrus* communities among sites in two different seasons as well as at the sites over seasons. Nevertheless, the analysis among *Dactylogyrus* infracommunities (i.e., host individuals at a given site and within a given season) showed that a nested pattern at the infracommunity level is uncommon. The parasite species richness of a given fish host species may also determine the structure of parasite communities, i.e., parasite infracommunities of fish host species with high parasite species richness showed a nested structure more often than parasite infracommunities of fish host species with low parasite species richness (Šimková *et al.*, 2003). Moreover, host specificity may also play a role (Matějusová *et al.*, 2000; Valtonen *et al.*, 2001).

Poulin and Guégan (2000) found another non-random structural pattern in fish ecto-parasite assemblages, i.e., antinestedness. This pattern is a situation where parasite species are always absent from infracommunities richer than the most depauperate one. The authors showed that antinested patterns are as common as nested patterns in ectoparasite communities of marine fish; antinested assemblages were characterized by lower prevalence and intensity of parasite infection than nested assemblages. An anti-nested pattern was also observed in some endoparasite communities of European bitter-ling (*Rhodeus sericeus*), a freshwater fish species of small body size with low parasite species richness relative to other common cyprinid fish species that have high ectopar-asite and endoparasite species richness (Šimková *et al.*, 2003). Poulin and Guégan (2000) proposed that a nestedness-antinestedness continuum might best illustrate the spatial organization of parasite communities, from coexistence to competitive exclusion. Parasite species found in nested assemblages have interspecific aggregation reduced relative to intraspecific aggregation, facilitating species coexistence, as proposed by the aggregation model of coexistence. On the other hand, parasite species in antinested assemblages are interspecifically aggregated, which leads to species exclusion.

Morand *et al.* (2002a) showed that epidemiology can explain order in ectoparasite assemblages of fish. They demonstrated a link between a nested pattern and distribution of parasite prevalence. Whereas a nested structure leads to a unimodal distribution of ectoparasites, a non-nested structure leads to a bimodal distribution. They proposed that the nested pattern is simply the result of demographic characteristics of each ectoparasite species in an assemblage.

Null model analyses

A major problem in testing various ecological models is the establishment of null models with which findings can be compared (see Rohde, 2005). Gotelli and Rohde (2002) tested for non-randomness in the structure of parasite communities on the heads and gills of 45 marine fish species using null model analyses. Data were organized as presence-absence matrices, in which each row represented a different parasite species and each column a different individual host. Four indices were used, i.e., the number of species combinations, the number of checkerboard species pairs, the C-score and the variance (V) ratio (for details see Gotelli & Rohde, 2002). The observed indices were compared with the indices of 5000 randomly selected null communities. Meta-analyses were conducted by calculating standardized effect sizes (SES = number of standard deviations in which the observed matrix is either above or below the mean index calculated) for each matrix. The null hypothesis is that the average SES for the set of matrices is zero. Fewer combinations of parasite species were supported throughout the study than expected on a random basis. This was the case even when empty sites were considered. Nevertheless, in almost all analyses, the null hypothesis could not be rejected. In other words, patterns of co-occurrence did not differ significantly from patterns due to random colonization and extinction, i.e., there is evidence neither for significant effects of interspecific competition nor for facilitation.

Conclusions

We conclude that ectoparasite communities on the gills and skin of fish are non-saturated, with many vacant niches, which suggests that these fish ectoparasites live under nonequilibrium conditions. Such communities, often with several congeneric species, are not structured by interspecific competition, but show a high degree of aggregation. The structure of ectoparasite communities especially in freshwater fish is affected by temporal and spatial variability, which promotes the coexistence of potentially competitive species. Niche segregation in congeneric ectoparasites is linked to the morphology of their attachment apparatus. Congeners with morphologically similar attachment apparatus tend to occupy the same or adjacent niches within a host. Co-occurrence of congeners is facilitated by reinforcement of reproductive barriers due to morphological differences in copulatory organs. Ectoparasite communities of fish may show non-random structural patterns, the likely result of the demography of species living in these communities. Many congeneric ectoparasite species of fish are specialists, adapted to certain host species; host specificity therefore also represents an important factor involved in structuring communities. Using null model analysis, neither interspecific competition nor facilitation was found to affect the structure of ectoparasite communities of marine fish.

Acknowledgments

AŠ was funded by the Czech Science Foundation, European Center of Excellence, Project No. P505/12/G112.

References

Aho, J. M., & Bush, A. O. (1993). Community richness in parasites of freshwater fishes from North America. In R. E. Ricklefs & D. Schluter (Eds.), *Species Diversity in Ecological Communities: Historical and Geographical Perspectives* (pp. 185–193). Chicago: University of Chicago Press.

Butlin, R. (1989). Reinforcement of premating isolation. In D. Otte & J. A. Endler (Eds.), *Speciation and Its Consequences* (pp. 158–179). Sunderland, MA: Sinauer Associates Inc.

Cornell, H. V., & Lawton, J. H. (1992). Species interactions, local and regional processes, and limits to the richness of ecological communities: a theoretical perspective. *Journal of Animal Ecology*, **61**, 1–12.

Culter, A. H. (1998). Nested patterns of species distribution: processes and implications. In M. L. McKinney & A. J. Drake (Eds.), *Biodiversity Dynamics. Turnover of Populations, Taxa, and Communities* (pp. 212–231). New York: Columbia University Press.

Gotelli, N. J., & Rohde, K. (2002). Co-occurrence of ectoparasites of marine fishes: null model analysis. *Ecology Letters*, **5**, 86–94.

Guégan, J.-F., & Huguény, B. (1994). A nested parasite species subset pattern in tropical fish: host as major determinant of parasite infracommunity structure. *Oecologia*, **100**, 184–189.

Holmes, J. C., & Price, P. W. (1986). Communities of parasites. In: J. Kikkawa & D. J. Anderson (Eds.), *Community Ecology: Pattern and Process* (pp. 186–213). Oxford: Blackwell Scientific.

Huguény, B., & Guégan, J.-F. (1997). Community nestedness and the proper way to assess statistical significance by Monte-Carlo tests: some comments on Worthen and Rohde's (1996) paper. *Oikos*, **80**, 572–574.

Hutchinson, G. E. (1959). Homage to Santa Rosalia, or why are there so many kinds of animals? *The American Naturalist*, **93**, 145–159.

Ives, A. R. (1988). Aggregation and the coexistence of competitors. *Annales Zoologici Fennici*, **25**, 75–88.

Ives, A. R. (1991). Aggregation and coexistence in a carrion fly community. *Ecological Monographs*, **61**, 75–94.

Jaenike, J., & James, A. C. (1991). Aggregation and the coexistence of mycophagous *Drosophila*. *Journal of Animal Ecology*, **60**, 913–928.

Jeffries, M. J., & Lawton, J. H. (1984). Enemy free space and the structure of ecological communities. *Biological Journal of the Linnean Society*, **23**, 269–286.

Kadlec, D., Šimková, A., & Gelnar M. (2003). The microhabitat distribution of two *Dactylogyrus* species parasitizing the gills of the barbel, *Barbus barbus*. *Journal of Helminthology*, **77**, 317–325.

Koskivaara, M., & Valtonen E. T. (1992). *Dactylogyrus* (Monogenea) communities on the gills of roach in three lakes in Central Finland. *Parasitology*, **104**, 263–272.

Koskivaara, M., Valtonen, E. T., & Vuori, K.-M. (1992). Microhabitat distribution and coexistence of *Dactylogyrus* species (Monogenea) on the gills of the roach. *Parasitology*, **104**, 273–281.

Matějusová, I., Morand, S., & Gelnar, M. (2000). Nestedness in assemblages of gyrodactylids (Monogenea: Gyrodactylidea) parasitising two species of cyprinid – with reference to generalists and specialists. *International Journal for Parasitology*, **30**, 1153–1158.

Matějusová, I., Šimková, A., Sasal, P., & Gelnar, M. (2003). Microhabitat distribution of *Pseudodactylogyrus anguillae* and *Pseudodactylogyrus bini* among and within gill arches of the European eel (*Anguilla anguilla* L.). *Parasitology Research*, **89**, 290–296.

Morand, S., Poulin, R., Rohde, K., & Hayward, C. J. (1999). Aggregation and species coexistence of ectoparasites of marine fishes. *International Journal for Parasitology*, **29**, 663–672.

Morand, S., Rohde, K., & Hayward, C. (2002a). Order in parasite communities of marine fish is explained by epidemiological processes. *Parasitology*, **124**, S57–S63.

Morand, S., Šimková, A., Matějusová, I., *et al.* (2002b). Investigating patterns may reveal processes: evolutionary ecology of ectoparasitic monogeneans. *International Journal for Parasitology*, **32**, 111–119.

Patterson, B. D., & Atmar, W. (1986). Nested subsets and the structure of insular mammalian faunas and archipelagos. *Biological Journal of the Linnean Society*, **28**, 65–82.

Poulin, R. (1997). Parasite faunas of freshwater fish: the relationship between richness and the specificity of parasites. *International Journal for Parasitology*, **27**, 1091–1098.

Poulin, R. (2007). *Evolutionary Ecology of Parasites*. Princeton, NJ: Princeton University Press.

Poulin, R., & Guégan, J. F. (2000). Nestedness, anti-nestedness, and the relationship between prevalence and intensity in ectoparasite assemblages of marine fish: a spatial model of species coexistence. *International Journal for Parasitology*, **30**, 1147–1152.

Rohde, K. (1977). A non-competitive mechanism responsible for restricting niches in parasites. *Zoologishe Anzeiger*, **199**, 164–172.

Rohde, K. (1979). A critical evaluation of intrinsic and extrinsic factors responsible for niche restriction in parasites. *The American Naturalist*, **114**, 648–671.

Rohde, K. (1989). Simple ecological systems, simple solutions to complex problems? *Evolutionary Theory*, **8**, 305–350.

Rohde, K. (1991). Intra-and inter-specific interactions in low density populations in resource-rich habitat. *Oikos*, **60**, 91–104.

Rohde, K. (2005). *Nonequilibrium Ecology*. Cambridge: Cambridge University Press.

Rohde, K., & Hobbs, R. P. (1986). Species segregation: competition or reinforcement of reproductive barriers? In M. Cremin, C. Dobson, & E. Noorhouse (Eds.), *Parasite Lives. Papers on Parasites, their Hosts and their Association to Honour J. F. A. Sprent* (pp. 189–199). St Lucia: University of Queensland Press.

Rohde, K., Worthen, W. B., Heap, M., Huguény, B., & Guégan, J.-F. (1998). Nestedness in assemblages of metozoan ecto- and endoparasites of marine fish. *International Journal of Parasitology*, **28**, 543–549.

Shaw, D. J., & Dobson, A. P. (1995). Patterns of macroparasite abundance and aggregation in wildlife populations: a quantitative review. *Parasitology*, **111**, S111–S133.

Shorrocks, B., & Rosewell, J. (1986). Guild size in drosophilids: a simulation model. *Journal of Animal Ecology*, **55**, 527–541.

Šimková, A., Desdevises, Y., Gelnar, M., & Morand, S. (2000). Co-existence of nine gill ectoparasites (*Dactylogyrus*: Monogenea) parasitizing roach (*Rutilus rutilus* L.): history and present ecology. *International Journal for Parasitology*, **30**, 1177–1188.

Šimková, A., Gelnar, M., & Sasal. P. (2001a). Aggregation of congeneric parasites (Monogenea: *Dactylogyrus*) among gill microhabitats within one host species (*Rutilus rutilus* L.). *Parasitology*, **123**, 599–607.

Šimková, A., Sasal, P., Kadlec, D., & Gelnar, M. (2001b). Water temperature influencing dacty-logyrid species communities in roach, *Rutilus rutilus*, in the Czech Republic. *Journal of Helminthology*, **75**, 373–393.

Šimková, A., Desdevises, Y., Gelnar, M., & Morand, S. (2001c). Morphometric correlates of host specificity in *Dactylogyrus* species (Monogenea) parasites of European Cyprinid fish. *Parasitology*, **123**, 169–177.

Šimková, A., Gelnar, M., & Morand, S. (2001d). Order and disorder in ectoparasite communities: the case of congeneric gill monogeneans (*Dactylogyrus* spp.). *International Journal for Parasitology*, **31**, 1205–1210.

Šimková, A., Ondračková, M., Gelnar, M., & Morand, S. (2002). Morphology and coexistence of congeneric ectoparasite species: reinforcement of reproductive isolation? *Biological Journal of the Linnean Society*, **76**, 125–135.

Šimková, A., Goüy de Bellocq, J., & Morand, S. (2003). The structure of host-parasite communities: order and history. In C. Combes & J. Jourdane (Eds.), *Taxonomy, Ecology and Evolution of Metazoan Parasites* (pp. 237–257). Livre hommage à Louis Euzet. Perpignan: Presses Universitaires de Perpignan (Collection Etudes).

Šimková, A., Verneau, O., Gelnar, M., & Morand, S. (2006). Specificity and specialization of congeneric monogeneans parasitizing Cyprinid fish. *Evolution*, **60**, 1023–1037.

Valtonen, E. T., Pulkinnen, K., Poulin, R., & Julkunen, M. (2001). The structure of parasite component communities in brackish water fishes of the northeastern Baltic Sea. *Parasitology*, **122**, 471–481.

Worthen, W. R., & Rohde, K. (1996). Nested subset analyses of colonization-dominated communities: metazoan ectoparasites of marine fishes. *Oikos*, **75**, 471–478.

Zelmer, D. A., & Arai, H. P. (1998). The contributions of host age and size to the aggregated distribution of parasites in yellow perch, *Perca flavescens*, from Garner Lake, Alberta, Canada. *Journal of Parasitology*, **84**, 24–28.

7 Ectoparasites of small mammals: interactive saturated and unsaturated communities

Boris R. Krasnov

Introduction

Parasites of different species often co-occur on a host individual or host population forming a community. Spatial distribution of parasite communities is fragmented among host individuals, among host species within a location, and among locations. To distinguish between scales, a hierarchical terminology has been proposed (Esch *et al.*, 1990; Combes, 2001; Poulin, 2007). In this chapter, I will refer to an assemblage of parasites of all species infesting an individual host as an infracommunity, to an assemblage of parasites of all species infesting a host population as a component community and to an assemblage of parasites of all species infesting a host community as a compound community.

There are at least two principal differences between infracommunities and communities at higher hierarchical levels. First, the former are short-living by definition, while the latter persist much longer. Second, parasite species in infracommunities may exert selective pressures on each other, which then induce the selection of traits that limit competition by separating niches (Holmes & Price, 1986). In contrast, interspecific interactions in component and compound communities are less likely. It is thus not surprising that studies of parasite community structure were focused mainly on infracommunities, while component and compound communities have received less attention.

In this section, I will demonstrate that ectoparasite communities at different hierarchical scales (infracommunities, component and compound communities) represent predictable sets of species rather than stochastic assemblages. However, I will show that these communities can be both saturated and non-saturated and, thus, provide examples of both equilibrium and nonequilibrium conditions. I will also discuss possible forces shaping the structure of these communities. I will take advantage of a series of recent comparative studies on the communities of fleas (Insecta: Siphonaptera) and gamasid mites (Acari: Parasitiformes) parasitic on small mammals, to address some fundamental questions about structure of ectoparasite communities. My aim is not to provide an extensive and exhaustive review, but rather to highlight main patterns and the key

The Balance of Nature and Human Impact, ed. Klaus Rohde. Published by Cambridge University Press.
© Cambridge University Press 2013.

processes that are the likely reasons for these patterns. In addition, I will intentionally limit this chapter to consideration of structural patterns in ectoparasite communities per se and will not consider patterns of structure in host-ectoparasite networks (e.g., Graham *et al.*, 2009; Fortuna *et al.*, 2010).

Ectoparasite co-occurrence: aggregative structure is a rule

Although a variety of patterns of community structure have been suggested (Diamond, 1975; Hanski, 1982; Patterson & Atmar, 1986; Fox & Brown, 1993), the fundamental question, when the pattern of the community organization is considered, is as follows: whether frequency of co-occurrence of different species in a real community differs from that expected in a community with a random species assemblage. If species co-occur more often than expected by chance (that is, they are positively associated), then the assemblage is said to be aggregatively structured, whereas if species co-occur less frequently than expected by chance (that is, they are negatively associated), then the assemblage is segregatively structured. The answer to this question can be found using null model analyses. The null model analysis compares frequencies of co-occurrences of species (ectoparasites) across sites (host individuals or populations or species) with those expected by chance, i.e., derived from randomly assembled species (=parasites) × site (=hosts) matrices. These frequencies are evaluated by one or more of co-occurrence metrics such as the C-score, the number of checkerboard species pairs, the number of species combinations, or the variance ratio (V-ratio) (see Gotelli, 2000, for details).

Pioneering study with application of the null model analysis to parasite communities has been carried out by Gotelli and Rohde (2002). They found little evidence for non-random species co-occurrence patterns in parasite assemblages of marine fish. However, investigations of species co-occurrences in ectoparasites exploiting small mammals produced contrasting results. Krasnov *et al.* (2006a) studied the community structure of fleas parasitic on small mammals and its temporal variation. It appeared that in many, albeit not all, cases the observed co-occurrence indices differed significantly from the null expectations. Furthermore, the results of this study demonstrated two important patterns, namely (a) flea assemblages on small mammalian hosts were structured at some times, whereas they appear to be randomly assembled at other times; and (b) whenever non-randomness of flea co-occurrences was detected, it suggested aggregation but never segregation of flea species.

Although the results of Krasnov *et al.* (2006a) concerned several different host species and flea assemblages of somewhat different species composition, it was unclear whether aggregative structure of communities is a rule for ectoparasites of terrestrial hosts or, alternatively, is characteristic of only a restricted set of conditions; that is, a single ectoparasite taxon (fleas) in a single geographic region. Other ectoparasite-host associations demonstrated similar patterns. Tello *et al.* (2008) investigated co-occurrence of sterblid batflies (Sterblidae) on a bat (Chiroptera) host. Their analyses showed strong evidence for interspecific aggregation, rather than for segregation of streblid species. The same was found by Presley (2007, 2011) for batflies (Streblidae and Nycteribiidae) and bat bugs (Polyctenidae) on bats.

To test whether ectoparasite community structure is affected by either parasite or host biology, Krasnov *et al*. (2010) compared the occurrence and pattern of community organization among different ectoparasite taxa (ixodid ticks, gamasid mites and fleas) exploiting the same host species and among different host species harboring the same parasite taxon using data on ectoparasites of rodent hosts from different continents. Again, whenever non-randomness of parasite co-occurrences was found, it indicated positive but not negative co-occurrences of parasite species. However, the frequency of the detection of non-randomness of parasite co-occurrences differed among parasite taxa as well as among host species independent of parasite taxon (Krasnov *et al*., 2010).

Studies by Krasnov *et al*. (2006a, 2010), Presley (2007, 2011) and Tello *et al*. (2008) were focused on ectoparasite co-occurrence in infracommunities, i.e., at the lowest hierarchical level. What happens at higher levels? Do structural patterns found for ectoparasite infracommunities apply also for component or compound communities? The results of Krasnov *et al*. (2011a) suggested that this is the likely case at least for fleas parasitic on small mammals. When flea co-occurrences in co-habitating host species (i.e., in component communities) from compound communities in four biogeographic realms were examined, patterns of co-occurrences on the same host species again indicated aggregation but not segregation of flea species. Furthermore, the degree of non-randomness of the entire flea community was similar among biogeographic realms.

The main conclusion from the above-cited studies is that aggregative structure of ectoparasites is a general rule independently of ectoparasite taxon, host taxon, geographic location and scale of consideration. Positive relationships between ectoparasite species have been supported by the results of other studies. For example, patterns of interspecific interactions in ectoparasite communities could be inferred from patterns of abundance. Indeed, negative relationships between the abundances of species co-occurring in a community can indicate competitive relationships between these species, while positive relationships would be indicative of facilitation. Indeed, in some flea-host associations, it was found that hosts highly infested with one flea species were also highly infested with other flea species (Faulkenberry & Robbins, 1980; Brinkerhoff *et al*., 2006).

Furthermore, examination of how the overall abundance and diversity of ectoparasite communities affect the abundance of individual parasite species in these communities would allow one to understand the potential role of diffuse competition in communities because ectoparasites use the same resource. For example, the decrease in the abundance of a parasite with an increase in the abundance of all other co-occurring parasites would suggest the occurrence of diffuse competition (Bock *et al*., 1992). The latter can also be revealed by negative relationships between the abundance of a given parasite species and the species richness or any other measure of diversity of the entire parasite assemblage, given that a higher number of species leads to more intense competition (MacArthur, 1972). On the other hand, positive relationships between the abundance of a given species and either the abundance of other co-occurring species or their diversity or both would indicate facilitation among parasite species or heterogeneity of transmission. Krasnov *et al.* (2005a) applied this approach to a large dataset of fleas and small mammalian hosts from different geographic regions and found that (a) the abundance of a given flea species correlated positively with the total abundance of all other co-occurring flea species in the community and (b) the abundance of any given flea species correlated negatively with diversity of the flea community. In general, the abundance of a given flea species was highest in assemblages consisting of few species of limited taxonomic diversity. While this supported the existence of some form of negative interactions among species, it also supported the occurrence of facilitation mediated via the host.

To test whether similar patterns are characteristic of other ectoparasite taxa, Krasnov *et al.* (2009) used census data on gamasid mites parasitic on small mammals and assessed how the abundance of individual mite species is influenced by the abundance and diversity of other mite species on the same host. The abundance of individual mite species appeared to be generally positively correlated with the combined abundances of all other mite species in the component community. In contrast, there were generally no consistent relationships between the abundance of individual mite species and diversity of the community in which they occur. Instead, an increase in mite diversity was accompanied by an increase of abundance in some species and decrease of abundance in other species. This suggested that both positive and negative interactions might exist in mite communities.

Interestingly, in aquatic host-parasite associations (ectoparasites on fish hosts), positive associations between parasite species were found to be much more common than negative ones (Rohde *et al.*, 1994, 1995). Rohde (1993) pointed out that heterogeneity of transmission alone can bring about negative or positive associations.

Nestedness in ectoparasite communities

Another possible departure from random community assembly is the nested pattern. This is a pattern in which species comprising depauperate assemblages constitute non-random subsets of the species occurring in successively richer assemblages (e.g., Patterson & Atmar, 1986). Studies of nestedness in parasite communities have provided contradictory results (e.g., Guégan & Hugueny, 1994; Worthen & Rohde, 1996; Poulin & Guégan, 2000; Poulin & Valtonen, 2001, 2002). In part, these contradictions stem from the type of statistical technique used. Another reason for the conflicting results might be the complicated relationships between extrinsic and intrinsic factors that affect the species composition of parasite communities (Poulin & Valtonen, 2001). In addition, earlier nestedness analyses may have focused on the wrong spatial scale, i.e., infracommunities, whereas component communities have received less attention (but see Guégan & Kennedy, 1996; Calvete *et al.*, 2004), although this higher hierarchical level is more relevant to nestedness analyses.

Nestedness analysis has rarely been applied to terrestrial ectoparasite communities. Nevertheless, the available studies demonstrated that at least some ectoparasite communities are characterized by nested patterns. Krasnov *et al.* (2005b) searched for the occurrence of nested patterns in flea assemblages among host populations of the same species across the species' geographic range. A flea community was considered to be structured if it departed from random assembly in either direction, i.e., towards either nestedness or antinestedness. As defined above, nestedness occurs if flea assemblages can be arranged in such a way that depauperate assemblages consist of proper subsets of progressively richer ones. The assemblages are considered as antinested if species are always absent from assemblages richer than the most depauperate one in which they occur (see Poulin & Guégan, 2000). The organization of flea assemblages across host populations within host species was found to form a continuum among host species from true nestedness to true antinestedness. Furthermore, the results of this study showed that the structure of flea assemblages in mammalian hosts might be driven to a large extent by features of host biology. The study of nestedness of ectoparasite (batflies and bat bugs) assemblages on bats carried out by Presley (2011) also demonstrated that parasite communities on some, albeit not other, hosts exhibited nestedness.

To test whether nested patterns in ectoparasite assemblages differ between ectoparasite taxa and scale of consideration, Krasnov *et al.* (2011b) studied fleas and gamasid mites exploiting small mammals across three scales, namely (a) local spatial, that is, across different locations within the same region; (b) local temporal, that is, within the same location across different times and (c) regional, that is, across distinct geographic regions. It was found that the organization of flea and mite assemblages across host populations within host species at smaller and larger spatial scales, as well as at temporal scales, was characterized by nestedness. However, the nested pattern estimated via metric NODF (Almeida-Neto *et al.*, 2008) was found to (a) differ between ectoparasite taxa (at least, at the local spatial scale) and (b) be scale-dependent, being the lowest at the regional scale. Scale-dependence of the degree of nestedness indicated that ectoparasite communities at different scales are governed by different mechanisms. Nestedness across insular or fragmented habitats of free-living organisms is commonly thought to result from differential colonization/extinction dynamics among species (Patterson & Atmar, 1986). The same can be true for parasites (Poulin & Valtonen, 2001). However, proximate mechanisms affecting colonizations and extinctions of parasites seem to be different at different scales. At local scale(s), colonization/extinction dynamics is likely to be ruled by epidemiological processes (Morand *et al.*, 2002). At regional scale, colonization/extinction dynamics is influenced by biogeographic and environmental processes (González & Oliva, 2009; Rohde, 2010).

Saturation of ectoparasite communities: the scale matters

Patterns in local communities are governed not only by local (competition, predation and habitat heterogeneity) but also by regional and historical processes (long-distance migration and speciation) (Gaston & Blackburn, 2000). The relative importance of local and regional processes in governing local species composition can be inferred from examination of the relationship between local and regional species richness (Srivastava, 1999).

If, for example, regional processes strongly control local communities by, for example, dispersal limiting local species richness, then the relationship between local and regional species richness will be linear (Cornell & Lawton, 1992). Local communities are, thus, unsaturated (species are often absent from suitable habitats) and exhibit "proportional sampling" of the available regional species pool. If, however, local processes play the main role in structuring local communities and impose upper limits on the number of species that are able to coexist, then local species richness will approach an asymptote with an increase in regional richness (Cornell & Lawton, 1992). At higher regional species richness, local richness becomes independent of regional richness. Local communities demonstrating a curvilinear relationship of local versus regional species richness are considered to be saturated with species (Guégan et al., 2005; but see Rohde, 1998).

Testing the relationship between local and regional species richness is, at first glance, rather straightforward and can be carried out using regression analysis (e.g., Oberdorff et al., 1998). However, some methodological problems arise, so the use of local/regional richness plots to test for saturation of diversity has been strongly criticized (Srivastava, 1999; Hillebrand, 2005). Nevertheless, the use of regional to local diversity regressions remains widespread (Heino et al., 2003; Calvete et al., 2004; Karlson et al., 2004). One of the most important methodological issues is a precise definition of borders for local and regional communities, which is sometimes self-evident for freshwater organisms, but it is much more difficult for terrestrial or marine organisms. However, for ectoparasites the definition of a community at the lowest hierarchical scale is relatively easy. This is the infracommunity. Obviously, the next hierarchical level is the component community. Finally, all component communities within a given host species represent either a regional parasite community or a parasite fauna (Poulin, 2007). Although Srivastava (1999) argued that in the case of parasites, an equivalent of "regional" species richness is the parasite fauna, I believe that component community richness can be considered as "regional" in relation to infracommunity richness. This is because the species pool of a component community contains all species that can colonize an infracommunity, assuming the absence of competitive exclusion. Dispersal of species within a component community may be slow but, nevertheless, it is much more frequent than host-switching (equivalent to dispersal between regions; see Srivastava, 1999).

Analyses of local versus regional species richness in helminth parasites have revealed that saturated and unsaturated communities are equally common (e.g., Morand et al., 1999; Calvete et al., 2004). In the only study of ectoparasites of terrestrial mammals, Krasnov et al. (2006b) investigated this relationship in communities of fleas on small mammalian hosts at two different spatial scales: between the richness of infracommunities and that of component communities, and between the richness of component communities and that of the entire regional species pool. At both spatial scales, consistent curvilinear relationships between species richness of the more "local" communities and richness of the more "regional" communities were found, suggesting that the number of species in species-rich infracommunities was independent of the species richness of the component community of which they were part. Thus, at first glance, the flea infracommunities were "saturated" and vacant niches seem to be generally unavailable in these communities. The observed pattern may arise because some species can be eliminated or not allowed to invade local communities due to some ecological constraints such as negative interactions among species in an infracommunity (Srivastava, 1999; Calvete *et al.*, 2004).

However, if negative competitive interactions among flea species in an infracommunity were indeed important, one would expect density compensation in species-poor infracommunities (Cornell, 1993). Consequently, demonstrating the existence or absence of saturation in parasite assemblages requires the additional investigation of interspecific interactions (Guégan *et al.*, 2005). To test for this, Krasnov *et al.* (2006b) assessed the relationship between mean flea abundance per host individual and richness of the "local" flea community. There was no strong evidence for density compensation at the infracommunity level, although its existence at the component community level appeared likely.

Consequently, negative interspecific interactions appear to be not the case for flea infracommunities, suggesting that the curvilinear relationship between infracommunity and component community species richness may occur for reasons other than "saturation". Indeed, Rohde (1998) demonstrated that curvilinearity in the local versus regional species richness relationship may be caused by processes other than species interactions within a local community. In particular, this curvilinearity may be a consequence of the differential likelihood of parasite species occurring in an infracommunity because of different transmission rates and lifespans (Rohde, 1998). In the case of fleas, these reasons can also be related to differential abiotic preferences of either imago or larval fleas of different species that contribute to the elimination of some flea species from some infracommunities (Krasnov *et al.*, 2001). All the above indicate that flea infracommunities are governed by processes acting at higher than "local" levels, and that further species could possibly be added over evolutionary time (Rohde, 1998).

The relationship between local and regional flea richness appeared to be the same at the larger scale as at the smaller scale; in other words, the relationship between richness of component communities and that of the regional flea pool seems to be similar to that found for infracommunities versus component communities. However, the absence of a relationship between mean flea abundance and component community species richness suggested the existence of density compensation. Therefore, component communities appeared to be saturated. The causes of this saturation are likely to be some intrinsic limiting factors that may play an important role in shaping flea component communities. One of the common factors responsible for community saturation is negative interspecific interactions such as competition (Cornell, 1993). Although direct interspecific competition between flea imagos within a host population seems not to be the case (see above and below), such competition can occur among larval fleas (Krasnov *et al.*, 2005c; see below). Thus, similar patterns in the relationships between "local" and "regional" species richness in the same host-parasite system but at different spatial scales may arise because of different mechanisms. This could be one explanation for the contrasting relationships reported between local and regional species richness in earlier studies of different host-parasite systems.

Mechanisms of aggregative community structure

The results of many studies described in the previous subchapters unequivocally showed that ectoparasite species co-occurred more frequently than expected by chance, so that different parasite species were aggregated in some host individuals or species. Obviously, there should be some mechanisms producing this aggregation.

Many sources of heterogeneity among host individuals such as differences in grooming abilities, immunocompetence, and age could cause aggregative patterns of ectoparasites. Although Presley (2011) argued that between-host variation in immunocompetence cannot be the case due to lack of proof, recent results of Krasnov *et al.* (2011c) demonstrated that this explanation is plausible. Indeed, the frequency of detection of positive flea co-occurrences was significantly higher in male than in female hosts. Males are known to be less immunocompetent than females due to the immunosuppressive effect of androgens (Folstad & Karter, 1992; Zuk, 1996). Furthermore, ectoparasite co-occurrence could arise from immunodepression in a host subjected to multiple challenges from a variety of parasites (Bush & Holmes, 1986; Cox, 2001). As a result, the effectiveness of energy allocation to immune defense decreases with an increase in the diversity of parasite attacks (Jokela *et al.*, 2000). Another, not necessarily alternative, explanation of the observed aggregative pattern in ectoparasites species co-occurrences in the infracommunities is heterogeneity of spatial behavior of host individuals (residents versus transients; Krasnov *et al.*, 2002).

Regarding component communities, some host species may represent better habitats for multiple ectoparasite species than other host species due to some of their inherent characters. Obviously, characters that make a host species preferred over other host species by a hematophagous parasite could be the same for different taxa of hematophages. These are characters that allow a parasite either to obtain more food or to obtain food of higher quality, or that make food acquisition easier (Kelly & Thompson, 2000). In addition, the main habitat for many ectoparasite taxa is not the body of a host, but its shelter (burrow or nest) where ectoparasites spend most of their lives and where their pre-imaginal stages develop. Thus, between-host difference in the architecture and microclimate of shelters can be another reason for ectoparasite aggregation in some but not other host species. Community organization of ectoparasites may also be affected by the social structure of their hosts. Host group-living can facilitate the exchange of ectoparasites and thus lead to more frequent co-occurrence of different parasites on the same host individual.

An additional mechanism by which multiple species in a community may coexist is reduction in the overall intensity of competition via aggregated utilization of resources (Hartley & Shorrocks, 2002). This model, known as the aggregation model of coexistence, states that if competing species are distributed such that interspecific aggregation is reduced relative to intraspecific aggregation then species coexistence is facilitated. The only attempt to test for the aggregation model of coexistence in relation to ectoparasite communities has been undertaken by Krasnov *et al.* (2006c) again using data on fleas parasitic on small mammals. Intraspecific aggregation was stronger than interspecific aggregation in all component communities of fleas. In other words, flea assemblages were not dominated by negative interspecific associations, and the level of interspecific aggregation in flea assemblages was reduced in relation to the level of intraspecific aggregation, thus facilitating flea coexistence. Consequently, aggregation of conspecific flea individuals may limit their own population growth to such an extent that the remaining resources appear to be sufficient to support other flea species (Hartley & Shorrocks, 2002).

Variation in community structure among ectoparasite taxa

It has been proposed that the level of vagility and/or dispersal ability may determine whether animals are subject to structuring mechanisms (Rohde, 2005). As a result, communities of large-bodied and/or highly vagile taxa are predominantly non-random, whereas communities of small-bodied and/or weakly vagile taxa predominantly represent random assemblages (Gotelli & McCabe, 2002; Gotelli & Rohde, 2002; Rohde, 2005). One of the reasons for this difference is that ecological niches in the latter are not saturated and their population densities are chronically low, so that interspecific interactions are weak and do not have much influence on community structure.

Applying these ideas to ectoparasites, the effect of vagility on the occurrence of non-random pattern in community organization may be realized in the difference between temporary and periodic ectoparasites (*sensu* Lehane, 2005). Temporary ectoparasites are largely free-living and visit the host for long enough to take a blood meal (e.g., ixodid ticks). Periodic ectoparasites spend a considerably longer time on the hosts than is required merely to obtain a blood meal but nevertheless spend a significant amount of time off-host (e.g., most fleas and gamasid mites). Obviously, temporary ectoparasites are more vagile than periodic ectoparasites. Consequently, the non-randomness could be expected to be found in the communities of the former rather than in those of the latter. However, the results of Krasnov *et al.* (2010) demonstrated that, contrary to these expectations, the frequency of the detection of community non-randomness was high in fleas and low in ticks, although as predicted it was low in mites. This difference among ectoparasite taxa could be caused by differences in life history characteristics of the ectoparasites other than the vagility and/or dispersal ability such as necessity of blood feeding, temporal pattern of feeding and relationships with off-host environment.

Concluding remarks

It is my hope that the review of various, albeit not yet numerous, studies presented in this chapter succeeded in convincing the reader that ectoparasite communities on small mammalian hosts do not represent random species assemblages. There are interactions, which, however, are predominantly positive either as a product of apparent facilitation resulting from host behavior, immunology and ecology or due to various parasite characteristics (Presley, 2011). Manifestation of structure in parasite communities thus strongly depends on the life history of the parasites and may be affected, to a great extent, by the life history of the host and scale of consideration.

In some ectoparasites (e.g., ticks) all developmental stages are hematophagous, while in other ectoparasites (e.g., fleas) larvae, in contrast to imagos, are not parasitic. These biological differences between larval and imago parasites can produce a striking difference in mechanisms prevailing in their communities (competition versus facilitation). Indeed, the only experimental study of interspecific interactions between flea larvae demonstrated strong and asymmetric interspecific competition (Krasnov *et al.*, 2005c), so that facilitation-governed communities of imago fleas are the source of competition-governed communities of larval fleas and vice versa. This suggests that an ectoparasite community in its entirety, i.e., comprising both imagos and pre-imagos, is governed by complicated rules which still remain to be revealed. Furthermore, ectoparasite communities present evidence of both equilibrium and nonequilibrium depending on parasite taxon, host biology and scale of consideration.

References

Almeida-Neto, M., Guimarães, P., Guimarães, P. R., Loyola, R. D., & Ulrich, W. (2008). A consistent metric for nestedness analysis in ecological systems: reconciling concept and measurement. *Oikos*, **117**, 1227–1239.

Bock, C. E., Cruz, A., Grant, M. C., Aid, C. S., & Strong, T. R. (1992). Field experimental evidence for diffuse competition among southwestern riparian birds. *The American Naturalist*, **140**, 515–528.

Brinkerhoff, R. J., Markeson, A. B., Knouft, J. A., Gage, K. L., & Montenieri, J. A. (2006). Abundance patterns of two *Oropsylla* (Ceratophyllidae: Siphonaptera) species on black-tailed prairie dog (*Cynomys ludovicianus*) hosts. *Journal of Vector Ecology*, **31**, 355–363.

Bush, A. O., & Holmes, J. C. (1986). Intestinal helminths of lesser scaup ducks: patterns of association. *Canadian Journal of Zoology*, **64**, 132–141.

Calvete, C., Blanco-Aguiar, J. A., Virgós, E., Cabezas-Díaz, S., & Villafuerte, R. (2004). Spatial variation in helminth community structure in the red-legged partridge (*Alectoris rufa* L.): effects of definitive host density. *Parasitology*, **129**, 101–113.

Combes, C. (2001). *Parasitism. The Ecology and Evolution of Intimate Interactions*. Chicago: University of Chicago Press.

Cornell, H. V. (1993). Unsaturated patterns in species assemblages: the role of regional processes in setting local species richness. In R. F. Ricklefs & D. Schluter (Eds.), *Species Diversity in Ecological Communities: Historical and Geographical Perspectives* (pp. 243–252). Chicago, IL: University of Chicago Press.

Cornell, H. V., & Lawton, J. H. (1992). Species interactions, local and regional processes, and limits to richness of ecological communities: a theoretical perspective. *Journal of Animal Ecology*, **61**, 1–12.

Cox, F. E. G. (2001). Concomitant infections, parasites and immune responses. *Parasitology*, **122**, S23–S38.

Diamond, J. M. (1975). Assembly of species communities: chance or competition. In M. L. Cody & J. M. Diamond (Eds.), *Ecology and Evolution of Communities* (pp. 342–444). Cambridge, MA: Harvard University Press.

Esch, G. W., Shostak, A. W., Marcogliese, D. J., & Goater, T. M. (1990). Patterns and processes in helminth parasite communities: an overview. In G. W. Esch, A. O. Bush & J. M. Aho (Eds.), *Parasite Communities: Patterns and Processes* (pp. 1–19). London: Chapman and Hall.

Faulkenberry, G. D., & Robbins, R. G. (1980). Statistical measures of interspecific association between the fleas of the gray-tailed vole, *Microtus canicaudus* Miller. *Entomological News*, **91**, 93–101.

Folstad, I., & Karter, A. J. (1992). Parasites, bright males, and the immunocompetence handicap. *The American Naturalist*, **139**, 603–622.

Fortuna, M. A., Stouffer, D. B., Olesen, J. M., *et al.* (2010). Nestedness versus modularity in ecological networks: two sides of the same coin? *Journal of Animal Ecology*, **79**, 811–817.

Fox, B. J., & Brown, J. H. (1993). Assembly rules for the functional groups in North American desert rodent communities. *Oikos*, **67**, 358–370.

Gaston, K. J., & Blackburn, T. M. (2000). *Pattern and Process in Macroecology.* Oxford: Blackwell Science.

González, M. T., & Oliva, M. E. (2009). Is the nestedness of metazoan parasite assemblages of marine fishes from the southeastern Pacific coast a pattern associated with the geographical distributional range of the host? *Parasitology*, **136**, 401–409.

Gotelli, N. J. (2000). Null model analysis of species co-occurrence patterns. *Ecology*, **81**, 2606–2621.

Gotelli, N. J., & McCabe, D. J. (2002). Species co-occurrence: a meta-analysis of J. M. Diamond's assembly rules model. *Ecology*, **83**, 2091–2096.

Gotelli, N. J., & Rohde, K. (2002). Co-occurrence of ectoparasites of marine fishes: a null model analysis. *Ecology Letters*, **5**, 86–94.

Graham, S. P., Hassan, H. K., Burkett-Cadena, N. D., Guyer, C., & Unnasch, T. R. (2009). Nestedness in ectoparasite-vertebrate host networks. *PLoS ONE*, **4**, e7873. doi:10.1371/journal.pone.0007873.

Guégan, J.-F., & Hugueny, B. A. (1994). Nested parasite species subset pattern in tropical fish host as major determinant of parasite infracommunity structure. *Oecologia*, **100**, 184–189.

Guégan, J.-F., & Kennedy, C. R. (1996). Parasite richness/sampling effort/host range: the fancy three-piece jigsaw puzzle. *Parasitology Today*, **12**, 367–369.

Guégan, J.-F., Morand, S., & Poulin, R. (2005). Are there general laws in parasite community ecology? The emergence of spatial parasitology and epidemiology. In F. Thomas, J.-F. Guégan & F. Renaud (Eds.), *Parasitism and Ecosystems* (pp. 22–42). Oxford: Oxford University Press.

Hanski, I. (1982). Communities of bumblebees: testing the core-satellite hypothesis. *Annales Zoologici Fennici*, **19**, 65–73.

Hartley, S., & Shorrocks, B. (2002). A general framework for the aggregation model of coexistence. *Journal of Animal Ecology*, **71**, 651–662.

Heino, J., Muotka, T., & Paavola, R. (2003). Determinants of macroinvertebrate diversity in headwater streams: regional and local influences. *Journal of Animal Ecology*, **72**, 425–434.

Hillebrand, H. (2005). Regressions of local on regional diversity do not reflect the importance of local interactions or saturation of local diversity. *Oikos*, **110**, 195–198.

Holmes, J. C., & Price, P. W. (1986). Communities of parasites. In J. Kikkawa & D. J. Anderson (Eds.), *Community Ecology: Patterns and Processes* (pp. 187–213). New York: Blackwell Science.

Jokela, J., Schmid-Hempel, P., & Rigby, M. C. (2000). Dr. Pangloss restrained by the Red Quinn – steps towards a unified defence theory. *Oikos*, **89**, 267–274.

Karlson, R. H., Cornell, H. V., & Hughes, T. P. (2004). Coral communities are regionally enriched along an oceanic biodiversity gradient. *Nature*, **429**, 867–870.

Kelly, D. W., & Thompson, C. E. (2000). Epidemiology and optimal foraging: modeling the ideal free distribution of insect vectors. *Parasitology*, **120**, 319–327.

Krasnov, B. R., Khokhlova, I. S., Fielden, L. J., & Burdelova, N. V. (2001). The effect of air temperature and humidity on the survival of pre-imaginal stages of two flea species (Siphonaptera: Pulicidae). *Journal of Medical Entomology*, **38**, 629–637.

Krasnov, B. R., Khokhlova, I. S., & Shenbrot, G. I. (2002). The effect of host density on ectoparasite distribution: an example with a desert rodent parasitized by fleas. *Ecology*, **83**, 164–175.

Krasnov, B. R., Mouillot, D., Shenbrot, G. I., Khokhlova, I. S., & Poulin, R. (2005a). Abundance patterns and coexistence processes in communities of fleas parasitic on small mammals. *Ecography*, **28**, 453–464.

Krasnov, B. R., Shenbrot, G. I., Khokhlova, I. S., & Poulin, R. (2005b). Nested pattern in flea assemblages across the host's geographic range. *Ecography*, **28**, 475–484.

Krasnov, B. R., Burdelova, N. V., Khokhlova, I. S., Shenbrot, G. I., & Degen, A. A. (2005c). Pre-imaginal interspecific competition in two flea species parasitic on the same rodent host. *Ecological Entomology*, **30**, 146–155.

Krasnov, B. R., Stanko, M., & Morand, S. (2006a). Are ectoparasite communities structured? Species co-occurrence, temporal variation and null models. *Journal of Animal Ecology*, **75**, 1330–1339.

Krasnov, B. R., Stanko, M., Khokhlova, I. S., *et al.* (2006b). Relationships between local and regional species richness in flea communities of small mammalian hosts: saturation and spatial scale. *Parasitology Research*, **98**, 403–413.

Krasnov, B. R., Stanko, M., Khokhlova, I. S., *et al.* (2006c). Aggregation and species coexistence in fleas parasitic on small mammals. *Ecography*, **29**, 159–168.

Krasnov, B. R., Vinarski, M. V., Korallo-Vinarskaya, N. P., Mouillot, D., & Poulin, R. (2009). Inferring associations among parasitic gamasid mites from census data. *Oecologia*, **160**, 175–185.

Krasnov, B. R., Matthee, S., Lareschi, M., Korallo-Vinarskaya, N. P., & Vinarski, M. V. (2010). Co-occurrence of ectoparasites on rodent hosts; null model analyses of data from three continents. *Oikos*, **119**, 120–128.

Krasnov, B. R., Shenbrot, G. I., & Khokhlova, I. S. (2011a). Aggregative structure is the rule in communities of fleas: null model analysis. *Ecography*, **34**, 751–761.

Krasnov, B. R., Stanko, M., Khokhlova, I. S., *et al.* (2011b). Nestedness and beta-diversity in ectoparasite assemblages of small mammalian hosts: effects of parasite affinity, host biology and scale. *Oikos*, **120**, 630–639.

Krasnov, B. R., Stanko, M., Matthee, S., *et al.* (2011c). Male hosts drive infracommunity structure of ectoparasites. *Oecologia*, **166**, 1099–1100.

Lehane, M. (2005). *The Biology of Blood-Sucking in Insects* (2nd edn). Cambridge: Cambridge University Press.

MacArthur, R. H. (1972). *Geographical Ecology*. New York: Harper and Row.

Morand, S., Poulin, R., Rohde, K., & Hayward, C. (1999). Aggregation and species coexistence of ectoparasites of marine fishes. *International Journal for Parasitology*, **29**, 663–672.

Morand, S., Rohde, K., & Hayward, C. (2002). Order in ectoparasite communities of marine fish is explained by epidemiological processes. *Parasitology*, **124**, S57–S63.

Oberdorff, T., Hugueny, B., Compin, A., & Belkessam, D. (1998). Non-interactive fish communities in the coastal streams of North-western France. *Journal of Animal Ecology*, **67**, 472–484.

Patterson, B. D., & Atmar, W. (1986). Nested subsets and the structure of insular mammalian faunas and archipelagos. *Biological Journal of the Linnean Society*, **28**, 65–82.

Poulin, R. (2007). *Evolutionary Ecology of Parasites: From Individuals to Communities* (2nd edn). Princeton, NJ: Princeton University Press.

Poulin, R., & Guégan, J.-F. (2000). Nestedness, antinestedness, and relationship between prevalence and intensity in ectoparasite assemblages of marine fish: a spatial model of species co-existence. *International Journal for Parasitology*, **30**, 1147–1152.

Poulin, R., & Valtonen, E. T. (2001). Nested assemblages resulting from host-size variation: the case of endoparasite communities in fish hosts. *International Journal for Parasitology*, **31**, 194–1204.

Poulin, R., & Valtonen, E. T. (2002). The predictability of helminth community structure in space: a comparison of fish populations from adjacent lakes. *International Journal for Parasitology*, **30**, 1235–1243.

Presley, S. J. (2007). Streblid bat fly assemblage structure on Paraguayan *Noctilio leporinus* (Chiroptera: Noctilionidae): nestedness and species co-occurrence. *Journal of Tropical Ecology*, **23**, 409–417.

Presley, S. J. (2011). Interspecific aggregation of ectoparasites on bats: importance of hosts as habitats supersedes interspecific interactions. *Oikos*, **120**, 832–841.

Rohde, K. (1993). *Ecology of Marine Parasites* (2nd edn). Wallingford: CAB International.

Rohde, K. (1998). Is there a fixed number of niches for endoparasites of fish? *International Journal for Parasitology*, **28**, 1861–1865.

Rohde, K. (2005). *Nonequilibrium Ecology*. Cambridge: Cambridge University Press.

Rohde, K. (2010). Marine parasite diversity and environmental gradients. In S. Morand and B. R. Krasnov (Eds.), *The Biogeography of Host-Parasite Interactions* (pp. 73–88). Oxford: Oxford University Press.

Rohde, K., Hayward, C., Heap, M., & Gosper, D. (1994). A tropical assemblage of ectoparasites: gill and head parasites of *Lethrinus miniatus* (Teleostei, Lethrinidae). *International Journal for Parasitology*, **24**, 1031–1053.

Rohde, K., Hayward, C., & Heap, M. (1995). Aspects of the ecology of metazoan ectoparasites of marine fishes. *International Journal for Parasitology*, **25**, 945–970.

Srivastava, D. (1999). Using local-regional richness plots to test for species saturation: pitfalls and potentials. *Journal of Animal Ecology*, **68**, 1–16.

Tello, J. S., Stevens, R. D., & Dick, C. W. (2008). Patterns of species co-occurrence and density compensation: a test for interspecific competition in bat ectoparasite infracommunities. *Oikos*, **117**, 693–702.

Worthen, W. B., & Rohde, K. (1996). Nested subsets analyses of colonization-dominated communities: metazoan ectoparasites of marine fishes. *Oikos*, **75**, 471–478.

Zuk, M. (1996). Disease, endocrine-immune interactions, and sexual selection. *Ecology*, **77**, 1037–1042.

8 A macroecological approach to the equilibrial vs. nonequilibrial debate using bird populations and communities

Brian McGill

Introduction

The debate about whether ecological systems behave in an equilibrial or nonequilibrial fashion is a thread that runs through the entire history of ecology. The debate between Clements's super-organism (Clements, 1936) vs. Gleason's individualistic assortment (Gleason, 1926) is one of the earliest well-known examples. The density-dependent (Nicholson & Bailey, 1935) vs. density-independent (Davidson & Andrewartha, 1948) regulation debate was very explicitly about equilibrial vs. nonequilibrial forces. This debate was so intense and long-lasting that a peace-making conference was called at Cold Spring Harbor (Cold Spring Harbor Symposium on Quantitative Biology (22nd), 1957). One could argue that the MacArthurian development of community ecology (MacArthur, 1968) and the subsequent backlash against it were also in this vein. The heated debate about null-models (Connor & Simberloff, 1979; Diamond, 1975) was basically sparked by a bold assertion of a specific version of the nonequilibrial view (namely complete randomness) as a challenge to the then prevailing equilibrial viewpoint. Most recently, niche vs. neutral theory is a battle between two very specific versions of the equilibrial and nonequilibrial theories respectively. Thus although the specific debate seems to mutate every couple of decades, the dominant points of contention in ecology have all had an underlying theme of equilibrial vs. nonequilibrial concepts at the center for close to 100 years now.

The noted historian of ecology, Sharon Kingsland, follows this recurring debate and interprets it as a battle of modelers, who like to find regular patterns, vs. field ecologists, who experience the variation in nature, (Kingsland, 1995). But I think this is perhaps imprecise. It is true that many of the earliest models introduced into ecology such as the Verlhulst-Pearl logistic equation (Verhulst, 1838) and the Lotka-Volterra competition equations (Lotka, 1925; Volterra, 1927) were highly deterministic, strong equilibrium, differential equation models. However, probabilistic, nonequilibrial models have been known in ecology and evolution at least since the 1920s (Arrhenius, 1921; Yule, 1924).

The Balance of Nature and Human Impact, ed. Klaus Rohde. Published by Cambridge University Press.
© Cambridge University Press 2013.

And the discovery of chaos by ecologists has meant that we have known that even seemingly deterministic differential equation models can lead to, at a minimum, a different kind of equilibrium than originally conceptualized (and many would argue to a nonequilibrial system). For the 2000s the most cutting-edge theoretical development has occurred within the nonequilibrial neutral theory model (Bell, 2000, Hubbell, 2001) using stochastic process theory (Etienne & Olff, 2004; McKane *et al.*, 2004; Etienne, 2005; Black & McKane, 2012). Equally field ecologists have found both equilibrial and nonequilibrial dynamics in their field systems, often depending on which organisms they study.

To my mind the debate between equilibrial and nonequilibrial viewpoints has always been a bit overblown. This is in part because it seems ecologists need both toolkits in their bag. Not only because one can find some specific systems lying at the extremes that are clearly equilibrial or nonequilibrial, but because most systems are likely to show aspects of both. Secondly, this has always seemed to me a poor topic for theoretical debate (where most of the debate seems to have occurred). This is fundamentally an empirical question and deserves empirical answers.

My personal bias and background is to adopt a macroecological approach to addressing this empirical question. This to me seems ideally suited to an empirical examination of the equilibrial/nonequilibrial debate. Macroecologists analyze large datasets using advanced statistical and ecoinformatics approaches that allow them to make generalizations across many species and many locations (i.e., ecological communities) spanning large areas. We have known since the 1950s that there are systems that clearly demonstrate equilibrial dynamics such as Gause's paramecia (Gause, 1934) and Nicholson and Bailey's blowflies (Nicholson & Bailey, 1935) and others that clearly demonstrate nonequilibrial dynamics such as Andrewartha's thrips (Davidson & Andrewartha, 1948). What is needed is an approach that addresses many systems and many species simultaneously with the same methods to allow for generalization – at least to a broad class of organisms. This is the macroecological approach which I adopt in this chapter.

One of the drawbacks of the macroecological approach is that it is data hungry, demanding large volumes of data collected in a uniform fashion. Questions of equilbrial vs. nonequilibrial dynamics further require that the data be collected over a long enough period of time to observe population dynamics (ideally over a time extending over multiple generations of the organisms studied). One dataset that meets these requirements is the North American Breeding Bird Survey (Sauer *et al.*, 1997; Patuxent Wildlife Research Center, 2001). This survey is run by government agencies in the USA and Canada using thousands of routes every year. Each route is 25 miles (approximately 40 kilometers) long along rural and wilderness roads with a stop every half mile (approximately 0.8 km for 50 stops total). The observers count every bird seen or heard at each stop. The observers are screened for advanced birding skills. The same routes are run every year in late May or early June to capture peak breeding season. Routes with unusual weather conditions or errors indicative of poor-quality observers are noted. Several hundred species of birds are identified each year. The program has been run since 1966 (with more routes coming online over time). In short this is an ideal dataset for a macroecological approach to the equilibrium/nonequilibrium question.

Fortunately, birds are not just a group of organisms with data available but they are an interesting group of organisms for the equilibrial/nonequilibrial question. Birds (along with mammals) are expected to be among the most equilibrial classes of organisms. Birds are homeothermic, long-lived, and make high parental investment in offspring resulting in relatively low juvenile mortality and small numbers of offspring. The fact that many birds are vagile enough to migrate seasonally also allows escape from the worst environments. These traits all suggest that birds should be buffered from climatic variability and have a high degree of inertia in their population dynamics relative to other classes of organisms. Thus findings of equilibrial dynamics in birds would largely be unsurprising – if any organisms are going to be equilibrial, it would be birds and mammals. Conversely, any findings of nonequilibrial behavior would be indicative that nonequilibrial processes must be very common and even perhaps approaching universal in the ecological world.

Are populations equilibrial?

To empirically test for the presence of equilibrial processes in bird populations, one needs a clear definition of what constitutes equilibrial conditions and processes. There is no exact definition agreed upon by all (as will be obvious even by the varying definitions in this book). Here I abandon theoretical models and mathematical definitions and adopt a definition that hopefully both matches the intuitive sense of equilibrium and is empirically focused (i.e., data driven). Thus I define an equilibrial population as one where:

1. there is an equilibrium population size (aka a setpoint) N^*.
2. Low CV – most observation times the population size is close to N^* (N_t near N^* usually) or, equivalently, variation around N^* is small in comparison to N^*, i.e., the coefficient of variation defined as σ/N^* is small.
3. Stationarity – N^* is constant over time (this assumption of constant mean when combined with assumptions that variance and all higher moments of N^* is constant over time is formally known to mathematicians as stationarity).
4. Density dependence dominates – N_t shows density-dependent population dynamics (growing populations when $N_t < N^*$ and shrinking populations when $N_t > N^*$) and this is the primary driver of year-to-year variability in growth rate.

Condition 1 cannot be tested directly, being more of a definition to enable 2–4, but the remaining three can be directly tested.

In this chapter I use the following methods. I downloaded the North American Breeding Bird Survey data (Patuxent Wildlife Research Center, 2001). I took all routes spanning the 30 year interval 1978–2007 that were marked as reliably observed for all 30 years (weather and observer quality adequate). This resulted in 131 routes. Similar results were obtained for the 44 routes rated good for 40 years (1968–2007), but the number of such routes becomes very small and so is not reported further here. Each route had between 22 and 115 species (mean of 60 species). Thus each species present on each route gave a 30 year time series for a total of 13 747 time series. Figure 8.1 gives examples of these time series. Because most species are rare on most routes (McGill &

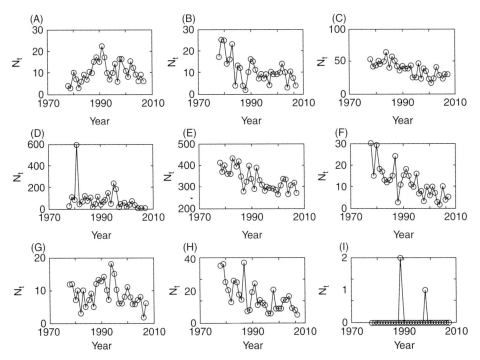

Figure 8.1. Random time series. Nine of these were chosen at random and plotted from 1977 to 2007. The series shows a wide variety of patterns and considerable noise (variation). The series in (I) is most typical (mostly zero abundances), while series A–H are filtered to have higher abundances that will not suffer from discrete counting problems and are randomly chosen from the 4400 species/site combinations that had a mean abundance of at least five individuals.

Collins, 2003; McGill *et al.*, 2007) (also see Figure 8.2A), most of these time series consist primarily of zeros with an occasional abundance of one or two, giving a very low average abundance (often averaging an abundance less than one per route). The commonness of this type of time series is shown by the fact that the median abundance is only 1.60 individuals. Over 30% of the route/species combinations had at least 25 of the 30 years as zero and a peak abundance less than or equal to 3, while over 60% were zero for at least half of the 30 years. Figure 8.1(I) is an example of this most common form of time series. However, when the average abundance is very small, it is hard to separate true process noise from the integer nature of abundance (only abundances of 0, 1, 2, etc. can occur). For this reason, depending on the test, some of these very low abundance time series are a priori removed. Figure 8.1A–H give examples of series screened for higher average abundances but otherwise random.

Testing criterion 2 – low CV

Note that the systems being observed were in existence long before the observations started in 1978. Thus equilibrial systems should have already reached the equilibrium N^* and just be expressing variation around this point N^*. This is in contrast to many

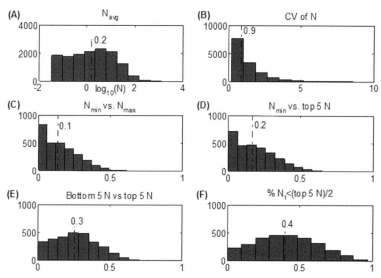

Figure 8.2.   There were 13 747 site/species pairs that had at least one bird one year. (A) A histogram of the
average abundance over the 30 years from 1978 to 2007 for each of these site/species pairs is
plotted on a $\log_{10}$ scale. The median over site/species pairs for average abundance over years was
1.60. (B) The coefficient of variation (CV) in N over the 30 years for the same 13 747 site/species
pairs. The median CV was 0.86 (86% of the mean). To avoid problems of observation error and
discrete lattice effects (Henson *et al.*, 1998), analyses C–F were performed on the 2863 species/site
pairs that had a mean abundance over time of at least 10 individuals. This is highly conservative
as all four distributions C–F were much further right-skewed when all site/species pairs were
included. (C) The proportion of the lowest observed abundance vs. the largest abundance. The
median had a lowest abundance that was 30% of the highest abundance. (D) The proportion of
the lowest observed abundance vs. the average of the five highest abundances over the 30 years.
The median was 0.2 (20%). (E) The proportion of the average of the five lowest abundances vs. the
average of the five highest abundances. The median was 0.3 (30%). (F) The fraction of years the
abundance was below half (50%) of the average of the 5 highest years. The median was 0.4 (40%)
or about 12 of the 30 years.

textbook examples of density-dependent growth models such as the logistic or Ricker
equations where the population starts small, goes through an inflection point of an
S-shaped curve and only eventually reaches equilibrium. The implication of this for
the BBS data is that if the systems are equilibrial we should assume that they are already
close to N^* and merely expressing their innate level of variation around the equilibrium.
If the standard deviation of this variability, σ, is large relative to N^*, the system will have
such large variance that it spends little time around N^*. This can be formally measured
using the coefficient of variation, $CV = \sigma/\mu$, or in these systems $CV = \sigma/N^*$. If the CV is
small, say around 0.10–0.20 (i.e., 10–20%), then the system may be noisy but never
deviates far from N^*. On the other hand if the CV is large then it will spend little time
around N^*, i.e., most observed N_t will be far from N^*. Across all species and sites, the
median CV is 0.86 (86%), with 95% of the species/site combinations falling in the range
0.19–3.75 or 19% to 375%. Thus all but 2.5% of the time series are experiencing at least

20% CV, and half are experiencing an enormous standard deviation that is at least 86% of the mean value. Filtering the time series to the 2863 site/species combinations with larger mean abundances (mean abundance of at least 10 individuals) that therefore largely stay away from 0 only produces a larger CV (median = 2.305 or 235% with 95% in the range 0.82–5.04 or 82–504%). Thus populations with large enough N^* to allow for significantly sized swings are almost all experiencing enormous variability in population size.

Several less traditional, less parameteric tests are also shown in Figure 8.2 (all on the 2863 site/species combinations with an average abundance of at least 10 – results are even more extreme when all populations are included). Figure 8.2C compares the smallest observed N_t in one time series for a given site/species combination (N_{min}) with the largest N_t (N_{max}) and in half the populations N_{min} is less than 10% of N_{max}. To avoid the possibility of being influenced by a single unusually high N_{max}, Figure 8.2D compares N_{min} with an average of the five highest N_t (N_{5max}) and finds similar results – over half the populations drop to less than 20% of N_{5max}. Figure 8.2E compares the mean of the five lowest N_t (N_{5min}) with N_{5max} and finds that half of the populations have N_{5min} less than 30% of N_{5max}. Finally Figure 8.2F shows that in single-species populations, most observations are of abundances less than half of N_{5max}. It is worth noting that in simple stochastic population models, increasing the variance in the noise enough can qualitatively change the model from one where most observations center on the equilibrium point to one where most observations are close to zero (Roughgarden, 1979). Thus by any evaluation, these time series are showing enormous amounts of fluctuation. It may be possible to declare an N^* value by taking the mean across N_t for all 30 years, but the population is fluctuating wildly around this value, often reaching or coming close to zero and often coming close to or exceeding double N^*. These bird populations are clearly not equilibrial by criterion 2.

Testing criterion 3 – stationarity

Full stationarity demands that the mean, variance, and higher moments (skew, kurtosis, etc.) of the distribution of N_t remain constant over time. Here I test only the most basic criterion that the mean, i.e., N^*, remains constant over time. A slightly less stringent criterion was used to filter time series for this test, namely that the mean abundance over 30 years, N^*, must be greater than or equal to five individuals, giving 4400 population time series. For each time series, a linear model as a function of time, $N_t = a + b * t$, and a quadratic model, $N_t = a + b * t + c * t^2$, were fit using OLS. Note that OLS regression fits a line through the mean, so if N^* (mean of N_t) was constant then the linear regression should have a slope b that is not significantly different from zero giving a horizontal line. If the slope was significantly different from zero, I identified four different types of non-stationarity: linear sloping up ($b > 0$), linear sloping down ($b < 0$), quadratic convex up ($c < 0$) and quadratic convex down ($c > 0$). I used AIC to choose whether a model was quadratic or linear. AIC assesses goodness of fit but penalizes for the extra parameters in the quadratic. Traditionally the model (linear or quadratic) with the lower AIC value is chosen. However, to be conservative, I required $AIC_{quadratic} > AIC_{linear} + 1$. This conservative test had the result of declaring more models linear which in turn increased the

Table 8.1. Analysis of whether times series representing one species at one site are horizontal (i.e., N^* is constant or stationary) or if N^* is trending up or down in a linear or quadratic fashion.

Time series type	Count	Frequency
Horizontal	1198	27.2%
Linear sloping up	713	16.2%
Linear sloping down	720	16.4%
Quadratic convex up	998	22.7%
Quadratic convex down	771	17.5%

number of models that were horizontal. Despite this, I found that only a small minority (less than 1/3) could not statistically ($p < 0.05$) reject a flat or horizontal ($b = 0$) model (Table 8.1). In the majority of cases, b was significantly different from zero, with the best model roughly evenly spread across the four different models of nonstationarity, with a slight preponderance of quadratic convex up (i.e., a hump or mountain shape). While test 2 demonstrated that the data were not equilibrial on a short timescale (year to year), this test shows that the systems are not equilibrial on longer (decades) timescales.

Testing criterion 4 – density dependence dominates

A strong defender of equilibrial models could write off tests of 2 and 3 by saying the system is inherently equilibrial in the sense of being density-dependent but is merely being forced by external factors to have very high variability and to have long-term trends. Here I test this directly by measuring density dependence. To do this I plot the year-on-year discrete growth rate, $\lambda_t = N_{t+1}/N_t$ vs. N_t. This is sometimes called a recruitment curve. If density dependence is occurring this plot should have a negative slope, causing the highest growth rate to be at $N_t = 0$ with the annual population growth λ_t decreasing down to a value of 1 (no change in population size) at the equilibrium value of N_t, i.e., N^*, and then becoming negative for $N_t > N^*$. Five examples of such plots are shown in Figure 8.3A–E. Calculating λ_t requires an invalid division by zero when $N_t = 0$. To avoid this I filtered species/site combinations only to populations where the average abundance $\bar{N} \geq 5$ and $N_t > 0$ for all t, giving 3276 time series. As the randomly chosen examples in Figure 8.3 suggest, there were broad signs of density-dependent regulation occurring. In this case I regressed $\log(\lambda_t)$ vs. N_t, effectively fitting a density-dependence model that was exponential since there were clear deviations from linearity of a convex down nature (as seen in the examples in Figure 8.3 and the overall goodness of fit was improved by doing this). This is equivalent to assuming the Ricker model of density-dependent population growth ($N_{t+1} = N_t \lambda_{max} e^{-cN_t}$), but my motivations were just to assess density dependence in a purely phenomenological manner without any specific model assumed. On first inspection, the density-dependent regulation model appears to hold up well; most regressions (2904 out of 3276 or 88.6%) were significant at $p < 0.05$ and all but five had $c < 0$ (i.e., negative density dependence). However, a glance at the examples in Figure 8.3 suggests that there is a great deal of noise around the density

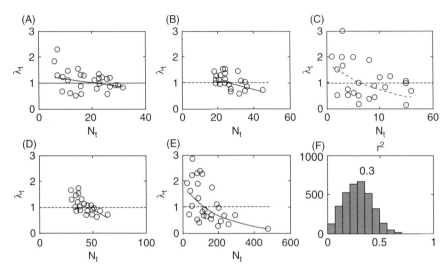

Figure 8.3. Fits of density-dependence (recruitment curve) for five random sites/species pairs (A–E). (F) Histogram of the r^2 with a median of 0.3 (30% of variance explained). Histogram is across 3276 species/site pairs with a mean abundance > 5 and no zeros in the time series since zeros lead to infinite growth rates and are problematic. NB: two points with very high growth rates (greater than 3) at very low abundances are truncated in panel (C).

dependence. This is confirmed by examining the r^2 value (percent of variance explained). The median r^2 is 0.30 (30% of variance explained), with 95% of the cases falling in the range 0.06–0.57. Thus while there is negative density dependence evident in the birds, density dependence explains less than half of the year-to-year variability in almost all cases. Whether this should be considered as support for the density-dependent model seems debatable.

In summary, populations of birds in North America show way too much short-term (yearly) variability in abundance to be considered equilibrial and they usually show long-term trends of change in equilibrium population size, N^*. Density dependence does appear to play a role in regulation of bird abundance, but it explains less than half of the variation. Thus, birds, presumed to be one of the most equilibrial classes of organisms, mostly contradicted the equilibrial model.

Are communities equilibrial?

In the last section I started with an empirical/data-driven definition of equilibrial populations and then tested these using the Breeding Bird Survey. I will now repeat this to see whether communities instead of populations are equilibrial or nonequilibrial. It is by now well understood that even small communities of three species can show chaotic behavior. Thus, even highly deterministic models can generate seemingly noisy, high-variance data without obvious patterns. This was a central limitation of early theoretical ecology when models required a true equilibrium point (with negative eigenvalues of the Jacobian matrix) to be considered stable (May, 1971). With hindsight, this is an overly restrictive condition, and studies of stability of communities have increasingly moved to data-driven, intuitive metrics such as coefficient of variation of total population size or total biomass (Pimm, 1984). This conforms with my approach in this chapter.

For this chapter, I define a community as being equilibrial if:

1. aggregate properties constant – aggregate measures of the community such as species richness, S, total abundance summed across species in the community, N_{total}, total biomass of the community, M, and total energy flux in the community E should be close enough to constant to have low coefficients of variation (CV);
2. community composition constant – the list of species present or absent and their relative abundances should show little variation; in practice this is measured by using some measure of distance between communities such as Bray-Curtis dissimilarity, Jaccard dissimilarity etc.;
3. strong species interactions – there should be a signal, perhaps through correlations of time series, that species are interacting strongly with each other.

Test of criterion 1 – aggregate properties constant

There are at least two different reasons to expect aggregate properties of communities to be more constant than the properties of individual populations (Lehman & Tilman, 2000; Mikkelson *et al.*, 2011):

- statistical averaging – also known as the portfolio effect; this suggests that the sum or mean of random variables, so long as the variables are independent, will have a lower variance than the individual variables as positive and negative noise cancel each other out;
- compensatory dynamics – also known as negative covariances; if a guild of competitive species is observed, it is expected that when one species does well, another will do worse, leading to net effects closer to zero (Houlahan *et al.*, 2007). A more specific version of this is that if the species are sharing a common resource, then the total resource availability can be divided among the species in varying fashion, but the signature of the constant resource should show at the community level. This idea has recently been presented as the idea of zero-sum community dynamics, which specifically suggests that energy flux (directly tied to the mechanism of resource consumption) should be more constant than abundance or body mass (Ernest *et al.*, 2008).

It is easy to test the idea that aggregate communities should be more constant. There are 131 communities used in this study, with mean species richness, S, across communities, of 60 species. The number of species at a site (Figure 8.4) is clearly more constant than total number of individuals, total biomass and total energy flux of the bird community (Figures 8.4B,C,D). Specifically, across the 131 communities the median CV for richness, S, is 0.076 (7.6%); for total abundance, N_{total}, it is 0.174 (17%); for total biomass, M, median CV is 0.376 (38%); and for total energy flux, E, is 0.261 (26%). Note that biomasses were based on species averages from Sibley (Sibley, 2000), and energy flux

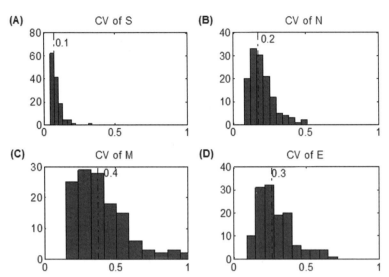

Figure 8.4.   Coefficient of variation (CV) of aggregate community properties. (A) CV of species richness, S. (B) CV of total number of individuals, N. (C) CV of total community biomass, M. M is calculated using species means and masses from Sibley (2000). (D) CV of energy flux in communities, E. E is calculated using allometric relationships with $E \propto M^{3/4}$ with M as in (C).

was based on species biomass raised to the ¾ power (no individual-level measurements of these values are available for the Breeding Bird Survey). Thus all four community-level aggregate properties (S, N, M, E) appear much more constant than population-level variation in abundance of individual species. The total numbers of species and individuals, especially, appear to approach what one might consider equilibrial (mean CV < 20% of community value). Total energy flux, E, which integrates changes in both biomass and abundance, had a lower CV than body mass, M, but higher than total abundance, N, thereby providing mixed support for the zero-sum idea. However, most time series of S, N, M, and E through time in a single community do show non-stationarity (specifically trends up or down over the 30 years). For S, 80 of the 131 communities had a slope with respect to time that was significantly different from zero (can reject the null hypothesis of no change in S over time at $p < 0.05$) with 56 communities trending up and 24 trending down. For N, 78 communities showed a trend with 31 up and 47 down. For M, 62 communities showed a trend with 56 up and just 6 down. The bias towards communities increasing in total biomass is striking and could be consistent with the idea that meso-predators are increasing at the expense of small birds. Finally, for E, 76 communities showed a significant trend with 54 up and 22 down.

In summary, aggregate community properties show much less year-to-year variability than population variables, with S and N especially showing small enough CV that one would be tempted to call them equilibrial. However, over longer timescales it seems that a majority of communities still experience significant trends up or down of these aggregate community properties, presumably due to external forcing from trends such as climate change (e.g., poleward shifts in regions: Devictor *et al.*, 2012) or human influence.

Test of criterion 2 – community composition constant

Analyzing communities provides the interesting possibility that aggregate properties can remain fixed while the internal details are constantly changing. This idea has precedence in the microscopic vs. macroscopic properties of statistical mechanics (McQuarrie & Allan, 2000), the "dynamic equilibrium" of species richness despite changing species composition in the theory of island biogeography (MacArthur & Wilson, 1967) and in the aforementioned studies finding "zero-sum" behavior in energy flux and species diversity (Ernest *et al.*, 2008). The previous section suggested that the aggregate properties of S and N_{total} had low year-on-year variability but showed long-term trends. Here I test whether the internal dynamics of community composition (presence/absence and abundance of species) is constant. Computational limits require looking at just 20 of the 131 communities (chosen randomly). For each community, the similarity of the community present in 1978 was compared with the community present in 1979, 1980, ... 2008. Similarity was measured using Spearman rank correlation to avoid issues of non-normality of abundances. Figure 8.5A shows the results of the 20 communities and a mean local-regression line fit through the 20 communities. It appears that something very

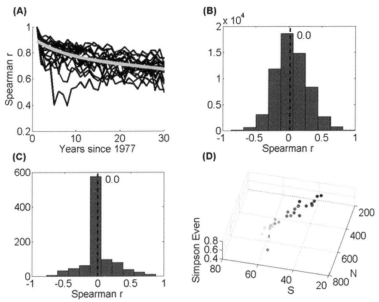

Figure 8.5.   (A) A plot of the decreasing Spearman correlation (similarity in community composition) with increasing time spans. (B) A histogram of the Spearman correlations between pairs of species co-occurring in one community. (C) A histogram of the Spearman correlations between N_{total} for different communities in the Northeastern US over time. (D) A simple 3-D ordination of one bird community. The lightest dot represents 1978, with increasingly darker colors representing succeeding years. This plot shows this route is experiencing a clear trend of both decreasing N_{total} and S. Although not completely clear in this one view of the 3-D data, the evenness bounces up and down throughout the years with no clear pattern.

close to exponential decay of similarity is occurring, with community similarity dropping off quickly in the first few years, and then slowing down but continuing to decay with time. Despite slowing down there is no obvious asymptote at 30 years (nor is there in an analysis of routes with 40 good years of observation). This suggests that internal community structure is changing constantly over time, despite the relative constancy of aggregate properties. This is different from the equilibrium model, where communities that are perturbed away from the original internal composition gradually return to that original composition.

Test of criterion 3 – strong species interactions

Just as one expects density-dependent regulation to be occurring in individual populations, one would expect species interactions to be a dominant process with a strong signal in equilibrial communities. Specifically competition ($-/-$) and predation ($-/+$) interactions should show up. This can be examined empirically by looking at the correlation of time series between pairs of species in the same community. If competition dominates, then the two time series should be negatively correlated (increased abundance of one coincides with decreases in the abundance of the other) (Veech, 2006; Houlahan et al., 2007).

Computational requirements again become intense, and so the pairwise correlation of all species co-occurring in a community (one BBS route) were analyzed for 10 communities. A histogram of all Spearman correlation coefficients (not including self-correlation) across all 10 communities is given in Figure 8.5B. Note that the correlation coefficients are strongly centered on 0 with a median of $\rho = 0.022$. One must be cautious in interpreting these results because mathematical constraints on correlation matrices prevent a matrix from being filled with all negative correlations. Indeed for large matrices (i.e., with many species), the lowest the average correlation can be is just below zero (Brown et al., 2004; Ranta et al., 2008) . Intuitively this is because if species A is strongly competitive (with strong negative correlation) with species B and C, then B and C must be positively correlated – or colloquially, the enemy of my enemy is my friend. Likewise, although predation creates both a positive and a negative entry in a covariance matrix (providing balance around a mean correlation of zero), apparent competition among species sharing a predator will also tend to produce too many negative covariances. However, even though the mean is constrained mathematically, the variance is not. One can still have strongly negatively correlated species as long as they are balanced out by strongly positively correlated species. However, this is not what is observed – most species are weakly interacting with almost no strong correlations in either direction (Figure 8.5B) with 95% of all interactions having correlations falling between $\rho = -0.41$ and $\rho = 0.55$ and echoing similar previous findings (Veech, 2006; Houlahan et al., 2007). Although it is an established and not surprising pattern that most species have weak interactions, it is surprising how few very strong interactions there are.

An alternative expectation is that all of the species would be strongly positively correlated if environmental drivers (e.g., good years vs. bad years and resource pulses) were trumping internal species interactions. There is no upper limit on the amount of positive correlations in a community (mathematically one could have every species with a correlation of +1 with every other species). However, this is not observed either in Figure 8.5B. This lack of year-to-year correlation due to strong external weather conditions is also supported by an analysis of how N_{total} is correlated between sites in a region (here the North-Eastern USA and adjacent Canada defined as north of latitude 40°N and east of longitude 80°W). The mean correlation between N_{total} for sites within a single region is $\rho = 0$, with 95% of all sites showing ρ between $(-0.50, 0.59)$. This leaves the suggestion that most of the correlations are weak and driven by fairly random forces.

Global change – what to measure?

The frequent occurrence of nonequilibrial signatures, even in one of the classes of organisms that might be expected to show a high level of equilibrial patterns, raises serious challenges for measuring human impacts on populations and communities. How does one separate natural "background" variability from human-caused changes (both of which contribute to the nonequilibrial dynamics described here)? This is, for example, an ongoing central issue in the debate on climate change; skeptics claim that recently observed trends of increasing temperature are well within the range of normal variability. Several possible solutions emerge from the analysis in this chapter. Specifically, population-level measures and compositions of communities seem to experience very high variability and thus would require very long time series to confidently deduce changes. In contrast, aggregate community measures such as S and N_{total} and long-term trends in them seem to be indicative of meaningful changes in the community structure. A simple technique that is easy to use and takes advantage of these facts is to do an ordination in 3-D space given by species richness, S, total abundance, N, and Shannon Evenness (McGill, 2011). This allows for quick and simple examination of long-term trends in low-varying aggregate community measures. Slightly different measures are needed when sample sizes are not constant (McGill, 2011). An example is shown in Figure 8.5D.

Conclusion

This chapter examined whether equilibrial or nonequilibrial models of population trends and community structure are the best fit for birds of North America. Birds are expected to be one of the groups of organisms most likely to exhibit equilibrial behavior, thus any finding of nonequilibrial dynamics may be general across many groups of organisms. I adopted a macroecological approach (analyzing a large dataset spanning many species, sites and years). The findings were mixed, with some support for both models, but the overall gestalt is that there appears to be more support for the nonequilibrial approach. Specifically, individual populations (single species at a single site or BBS route) showed extremely high variability year-to-year, often with a coefficient of variation larger than one. It seems most populations range from close to zero to double the equilibrium population size and spend little time close to N^*. Moreover most (72.8%) of the bird populations show some form of nonstationarity with a clear trend up or down in population size over 30 years. There does seem to be some support for negative density-dependent regulation roughly in the form found in the Ricker model, but this model explains less than half of the variation that occurs from year to year. Similarly, for communities, there is considerable turnover in community composition over 30 years with no sign of community turnover asymptoting to a constant level. And there is no sign of strong species interactions, nor of strong environmentally driven spatial synchrony in variation of community size (N_{total}). The aggregate community measures of richness, S, and total abundance, N_{total}, do appear to come close to equilibrial in having relatively low amounts of variability, but they still usually show long-term trends up or down in contradiction of equilibrial assumptions.

Overall, we see that populations have almost no sign of equilibrial dynamics except for a weak density-dependence. Likewise, communities appear mostly nonequilibrial except that S and N_{total} appear to have fairly low variability despite usually showing long-term trends and hiding ongoing turnover of internal community stucture. As noted in the introduction of this chapter, as well as much of this book, nonequilibrial models and ideas have been around for a long time. And it is a myth that the mathematics for equilibrial systems is easier or more useful. However, equilibrial systems remain the dominant paradigm in most ecologists' minds. This chapter shows that this is not really justified, even for a taxon such as birds. Climatologists and hydrologists have begun talking about how in a world of climate change "stationarity is dead", meaning that even probability distributions of variables like hottest day or peak flowering time have begun to change. These scientists have begun to seriously explore the implications of this fact (Milly *et al.*, 2008; Craig, 2010). Given the mostly nonequilibrial patterns of birds, it is time for ecologists to join this exploration.

References

Arrhenius, O. (1921). Species and area. *The Journal of Ecology*, **9**, 95–99.

Bell, G. (2000). The distribution of abundance in neutral communities. *The American Naturalist*, **155**, 606–617.

Black, A. J., & McKane, A. J. (2012). Stochastic formulation of ecological models and their applications. *Trends in Ecology & Evolution*, **27**, 337–345.

Brown, J., Bedrick, E., Ernest, S., Cartron, J., & Kelly, J. (2004). Constraints on negative relationships: mathematical causes and ecological consequences. In M. L. Tapper & S. R. Lele (Eds.), *The Nature of Scientific Evidence: Statistical, Philosophical, and Empirical Considerations* (pp. 298–323). Chicago, IL: University of Chicago Press.

Clements, F. E. (1936). Nature and structure of the Climax. *Journal of Ecology*, **24**, 252–284.

Cold Spring Harbor Symposium on Quantitative Biology (22nd symposium) (1957). *Population Studies: Animal Ecology and Demography*. Cold Spring Harbor, NY: Cold Spring Harbor Labs.

Connor, E. F., & Simberloff, D. (1979). The assembly of communities: chance or competition. *Ecology*, **60**, 1132–1140.

Craig, R. K. (2010). Stationarity is dead – long live transformation: five principles for climate change adaptation law. *Harvard Environmental Law Review*, **34**, 9–75.

Davidson, J., & Andrewartha, H. G. (1948). The influence of rainfall, evaporation, and atmospheric temperature on the fluctuations in the size of a natural populatin of *Thrips imaginis*. *Journal of Animal Ecology*, **17**, 200–222.

Devictor, V., van Swaay, C., Brereton, T., *et al.* (2012). Differences in the climatic debts of birds and butterflies at a continental scale. *Nature Climate Change*, **2**, 121–124.

Diamond, J. M. (1975). Assembly of species communities. In M. L. Cody & J. M. Diamond (Eds.), *Ecology and Evolution of Communities* (pp. 342–444). Cambridge, MA: Harvard University Press.

Ernest, S. K. M., Brown, J. H., Thibault, K. M., White, E. P., & Goheen, J. R. (2008). Zero sum, the niche, and metacommunities: long-term dynamics of community assembly. *The American Naturalist*, **172**, E257–E269.

Etienne, R. S. (2005). A new sampling formula for neutral biodiversity. *Ecology Letters*, **8**, 253–260.

Etienne, R. S., & Olff, H. (2004). A novel genealogical approach to neutral biodiversity theory. *Ecology Letters*, **7**, 170–175.

Gause, G. F. (1934). *The Struggle for Existence*. Dover 1971 reprint of 1934. New York: Williams & Wilkins.

Gleason, H. A. (1926). The individualistic concept of the plant association. *Torrey Botanical Club*, **53**, 7–26.

Henson, S. M., Cushing, J. M., Costantino, R. F., *et al.* (1998). Phase switching in population cycles. *Proceedings of the Royal Society of London B*, **265**, 2229–2234.

Houlahan, J. E., Currie, D. J., Cottenie, K., *et al.* (2007). Compensatory dynamics are rare in natural ecological communities. *Proceedings of the National Academy of Sciences of the USA*, **104**, 3273.

Hubbell, S. P. (2001). *A Unified Theory of Biodiversity and Biogeography.* Princeton, NJ: Princeton University Press.

Kingsland, S. E. (1995). *Modelling Nature: Episodes in the History of Population Ecology* (2nd edn). Chicago, IL: University of Chicago Press.

Lehman, C. L., & Tilman, D. (2000). Biodiversity, stability, and productivity in competitive communities. *The American Naturalist*, **156**, 534–552.

Lotka, A. J. (1925). *Elements of Mathematical Biology.* New York: Williams & Wilkins.

MacArthur, R. H. (1968). The theory of the niche. In R. Lewontin (Ed.), *Population Biology and Evolution* (pp. 159–176). Syracuse, NY: Syracuse University Press.

MacArthur, R. H., & Wilson, E. O. (1967). *The Theory of Island Biogeography.* Princeton, NJ: Princeton University Press.

May, R. M. (1971). Stability in multispecies community models. *Mathematical Bioscience*, **12**, 59–79.

McGill, B., & Collins, C. (2003). A unified theory for macroecology based on spatial patterns of abundance. *Evolutionary Ecology Research*, **5**, 469–492.

McGill, B. J. (2011). Species abundance distributions. In A. E. Magurran & B. J. McGill (Eds.), *Biological Diversity: Frontiers in Measurement and Assessment* (pp. 105–122). Oxford: Oxford University Press.

McGill, B. J., Etienne, R. S., Gray, J. S., *et al.* (2007). Species abundance distributions: moving beyond single prediction theories to integration within an ecological framework. *Ecology Letters*, **10**, 995–1015.

McKane, A. J., Alonso, D., & Sole, R. V. (2004). Analytic solution of Hubbell's model of local community dynamics. *Theoretical Population Biology*, **65**, 67–73.

McQuarrie, D. A., & Allan, D. (2000). *Statistical Mechanics*. Sausalito, CA: University Science Books.

Mikkelson, G., McGill, B., Beaulieu, S., & Beukema, P. (2011). Multiple links between species diversity and temporal stability in bird communities across North America. *Evolutionary Ecological Research*, **13**, 361–372.

Milly, P. C. D., Betancourt, J., Falkenmark, M., *et al.* (2008). Stationarity is dead: whither water management? *Science*, **319**, 573–574.

Nicholson, A. J., & Bailey, V. A. (1935). The balance of animal populations I. *Proceedings of the Zoological Society, London*, **3**, 551–598.

Patuxent Wildlife Research Center (2001). Breeding Bird Survey FTP site. Available at: ftp://www. mp2-pwrc.usgs.gov/pub/bbs/Datafiles/.

Pimm, S. L. (1984). The complexity and stability of ecosystems. *Nature*, **307**, 321–326.

Ranta, E., Kaitala, V., Fowler, M. S., *et al.* (2008). Detecting compensatory dynamics in competitive communities under environmental forcing. *Oikos*, **117**, 1907–1911.

Roughgarden, J. (1979). *Theory of Population Genetics and Evolutionary Ecology: An Introduction.* New York: MacMillan.

Sauer, J. R., Hines, J. E., Gough, G., Thomas, I., & Peterjohn, B. G. (1997). *The North American Breeding Bird Survey Results and Analysis*. Laurel, MD: USGS Patuxent Wildlife Research Center.

Sibley, D. A. (2000). *National Audubon Society. The Sibley Guide to Birds*. New York: Alfred A. Knopf.

Veech, J. A. (2006). A probability-based analysis of temporal and spatial co-occurrence in grassland birds. *Journal of Biogeography*, **33**, 2145–2153.

Verhulst, P. F. (1838). Notice sur la loi que la population suit dans son accroissement. *Correspondance Mathematique et Physique*, **10**, 113–121.

Volterra, V. (1927). Variazioni e fluttuazioni del numero d'individui in specie animali conviventi. *Mem. R. Accad. Naz. dei Lincei. Ser.* VI, vol. **2**. (**Volterra, V.** (1926). Fluctuations in the abundance of a species considered mathematically. *Nature*, **118**, 558–560).

Yule, G. U. (1924). A mathematical theory of evolution based on the conclusions of Dr J C Willis. *Philosophical Transactions of The Royal Society B-Biological Sciences*, **213**, 21–87.

Part III

Equilibrium and Nonequilibrium on Geographical Scales

9 Island flora and fauna: equilibrium and nonequilibrium

Lloyd W. Morrison

In this chapter, I evaluate the concept of equilibrium in insular systems. I focus on species numbers rather than population abundances, because many studies of island biotas have been motivated by the ideas of MacArthur and Wilson (1967), who focused on species numbers. A useful system for thinking about equilibrium on islands was presented by Rey (1984) and further expanded and developed by Whittaker (1998). In this system, an island is cross-classified according to two characteristics: (1) whether it is in equilibrium or not, and (2) whether the species composition is static or dynamic. In an equilibrium condition, immigration and extinction rates are balanced, and species are neither gained nor lost. In a nonequilibrium condition, immigrations exceed extinctions, or vice versa, and species number increases or decreases over time. A static condition implies no change in the identities of species present, whereas a dynamic condition describes changes in species composition, due to turnover. I make reference to this system, with some modifications, in describing a number of different island scenarios. I agree with Whittaker that it is better to think of these characteristics in terms of continua rather than discrete categories, although the use of categories facilitates the comparison of various studies.

Equilibrium vs. nonequilibrium states of islands

Stasis

An island in stasis ("static equilibrium" in Whittaker, 1998) experiences neither immigrations nor extinctions; the species composition remains unchanged over time (Figure 9.1a). The classic example of this scenario derives from the work of Lack (1969, 1976) on birds. Lack assumed dispersal was not limiting, and the main factor determining whether a given species would be present on an island was whether the appropriate habitat type existed. Once all niche spaces were occupied, the island would be closed to further colonization. Other studies, many with birds, have found a relative stasis in community composition, with very few immigrations or extinctions (e.g., Walter, 1998; Foufopoulos & Mayer, 2007). In this relatively deterministic view, observed turnover of species has often been attributed to anthropogenic effects, which Lack acknowledged could lead to changes in the fauna.

The Balance of Nature and Human Impact, ed. Klaus Rohde. Published by Cambridge University Press.
© Cambridge University Press 2013.

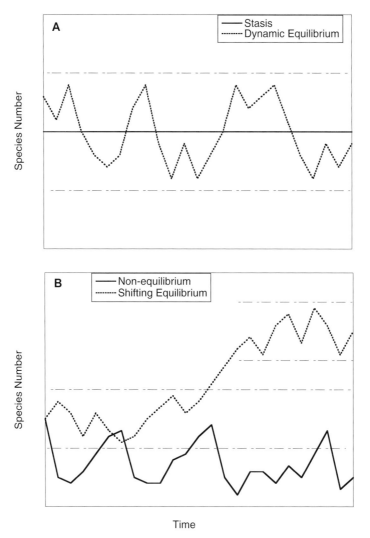

Figure 9.1.   Generalized equilibrium vs. nonequilibrium scenarios. (A) Stasis resulting from constant species composition over time. Dynamic equilibrium characterized by stochasticity in species number within some acceptable range of variability (indicated by the horizontal dashed lines). (B) Nonequilibrium characterized by deviations in species numbers beyond an acceptable range of variability (e.g., frequent disturbance events resulting in episodic extinctions). Shifting equilibrium, shown here as a unidirectional change to a higher species number. The second set of horizontal dashed lines represents the new acceptable range of variability.

Dynamic equilibrium

The theory of island biogeography (MacArthur & Wilson, 1967) posits that the number of species on an island is a function of immigration and extinction (and speciation, although this is not explicitly considered in studies conducted in ecological time). When immigration and extinction rates are equal, the island will be at equilibrium and the species number will be determined by the intersection of these two rates. The identities of species present will change over time, a phenomenon known as turnover, and the rate of turnover will equal the immigration and extinction rates at equilibrium. This elegant theory has garnered much attention among ecologists, as it represents a general stochastic model that can be applied to any system, in contrast to earlier ideas that invoked specific historical factors.

Many studies have attempted to test various aspects of this theory, including the prediction of a dynamic equilibrium (i.e., a constancy of species numbers but with changing species composition). Yet disagreement has existed over whether a given data set supports an equilibrium condition or not (Simberloff, 1983). MacArthur and Wilson (1967) did not assume that each extinction would be balanced by an immigration, but that a certain amount of stochasticity would characterize the equilibrium. Thus the number of species would be expected to vary around a long-term mean species number (Figure 9.1a). MacArthur and Wilson (1967) calculated that the ratio of the variance to the mean would be ~0.5 (closer to 1 in a "very unsaturated situation", as during the colonization process). Schoener (2010) refined MacArthur and Wilson's (1967) calculations of the variance/mean ratio, and determined that a value between 0 and 1 was consistent with MacArthur and Wilson's concept of a dynamic equilibrium. Such a variance/mean ratio threshold has rarely been applied to equilibrium studies, but many would have to be interpreted as evidence of a dynamic equilibrium if based on this criterion. Ultimately, many studies encompass so few surveys that it is difficult to accurately assess the variance.

Shifting equilibrium

Schoener (2010) suggests that an equilibrium may shift unidirectionally over time, due to basic changes in the processes underlying immigration or extinction rates (e.g., climate change). Clear examples of a shifting equilibrium are difficult to find, and what represents a shifting equilibrium to one may be interpreted as a nonequilibrium situation to another. Here I consider a shifting equilibrium to occur when the long-term mean species number increases (or decreases) unidirectionally to some higher (or lower) level, due to a permanent change in one or more factors affecting immigration or extinction rates (Figure 9.1b).

Nonequilibrium

In this scenario, immigration rates are greater than extinction rates, or vice versa, and an island gains or loses species over time. Nonequilibrium could describe either a unidirectional trend, or relatively large increases and decreases in species numbers beyond some acceptable limits of variability. This scenario could characterize islands that experience acute disturbance events resulting in a loss of species, followed by a slow gain in species back toward an equilibrium point (Figure 9.1b). If the disturbances are large or frequent, however, relative to the response time (i.e., dispersal ability) of the taxa in question, an equilibrium may rarely or never be reached. Whittaker (1995) argued that disturbance events affecting islands are more frequent than generally recognized, such as the impacts of hurricanes in the tropics. Although various non-equilibrium models have been created (e.g., Villa *et al.*, 1992; Russell *et al.*, 1995) they lack the elegant simplicity of MacArthur and Wilson's equilibrium model and have not been widely applied.

A classic example of apparent nonequilibrium is Abbott and Grant's (1976) study of the bird faunas on islands around Australia and New Zealand. Abbott and Grant reported a nonequilibrium condition for passerine species, and suggested that variability in the weather from year to year led to uncertainty in food resources, going on to declare that "nonequilibrium is to be expected at high latitudes with climate unpredictability". Unfortunately, their inferences came primarily from islands that had only been surveyed at two time periods, so it is not possible to determine how much variability characterized species numbers or if any regular oscillations were evident.

Whittaker (1998) distinguished between a dynamic and static nonequilibrium. The concept of a static nonequilibrium may seem counterintuitive (i.e., how can species composition be unchanging, yet species numbers be increasing or decreasing?). The determination of a static nonequilibrium is achieved by considering the system on two timescales: a short-term scale in which species identities are unchanging, and a longer timescale in which extinctions outnumber immigrations, or vice versa. The classic example is Brown's (1971) work on boreal mammals occupying mountaintop islands in the Great Basin, in which the islands were determined to be in a state of faunal relaxation. Brown posited that, since the Pleistocene, extinctions have occurred but no new immigrations, due to a changing climate that caused the intervening desert to become a barrier to dispersal. (Brown's interpretations have since been challenged by Lawlor (1998) and Grayson & Madsen (2000)). Whittaker and Fernández-Palacios (2007) also include a study of the birds of the lesser Antilles (Ricklefs & Bermingham, 2001) as an example of static nonequilibrium. Although there has been little change in the bird faunas in modern times, Ricklefs and Bermingham present evidence for either an abrupt increase in colonization or a mass extinction event *c.* half a million years ago.

Successional influences

New islands that rise out of the sea (e.g., Surtsey) or extant islands that are completely sterilized by a major volcanic eruption (e.g., Krakatau) represent somewhat specialized cases to apply classic island biogeographic theory. In the case of Krakatau, for example, the disturbance was so extreme that not only were all species eradicated, but the physical environment was altered so that it was no longer able to support most species that were present previously. Successional processes are at work following such a disturbance, and over time it is expected that species numbers will increase as the appropriate habitats become available or necessary biological interactions become established. Thus immigration rates will tend to exceed extinction rates over a long period of time (i.e., decades to centuries) but, in the absence of any additional extreme disturbances, they would be expected to eventually equalize. Such a case could obviously be described as nonequilibrium, but could also be considered as a type of shifting equilibrium. Whittaker (2004) summarized data for numerous taxa of Krakatau, finding a diversity of scenarios to exist. Some taxonomic groups appear to have achieved an equilibrium whereas others have not, and some appear static and others dynamic.

Thus, determining whether or not a particular island is in equilibrium (and if so, what type of equilibrium) is not simple, even if abundant data are available. Usually, an abundance of data is not available, which makes the determination even more difficult. Over the short term, the issue is how much variability in species numbers is acceptable to qualify as an equilibrium condition, and there is no widely accepted standard that has been consistently applied. Over the longer term, a larger problem has to do with whether one considers an equilibrium number of species to be a relatively permanent character-istic of an island, or a parameter that may change over time as the environment changes. The best examples of a static nonequilibrium suggest that the equilibrium number of species is a relatively immutable property of an island that does not change, even over evolutionary or geological time. In contrast, the idea of a shifting equilibrium considers that the equilibrium number of species may change due to underlying systematic changes in immigration or extinction rates. Thus it is critical to define an appropriate time period to allow an informative inference from any equilibrium study. Given enough time, all islands or habitat fragments are affected by geological change or evolution, and would have to be considered as nonequilibrium, although this is a rather trivial result. For the remainder of this chapter I focus on an ecological timescale.

Case study: small Bahamian islands

Attempts to compare and contrast a variety of equilibrium studies result in problems with issues of sampling intensity and temporal scale (e.g., how many surveys were conducted? How long were survey intervals? What was the total length of the investigation?). The most important question is: is the data record long enough to encompass the underlying environmental dynamics? If not, we are only seeing a portion of the overall picture. Unfortunately, the answer to this question for most studies is that it is almost certainly not long enough.

Thus, rather than attempting to scrutinize a broad range of studies, I focus on a summary of the findings from four taxa (plants, ants, web spiders, and lizards) inhabiting archipelagoes of small islands in the Bahamas. The data record for this overall body of research spans multiple decades, and many of the same islands have been surveyed for all four taxa. Given the apparent long-term variability and stochasticity in the environment, however, it is still unlikely that we have a complete understanding of the equilibrium (or nonequilibrium) dynamics for most taxa.

The Bahamian study system includes hundreds of small rocky islands in the northern and central Bahamas. The islands are physically stable, composed primarily of marine limestone with some fossilized sand deposits. The marine limestone substrate is riddled with crevices and galleries, so that in cross section it has a "Swiss cheese" appearance. All of the plants on the survey islands are long-lived perennials.

Ants and plants

The ants and plants have been surveyed in archipelagoes at Andros, the Abacos, and the Exumas. The data span almost two decades (1990–2007) for some islands. Annual surveys were conducted over the first 5 years in the Exumas, and it was discovered that very low rates of turnover characterized both plants ($\leq 1\%$/year; Morrison, 1997) and ants ($\leq 5\%$/year; Morrison, 1998). Because so few immigrations or extinctions occurred, after 1994 surveys were conducted at longer (4- or 5-year) intervals. After a decade of data from the Exumas and Andros, I observed similar low rates of turnover for plants (Morrison, 2003), and somewhat lower turnover rates for ants (Morrison, 2002), than in the first 5 years.

After the first 5 years, I had concluded that the ant data appeared to be consistent with the idea of a dynamic equilibrium, although it was not very dynamic since species compositions changed very little (Morrison, 1998). After a decade, however, even though immigrations usually exceeded extinctions, there were so few immigrations or extinctions that it appeared the ants fell closer to a stasis, or static equilibrium classification, at least for the time period in question (Morrison, 2002). This was also my determination for plants, although I did speculate that over longer time periods the system may be nonequilibrium, if periodic hurricanes caused large extinction events (Morrison, 2003).

Continued surveys extended this data set into a second decade and included a third archipelago (the Abacos). This decade revealed a relatively dramatic change, for both the ants and plants, in both the Exumas and Andros. Whereas immigration rates were often somewhat higher than extinction rates in the first decade, extinction rates exceeded immigration rates, by as much as an order of magnitude, in the second decade (Morrison, 2010a, 2010b). Although data were available for the Abacos only in the second decade, the same result was obvious for both ants and plants: a preponderance of extinctions (Figure 9.2).

What caused this change in the immigration/extinction dynamics? One potential contributing factor is that four hurricanes affected the study archipelagos near the end of the first decade and early part of the second decade (Floyd in 1999, Michelle in 2001, and Frances and Jeanne in 2004). The timing of the surveys relative to the hurricanes, however, indicates that acute effects of these storms did not cause most of the extinctions, for either taxon. All of the ants surveyed nest in subterranean galleries in the marine limestone, and would probably be able to survive complete inundation of the islands in air-filled pockets within the substrate. Similarly, the roots of the plants are protected by the limestone and although much of the above-ground biomass may be broken or removed by wind and wave action (Spiller et al., 1998), the plants are able to regenerate from the protected root system (Morrison & Spiller, 2008). Yet indirect effects of hurricanes, such as increased herbivory (Spiller & Agrawal, 2003) and decreased nutrient availability from scouring of topsoil, may have been responsible. Additionally, a long-term (quarter-century) increase in temperature and decrease in the number of days with rain was observed in the region between 1983 and 2007 (Morrison, 2010b).

After two decades of data, I re-characterized both the ants and plants as being nonequilibrium (Morrison, 2010a, 2010b). This interpretation was not based on the relative

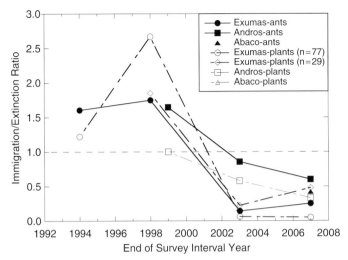

Figure 9.2. The ratio of immigrations to extinctions for ants (solid symbols) and plants (open symbols) in three Bahamian archipelagoes over time. The ratio was calculated by dividing the total number of immigrations by the total number of extinctions occurring in each survey interval. The point is plotted according to the ending year of the interval. Two sets of islands were surveyed for plants in the Exumas (see Morrison, 2010b). The horizontal dashed line represents an equilibrium between immigrations and extinctions. (Modified from figures in Morrison, 2010a, 2010b.)

constancy of species numbers or the variance/mean ratio (a strict interpretation of which would result in a verdict of dynamic equilibrium). The ultimate categorization of non-equilibrium resulted from three lines of evidence: (1) the underlying immigration and extinction rates changed. This was true for all archipelagoes, for both ants and plants, and for all survey intervals within each respective decade. (2) Potential environmental drivers have been identified. Chronic effects of increased hurricane activity or climate change (which may affect hurricane frequency and intensity) could underlie the changing rates of immigration and extinction. Although the observed climatic change was monotonic over the entire two-decade period, it is possible that the stress from a warmer, drier environment only reached levels that affected species occurrences during the second decade. (3) Piecing together a regional record including other related studies supports, at least for the ants, a potential long-term nonequilibrium cycle associated with hurricane activity, as described below.

Comparing records of ants on 21 islands near Puerto Rico surveyed in 1981–1983 to data obtained 18 years previously, Torres and Snelling (1997) found evidence of non-equilibrium, with 4.7 more immigrations than extinctions occurring. These results combined with the Bahamas data suggest a regional increase in ant species on small islands in the northern Caribbean-Bahamas region from the mid-1960s to the 1990s, and a decrease afterwards. Generalized hurricane activity in the North Atlantic was unusually low in the 1970s and 1980s, but has increased greatly since 1995 (Goldenberg *et al.*, 2001; Nyberg *et al.*, 2007). Thus it is possible that the surveys conducted in this region prior to the 1990s recorded a slow rebound in species numbers from more depauperate faunas that existed in the first half of the century due to greater hurricane activity. Likewise, the decline documented in the 1990s may reflect increased hurricane activity beginning in that decade. In the Exumas, no major hurricanes struck between 1932 and 1996 (Spiller *et al.*, 1998). In contrast, four hurricanes affected these study archipelagoes between 1999 and 2004, although storms varied in intensity across archipelagoes (Morrison 2010a). Thus, over the long term (i.e., decades to centuries) there may be an oscillation in species numbers driven by cycles of hurricane activity.

Web spiders and lizards

In this same island system, lizards (primarily *Anolis sagrei*) were found to have very low turnover rates ($\leq 1\%$/yr) in the absence of major disturbances, appearing almost static over time periods of up to a decade (Schoener, 1986, 1991). Web spiders, on the other hand, revealed relatively high turnover rates ($\sim 33\%$/yr; Schoener, 1991), even in the absence of major disturbances. The spider populations that exhibited turnover, however, were primarily small in size, and repeatedly went extinct and re-immigrated. Larger populations of spiders tended to be persistent over the 5-year period of the study (Schoener & Spiller, 1987). Thus the web spiders represent a sort of hybrid situation with respect to turnover, with a permanent fraction that was static, and another fraction that was dynamically changing.

It is of interest to compare the acute effects of hurricanes on the plants and ants versus the spiders and lizards in this same system. Unlike the plants and ants, which are protected from storms by the island substrate, the web spiders and lizards are vulnerable to the winds and waves of hurricanes. Hurricane Lili in 1996 swept away every lizard and web spider from a group of islands on the west side of Great Exuma (Spiller *et al.*, 1998). Yet web spiders, which disperse by ballooning, required only 1 year to recolonize the islands. In contrast lizards, which disperse by floating or rafting, have recolonized only a few islands. Schoener (2010) considers that the attainment of equilibrium for lizards on these islands may require a longer period of time than the intervals between such hurricanes. If so, an equilibrium species number may never or only rarely be reached, a case of nonequilibrium due to frequent disturbance relative to the dispersal ability of the lizards.

In 1999 Hurricane Floyd decimated the web spiders from a group of small islands near Abaco and, as was the case in the aftermath of Lili in the Exumas, species numbers recovered to pre-disturbance levels in 1 year (Schoener & Spiller, 2006). Hurricane Floyd also apparently swept the adult lizards from most of these small islands, but their eggs remained on many of the islands, from which new populations were established (Schoener *et al.*, 2001). On some islands no life stages remained, and recolonization apparently occurred overwater. Dispersal in the Abaco archipelago was greater than that in Great Exuma because of shorter distances separating the small islands from larger source islands. Consideration of the effects of three hurricanes (Lili, Floyd, and Michelle) on lizard populations reveals that the impacts vary depending upon the strength of the storm and the time of year in relation to the lizard reproductive cycle (Schoener *et al.*, 2004).

Thus, in this Bahamian system, the spiders appear to be the closest to the classic idea of a dynamic equilibrium. Although this equilibrium may be lost for short periods following hurricane-force disturbances, it is quickly regained. The lizards, plants and ants could all be characterized as nonequilibrium, at least over the longer term for most archipelagoes. In the short term, however, they may appear to be in stasis. Lizards may or may not be completely wiped out by hurricanes, depending upon the intensity and timing. Furthermore, distance to larger source islands is an important factor determining recolonization rate for lizards, and thus lizards on very close islands may resemble more of a dynamic equilibrium, whereas lizards on more distant islands could be characterized as nonequilibrium over the long term. In contrast to lizards and web spiders, plants and ants do not experience extinction pulses from hurricanes, although a direct hit from the strongest (category 5) hurricanes has not been evaluated. The chronic effects of hurricanes, however, may result in higher extinction rates and decreasing species richness for plants and ants, resulting in a long-term nonequilibrium.

Characteristics related to equilibrium

Based on the Bahamas system and other studies of island biotas, several general characteristics relating to equilibrium can be elucidated. (1) Equilibrium is less likely when disturbances (e.g., hurricanes) are intense or frequent. (2) Equilibrium is more likely (or more rapidly re-attained) for taxa with higher dispersal rates (e.g., web spiders vs. lizards), or for closer islands that may be more rapidly recolonized following a disturbance (e.g., lizards at Abaco vs. Great Exuma). (3) Equilibrium is more likely for taxa that have greater resistance to disturbance events (e.g., ants and plants protected by the substrate). (4) Equilibrium is less likely when there is a broad-scale, pervasive change in the environment that alters immigration or extinction rates (e.g., climate change, or an increase in hurricane frequency or intensity). (5) Some species (or populations) in an assemblage may be relatively static, whereas others are more dynamic (e.g., Bahamian web spiders). Thus attempting to classify different types of equilibria may not be as informative as focusing on the component species populations with respect to turnover. (6) The determination of an equilibrium or nonequilibrium state may depend on (i.e., be an artifact of) the number of surveys and time period observed.

Human impacts

Human activities have resulted in dramatic changes to the species assemblages of islands. Humans have introduced many species to islands (including competitors, predators and pathogens of native species) and greatly modified insular habitats. The majority of documented extinction events over the past 500 years have been on islands, as opposed to mainlands, for both terrestrial vertebrates and plants. Furthermore, predation, either alone or in concert with other factors (e.g., habitat destruction), has been responsible for the vast majority of extinction events (Sax & Gaines, 2008).

A complete review of the effects of human activities on islands is beyond the scope of this chapter. It is important to note, however, that implications for equilibrium vary by taxonomic group. For example, in a literature survey Sax *et al.* (2002) determined how species numbers have changed for vascular plants and land birds on oceanic islands following the establishment of non-native species. They estimated that species richness has remained relatively unchanged for land birds, but for vascular plants it has approximately doubled. Even though species richness of birds on the islands of their survey did not change, the identities of the birds changed dramatically, with over 50% of native species going extinct and being replaced by non-natives. In contrast, relatively few native plants have gone extinct, and the increase in plant species richness is thought to be due primarily to human-assisted dispersal.

An emerging threat to insular species assemblages is the current rate of climate change. The distributions of many organisms are shifting in latitude or elevation relatively rapidly in response to the changing climate (Chen *et al.*, 2011). Species inhabiting many islands, however, may be unable to shift their ranges, and simply go extinct. A study of the reptiles inhabiting Mediterranean islands found elevated extinction rates associated with historical climate change, and suggests the types of patterns that may be observed due to future climate change (Foufopoulos *et al.*, 2011). Relatively few studies have documented effects of climate change on island biotas in ecological time. A significant climate signal was found in a dataset spanning 23 years for the butterfly fauna on Hilbre Island in the UK (Dennis *et al.*, 2010). The changes in the ant and plant compositions of small Bahamian islands may also be due, to some extent, to climate change. A warming climate, along with higher sea levels and increased intensities of tropical cyclones, has important implications for the species assemblages of many islands. This represents an intriguing field for future studies.

References

Abbott, I., & Grant, P. R. (1976). Nonequilibrial bird faunas on islands. *The American Naturalist*, **110**, 507–528.

Brown, J. H. (1971). Mammals on mountaintops: nonequilibrium insular biogeography. *The American Naturalist*, **105**, 467–478.

Chen, I.-C., Hill, J. K., Ohlemuller, R., Roy D. B., & Thomas, C. D. (2011). Rapid range shifts of species associated with high levels of climate warming. *Science*, **333**, 1024–1026.

Dennis, R. L., Dapporto, H. L., Sparks, T. H., *et al.* (2010). Turnover and trends in butterfly communities on two British tidal islands: stochastic influences and deterministic factors. *Journal of Biogeography*, **37**, 2291–2304.

Foufopoulos, J., Kilpatrick, A. M., & Ives, A. R. (2011). Climate change and elevated extinction rates of reptiles from Mediterranean islands. *The American Naturalist*, **177**, 119–129.

Foufopoulos, J., & Mayer, G. C. (2007). Turnover of passerine birds on islands in the Aegean Sea (Greece). *Journal of Biogeography*, **34**, 1113–1123.

Goldenberg, S. B., Landsea, C. W., Mestas-Nunez, A. M., & Gray, W. M. (2001). The recent increase in Atlantic hurricane activity: causes and implications. *Science*, **293**, 474–479.

Grayson, D. K., & Madsen, D. B. (2000). Biogeographic implications of recent low-elevation recolonization by *Neotoma cinera* in the Great Basin. *Journal of Mammology*, **81**, 1100–1105.

Lack, D. (1969). The numbers of bird species on islands. *Bird Study*, **16**, 193–209.

Lack, D. (1976). *Island Biology Illustrated by the Land Birds of Jamaica*. Oxford: Blackwell Scientific.

Lawlor, T. E. (1998). Biogeography of great basin mammals: paradigm lost? *Journal of Mammology*, **79**, 1111–1130.

MacArthur, R. H., & Wilson, E. O. (1967). *The Theory of Island Biogeography*. Princeton, NJ: Princeton University Press.

Morrison, L. W. (1997). The insular biogeography of small Bahamian cays. *Journal of Ecology*, **85**, 441–454.

Morrison, L. W. (1998). The spatiotemporal dynamics of insular ant metapopulations. *Ecology*, **79**, 1135–1146.

Morrison, L. W. (2002). Island biogeography and metapopulation dynamics of Bahamian ants. *Journal of Biogeography*, **29**, 387–394.

Morrison, L. W. (2003). Plant species persistence and turnover on small Bahamian cays. *Oecologia*, **136**, 51–62.

Morrison, L. W. (2010a). Disequilibrial island turnover dynamics: a 17-year record of Bahamian ants. *Journal of Biogeography*, **37**, 2148–2157.

Morrison, L. W. (2010b). Long-term non-equilibrium dynamics of insular floras: a 17-year record. *Global Ecology and Biogeography*, **19**, 663–672.

Morrison, L. W., & Spiller, D. A. (2008). Patterns and processes in insular floras affected by hurricanes. *Journal of Biogeography*, **35**, 1701–1710.

Nyberg, J., Malmgren, B. A., Winter, A., *et al.* (2007). Low Atlantic hurricane activity in the 1970s and 1980s compared to the past 270 years. *Nature*, **447**, 698–702.

Rey, J. R. (1984). Experimental tests of island biogeographic theory. In D. R. Strong, Jr., D. Simberloff, L. G. Abele & A. B. Thistle (Eds.), *Ecological Communities: Conceptual Issues and the Evidence* (pp. 101–112). Princeton, NJ: Princeton University Press.

Ricklefs, R. E., & Bermingham, E. (2001). Nonequilibrium diversity dynamics of the Lesser Antillean avifauna. *Science*, **294**, 1522–1524.

Russell, G. J., Diamond, J. M., Pimm, S. L., & Reed, T. M. (1995). A century of turnover: community dynamics at three timescales. *Journal of Animal Ecology*, **64**, 628–641.

Sax, D. F., & Gaines, S. D. (2008). Species invasions and extinction: the future of native biodiversity on islands. *Proceedings of the National Academy of Sciences of the USA*, **105**, 11490–11497.

Sax, D. F., Gaines, S. D., & Brown, J. H. (2002). Species invasions exceed extinctions on islands worldwide: a comparative study of plants and birds. *The American Naturalist*, **160**, 766–783.

Schoener, T. W. (1986). Patterns in terrestrial vertebrate versus arthropod communities: do systematic differences in regularity exist? In J. M. Diamond, & T. J. Case (Eds.), *Community Ecology* (pp. 556–586). New York: Harper & Row.

Schoener, T. W. (1991). Extinction and the nature of the metapopulation: a case study. *Acta Oecologica*, **12**, 53–75.

Schoener, T. W. (2010). The MacArthur-Wilson equilibrium model: a chronicle of what it said and how it was tested. In J. B. Losos & T. J. Ricklefs (Eds.), *The Theory of Island Biogeography Revisited* (pp. 52–87). Princeton, NJ: Princeton University Press.

Schoener, T. W., & Spiller, D. A. (1987). High population persistence in a system with high turnover. *Nature*, **330**, 474–477.

Schoener, T. W., & Spiller, D. A. (2006). Nonsynchronous recovery of community characteristics in island spiders after a catastrophic hurricane. *Proceedings of the National Academy of Sciences of the USA*, **103**, 2220–2225.

Schoener, T. W., Spiller, D. A., & Losos, J. B. (2001). Natural restoration of the species-area relation for a lizard after a hurricane. *Science*, **294**, 1525–1528.

Schoener, T. W., Spiller, D. A., & Losos, J. B. (2004). Variable ecological effects of hurricanes: the importance of seasonal timing for survival of lizards on Bahamian islands. *Proceedings of the National Academy of Sciences of the USA*, **101**, 177–181.

Simberloff, D. (1983). When is an island community in equilibrium? *Science*, **220**, 1275–1277.

Spiller, D. A., & Agrawal, A. A. (2003). Intense disturbance enhances plant susceptibility to herbivory: natural and experimental evidence. *Ecology*, **84**, 890–897.

Spiller, D. A., Losos, J. B., & Schoener, T. W. (1998). Impact of a catastrophic hurricane on island populations. *Science*, **281**, 695–697.

Torres, J. A., & Snelling, R. R. (1997). Biogeography of Puerto Rican ants: a non-equilibrium case? *Biodiversity and Conservation*, **6**, 1103–1121.

Villa, F., Rossi, O., & Sartore, F. (1992). Understanding the role of chronic environmental disturbance in the context of island biogeographic theory. *Environmental Management*, **16**, 653–666.

Walter, H. S. (1998). Driving forces of island biodiversity: an appraisal of two theories. *Physical Geography*, **19**, 351–377.

Whittaker, R. J. (1995). Disturbed island ecology. *Trends in Ecology and Evolution*, **10**, 421–425.

Whittaker, R. J. (1998). *Island Biogeography*. Oxford: Oxford University Press.

Whittaker, R. J. (2004). Dynamic hypotheses of richness on islands and continents. In M. V. Lomolino & L. R. Heaney (Eds.), *Frontiers of Biogeography: New Directions in the Geography of Nature* (pp. 211–231). Sunderland, MA: Sinauer Associates.

Whittaker, R. J., & Fernández-Palacios, J. M. (2007). *Island Biogeography: Ecology, Evolution, and Conservation*. Oxford: Oxford University Press.

10 The dynamic past and future of arctic vascular plants: climate change, spatial variation and genetic diversity

Christian Brochmann, Mary E. Edwards and Inger G. Alsos

Introduction

The Arctic lies at a global extreme of human impact as well as climatic gradients. Human population density is low. Climate precludes agriculture in all areas but the southernmost ones, and hunting and reindeer husbandry have been the traditional ways of life. More recently coal, oil and gas exploitation and associated infrastructure have affected the vegetation locally and led to species introductions (Walker & Everett, 1987; Forbes *et al.*, 2001). Long-range pollution from industrial regions has reached the Arctic and is accumulating in some animals. Even so, except for Antarctica, the arctic tundra remains the least disturbed major world biome (CAFF, 2001). Short growing seasons, low temperature means and extremes, and high inter-annual variability act as strong environmental filters that allow only the hardiest species to survive. The region has experienced extreme climate changes in the past and is expected to do so in the future. Our focus of this chapter, therefore, is also on past plant responses to rapid climate change, since this history provides a basic framework to help understand the future.

The Arctic comprises relatively low-diversity ecosystems with a dramatic and relatively well-documented history of past responses to repeated climatic oscillations. New paleoecological evidence, recent advances in molecular studies, and predictive modeling of arctic species ranges have greatly increased our understanding of how arctic plants may respond to future environmental change, from individual species to communities. On a short timescale (< 100 years), the species composition of arctic plant communities is often remarkably stable. This partly reflects perennial strategies and clonal growth, enabling survival of fluctuating inter-annual climatic conditions (de Witte & Stöcklin, 2010; Jónsdóttir, 2011). In the longer term, however, plant communities and species ranges have been highly dynamic. During the Late Pleistocene, some regions experienced a change from complete ice cover to boreal forest within a few thousand years, with accompanying transformation of vegetation and species range shifts (Miller *et al.*, 2010a).

Here we briefly describe the key features of the modern arctic vegetation and flora, review the consequences of past changes driven primarily by long-term climate variation, and anticipate possible responses to future climate warming.

Arctic flora and vegetation

The arctic flora features notorious taxonomic complexities arising from frequent hybridization, reticulate evolution via genome doubling (polyploidy), inbreeding/asexuality and the occurrence of widespread species with complex morphological variation. Progress towards a unified arctic-wide taxonomy was long hampered by divergent traditions among European, Russian, and North American botanists. The same species was often known under different names in different parts of the Arctic, or the same name was applied to different species. A recent major achievement is the publication of the first consensus checklist for all arctic vascular plant species, the Pan-Arctic Flora checklist (PAF; Elven *et al.*, 2011), in which many plant groups are also revised taxonomically based on recent molecular studies. The PAF checklist accepts about 2200 taxa at the species or subspecies level. Many of these are, however, boreal or temperate alpine taxa that have marginal arctic occurrences, which leaves ~1000 species and subspecies as part of the regular arctic flora. Only a subset of these, ~200 taxa, are arctic specialists, i.e., with their ranges mainly limited to the arctic region. The proportion of widely distributed species increases towards the north, and more than 90% of high arctic species have a circumpolar distribution (Elven *et al.*, 2011).

Arctic vegetation patterns are strongly related to summer temperature through its effects on nutrient availability and length of growing season (Chapin, 1983). There are clear thermal limits to species distributions (Young, 1971; Karlsen & Elvebakk, 2003; Elven *et al.*, 2011). Arctic vegetation can be described in terms of five zones in which both plant cover and species richness decrease from south to north (Figure 10.1). A mean July temperature difference of only 1–2°C corresponds to a large difference in terms of vertical and horizontal structure of plant cover, dominance of major functional types, phytomass, net annual production and species diversity (Callaghan *et al.*, 2004; Walker *et al.*, 2005). In the warmest, southernmost zone (E), 200–500 species are found in regions of the size of, for example, St. Lawrence Island or South Island of Novaya Zemlya, whereas fewer than 50 species occur in regions of similar sizes in the northernmost zone (A), the polar desert (Young, 1971; Walker *et al.*, 2005). In the polar desert, July temperatures range from 1 to 3°C, and vascular plant cover is < 5%, with individuals barely protruding above the moss layer (Walker *et al.*, 2006). One zone further south, the July temperature is 4–5°C, and the vascular plant cover increases to 5–25%.

Summer temperature is the most important factor structuring arctic vegetation on large geographical scales, but arctic ecosystems show strong resilience to temperature change on regional and landscape scales, where other factors such as topography, snow cover, moisture and biotic interactions may become more important (Jónsdóttir, 2005; Armbruster *et al.*, 2007). In the high-arctic archipelago of Svalbard ("Spitsbergen"; Figure 10.1), where the mean annual temperature has increased by 1.8°C from 1912 to

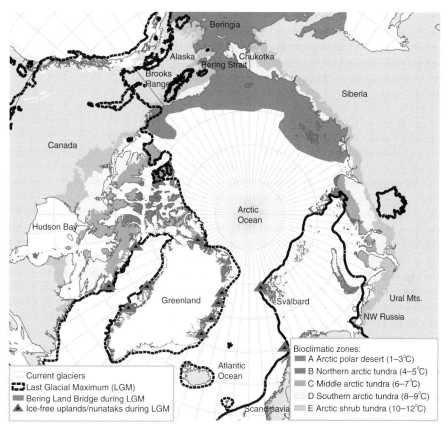

Figure 10.1.   Bioclimatic zones (http://www.arcticatlas.org/maps/catalog/index.shtml) and glaciation in the Arctic. Glacial limits are according to Ehlers and Gibbard (2004) for Europe and Dyke *et al.* (2003) for North America. Ice-free uplands and nunataks are according to Brochmann *et al.* (2003). See plate section for color version.

2002, vegetation analyses revealed only minor climate-change-related differences after 70 years (Prach *et al.*, 2010). Similarly, only minor changes in vegetation were found over 40 years in low-arctic southeastern Greenland (Daniëls *et al.*, 2011, but see also Callaghan *et al.*, 2011, who detected some vegetation change over 40 years in western Greenland). This slow reaction to environmental change may be explained by the long life span and generation time of most arctic plant species, their predominantly clonal reproduction, low flowering frequency and irregular seed set (Jónsdóttir, 2011; Müller *et al.*, 2011). The majority of arctic species are able to reproduce both sexually and asexually (Murray, 1987; Brochmann & Steen, 1999). Most of them are believed to have life spans of some decades (Jónsdóttir, 2011), and some clonal individuals may be several thousand years old (de Witte & Stöcklin, 2010; Jónsdóttir, 2011). Thus, individual plants may have survived more extreme climate fluctuations in their lifetime than those experienced in the past century and possibly the average change anticipated for the coming 100 years.

Biotic factors such as competition and herbivory do affect community stability also in the Arctic, in spite of the typically fragmented vegetation cover (Gough, 2006; Jónsdóttir, 2005). However, the few cases where herbivory is known to cause cascading effects on a local ecosystem are directly or indirectly linked to human interference (Jónsdóttir, 2005). The most well-known example is the overgrazing of salt marshes by geese in the Hudson Bay area (Jefferies *et al.*, 2006).

The past

Changes in arctic climate and geography

The arctic tundra as we know it today is young biogeographically yet has a dynamic evolutionary history. It formed only 2–3 million years ago, when the Pliocene forests that stretched to the coast of the Arctic Ocean gave way to tundra in response to the planetary cooling at the transition to the Pleistocene period. During the Pleistocene, with its marked temperature fluctuations driven by orbital cycles that show increasing amplitudes over the past ~1 Myr, extensive ice sheets have advanced and retreated in the northern high latitudes. Large areas of the western Eurasian and North American Arctic and adjacent regions have been periodically glaciated, while ice cover has tended to be less extensive in eastern Siberia and northwest North America (Figure 10.1). During times of lowered sea level associated with increased global ice volumes, the exposed coastal shelves united easternmost Siberia and Chukotka with Alaska and northwest Canada into a sub-continental area known as Beringia (Hopkins, 1982).

During the coldest and driest intervals of glacial periods, the vegetation of unglaciated arctic regions was characterized by a suite of largely herbaceous cold- and drought-tolerant taxa, while taxa with more mesic requirements, including many now widespread shrub species, had restricted distributions (Lamb & Edwards 1988; Bigelow *et al.*, 2003). The Pleistocene period has thus seen two kinds of distributional rearrangements of plant species: within- and between-region shifts in abundance resulting from climate-controlled changes in habitat availability, and large-scale recolonization events in the wake of receding ice sheets at the termination of glacial periods, the last occurring between ~15 000 and 6000 yr BP.

Climate-driven speciation

The current arctic flora is composed of a mixture of lineages resulting from immigration from southern mountains and steppes, *in situ* Pleistocene speciation, and probably *in situ* survival of some Pliocene forest elements (Murray, 1995). Pliocene macrofossils of the purple saxifrage *Saxifraga oppositifolia* from the Canadian Arctic and north Greenland as well as its plastid DNA variation suggest that this species was among the first plants to colonize the Arctic approximately 3 million years ago (Abbott *et al.*, 2000). Several other arctic plant species, such as netleaf willow *Salix reticulata*, black crowberry *Empetrum nigrum*, bog blueberry *Vaccinium uliginosum*, arctic white heather *Cassiope tetragona*, and mountain avens *Dryas octopetala*, have also been recorded as late Pliocene fossils in Greenland (Bennike, 1990) or Arctic Canada (Matthews & Ovenden, 1990).

In several other arctic plant groups, such as poppies (*Papaver*), whitlow-grasses (*Draba*), and mouse-ear chickweeds (*Cerastium*), major episodes of speciation have occurred more recently, probably within the period of the major ice ages (< 1 million years; reviewed in Brochmann *et al.*, 2003, 2004; Brochmann & Brysting, 2008). Many arctic plant species have now been sequenced for DNA regions commonly used for phylogenetic reconstruction. In many cases, little within-genus variation or none at all has been observed, insufficient for phylogenetic resolution and suggestive of a very young flora (Sønstebø *et al.*, 2010).

Pleistocene speciation in arctic plants has occurred both at the diploid level and via polyploidization (genome doubling), which in most studied cases has originated after hybridization between divergent lineages. Many studies have documented the importance of recent, recurrent and successively higher-level polyploid evolution in arctic plants, associated with range expansions induced by climatic shifts and mixing of divergent lineages (reviewed in Brochmann *et al.*, 2004). The frequency and level of polyploidy increase towards the north; 73% of the arctic specialist taxa are polyploids, and more than half of these are hexaploids (i.e., with six sets of chromosomes) or even more highly polyploid. For arctic specialists with restricted distributions, the frequency of diploids is much higher in unglaciated Beringia than in the heavily glaciated Atlantic area (Figure 10.1), suggesting that polyploids are most successful in colonizing after deglaciation. Most important in the context of climate change is that hybridization followed by genome doubling results in highly heterozygous genomes, which act to buffer against genetic depauperation through periods of inbreeding and long-distance (re-)colonization. A polyploid combines and preserves, in modified form, all the genomes of its original diploid progenitor species. As the average ploidy level of the ~1000 species in the regular arctic flora is pentaploid (five sets of chromosomes), these arctic species thus represent many more species (perhaps up to 2500) in terms of genetic or "ancestral species" diversity, that have been combined into the current members of the flora in response to past climate change.

There is abundant molecular evidence for recurrent formation of arctic polyploid species from more or less divergent diploid or low-polyploid progenitors. Taken together with the high frequency of self-pollinating and asexual reproductive systems in arctic plants (Brochmann & Steen, 1999) and the reshuffling of populations during climate shifts, this provides an explanation for the intricate taxonomy of many arctic plant groups (e.g., Brochmann & Elven, 1992; Jørgensen *et al.*, 2006; Guggisberg *et al.*, 2006, 2009; Brysting *et al.*, 2007; Carlsen *et al.*, 2009, 2010; Jordon-Thaden *et al.*, 2010).

There is also increasing molecular evidence for recent and rapid speciation at the diploid level in several arctic plant groups, possibly driven by genetic drift in small inbreeding populations. In arctic *Draba* spp. rapid development of sterility barriers has resulted in numerous (cryptic) biological species within single diploid taxonomic species, to an extent unknown in other floras (Brochmann *et al.*, 1993; Grundt *et al.*, 2006; Skrede *et al.*, 2008). The sterility barriers do not appear to be accompanied by morphological or ecological differentiation. The plants are predominantly, although not exclusively, self-pollinating, which provides instantaneous isolation from other lineages and thereby facilitates the accumulation of hybrid incompatibilities. In spite of the recent formation of these cryptic species, genetic mapping and quantitative trait locus (QTL) analyses have shown that the evolution of hybrid sterility involves multiple loci and genetic mechanisms (Skrede *et al.*, 2008). During periods of range expansions, such cryptic species can avoid hybridization with other expanding lineages and thus develop into taxonomically distinct species over time.

Apart from the genetic diversity acquired via their polyploid hybrid origins, many arctic plant species show considerable variation both within and among populations (Alsos *et al.*, 2007, 2012; Brochmann & Brysting, 2008). Genetic diversity varies with breeding system, frequency of inter- and intraspecific hybridization, and history as related to refugial isolation and range shifts. Migration in response to fluctuating ice extent and climate change is likely to have been an important driver of arctic plant diversification.

Arctic phylogeography

The phylogeographic history of many arctic plant species and species complexes is now known in some detail (summarized in Brochmann *et al.*, 2003; Brochmann & Brysting, 2008; P.B. Eidesen & C. Brochmann, unpubl.). Most arctic plants show some degree of geographic structuring with more or less distinct genetic groups. In a comparative study of 17 widespread northern plant species, genetic structure, diversity and distinctiveness were estimated from amplified fragment length polymorphisms (AFLPs) in some 8000 individual plants (P.B. Eidesen & C. Brochmann, unpubl.; cf. Eidesen, 2007). The results were spatially extrapolated into summary composite maps using GIS, showing that large-scale common phylogeographic patterns do exist and that the Pleistocene glaciations had a major impact on overall genetic diversity. Gene flow across the Arctic has been most severely hampered by long-standing physical barriers, i.e., the Atlantic Ocean, the Greenlandic ice cap, the Ural mountains, and the lowlands between the European Arctic and southern alpine areas (Figure 10.1). Overall high levels of genetic diversity are found only around the Bering Strait, corroborating the role of this mainly unglaciated area as a major refugium during the Pleistocene (Hultén, 1937; Abbott & Brochmann, 2003). In contrast, the genetically most distinct populations (measured as genetic distinctiveness, i.e., occurrence of rare alleles) are located in unglaciated east and west Siberia. Beringia harbors higher genetic diversity but less distinctiveness, suggesting higher gene flow out of this refugium than from the Siberian ones.

In some arctic plants, repeated bottlenecks during colonization after the last glaciation have resulted in extremely depauperate and virtually identical populations over vast northern areas. The alpine rock-cress *Arabis alpina* (Ehrich *et al.*, 2007) and the glacier buttercup *Ranunculus glacialis* (Schönswetter *et al.* 2003) consist of a single or very few genotypes in the north but harbor considerable diversity in their populations in the European Alps, which served as sources for northward colonization. A similar pattern is found for the European populations of the pygmy buttercup *Ranunculus pygmaeus*, but notably in this case the variable source populations are found in the Arctic itself, in the Polar Urals, and the species is variable also in other arctic areas (Schönswetter *et al.*,

2006). The study of *Arabis alpina* also demonstrated conspicuously different genetic consequences of Pleistocene range shifts between the Arctic (latitudinal recolonization over vast distances, causing loss of diversity), the Alps (vertical shifts over short distances involving many peripheral glacial populations, maintaining diversity), and the East African high mountains (highly divergent present-day refugial populations in different isolated mountains, high levels of diversity; Ehrich *et al.*, 2007).

Range shifts following climate change in other arctic plants have not, however, followed the leading-edge model of genetic depauperation through repeated bottlenecks during colonization (cf. Hewitt, 1996). Many cold-adapted species simply shifted their large distributions in response to the glacial cycles, and many of them were probably more widely distributed during cold periods, rather than contracted into smaller refugia (e.g., the dwarf willow *Salix herbacea*; Alsos *et al.*, 2009). In some species, the northern populations are genetically more variable than the southern ones, where they are restricted to refugia under the current climate (e.g., the nodding saxifrage *Saxifraga cernua*, Bauert *et al.*, 1998; Gabrielsen & Brochmann, 1998; Kjølner *et al.*, 2004; *Dryas octopetala*, Skrede *et al.*, 2006; *Vaccinium uliginosum*, Alsos *et al.*, 2005; Eidesen *et al.*, 2007a).

High genetic diversity in arctic plants is not only found in areas little affected by the major Pleistocene ice sheets, such as Beringia, but also in areas that were colonized postglacially by several intraspecific lineages. *Dryas octopetala*, for example, which consists of two distinct genetic groups in Eurasia, has high diversity in northern Siberia and the Urals, and also in secondary contact zones in northern Scandinavia, Svalbard, and East Greenland (Skrede *et al.*, 2006).

Historical evidence for high mobility of arctic plants

The traditional view of low, even lost, dispersal ability of arctic plants (Hultén, 1937; Dahl, 1963) can now be confidently rejected. High dispersal ability has long been suspected based on the fossil record, which suggests rapid range shifts following climate changes (e.g., Birks, 2008). A wealth of molecular data has now demonstrated that the arctic flora is highly mobile, with long-distance colonization occurring at higher rates than previously envisioned and from many source regions (reviewed in Brochmann & Brysting, 2008; cf. also Westergaard *et al.*, 2010; Popp *et al.*, 2011). It appears increasingly clear that at least part of the arctic flora as we know it today has been selected for mobility during the glacial cycles.

Although physical barriers such as the North Atlantic Ocean, not unexpectedly, have hampered gene flow in arctic plants more than continuous land masses (P.B. Eidesen & C. Brochmann, unpubl.), recent cross-oceanic dispersal, probably with birds or with wind across winter sea ice, has occurred in many species. This has been demonstrated in several studies at the entire circumpolar scale, and at the regional scale for the isolated high-arctic Svalbard archipelago, which today is also heavily glaciated. In particular, Alsos *et al.* (2007) showed that long-distance colonization of Svalbard has occurred frequently and from several source regions. The genetic effect of restricted colonization (the founder effect) was strongly correlated with the temperature requirements of the species, showing that the hardiest species have established in the archipelago more frequently than the less hardy ones, regardless of their dispersal adaptations (Figure 10.2A).

Svalbard is one of the most isolated arctic archipelagos and well explored with respect to biodiversity and glaciation and climate history. It was almost completely glaciated during the last glaciation (Landvik *et al.*, 1998, 2003), suggesting that most plant species occurring there today must have immigrated via cross-oceanic dispersal after the last glaciation. It thus forms a suitable model system for the study of arctic plant dispersal and establishment. Alsos *et al.* (2007) found the predominant source for colonization of Svalbard to be the most distant region, northwestern Russia. In eight of the nine species studied, multiple propagules resulting in successful establishment (a minimum of 6–38) were necessary to bring the observed genetic diversity to Svalbard, implying that many more propagules actually reached the archipelago. In *Dryas octopetala*, for example, the genetic diversity in Svalbard was similar to that in the source regions, and it is likely that many thousand propagules of this species were dispersed there. In *Vaccinium uligino-sum*, two genetically distinct types of Svalbard populations were identified; one originating from northwestern Russia and one originating from East Greenland.

At the circumpolar scale, Hultén (1937) suggested that arctic plants radiated east- and westward from Beringia and reached a full circumpolar distribution before the onset of the Pleistocene glaciations. He proposed that during each glacial cycle the arctic plants persisted in Beringia, whereas their distribution ranges were repeatedly fragmented and reformed elsewhere. However, he was convinced that arctic plants spread very slowly and that they were unable to reform their full distribution from Beringia during each interglacial period, and that they therefore had to rely on several additional glacial refugia. The scenario inferred based on plastid DNA variation in *Saxifraga oppositifolia* fits nicely with Hultén's hypothesis (Abbott *et al.*, 2000; Abbott & Brochmann, 2003). Macrofossils demonstrate its presence in north Greenland already 3 million years ago, and it contains two divergent cpDNA lineages, one broadly Beringian and one broadly Atlantic, which overlap in northern Greenland and northern Siberia.

However, studies of *Cassiope tetragona* (Eidesen *et al.*, 2007b), *Rubus chamaemorus* (Ehrich *et al.*, 2008), and *Saxifraga rivularis* (Westergaard *et al.*, 2010) gave surprisingly different results. The two first-mentioned species have also been found as early macro-fossils in northern Greenland (2.5–2.0 Myr old; Bennike & Böcher, 1990). In contrast to *Saxifraga oppositifolia*, however, no plastid DNA variation at all was found in *Cassiope tetragona*, suggesting a much more recent history. The molecular data are consistent with a Beringian origin of ssp. *tetragona* and westward expansion into northern Siberia at least one glacial cycle ago, but the eastward expansion probably occurred as late as in the current interglacial. The AFLP data are in accordance with a leading-edge model of recent colonization from Beringia throughout Canada and Greenland and across the Atlantic into Svalbard and Scandinavia, demonstrating exceptional ability for rapid long-distant colonization. Thus, *C. tetragona*, and probably *Rubus chamaemorus* as well (Ehrich *et al.*, 2008), most likely expanded from Beringia several times, but the earlier emigrants became extinct. In *Saxifraga rivularis*, the extreme Beringian-Atlantic dis-junction must have formed at least twice (Westergaard *et al.*, 2010). The first expansion from Beringia occurred at least one glacial cycle ago, followed by allopatric

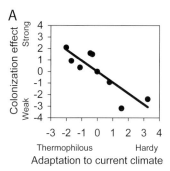

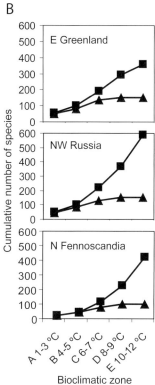

Figure 10.2.   (A) Index of the genetic effect of restricted colonization of Svalbard for nine species versus an index of adaptation to the current climate in Svalbard. The axes are principal components summarizing three measures of climatic adaptations and six quantities related to the effect of restricted colonization based on genetic (AFLP) data for 4439 samples. Thus, the hardiest species have established in the archipelago more frequently than the less hardy ones, regardless of their mechanism of dispersal, suggesting that establishment rather than dispersal is the limiting factor. (B) Number of species that may colonize the geographically isolated Svalbard archipelago from adjacent land masses following climate warming. The graphs show cumulative number of species in successively warmer bioclimatic zones in the source regions (squares) and the cumulative number of these species which are present in Svalbard today (triangles). Mean July temperature is given for each bioclimatic zone. Most of Svalbard's current flora belongs to zones A–C. A summer temperature increase of 2–4°C could shift Svalbard towards zones D and E. The gap between the lines at bioclimatic zones D and E represents the high numbers of potential colonizing species. Reprinted after Alsos *et al.* (2007) with permission from *Science*.

differentiation into one Beringian and one Atlantic subspecies. However, a new expansion from Beringia most likely occurred via several long-distance dispersals in the current interglacial, resulting in colonization of the western Atlantic region by the Beringian subspecies.

Thus, the fairly species-poor arctic flora is likely to be among the best adapted ones to environmental change, through long-term selection for high mobility, high potential for successful establishment in the low-competitive arctic environment, and buffering against inbreeding- and bottleneck-induced gene loss via polyploidy. The fossil record and the historical imprints kept in modern and permafrost-preserved ancient DNA molecules (Sønstebø et al., 2010) suggest that many arctic species had their largest Pleistocene ranges during cold periods and were confined to smaller refugial (sensu Stewart et al., 2010) ranges during warm periods. Genetic diversity must have been lost during range contractions via loss of populations, but this was counteracted not only by accumulation of new diversity during refugial isolation, but also by mixing of divergent refugial lineages during rapid large-scale recolonizations, resulting in new diversity hotspots and novel gene combinations, in many cases preserved through polyploid hybrid speciation.

The future

The pace and scale of climate change

The regional near-surface air temperature in the Arctic is rising at two to four times the global average rate (IPCC, 2007; Bekryaev et al., 2010; Miller et al., 2010a). Temperatures are expected to exceed those of the warmest periods of the Holocene as well as the last interglacial climate optimum about 120 000 years ago (Frenzel et al., 1992; Miller et al., 2010b). This is expected to have a large impact on both the physical environment and the biota of the arctic region (ACIA, 2006). However, Beaumont et al. (2011) point out that at arctic and boreal latitudes, despite high rates of warming, climate will tend to remain within the current envelope of variation (as expressed by the 2-σ value of monthly temperature patterns for 1961–1990) for longer than for regions nearer the equator, as northern climates are inherently more variable. Furthermore, latitudinal ranges of species are positively correlated with the annual range of temperatures experienced – that is, northern species tend to occupy large ranges (Stevens, 1989). In the past, some elements of the arctic flora persisting in nunataks and other unglaciated northern areas must have experienced major, rapid climate changes during past glacial periods (Dansgaard et al., 1993). Their Pleistocene genetic inheritance may have pre-adapted them to be tolerant of rapid climate change (Crawford & Abbott, 1994), as discussed in the previous section of this chapter. Taken together, these characteristics suggest potentially high resilience of the arctic flora and vegetation in the face of climate change, but future warming is nevertheless likely to transform many aspects of arctic ecosystems.

July temperatures in the Arctic are low, and even a small change in temperature may cause a several-fold change in warmth available to plant growth. This may result in major changes in vegetation structure, plant productivity, phytomass, species diversity and shift in altitudinal and zonal vegetation boundaries (Walker *et al.*, 2005). The expected increase in arctic July temperatures is 1.1–5.2°C (mean 2.5°C) by the end of this century (Christensen *et al.*, 2007). Assuming that dispersal rates and biotic interactions stay as they were in the past, this would imply a northwards shift in bioclimatic zonation by one to three zones (Figure 10.1). However, as the rate of the current warming exceeds the century-scale warming at the Pleistocene-Holocene transition (Post *et al.*, 2009), a higher dispersal rate may be required to track the future climate niche. The mean velocity of climate change (geographic shifts of isotherms over time) is regionally variable, and higher in the Low Arctic than in most other regions (see Figure 1C in Burrows *et al.*, 2011). Here, species would have to move about 50 km per decade to track their niches (Burrows *et al.*, 2011).

Northward expansion of high-shrub tundra has already been observed in many places (Myers-Smith *et al.*, 2011; Elmendorf *et al.* 2012a). While development of shrubs may be affected by changes in grazing pressure (Olofsson *et al.*, 2009), there is strong experimental support for climate being the driver of the current arctic shrub expansion (Walker *et al.*, 2006; Elmendorf *et al.*, 2012b). Dominance of high-shrub tundra during the orbital-driven early-Holocene summer thermal maximum has also been inferred for Beringia from fossil data (Edwards *et al.*, 2005) . Tall shrubs function as a low canopy layer, changing growth conditions, particularly for light, with negative competitive effects on low-growing arctic species.

The crucial questions are, therefore, (i) whether cold-adapted species are able to track such fast changes, (ii) whether they can cope with increased competition from more thermophilous species, and (iii) whether enough suitable habitat will be left to sustain viable populations with sufficient genetic diversity over time. These questions are addressed separately in the next sections.

Estimates of dispersal ability and species saturation

The median and mean dispersal distances estimated for arctic vascular plants are 460 and 570 km, respectively (Hoffmann, 2012). Assuming that they all arrived within the first 1000 years of adequate climate, this would give a dispersal rate of 46–57 km per century, which is sufficient to track current climate change. However, these estimates are based on current distribution of species on arctic islands and the nearby mainland, and do not include a timescale. Without fossil records of each species, we only know that most of them must have arrived after the last glacial maximum, thus any time between ~15 000 yr BP and the present time, giving dispersal rates of 0.3–57 km per century. This may be insufficient to track current rates of change.

While fossil and molecular studies provide strong evidence for high mobility of the arctic flora, neither the arctic archipelago Svalbard (Alsos *et al.*, 2007) nor other arctic islands (Hoffmann, 2012) appear to be fully saturated with species. However, the role of dispersal in lack of saturation is not clear. Alsos *et al.* (2007) showed that the genetic effect of restricted colonization (the founder effect) was strongly correlated with the temperature requirements of the species, as the hardiest species have established in the archipelago more frequently than the less hardy ones, regardless of their mechanism of dispersal (Figure 10.2A). This result suggests that it is not cross-oceanic dispersal, but rather establishment involving germination, survival, and local reproduction, which is the main limiting process for colonization of Svalbard. This interpretation is supported by the observation that 80–90% of the most cold-adapted species (1–5°C July temperature; bioclimatic zones A and B) that occur in Greenland and the two continental source regions (northwestern Russia and Scandinavia) are currently present in Svalbard, whereas only 40–60% of the species limited to 6–7°C July temperature (bioclimatic zone C) are present in Svalbard (Figure 10.2B).

In contrast, Hoffmann (2012) infers a lack of saturation caused by dispersal limitations. He concludes that "species numbers on the arctic island may increase if migration becomes complete and may not necessarily be linked to climate change". Hoffmann mentions the possibility that many species may already be at their northern limit and thus cannot establish further north. He argues, however, that this is not likely as his dispersal model takes the bioclimatic zone of the species into account: if dispersal is unlimited, species should be present on all islands on which their northernmost bioclimatic zones occur. Here we will argue that the simplistic bioclimatic zone concept does not encompass all aspects of a species' niche. In the three northernmost bioclimatic zones across the Arctic (Figure 10.1), 17–24% of the taxa only have scattered occurrences and 41–45% are rare (Elven et al., 2011), believed to be due to climate constraints (Bay, 1997; Engelskjøn et al., 2003; Solstad et al., 2010). Thus, we think that restricted establishment may partly explain the lack of saturation also in Hoffmann's (2012) study. Still, there seem to be other factors limiting distribution of arctic species, and dispersal may be one of them, especially in the Low Arctic where species ranges are more restricted than in the High Arctic (Elven et al., 2011).

Colonization and competitive interactions

A summer temperature increase of 2–4°C could shift high-arctic Svalbard, which currently has a total flora of 165 vascular plant species, towards the warmer bioclimatic zones D and E (Figure 10.1). A simple calculation, assuming that bioclimatic zones sufficiently represent species niches and that establishment rather than dispersal is the main limiting factor, shows that Svalbard can experience colonization by hundreds of new plant species currently occurring in these bioclimatic zones in Greenland, northwestern Russia, and Scandinavia (Figure 10.2B; Alsos et al., 2007). This is in accordance with modeled change of capacity for species richness, which predicts the highest increase in tundra (Sommer et al., 2010).

We can therefore predict that even the isolated Svalbard archipelago will experience increased species diversity with a warmer climate. Glacier retreat in the mountains will probably ensure continued habitat availability for the most cold-adapted species (Müller et al., 2012), whereas colonization by new low-arctic and boreal species can be expected in the lowlands, in particular in the inner warm fjord districts. Immigration of certain taxa, such as large willow shrubs *Salix* spp., may create entirely new habitats and lead to ecological cascade effects. Arctic islands such as Svalbard and mountain ranges such as the Polar Urals and Brooks Range may nevertheless be able to serve as critical refuges for the most cold-adapted species because they offer high mountain habitats, where taxa with low competitive ability can escape competition from low-arctic and boreal species. The situation is different from many continental arctic areas, which have large regions of subdued topography and where there is likely to be more homogeneous development of productive low-arctic shrub vegetation and widespread exclusion of less competitive arctic taxa.

Habitat availability and maintenance of genetic diversity

The rapid changes expected in the near future will directly affect only a few generations in long-lived arctic plant species. Thus, their immediate response is expected to be relocation rather than adaptation (Callaghan *et al.*, 2004; ACIA, 2006). However, maintaining genetic diversity is essential both for short-term adaptation to environmental change and for long-term survival (Reusch *et al.*, 2005). A number of arctic species are already experiencing reduction in their distributions, abundance and ability to exchange genes and individuals among populations (Cook *et al.*, 2012).

Although most arctic plant species are polyploids and preserve a variety of alleles inherited from their diploid ancestral species, this type of genetic variation is more or less "fixed" over short timescales; i.e., it is not normally released by recombination and thus not readily accessible for short-term adaptation to climate change (Brochmann *et al.*, 2004). With some exceptions (e.g., *Carex rufina*; Westergaard *et al.*, 2011), arctic species also contain the type of genetic diversity measured with standard population genetic approaches, which is most relevant for their ability for short-term adaptation. We recently estimated consequences of future climate change on geographic range and genetic diversity in 27 northern plant species, diploids as well as polyploids, based on amplified fragment length polymorphisms (AFLPs) in 9581 range-wide samples from 1200 populations (Alsos *et al.*, 2012). Range reduction was estimated using species distribution models, two global circulation models, and two emissions scenarios. We used a randomization procedure to estimate gene loss under increasing range loss and explored the effect of differences in species traits. In particular, we explored whether measures of genetic differentiation among populations (F_{ST}), which have been published for many thousand species, could be used as a proxy for predicting vulnerability to climate warming.

We predicted that 18 of these 27 northern species will lose close to half of their range by 2080 under at least one emission scenario and circulation model (Alsos *et al.*, 2012). New range area gained was in most cases considerably lower than the range area lost. In the worst-case scenario, range loss will lead to gene diversity loss in all 27 species, and one-third of them will lose more than 50%. Importantly, we found that vulnerability (in this specified sense) of a species to climate warming can be reasonably well predicted from its dispersal adaptation and/or by measures of population differentiation (F_{ST} estimates), which can aid in conservation prioritization. Species adapted to animal or wind dispersal will lose diversity at only half the rate of species with no particular adaptation to long-distance dispersal (although the latter certainly are able to occasionally disperse over vast areas as well; e.g., Westergaard *et al.*, 2010). Notably, the mean expected range reductions in these northern species are 1.2–1.8 times larger than that reported for temperate plants (Normand *et al.*, 2007; Morin *et al.*, 2008).

For many of the studied species, the effect of range reduction on loss of genetic diversity appeared to be independent of which particular part of the range that was lost. However, in some cases the actual gene loss can depend strongly on geography and on individual species histories as well. Species derived postglacially from alpine sources via northwards leading-edge colonization (e.g., *Arabis alpina*, Ehrich *et al.*, 2007; *Ranunculus glacialis*, Schönswetter *et al.*, 2003), which have high diversity in the south and hardly any in the north, will lose most of their diversity if their alpine populations go extinct. For *Salix herbacea*, on the other hand, a 50% range loss was estimated to cause only 5% gene diversity loss, because of its high dispersal ability combined with a history of broad-fronted postglacial colonization, which allowed high levels of diversity to be retained in most parts of its current range (Alsos *et al.*, 2009).

Conclusions

The Arctic represents a relatively simple and evolutionarily young system with a dramatic and quite well-known history of repeated responses to climatic oscillations. The species-poor arctic flora is likely to be among the best adapted to environmental change, through long-term selection for high mobility and buffering against inbreeding- and bottleneck-induced gene loss via genome doubling (polyploidy). The fossil record and the historical imprints in DNA molecules show that many arctic species had their largest Pleistocene ranges during cold periods and were confined to smaller (refugial) ranges during warm periods. Genetic diversity must have been lost during range contractions via loss of populations, but this was counteracted not only by accumulation of new diversity during refugial isolation, but also by mixing of divergent refugial lineages during rapid large-scale recolonizations, resulting in new diversity hotspots and novel gene combinations preserved through polyploid hybrid speciation. Today's arctic flora is, however, challenged by an even warmer climate over a very short time horizon. Many cold-adapted species will probably experience a decrease in range as well as in genetic diversity. Modeling predicts that 18 of 27 northern species will lose close to half of their range by 2080 under at least one emission scenario and circulation model. In the worst-case scenario, range loss will lead to gene diversity loss in all 27 species, and one-third of them will lose more than 50%. Vulnerability of a species to climate warming can be predicted from its dispersal adaptation and/or F_{ST} estimates, aiding conservation prioritization. Gene loss will, however, depend on individual species histories as well. Species derived postglacially from alpine sources via northwards leading-edge colonization, which have high diversity in the south and hardly any at all in the north, will lose most of their diversity if the alpine populations go extinct. Some northern areas, such as the currently heavily glaciated and isolated arctic archipelago of Svalbard, will probably increase in species diversity following glacier retreat and colonization by new low-arctic and boreal species, potentially leading to ecological cascade effects.

Acknowledgments

We thank current and previous members of our research groups as well as other collaborators for their invaluable contributions to many of the papers reviewed here, and for numerous discussions on these topics. I. S. Jónsdóttir is acknowledged for valuable comments on the manuscript.

References

Abbott, R. J., & Brochmann, C. (2003). History and evolution of the arctic flora: in the footsteps of Eric Hultén. *Molecular Ecology*, **12**, 299–313.

Abbott, R. J., Smith, L. C., Milne, R. I., *et al.* (2000). Molecular analysis of plant migration and refugia in the Arctic. *Science*, **289**, 1343–1346.

ACIA (2006). *Arctic Climate Impact Assessment – Scientific Report*. Cambridge: Cambridge University Press.

Alsos, I. G., Engelskjøn, T., Gielly, L., Taberlet, P., & Brochmann, C. (2005). Impact of ice ages on circumpolar molecular diversity: insight from an ecological key species. *Molecular Ecology*, **14**, 2739–2753.

Alsos, I. G., Alm, T., Normand, S., & Brochmann, C. (2009). Past and future range shift and loss of genetic diverstity in dwarf willow (*Salix herbacea* L.) inferred from genetics, fossils, and modelling. *Global Ecology and Biogeography*, **18**, 223–239.

Alsos, I. G., Ehrich, D., Thuiller, W., *et al.* (2012). Genetic consequences of climate change for northern plants. *Proceedings of the Royal Society of London B*, **279**, 2042–2051.

Alsos, I. G., Eidesen, P. B., Ehrich, D., *et al.* (2007). Frequent long-distance colonization in the changing Arctic. *Science*, **316**, 1606–1609.

Armbruster, W. S., Rae, D. A., & Edwards, M. E. (2007). Topographic complexity and terrestrial biotic response to high-latitude climate change: variance is as important as the mean. In J. B. Ørbæk, R. Kallenborn, I. Tombre, E. N. Hegseth, S. Falk-Petersen, & A. H. Hoel (Eds.). *Arctic-Alpine Ecosystems and People in a Changing Environment* (pp. 105–121). Berlin: Springer-Verlag.

Bauert, M. R., Kalin, M., Baltisberger, M., & Edwards, P. J. (1998). No genetic variation detected within isolated relict populations of *Saxifraga cernua* in the Alps using RAPD markers. *Molecular Ecology*, **7**, 1519–1527.

Bay, C. (1997). Floristical and ecological characterization of the polar desert zone of Greenland. *Journal of Vegetation Science*, **8**, 685–696.

Beaumont, L. J., Pitman, A., Perkins, S., *et al.* (2011). Impacts of climate change on the world's most exceptional ecoregions. *Proceedings of the National Academy of Sciences of the USA*, **108**, 2306–2311.

Bekryaev, R. V., Polyakov, I. V., & Alexeev, V. A. (2010). Role of polar amplification in long-term surface air temperature variations and modern arctic warming. *Journal of Climate*, **23**, 3888–3906.

Bennike, O. (1990). The Kap København Formation: stratigraphy and palaeobotany of a Plio-Pleistocene sequence in Peary Land, North Greenland. *Meddelelser om Grønland. Geoscience*, **23**, 1–85.

Bennike, O., & Böcher, J. (1990). Forest-tundra neighbouring the North Pole: plant and insect remains from the Plio-Pleistocene Kap København formation, North Greenland. *Arctic*, **43**, 331–338.

Bigelow, N. H., Brubaker, L. B., Edwards, M. E., *et al.* (2003). Climate change and arctic ecosystems: 1. Vegetation changes north of 55°N between the last glacial maximum, mid-Holocene, and present. *Journal of Geophysical Research-Atmospheres*, **108**, No. D19, 8170. doi10.1029/2002JD002558.

Birks, H. H. (2008). The Late-Quaternary history of arctic and alpine plants. *Plant Ecology & Diversity*, **1**, 135–146.

Brochmann, C., Borgen, L., & Stedje, B. (1993). Crossing relationships and chromosome numbers of Nordic populations of *Draba* (Brassicaceae), with emphasis on the *D. alpina* complex. *Nordic Journal of Botany*, **13**, 121–147.

Brochmann, C., & Brysting, A. K. (2008). The Arctic – an evolutionary freezer? *Plant Ecology and Diversity*, **1**, 181–195.

Brochmann, C., Brysting, A. K., Alsos, I. G., *et al.* (2004). Polyploidy in arctic plants. *Biological Journal of the Linnean Society*, **82**, 521–536.

Brochmann, C., & Elven, R. (1992). Ecological and genetic consequences of polyploidy in arctic *Draba* (Brassicaceae). *Evolutionary Trends in Plants*, **6**, 111–124.

Brochmann, C., Gabrielsen, T. M., Nordal, I., Landvik, J. Y., & Elven, R. (2003). Glacial survival or *tabula rasa*? The history of North Atlantic biota revisited. *Taxon*, **52**, 417–450.

Brochmann, C., & Steen, S. W. (1999). Sex and genes in the flora of Svalbard – implications for conservation biology and climate change. *Det Norske Videnskaps-Akademi. I. Matematisk Naturvitenskapelig Klasse, Skrifter, Ny Serie*, **38**, 33–72.

Brysting, A. K., Oxelman, B., Huber, K. T., Moulton, V., & Brochmann, C. (2007). Untangling complex histories of genome mergings in high polyploids. *Systematic Biology*, **56**, 467–476.

Burrows, M. T., Schoeman, D. S., Buckley, L. B., *et al.* (2011). The pace of shifting climate in marine and terrestrial ecosystems. *Science*, **334**, 652–655.

CAFF (2001). *Arctic Flora and Fauna. Status and Conservation*. Helsinki: Edita.

Callaghan, T. V., Björn, L. O., Chernov, Y., *et al.* (2004). Biodiversity, distributions and adaptations of arctic species in the context of environmental change. *Ambo*, **33**, 404–417.

Callaghan, T. V., Christensen, T. R., & Jantze, E. J. (2011). Plant and vegetation dynamics on Disko Island, West Greenland: snapshots separated by over 40 years. *Ambio*, **40**, 624–637.

Carlsen, T., Bleeker, W., Hurka, H., Elven, R., & Brochmann, C. (2009). Biogeography and phylogeny of *Cardamine* (Brassicaceae). *Annals of the Missouri Botanical Garden*, **96**, 215–236.

Carlsen, T., Elven, R., & Brochmann, C. (2010). The evolutionary history of Beringian *Smelowskia* (Brassicaceae) inferred from combined microsatellite and DNA sequence data. *Taxon*, **59**, 427–438.

Chapin, F. S. (1983). Direct and indirect effects of temperature on arctic plants. *Polar Biology*, **2**, 47–52.

Christensen, J. H., Hewitson, B., Busuioc, A., *et al.* (2007). Regional climate projections. In S. Solomon, D. Qin, M. Manning, Z. Chen, M. Marquis, K. B. Averyt, M. Tignor & H. L. Miller (Eds.), *Climate Change 2007: The Physical Science Basis. Contribution of*

Working Group I to the Fourth Assessment Report of the Intergovernmental Panel on Climate Change. Cambridge: Cambridge University Press.

Cook, J. A., Brochmann, C., Talbot, S. L., *et al.* (2012). Genetics. In *Arctic Biodiversity Assessment: Status and Trends in Arctic Biodiversity. Scientific Report to Conservation of Arctic Flora and Fauna (CAFF)*. Arctic Council (in press).

Crawford, R. M. M., & Abbott, R. J. (1994). Pre-adaptions of arctic plants to climate change. *Botanica Acta*, **107**, 271–278.

Dahl, E. (1963). Plant migration across the North Atlantic ocean and their importance for the palaeogeography of the region. In A. L. Löve & D. Löve (Eds.), *North Atlantic Biota and their History* (pp. 173–188). Oxford: Pergamon.

Daniëls, F. J., De Molenaar, J. G., Chytrý, M., & Tichý, L. (2011). Vegetation change in Southeast Greenland? Tasiilaq revisited after 40 years. *Applied Vegetation Science*, **14**, 230–241.

Dansgaard, W., Johnsen, S. J., Clausen, H. B., *et al.* (1993). Evidence for general instability of past climate from a 250-kyr ice-core record. *Nature*, **364**, 218–220.

de Witte, L. C., & Stöcklin, J. (2010). Longevity of clonal plants: why it matters and how to measure it. *Annals of Botany*, **106**, 859–870.

Dyke, A. S., Moore, A., & Robertson, L. (2003). *Deglaciation of North America*. Geological Survey of Canada, Open File 1574.

Edwards, M. E., Brubaker, L. B., Lozhkin, A. V., & Anderson, P. M. (2005). Structurally novel biomes: a response to past warming in Beringia. *Ecology*, **86**, 1696–1703.

Ehlers, J., & Gibbard, P. L. (2004). *Quaternary Glaciations – Extent and Chronology. Part I: Europe*. Amsterdam: Elsevier.

Ehrich, D., Alsos, I. G., & Brochmann, C. (2008). Where did the northern peatland species survive the dry glacials: cloudberry (*Rubus chamaemorus*) as an example. *Journal of Biogeography*, **35**, 801–814.

Ehrich, D., Gaudeul, M., Assefa, A., *et al.* (2007). Genetic consequences of Pleistocene range shifts: contrast between the Arctic, the Alps and the East African mountains. *Molecular Ecology*, **16**, 2542–2559.

Eidesen, P. B. (2007). *Arctic-alpine plants on the move: Individual and comparative phylogeographies reveal responses to climate change*. PhD Thesis, Natural History Museum, University of Oslo.

Eidesen, P. B., Alsos, I. G., Popp, M., *et al.* (2007a). Nuclear versus plastid data: complex Pleistocene history of a circumpolar key species. *Molecular Ecology*, **16**, 3902–3925.

Eidesen, P. B., Carlsen, T., Molau, U., & Brochmann, C. (2007b). Repeatedly out of Beringia: *Cassiope tetragona* embraces the Arctic. *Journal of Biogeography*, **34**, 1559–1574.

Elmendorf, S. C., Henry, G. H. R., Hollister, R. D., *et al.* (2012a). Plot-scale evidence of tundra vegetation change and links to recent summer warming. *Nature Climate Change*, **2**, 453–457.

Elmendorf, S. C., Henry, G. H. R., Hollister, R. D., *et al.* (2012b). Global assessment of experimental climate warming on tundra vegetation: heterogeneity over space and time. *Ecology Letters*, **15**, 164–175.

Elven, R., Murray, D. F., Razzhivin, V., & Yurtsev, B. A. (2011). *Checklist of the Panarctic Flora (PAF)*. Oslo: Natural History Museum, University of Oslo. Available at: http://nhm2.uio.no/paf [retrieved 25 April 2012].

Engelskjøn, T., Lund, L., & Alsos, I. G. (2003). Twenty of the most thermophilous vascular plant species in Svalbard and their conservation state. *Polar Research*, **22**, 317–339.

Forbes, B. C., Ebersole, J. J., & Strandberg, B. (2001). Anthropogenic disturbance and patch dynamics in circumpolar arctic ecosystems. *Conservation Biology*, **15**, 954–969.

Frenzel, B., Pécsi, M., & Velichko, A. A. (1992). *Atlas of Paleoclimates and Paleoenvironments of the Northern Hemisphere. Late Pleistocene – Holocene*. Stuttgart: Gustav Fischer Verlag.

Gabrielsen, T. M., & Brochmann, C. (1998). Sex after all: high levels of diversity detected in the arctic clonal plant *Saxifraga cernua* using RAPD markers. *Molecular Ecology*, **7**, 1701–1708.

Gough, L. (2006). Neighbor effects on germination, survival, and growth in two arctic tundra plant communities. *Ecography*, **29**, 44–56.

Grundt, H. H., Kjølner, S., Borgen, L., Rieseberg, L. H., & Brochmann, C. (2006). High biological species diversity in the arctic flora. *Proceedings of the National Academy of Sciences of the USA*, **103**, 972–975.

Guggisberg, A., Mansion, G., & Conti, E. (2009). Disentangling reticulate evolution in an arctic-alpine polyploid complex. *Systematic Biology*, **58**, 55–73.

Guggisberg, A., Mansion, G., Kelso, S., & Conti, E. (2006). Evolution of biogeographic patterns, ploidy levels, and breeding systems in a diploid-polyploid species complex of *Primula*. *New Phytologist*, **171**, 617–632.

Hewitt, G. M. (1996). Some genetic consequences of ice ages, and their role in divergence and speciation. *Biological Journal of the Linnean Society*, **58**, 247–276.

Hoffmann, M. H. (2012). Not across the North Pole: plant migration in the Arctic. *New Phytologist*, **193**, 474–480.

Hopkins, D. (1982). Aspects of the paleogeography of Beringia during the late Pleistocene. In D. M. Hopkins, J. V. Matthews, C. E. Schweger & S. B. Young (Eds.), *Paleoecology of Beringia* (pp. 3–28). New York: Academic Press.

Hultén, E. (1937). *Outline of the History of Arctic and Boreal Biota during the Quarternary Period*. New York: Lehre J. Cramer.

IPCC (2007). Core Writing Team, R. K. Pachauri & A. Reisinger (Eds.) *Climate Change 2007: Synthesis Report. Contribution of Working Groups I, II and III to the Fourth Assessment Report of the Intergovernmental Panel on Climate Change* (p. 104). Geneva: IPCC.

Jefferies, R. L., Jano, L. P., & Abraham, K. F. (2006). A biotic agent promotes large-scale catastrophic change in the coastal marshes of Hudson Bay. *Journal of Ecology* **94**, 234–242.

Jónsdóttir, I. S. (2005). Terrestrial ecosystems on Svalbard: heterogeneity, complexity and fragility from an arctic island perspective. *Biology and Environment: Proceedings of the Royal Irish Academy*, **105B**, 155–165.

Jónsdóttir, I. S. (2011). Diversity of plant life histories in the Arctic. *Preslia*, **83**, 281–300.

Jordon-Thaden, I., Hase, I., Al-Shehbaz, I., & Koch, M. A. (2010). Molecular phylogeny and systematics of the genus *Draba* (Brassicaceae) and identification of its most closely related genera. *Molecular Phylogenetics and Evolution*, **55**, 524–540.

Jørgensen, M. H., Elven, R., Tribsch, A., *et al.* (2006). Taxonomy and evolutionary relationships in the *Saxifraga rivularis* complex. *Systematic Botany*, **31**, 702–729.

Karlsen, S. R., & Elvebakk, A. (2003). A method using indicator plants to map local climatic variation in the Kangerlussuaq/Scoresby Sund area, East Greenland. *Journal of Biogeography*, **30**, 1469–1491.

Kjølner, S., Såstad, M., Taberlet, P., & Brochmann, C. (2004). Amplified fragment length polymorphism versus random amplified polymorphic DNA markers: clonal diversity in *Saxifraga cernua*. *Molecular Ecology*, **13**, 81–86.

Lamb, H. F., & Edwards, M. E. (1988). The Arctic. In B. Huntley, & T. Webb, III (Eds.), *Vegetation History. Handbook of Vegetation Science 7*. Dordrecht: Kluwer Academic.

Landvik, J. Y., Bondevik, S., Elverhøi, A., *et al.* (1998). The last glacial maximum of Svalbard and the Barents Sea area: ice sheet extent and configuration. *Quaternary Science Reviews*, **17**, 43–75.

Landvik, J. Y., Brook, E. J., Gualtieri, L., *et al.* (2003). Northwest Svalbard during the last glaciation: ice-free areas existed. *Geology*, **31**, 905–908.

Matthews, J. V., & Ovenden, L. E. (1990). Late Tertiary plant macrofossils from localities in Arctic/ Subarctic North America (Alaska, Yukon and Northwest Territories) – a review of the data. *Arctic*, **43**, 364–392.

Miller, G. H., Brigham-Grette, J., Alley, R. B., *et al.* (2010a). Temperature and precipitation history of the Arctic. *Quaternary Science Reviews*, **29**, 1679–1715.

Miller, G. H., Alley, R. B., Brigham-Grette, J., *et al.* (2010b). Arctic amplification: can the past constrain the future? *Quaternary Science Reviews*, **29**, 1779–1790.

Morin, X., Viner, D., & Chuine, I. (2008). Tree species range shifts at a continental scale: new predictive insights from a process-based model. *Journal of Ecology*, **96**, 784–794.

Müller, E., Cooper, E. J., & Alsos, I. G. (2011). Germinability of arctic plants is high in perceived optimal conditions but low in the field. *Botany*, **89**, 337–348.

Müller, E., Eidesen, P. B., Ehrich, D., & Alsos, I. G. (2012). Frequency of local, regional, and long-distance dispersal of diploid and tetraploid *Saxifraga oppositifolia* (Saxifragaceae) to arctic glacier forelands. *American Journal of Botany*, **99**, 459–471.

Murray, D. F. (1987). Breeding systems in the vascular flora of Arctic North America. In K. M. Urbanska (Ed.), *Differentiation Patterns in Higher Plants* (pp. 239–262). New York: Academic Press.

Murray, D. F. (1995). Causes of arctic plant diversity: origin and evolution. In F. S. Chapin III and C. Körner (Eds.), *Arctic and Alpine Biodiversity: Pattern, Causes and Ecosystem Consequences* (pp. 21–32). Berlin: Springer.

Myers-Smith, I. H., Forbes, B. C., Wilmking, M., *et al.* (2011). Shrub expansion in tundra ecosystems: dynamics, impacts and research priorities. *Environmental Research Letters*, **6**, 610–623.

Normand, S., Svenning, J.-C., & Skov, F. (2007). National and European perspectives on climate change sensitivity of the habitats directive characteristic plant species. *Journal of Nature Conservation*, **15**, 41–53.

Olofsson, J., Oksanen, L., Callaghan, T., *et al.* (2009). Herbivores inhibit climate-driven shrub expansion on the tundra. *Global Change Biology*, **15**, 2681–2693.

Popp, M., Mirré, V., & Brochmann, C. (2011). A single Mid-Pleistocene long-distance dispersal by a bird can explain the extreme bipolar disjunction in crowberries (*Empetrum*). *Proceedings of the National Academy of Sciences of the USA*, **108**, 6520–6525.

Post, E., Forchhammer, M. C., Bret-Harte, M. S., *et al.* (2009). Ecological dynamics across the Arctic associated with recent climate change. *Science*, **325**, 1355–1358.

Prach, K., Košnar, J., Klimešová, J., & Hais, M. (2010). High Arctic vegetation after 70 years: a repeated analysis from Svalbard. *Polar Biology*, **33**, 635–639.

Reusch, T. B. H., Ehlers, A., Hämmerli, A., & Worm, B. (2005). Ecosystem recovery after climatic extremes enhanced by genotypic diversity. *Proceedings of the National Academy of Sciences of the USA*, **102**, 2826–2831.

Schönswetter, P., Paun, O., Tribsch, A., & Niklfeld, H. (2003). Out of the Alps: colonization of Northern Europe by East Alpine populations of the Glacier Buttercup *Ranunculus glacialis* L. (Ranunculaceae). *Molecular Ecology*, **12**, 3373–3381.

Schönswetter, P., Popp, M., & Brochmann, C. (2006). Rare arctic-alpine plants of the European Alps have different immigration histories: the snowbed species *Minuartia biflora* and *Ranunculus pygmaeus*. *Molecular Ecology*, **15**, 709–720.

Skrede, I., Brochmann, C., Borgen, L., & Rieseberg, L. H. (2008). Genetics of intrinsic postzygotic isolation in a circumpolar plant species, *Draba nivalis* (Brassicaceae). *Evolution*, **62**, 1840–1851.

Skrede, I., Eidesen, P. B., Portela, R. P., & Brochmann, C. (2006). Refugia, differentiation and postglacial migration in arctic-alpine Eurasia, exemplified by the mountain avens (*Dryas octopetala* L.). *Molecular Ecology*, **15**, 1827–1840.

Solstad, H., Elven, R., Alm, T., *et al.* (2010). Karplanter: Pteridophyta, Pinophyta, Magnoliophyta. In J. A. Kålås, S. Henriksen, S. Skjelset and Å. Viken (Eds.), *Norsk Rødliste for Arter 2010 [The 2010 Norwegian red list for species]* (pp. 155–182). Trondheim: Artsdatabanken.

Sommer, J. H., Kreft, H., Kier, G., *et al.* (2010). Projected impacts of climate change on regional capacities for global plant species richness. *Proceedings of the Royal Society of London B*, **277**, 2271–2280.

Sønstebø, J. H., Gielly, L., Brysting, A. K., *et al.* (2010). Using next-generation sequencing for molecular reconstruction of past Arctic vegetation and climate. *Molecular Ecology Resources*, **10**, 1009–1018.

Stevens, G. (1989). The latitudinal gradient in geographical range: how so many species coexist in the tropics. *The American Naturalist*, **133**, 240–256.

Stewart, J. R., Lister, A. M., Barnes, I., & Dalén, L. (2010). Refugia revisited: individualistic responses of species in space and time. *Proceedings of the Royal Society of London B*, **277**, 661–671.

Walker, D. A., & Everett, K. R. (1987). Road dust and its environmental-impact on Alaskan taiga and tundra. *Arctic and Alpine Research*, **19**, 479–489.

Walker, D. A., Raynolds, M. K., Daniels, F. J. A., *et al.* (2005). The circumpolar arctic vegetation map. *Journal of Vegetation Science*, **16**, 267–282.

Walker, M. D., Wahren, C. H., Hollister, R. D., *et al.* (2006). Plant community responses to experimental warming across the tundra biome. *Proceedings of the National Academy of Sciences of the USA*, **103**, 1342–1346.

Westergaard, K. B., Alsos, I. G., Engelskjøn, T., Flatberg, K. I., & Brochmann, C. (2011). Trans-Atlantic genetic uniformity in the rare snowbed sedge *Carex rufina*. *Conservation Genetics*, **12**, 1367–1371.

Westergaard, K. B., Jørgensen, M. H., Gabrielsen, T. M., Alsos, I. G., & Brochmann, C. (2010). The extreme Beringian/Atlantic disjunction in *Saxifraga rivularis* (Saxifragaceae) has formed at least twice. *Journal of Biogeography*, **37**, 1262–1276.

Young, S. B. (1971). The vascular flora of St. Lawrence Island with special reference to floristic zonation in the arctic regions. *Gray Herbarium*, **201**, 11–104.

Part IV

Latitudinal Gradients

11 Latitudinal diversity gradients: equilibrium and nonequilibrium explanations

Klaus Rohde

General background, examples, exceptions

A central question in evolutionary ecology is: what are the reasons for differences in the abundance and diversity of organisms in different habitats and regions? Such differences are universal, i.e., it is highly unlikely that any two habitats will have the same number of species and organisms, and on a larger scale they are apparent between ecosystems and between latitudes, altitudes and different depths (in aquatic systems), as well as between different longitudes. By far the best-documented gradients are latitudinal ones, i.e., a very marked increase in diversity from high to low latitudes. An analysis of these gradients presents us with the opportunity to find a general explanation of the causes that determine diversity.

Latitudinal gradients in species diversity or better biodiversity were first described by Alexander von Humboldt in 1799 during his expedition in South America (Humboldt, 1808), and they have attracted the attention of numerous biologists since (for recent reviews and important papers see Rohde, 1992, 1999; Gaston, 2000; Willig 2001; Willig *et al.*, 2003; Hillebrand, 2004; Mittelbach *et al.*, 2007). They are known for most groups of plants and animals, although there are exceptions. A few examples are given in the following. According to Krebs (1985), the number of ant species at various localities is: Alaska 7 species, Utah 63, Cuba 101, and Brazil 222. Also according to Krebs, there are 22 species of snakes in Canada, 126 in the USA, and 293 in Mexico, and whereas a deciduous forest in Michigan, USA, contains 10–15 species on two hectares, a Malaysian tropical rainforest contains 227 in the same area. Wright (2002, references therein) gives even higher numbers for tropical rainforests: a $0.52 \, km^2$ plot in Borneo had 1175 species (with a diameter at breast height of at least 1 cm), and one hectare of Amazonian rainforests can have more than 280 species of trees in one hectare (with a diameter at breast height of at least 10 cm). In the oceans, latitudinal gradients are well documented for surface waters, but there is some evidence that they also occur in the deep sea (e.g., Lambshead *et al.*, 2002).

Among exceptions to the gradients are parasitoid hymenopterans on land and helminths (parasitic worms) of marine mammals, i.e., of cetaceans and pinnipeds, which have the greatest diversity outside the tropics. Also, many groups are composed of cold- and warm-adapted subgroups, the diversity of the former peaking at high latitudes. One can expect "typical" gradients therefore only if sufficiently large groups are considered.

Importantly, a gradient in species richness does not always lead to a gradient in community richness (Rohde & Heap, 1998).

Fossil evidence indicates that latitudinal diversity gradients have been in existence for long geological periods (Stehli *et al*., 1969; Crame, 2001), although their steepness underwent changes (e.g., Powell, 2007 for brachiopods in the late Paleozoic ice age, and Brayard *et al*., 2006 for Triassic ammonoids). Both latitudinal and longitudinal gradients in diversity have strongly increased through the Cenozoic, i.e., during the last 65 million years (Crame, 2001).

Explanations of latitudinal gradients in species diversity

There is still no consensus about the causes of latitudinal gradients in diversity. Numerous hypotheses have been proposed, and various authors have reviewed the evidence given for them (Pianka, 1966; Rohde, 1992, 1999, 2010; Gaston, 2000; Willig *et al*., 2003 ; Mittelbach *et al*., 2007). Gaston (2000), discussing global patterns in biodiversity in general including latitudinal gradients, writes that "no single mechanism adequately explains all examples of a given pattern", "that the patterns . . . may vary with spatial scale, that processes operating at regional scales influence patterns at local ones, and that the relative balance of causal mechanisms means that there will invariably be variations in and exceptions to any given pattern". Rohde (1992) discussed 28 hypotheses on mechanisms thought to be responsible for the gradients. He considered 12 of these to be circular (e.g., competition, predation), and 11 based on insufficient evidence (e.g., productivity, area); a further five are time hypotheses which assume that longer evolutionary time spans available for diversification in the tropics have led to greater diversity. Included among circular hypotheses listed by Rohde (1992) are those which give greater predation and/or competition in the tropics as an "explanation". That interactions are often greater at low latitudes has been shown by various authors (e.g., Schemske *et al*., 2009). However, it is likely that a greater number of species in the topics, whatever the reason for the greater diversity, will include a greater number of predators and competitors, but this may be a consequence and not an "explanation" of greater diversity, although conceivably increased predation and competition, once established, may accelerate the process of diversification. Any explanation of the gradients must incorporate a time component, because time is necessary to build up species numbers. However, there is no convincing evidence that the tropics have had a longer undisturbed history than other regions. A time element not dependent on longer undisturbed history in the tropics is an essential component of the hypothesis of effective evolutionary time (see below).

Many studies have shown that latitudinal gradients in temperature (or other measures of solar energy input, such as annual potential evapotranspiration, solar radiation, actual evapotranspiration) are best correlated with latitudinal diversity gradients of many taxa and in many habitats (e.g., Currie, 1991; Currie & Paquin, 1987; further references in

Rohde, 1992, 1999). For angiosperms, Francis and Currie (2003) demonstrated the primary importance of mean annual temperature (or annual potential evapotranspiration) and annual water deficit and their interaction for global diversity gradients.

It is likely that – in view of the generality of the gradients – some primary cause or causes must be involved, which – in view of the preceding paragraph – may be solar energy input (temperature) (e.g., Rohde, 1978a, 1978b, 1992, 1999).

Proceeding from this assumption, Rohde (1978a, 1978b, 1992), expanding on earlier work by Rensch, among others, proposed the hypothesis of "effective evolutionary time" (EET), according to which higher tropical diversity is determined by faster diversification rates directly caused by higher mutation rates, shorter generation times and faster speciation in warm environments, as well as by the time span under which ecosystems have existed under relatively undisturbed conditions (see below) (for forerunners see Rohde, 1978b, 1992 and 2009). However, several other factors, whose relative importance differs between taxa and habitats, contribute as well. Among the other important factors that have been suggested are area (e.g., Rosenzweig, 1995), heterogeneity of the habitats (e.g., Rahbeck & Graves, 2001), productivity (e.g., Rosenzweig, 1995), narrower niches in the tropics (e.g., Ben-Eliahu & Safriel, 1982), and midpoint of distribution (Collwell & Hurtt, 1994). The following examples show that at least some of these factors are indeed important.

Area

Terborgh (1973), Rosenzweig (1995) and Gorelick (2008), among others, suggested that the larger area in the tropics leads to greater species diversity. A study by Valdovinos *et al.* (2003) has indeed shown that area explains some of the great diversity of marine mollusks in the Southern Pacific. Analyzing the distribution of 629 mollusk species along the Pacific coast of South America, they demonstrated that richness increases sharply south of 42°S where shelf area also increases sharply (Figure 11.1A). This increase is not the result of artifacts produced by sampling in the larger area, but of an increased alpha diversity, as shown by the results of sampling in soft bottom habitats along the South American coast (Figure 11.1B). According to the authors, geographic isolation due to divergence of ocean currents and the formation of refuges during glaciations has facilitated diversification in this area. It should be stressed, however, that richness in the southernmost areas is much less than in the tropical zone, and on the Atlantic and Pacific shelves of North America the latitudinal gradient is more typical with lowest diversity at high latitudes (Roy *et al.*, 1998), although the peak of diversity is displaced somewhat to the north.

An effect of area has also been demonstrated for freshwater fish. There is a distinct increase in diversity of freshwater fish with area in the tropics, and in North America and Northern Eurasia. But as in the case of South American marine mollusks, diversity in warm waters is far greater than in the cold-temperate zones. An exception is Madagascar, a likely consequence of long isolation from a large continent (Rohde, 2005) (for further examples of the effects of area on diversity see Rosenzweig, 1995). In spite of these findings there are important reasons to reject the view that area can give a general

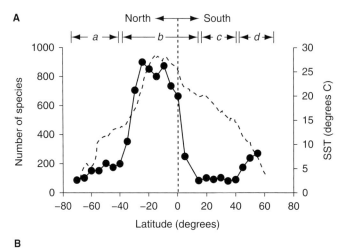

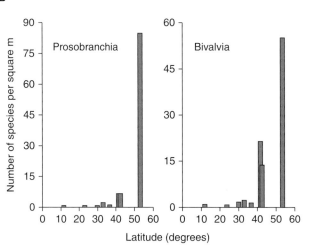

Figure 11.1.   (A) Latitudinal gradient in species diversity of molluscs on North and South American Pacific shelves. Points are selected in bands of 5° latitude. Mean latitudinal SST (sea surface temperature) indicated by segmented line. Note greatest diversity north of the equator, and a rise in diversity in the southernmost zone. (B) Local (alpha) diversity of prosobranch snails (left) and bivalves (right) in shallow soft bottom habitats. Note steep increase towards southern high latitudes. From Valdovinos, C., Navarete, S. A., & Marquet, P. A. (2003), with the permission of the authors and Wiley. Data for North American mollusks from Roy, K., Jablonski, D., & Valentine, J. W., *et al.* (1998).

explanation for latitudinal biodiversity gradients (Rohde, 1997, 1998). They include: (1) only in Africa are tropical land masses greater than cold-temperate ones; in Eurasia non-tropical land arcas are much larger than tropical ones; (2) the greatest diversity of many plant and animal groups is found in southeast Asia, which has a relatively small land area and small freshwater systems (e.g., freshwater fish; for details and references see Rohde, 1997, 1998); (3) the tropical Atlantic is not larger but smaller than the non-tropical Atlantic, but nevertheless has the greatest diversity of planktonic Foraminifera (reference in Rohde, 1997; further examples in Rohde, 1998).

Productivity

Rosenzweig and Sandlin (1997) explain the finding that some larger areas have fewer species than much smaller tropical ones by differences in productivity. However, even a cursory glance at some large-scale patterns shows that productivity is unlikely to be a major factor in determining latitudinal diversity gradients. For example, the large and highly productive Antarctic/Subantarctic Ocean has low diversity. In contrast, most tropical seas are not highly productive but possess very high species diversity (for details see Rohde, 1998). Waide *et al.* (1999), evaluating data from many plant and animal studies, have shown that the relationship between diversity and productivity is scale-dependent. At local and landscape scales (maximum 200 km), absence of or negative correlations are prevalent, at continental to global scales (more than 4000 km) there often are unimodal or positive relationships, but absence of or negative correlations occur as well. At regional scales (200–4000 km) an absence of or negative correlations are about as common as unimodal or positive relationships for plants, and somewhat more common for animals. Irigoien *et al.* (2004) examined the relationship between diversity and productivity (biomass) of plankton from different oceanic areas and found patterns consistent over all areas. Phytoplankton diversity is a unimodal function of its biomass; it has highest diversity at intermediate levels of biomass and lowest diversity during blooms. Zooplankton and phytoplankton diversity were not related, but diversity of zooplankton was found to be a unimodal function of its biomass. Rohde (1998) pointed out that increased productivity may be the result of greater diversity and not vice versa. For example, highly diverse coral reefs are very productive, but they have evolved in seas of low productivity (Rohde, 1998). And experiments have shown that an increase in biodiversity by a factor of two or three increased productivity by the same factor (Kareiva, 1994).

Spatial heterogeneity

Rahbeck and Graves (2001) provided a well-supported example of the effect of spatial heterogeneity on diversity. They analyzed the geographic ranges of 2869 bird species breeding in South America, that is, of almost one-third of the world's bird fauna. They considered the influence of climate, area, ecosystem diversity, and topography. Their analysis was conducted at 10 spatial scales (about 12 300 to 1 225 000 km^2 quadrat area). Regional variability in species richness correlated best with topography, precipitation, topography × latitude, ecosystem diversity, and cloud cover. Ranking of these factors depended on the scale used for analysis. Direct measures of ambient energy (mean and maximum temperature) were of secondary importance. Altogether, humid montane regions near the equator were most diverse. The authors conclude that a synergism between climate and coarse-scale topographic heterogeneity ultimately controls terrestrial diversity from the poles to the equator. It must not be forgotten, however, that temperature – although not giving good correlations with diversity as such – is the primary factor affecting climate.

There is no general trend of increasing heterogeneity towards low latitudes. Exceptions are the presence of a greater variety of habitats in tropical mountain ranges, because there we have tropical habitats at the base to cold habitats near the peaks (e.g., Körner, 2000). In South America the effect of spatial heterogeneity should be most marked, because the Andes extending along the continent provide an increasingly wide range of habitats towards the equator. Great complexity of coral reefs, often used as an example of the importance of heterogeneity for species richness, is the result of rather than the cause of great diversity, as pointed out by Rohde (1998, see above: productivity).

Narrower niches in the tropics

The view that greater species richness in the tropics leads to denser species packing is widespread (e.g., MacArthur & Wilson 1963; MacArthur, 1967, 1972). It is an underlying assumption of many of the hypotheses proposed to "explain" latitudinal diversity gradients. More generally, the assumption that greater species diversity must lead to denser species packing, i.e., narrower niches, is a central thesis of equilibrium ecology. An important aspect of a species' niche is its latitudinal range, studied by many authors in the context of Rapoport's rule. The rule claims that latitudinal ranges of organisms are generally smaller at low than at high latitudes and that this "explains" greater tropical diversity (Stevens, 1989). Evidence for the generality of the rule is controversial; it will not be discussed here (for a review see Rohde, 1999). Even where smaller latitudinal ranges in the tropics exist, they cannot give an "explanation" of increased diversity in the tropics: there is an alternative explanation: they could also be the result of faster evolution at low latitudes, which has led to closer species packing and hence smaller niches, including latitudinal ranges. Rohde (1998) suggested the existence of "two opposing trends: on the one hand, tropical species should have large latitudinal ranges because temperature conditions in the tropics are uniform over a much larger latitudinal range than at higher latitudes; on the other hand, newly evolved taxa (in particular subspecies, most of them in the tropics) that have little vagility and no dispersal stages, should have narrow ranges because they did not have the time to spread away from their place of origin even within the tropics". (In this context it is important to point out that Rapoport, after whom the rule is named, suggested it first for some tropical subspecies.) "Hence, Rapoport's rule should apply to recently evolved groups with little vagility, it should not apply to older and more vagile groups", but for neither case can it give an "explanation" of the latitudinal diversity gradient. Futhermore, Thorson's rule, according to which marine benthic invertebrates in the tropics generally have widely dispersing pelagic larvae, whereas benthic invertebrates at high latitudes often have larvae produced by viviparity, ovoviviparity and brooding, counteracts Rapoport effects at least in the oceans (see Rohde, 1989, 1992 for a discussion and references). That niche width is not generally narrower in the tropics was shown in the meta-analysis by Vázquez and Stevens (2004). For niche width of parasites see the review by Rohde (1989, references therein).

Mid-domain model

The mid-domain model was proposed by Colwell and Hurtt (1994) as an explanation of latitudinal diversity gradients. It assumes that species closer to the geographic midpoint of distribution can extend further away from the midpoint than species having their midpoint closer to some climatic or geographic boundary, leading to denser species packing around the midpoint. The hypothesis can be used as a null model which generates gradients in a completely stochastic manner. Colwell and Hurtt (1994) selected Madagascar as an ideal region to test this hypothesis, because it has clearly defined boundaries and a large endemic fauna, and indeed found support for the hypothesis. However, a study by Kerr *et al.* (2006) did not find support, a claim refuted by Lees and Colwell (2007). With regard to latitudinal gradients, it is clear that both terrestrial and marine regions of southeast Asia, centers of very high biodiversity, are not located in the center of a continent or an ocean. The mid-domain model can therefore not give an explanation for the very great diversity there. Nevertheless, the model may well explain some diversity patterns.

Time and area hypothesis and diversification rate hypothesis

Mittelbach *et al.* (2007) distinguish two major hypotheses, i.e., the "time and area" and the "diversification rate hypotheses". According to the former, tropical climates are older and larger, thereby permitting a higher degree of diversification. According to the latter, in the tropics diversification is faster for a number of reasons, which are: faster speciation due to more opportunities for reproductive isolation, faster molecular evolution and the increased importance of biotic interactions. Among the many mechanisms proposed to explain latitudinal variation in diversification rates are genetic drift, climate change, speciation types, area, physiological tolerances and dispersal limitation, evolutionary speed and biotic interactions. Rohde's (1978a, 1978b, 1992) hypothesis of effective evolutionary time, which will be discussed in the following, includes aspects of the time-area and diversification rate hypotheses in assuming direct effects of time spans under which systems have existed under more or less unchanged conditions, as well as direct effects of temperature on evolutionary speed.

Effective evolutionary time (EET)

Most hypotheses proposed to explain latitudinal gradients share the underlying assumption that habitats overall are saturated with species and that differences in species numbers should be explained by differences in the number of species that can be accommodated, determined by either area, productivity, spatial heterogeneity, climate, or a combination of these and possibly other factors. In other words, they are equilibrium explanations. Also, many authors have used positive correlations to find "explanations" of patterns, disregarding alternative explanations and ignoring historical events. For example, correlations between tree diversity and contemporary climate and, in particular, energy have been used to claim that explanations of tree diversity based on extant factors are sufficient and that historical processes or events are superfluous for explaining the patterns (references in McGlone, 1996). However, correlations cannot give an explanation of a pattern, although they may suggest an explanation. McGlone (1996) made this point when stating that diversity-energy correlations of tree diversity are strong only at regional scales, and cannot predict diversity at small plots within latitudinal bands, or between continents. Also, "tree diversity cannot have responded to global glacial-interglacial energy fluctuations because plant species cannot evolve that rapidly nor, in most areas of the world, can migration plausibly adjust floral diversity. Thus, contemporary climate or energy, while yielding excellent correlations with plant diversity, has no explanatory power". Contemporary climate merely acts as a surrogate for past climatic changes (see also Latham & Ricklefs, 1993; Francis & Currie, 1998). More generally, positive correlations between energy input (temperature) and diversity do not explain diversity patterns in the sense that temperature somehow limits diversity (see for example, the review by Hawkins *et al.*, 2003). Such correlations may also mean that temperature is involved in determining diversity via acceleration of evolutionary processes.

Based on these considerations, Rohde (1978a, 1978b, 1992, 1999) concluded that none of the explanations for latitudinal gradients in species diversity which assume saturation of habitats with species and equilibrium conditions, and higher "ceilings" of diversity in the tropics due to some limiting factor, is acceptable. Likewise, hypotheses based exclusively on evolutionary or ecological time do not give a satisfactory explanation, because tropical habitats are not generally older than cold-temperate ones; there were marked temperature changes in tropical seas in the geological past, possibly greater than in cold-temperate waters; and mass extinctions have occurred in tropical and cold-water environments (for discussion and references see Rohde, 1992). Rohde suggested that the best explanation for the gradients is that species saturation has not been reached, that higher energy input accelerates evolutionary speed by shortening generation times, increasing mutation rates and speeding up selection, and that evolutionary speed and the time under which communities have existed under relatively constant conditions (both determining the "effective evolutionary time") are responsible. In other words, more species have accumulated in the tropics because evolution there is faster. Essential evidence for this hypothesis is as follows:

(1) diversity has increased over evolutionary time;
(2) extant niche space is not filled, and an increase in diversity is likely (i.e., non-equilibrium conditions prevail);
(3) higher temperature increases speciation rates by (a) shortening generation times, (b) increasing mutation rates and (c) speeding up selection due to generally accelerated physiological processes (Q_{10});
(4) niches are not generally (or at least significantly) narrower in the tropics;
(5) when peaks in temperature are shifted away from the equator, diversity peaks are shifted as well.

There is evidence for all points.

(1) Marine benthic diversity, interrupted by several events of mass extinction, has markedly increased to the present (e.g., Jablonski, 1999; also Benton, 1995, 1998; Courtillot & Gaudemer, 1996; Jackson & Johnston, 2001; discussion in Rohde, 2005; review by Erwin, 2009, his figure 2). However, the degree of increase found depends to a large degree on the method used (see for example Alroy, 2010a). For instance, according to Alroy (2010a), who used standardization methods correcting

for sampling size, the "key factors" that determine diversity "are standing diversity and the dominance of onshore environments such as reefs. These factors combine to produce logistic growth patterns with slowly changing equilibrium values. There is no evidence of unregulated exponential growth across any long stretch of the Phanerozoic, and in particular there was no large Cenozoic radiation beyond the Eocene". And according to Alroy (2010b): "The fossil record demonstrates that each major taxonomic group has a consistent net rate of diversification and a limit to its species richness". Methods used by Alroy have been criticized on the basis that his (sampling-corrected) standardization approach gives less reliable estimates of diversity than directly interpreting the fossil record, as for example done by Hannisdal and Peters (2011), and the increase of diversity over time may be greater than found by him. According to Hannisdal and Peters (2011), "mutual responses to interacting Earth systems, not sampling biases, explain much of the observed covariation between Phanerozoic patterns of sedimentation and fossil biodiversity".

(2) Nonsaturation of habitats with species, i.e., the existence of vacant niches, has been demonstrated for many groups, including parasites, marine invertebrates, and insects (Rohde, 1998; Walker & Valentine, 1984; Lawton, 1984; discussions in Rohde, 1991, 2005; Chapter 6).

(3) That diversification rates (cladogenesis) are faster at low latitudes was for example shown by Storch *et al.* (2006), according to whom richness in the global distribution of birds is best correlated with range dynamics modulated by actual evapotranspiration; by Martin *et al.* (2007), according to whom latitudinal diversity gradients in marine invertebrates of 26 orders are best explained by the speed of cladogenesis followed by range expansion; and by Mullen *et al.* (2011), according to whom not clade age but higher diversification rates in the tropics explain species richness in a species-rich butterfly genus. Generation times are shorter for many groups of heterothermic and also some homeothermic animals (details in Rohde, 1993); increased mutation rates at low latitudes are discussed in the following chapter, "Effective evolutionary time and the latitudinal diversity gradient". It is well known that physiological processes are accelerated at higher temperatures. (For more detailed discussions see Rohde, 1992, 2005.)

(4) The meta-analysis by Vázquez and Stevens (2004) has shown that there is no evidence for generally narrower niches in the tropics and, even if future studies should demonstrate somewhat narrower niches in the tropics, it is not likely that they are many times narrower, as required for an explanation of the steep rise in species diversity near the equator.

(5) Shifts of maximum diversity are indeed often shifted in concordance with shifts in maximum temperatures (e.g., Figure 11.1A).

The hypothesis of effective evolutionary time incorporates elements of both the major hypotheses distinguished by Mittelbach *et al.* (2007), and is well in agreement with the metabolic theory of ecology (Brown *et al.*, 2004). Evidence that EET and not other evolutionary speed hypotheses give an explanation comes from molecular studies (see Chapter 12, "Effective evolutionary time and the latitudinal diversity gradient").

Wright *et al.* (2010), who provided molecular evidence for the hypothesis, have used the term "evolutionary speed hypothesis (ESH)" for it. I suggest not to use it, because it ignores the time component, which seems essential for any hypothesis that aims to explain not just latitudinal but for example also depth gradients, and because ESH does not distinguish between the hypothesis of effective evolutionary time and hypotheses that explain higher speciation speeds in the tropics by other mechanisms than direct temperature effects.

Acknowledgment

Some parts of this section are based on my earlier discussion in *Nonequilibrium Ecology*, Cambridge University Press, 2005. Used here with the permission of Cambridge University Press.

References

Alroy, J. (2010a). Geographical, environmental and intrinsic controls on Phanerozoic marine diversification. *Palaeontology*, **53**, 1211–1235.

Alroy, J. (2010b). The shifting balance of diversity among major marine animal groups. *Science*, **329**, 1191–1194.

Ben-Eliahu, M. N., & Safriel, U. N. (1982). A comparison between species diversity of polychaetes from tropical and temperate structurally similar rocky intertidal habitats. *Journal of Biogeography*, **9**, 371–390.

Benton, M. J. (1995). Diversification and extinction in the history of life. *Science*, **268**, 52–58.

Benton, M. J. (1998). Analyzing diversification through time: reply to Sepkoski and Miller. *Trends in Ecology & Evolution*, **13**, 201.

Brayard, A., Bucher, H., Escarguel, G., *et al.* (2006). The Early Triassic ammonoid recovery: paleoclimatic significance of diversity gradient. *Paleogeography, Paleoclimatology, Paleoecology*, **239**, 374–395.

Brown, J. H., Gillooly, J. F., Allen, A. P., Savage, Van M., & West, G. B. (2004). Toward a metabolic theory of ecology. *Ecology*, **85**, 1771–1789.

Colwell, R. K., & Hurtt, G. C. (1994). Nonbiological gradients in species richness and a spurious Rapoport effect. *The American Naturalist*, **144**, 570–59.

Courtillot, V., & Gaudemer, Y. (1996). Effects of mass extinctions on biodiversity. *Nature*, **381**, 146–148.

Crame, J. A. (2001). Taxonomic diversity gradients through geological time. *Diversity and Distributions*, **7**, 175–189.

Currie, D. J. (1991). Energy and large-scale patterns of animal and plant-species richness. *The American Naturalist*, **137**, 27–49.

Currie, D. J., & Paquin, V. (1987). Large scale biogeographical patterns of species richness of trees. *Nature*, **329**, 326–327.

Erwin, D. H. (2009). Climate as a driver of evolutionary change. *Current Biology*, **19**, R575–R583.

Francis, A. P., & Currie, D. J. (1998). Global patterns of tree species richness in moist forests: another look. *Oikos* **81**, 598–602.

Francis, A. P., & Currie, D. J. (2003). A globally consistent richness-climate relationship for angiosperms. *The American Naturalist*, **161**, 523–536.

Gaston, K. J. (2000). Global patterns in biodiversity. *Nature*, **405**, 220–227.

Gorelick, R. (2008). Species richness and the analytic geometry of latitudinal and altitudinal gradients. *Acta Biotheoretica*, **56**, 197–203.

Hannisdal, B., & Peters, S. E. (2011). Phanerozoic Earth system evolution and marine biodiversity. *Science*, **334**, 1121–1124.

Hawkins, B. A., Field, R., Cornell, H. V., *et al.* (2003). Energy, water, and broad-scale geographic patterns of species richness. *Ecology*, **84**, 3105–3117.

Hillebrand, H. (2004). On the generality of the latitudinal diversity gradient. *The American Naturalist*, **163**, 192–211.

Humboldt, A. von (1808). Ansichten der Natur mit wissenschaftlichen Erläuterungen, Tübingen (new edn. Eichborn, Frankfurt a. M. 2004).

Irigoien, X., Huisman, J., & Harris, R. P. (2004). Global biodiversity patterns of marine phytoplankton and zooplankton. *Nature*, **429**, 863–867.

Jablonski, D. (1999). The future of the fossil record. *Science*, **284**, 2114–2116.

Jackson, J. B. C., & Johnson, K. G. (2001). Measuring past biodiversity. *Science*, **293**, 2401–2404.

Kareiva, P. (1994). Diversity begets productivity. *Nature*, **368**, 686–687.

Kerr, J. T., Perring, M., & Currie, D. J. (2006). The missing Madagascan mid-domain effect. *Ecology Letters*, **9**, 149–159.

Körner, C. (2000). Why are there global gradients in species richness? Mountains might hold the answer. *TREE*, **15**, 513–514.

Krebs, C. J. (1985). *Ecology. The Experimental Analysis of Distribution and Abundance*. New York: Harper & Row.

Lambshead, P. J., Brown, D., Ferrero, C. J., *et al.* (2002). Latitudinal diversity patterns of deep-sea marine nematodes and organic fluxes: a test from the central equatorial Pacific. *Marine Ecology Progress Series*, **236**, 129–135.

Latham, R. E., & Ricklefs, R. E. (1993). Global patterns of tree species richness in moist forests: energy-diversity theory does not account for variation in tree species richness. *Oikos* **67**, 325–333.

Lawton, J. H. (1984). Non-competitive populations, non-convergent communities, and vacant niches: the herbivores of bracken. In D. R. Strong, Jr., D. Simberloff, L. G. Abele & A. B. Thistle (Eds.), *Ecological Communities: Conceptual Issues and the Evidence* (pp. 67–101). Princeton, NJ: Princeton University Press.

Lees, D. C., & Colwell, R. K. (2007). A strong Madagascan rainforest MDE and no equatorward increase in species richness: re-analysis of 'The missing Madagascan mid-domain effect', by Kerr J. T., Perring M., & Currie D. J (*Ecology Letters*, **9**, 149–159, 2006). *Ecology Letters*, **10**, E4–E8.

MacArthur, R. H. (1967). *The Theory of Island Biogeography*. Princeton, NJ: Princeton University Press.

MacArthur, R. H. (1972). *Geographic Ecology: Patterns in the Distribution of Species*. New York: Harper and Row.

MacArthur, R. H., & Wilson, E. O. (1963). An equilibrium theory of insular zoogeography. *Evolution*, **17**, 373–387.

Martin, P. R., Bonier, F., & Tewksbury, J. J. (2007). Revisiting Jablonski (1993): cladogenesis and range expansion explain latitudinal variation in taxonomic richness. *Journal of Evolutionary Biology*, **20**, 930–936.

McGlone, M. S. (1996). When history matters: scale, time, climate and tree diversity. *Global Ecology and Biogeography Letters*, **5**, 309–314.

Mittelbach, G. G., Schemske, D. W., Cornell, H. V., *et al.* (2007). Evolution and the latitudinal diversity gradient: speciation, extinction and biogeography. *Ecology Letters*, **10**, 315–331.

Mullen, S. P., Savage, W. K., Wahlberg, N., & Willmott, K. R. (2011). Rapid diversification and not clade age explains high diversity in neotropical Adelpha butterflies. *Proceedings of the Royal Society of London B*, **278**, 1777–1785.

Pianka, E. R. (1966). Latitudinal gradients in species diversity: a review of concepts. *The American Naturalist*, **100**, 33–46.

Powell, M. G. (2007). Latitudinal diversity gradients for brachiopod genera during late Paleozoic time: links between climate, biogeography and evolutionary rates. *Global Ecology and Biogeography*, **16**, 519–528.

Rahbeck, C., & Graves, G. R. (2001). Multiscale assessment of patterns of avian speciess richness. *Proceedings of the National Academy of Sciences of the USA*, **98**, 4534–4539.

Rohde, K. (1978a). Latitudinal differences in species diversity and their causes. I. A review of the hypotheses explaining the gradients. *Biologisches Zentralblatt*, **97**, 393–403.

Rohde, K. (1978b). Latitudinal gradients in species diversity and their causes. II. Marine parasitological evidence for a time hypothesis. *Biologisches Zentralblatt*, **97**, 405–418.

Rohde, K. (1989). Simple ecological systems, simple solutions to complex problems? *Evolutionary Theory*, **8**, 305–350.

Rohde, K. (1991). Intra- and interspecific interactions in low density populations in resource-rich habitats. *Oikos*, **60**, 91–104.

Rohde, K. (1992). Latitudinal gradients in species diversity: the search for the primary cause. *Oikos*, **65**, 514–527.

Rohde, K. (1993). *Ecology of Marine Parasites* (2nd edn). Wallingford: CAB International (Commonwealth Agricultural Bureau),

Rohde, K. (1997). The larger area in the tropics does not explain latitudinal gradients in species diversity. *Oikos*, **79**, 169–172.

Rohde, K. (1998). Latitudinal gradients in species diversity. Area matters, but how much? *Oikos*, **82**, 184–190.

Rohde, K. (1999). Latitudinal gradients in species diversity and Rapoport's rule revisited: a review of recent work, and what can parasites teach us about the causes of the gradients? *Ecography*, **22**, 593–613 (invited Minireview on the occasion of the 50th anniversary of the Nordic Ecological Society Oikos). Also published in T. Fenchel (Ed.), *Ecology 1999 – And Tomorrow* (pp. 73–93). University Lund: Oikos Editorial Office.

Rohde, K. (2005). *Nonequilibrium Ecology*. Cambridge: Cambridge University Press.

Rohde, K. (2009). *Effective Evolutionary Time*. Available at: http://krohde.wordpress.com/article/effective-evolutionary-time-xk923bc3gp4-11/[2009].

Rohde, K. (2010). Marine parasite diversity and environmental gradients. In S. Morand & B. Krasnoff (Eds.), *The Biogeography of Host-Parasite Interactions*. Oxford: Oxford University Press.

Rohde, K., & Heap, M. (1998). Latitudinal differences in species and community richness and in community structure of metazoan endo- and ectoparasites of marine teleost fish. *International Journal for Parasitology*, **28**, 461–474.

Rosenzweig, M. L. (1995). *Species Diversity in Space and Time*. Cambridge: Cambridge University Press.

Rosenzweig, M. L., & Sandlin, E. A. (1997). Species diversity and latitude: listening to area's signal. *Oikos*, **80**, 172–176.

Roy, K., Jablonski, D., & Valentine, J. W., *et al.* (1998). Marine latitudinal diversity gradients: tests of causal hypotheses. *Proceedings of the National Academy of Sciences of the USA*, **95**, 3699–3702.

Schemske, D. W., Mittelbach, G. G., Cornell, H. V., Sobel, J. M., & Roy, K. (2009). Is there a latitudinal gradient in the importance of biotic interactions? *Annual Review of Ecology, Evolution and Systematics*, **40**, 245–269.

Stehli, F. G., Douglas, D. G., & Newell, N. D. (1969). Generation and maintenance of gradients in taxonomic diversity. *Science*, **164**, 947–949.

Stevens, G. C. (1989). The latitudinal gradients in geographical range: how so many species co-exist in the tropics. *The American Naturalist*, **133**, 240–256.

Storch, D., Davies, R. G., Zajicek, S., *et al.* (2006). Energy, range dynamics and global species richness patterns: reconciling mid-domain effects and environmental determinants of avian diversity. *Ecology Letters*, **9**, 1308–1320.

Terborgh, J. (1973). On the notion of favourableness in plant ecology. *The American Naturalist*, **107**, 481–501.

Valdovinos, C., Navarete, S. A., & Marquet, P. A. (2003). Mollusk species diversity in the Southeastern Pacific: why are there more species towards the pole? *Ecography* **26**, 129–144.

Vázquez, D. P., & Stevens, R. D. (2004). The latitudinal gradient in niche breadth: concepts and evidence. *The American Naturalist*, **164**, E1–E19.

Waide, R. B., Willig, M. R., Steiner, C. F., *et al.* (1999). The relationship between productivity and species richness. *Annual Review of Ecology and Systematics*, **30**, 257–300.

Walker, T. D., & Valentine, J. W. (1984). Equilibrium models of evolutionary diversity and the number of empty niches. *The American Naturalist*, **124**, 887–899.

Willig, M. R. (2001). Latitude, common trends within. In S. Levin (Ed.), *Encyclopedia of Biodiversity* (vol. 3, pp. 701–714). New York: Academic Press.

Willig, M. R., Kaufman, D. M., & Stevens, R. D. (2003). Latitudinal gradients of biodiversity: pattern, process, scale, and synthesis. *Annual Review of Ecology, Evolution and Systematics*, **34**, 273–309.

Wright, S. J. (2002). Plant diversity in tropical forests: a review of mechanisms of plant coexistence. *Oecologia*, **130**, 1–14.

Wright S. D., Gillman L. N., Ross H. A., & Keeling D. J. (2010). Energy and tempo of evolution in amphibians. *Global Ecology and Biogeography*, **19**, 733–740.

12 Effective evolutionary time and the latitudinal diversity gradient

Len N. Gillman and Shane D. Wright

Introduction

The relationship between climate and biodiversity is perhaps the most widely recognized and extensively studied pattern of nature on earth. Attempts to explain this relationship and the attendant latitudinal gradient in diversity began more than 200 years ago (von Humboldt, 1808; Wallace, 1878) and indeed the number of theories that attempt to address this question appears to be accumulating at an ever increasing rate. However, one theory which has received relatively little attention began with the observation by Rensch (1959) that animals living in warmer tropical climates have shorter generation times than those living at higher latitudes. He suggested that because natural selection accumulates change with each generation, shorter generation times found among tropical fauna might increase the pace of natural selection and thereby the pace at which evolution progresses. A faster evolutionary speed in the tropics would therefore lead to the evolution of more species there than at higher latitudes over an equivalent period of time.

Evidence suggesting that mutations can be induced by high temperatures prompted Rohde (1978, 1992) to predict that not only might rates of selection increase with increasing ambient energy towards the tropics, but that rates of mutation may also be greater in lower-latitude climates. It is predicted that the combined effects of faster rates of mutation and faster rates of selection will lead to greater rates of diversification. Over an equivalent period of time, regions experiencing generally faster rates of genetic evolution will therefore generate and accumulate more species and greater species richness than regions where genetic evolution is slower. Fundamental to this hypothesis is the precept that the incumbent diversity of species within communities is not at an equilibrium number set by contemporary environmental conditions. Instead, it suggests that communities continue to accumulate species at rates that depend on climatic variables. This theory is therefore quite different to those, such as the energy-richness or more individuals hypothesis (Hutchinson, 1959; Brown, 1981; Wright, 1983), that suggest that diversity is limited by energetic capacity of the environment and that species origination and extinction are therefore held in balance with climate.

The Balance of Nature and Human Impact, ed. Klaus Rohde. Published by Cambridge University Press.
© Cambridge University Press 2013.

Rohde also proposed that the time over which speciation had been able to occur would have a direct bearing on the number of extant species. Thus, species richness is predicted to be dependent on both the rate of evolution and the time over which species have had to accumulate; he called this the effective evolutionary time hypothesis.

Testing for a thermal gradient in rates of genetic evolution among ectotherms

If ambient energy directly affects the rate of mutation then we would expect genetic evolution among ectotherms, such as plants and cold-blooded animals, to vary positively with ambient temperature, whereas among endotherms (birds and mammals) that maintain a relatively constant body temperature of 35–40°C while active, regardless of ambient temperature, the same relationship should not be evident (Allen *et al.*, 2006).

In 2003, Wright *et al.* examined 24 species of plants within Myrtaceae that differed in latitudinal distribution and found the first evidence that species occupying tropical climates had faster rates of genetic evolution than those occupying cooler climates at higher latitudes. They found that rates of genetic evolution in the ITS-ETS regions of rDNA of the tropical species were almost three times faster than those occupying temperate climates. Although this study lacked phylogenetically independent replication, confirmation of these results with well-replicated studies for a diversity of genes soon followed: replication across 86 angiosperm families (Davies *et al.*, 2004) and 45 phylogenetically independent pairs of gymnosperm and angiosperm tree species (Wright *et al.*, 2006; Gillman *et al.*, 2010).

The same pattern is apparent among ectothermic animals and microbes. Using 22 pairs of marine foraminifera that were dated to first appearance from fossil data, Allen *et al.* (2006) demonstrated an exponential increase in the rate of nuclear DNA evolution with increasing ocean temperature. Similarly, comparisons between 68 sister species pairs of teleost fish in which each member of the pair occurs at different depths or different latitudes revealed faster rates of evolution in both cytochrome *b* and ribosomal 12S and 16S genes for the species occurring in warmer waters (Wright *et al.*, 2011). Amphibians from 18 families of caudates and anurans (188 species) were also found to have faster rates of genetic evolution in the mitochondrial RNA genes, 12S and 16S, at both lower latitudes and lower elevations (Wright *et al.*, 2010).

Thus, empirical studies that have tested a range of nuclear and mitochondrial genes across a diversity of ectothermic taxa support the hypothesis that rates of genetic evolution are faster within species that occupy warmer habitats. Furthermore, the association between ambient temperature and rates of evolution comes from thermal dimensions that include both elevation and latitude in terrestrial environments and both depth and latitude in the marine environment. Although many variables vary with latitude, elevation and depth, temperature is the only one that varies consistently in the same way with all three. Taken together these results suggest a general link between thermal energy and the tempo of genetic evolution within ectotherms.

Plausible explanations

1. Metabolic rate

It has been suggested that metabolic rate influences mutagenesis within species either via the rate of cell division and consequential replication error in the germline, or via the rate of DNA damage due to the production of oxygen free-radicals (Martin & Palumbi, 1993). Metabolic rates increase with temperature towards the equator among ectotherms consistent with faster rates of genetic evolution in warmer environments (Allen *et al.*, 2006). However, body temperatures and metabolic rates in endotherms increase with latitude rather than decreasing (Anderson & Jetz, 2005). For example, a mammal living at $-10°C$ at a high polar latitude has a body temperature approximately $2.7°C$ warmer and a basal metabolic rate (BMR) approximately 40% higher than a tropical mammal of similar size living at $25°C$ (Clarke *et al.*, 2010). Thus, a positive association between metabolic rate and genetic evolution would predict microevolution for birds and mammals, in contrast to ectotherms, to be faster at higher latitudes, not slower.

Two studies that have tested for latitudinal variation in the rate of genetic evolution among birds failed to find statistically significant results (Bromham & Cardillo, 2003; Weir & Schluter, 2008), whereas two others involving endotherms report significant associations. The largest study involving endotherms was one using 131 independent sister pairs of mammal species from 10 orders and 29 families (Gillman *et al.*, 2009). In this study, rates of genetic evolution were found to be independently faster for species at both lower latitudes and lower elevations. The second of these, using 30 phylogenetically independent bird species pairs, also found faster rates in warmer environments (Gillman *et al.*, 2012). The latter two studies therefore show the opposite trend to that expected if BMR had a positive influence on rates of genetic evolution.

Early studies with limited data that tested for a direct relationship between BMR and rates of genetic evolution found positive correlations (Martin & Palumbi, 1993; Bleiweiss, 1998). However, studies using much larger data sets of mammals (61 species from 14 orders, Bromham *et al.*, 1996) and more generally of metazoans (> 300 species for 12 different genes, Lanfear *et al.*, 2007) have found no support for such a relationship. Basal metabolic rate would not, therefore, appear to be instrumental in controlling rates of genetic evolution. However, average annual metabolic rates may nonetheless be important (see discussion below).

2. Body mass

Average body mass has been found to correlate positively with latitude and inversely with temperature (Ashton *et al.*, 2000). Body mass also correlates inversely with rates of genetic evolution (Martin & Palumbi, 1993; Nunn & Stanley, 1998; Gillooly *et al.*, 2005) and therefore the slower rates of evolution reported in cooler climates towards the poles might be due to the tendency for average body size to increase with latitude. However, rates of genetic evolution were also faster for subsets of warmer climate species (occurring at a lower latitude, elevation or depth) of mammals and fishes that are heavier than their cold climate sisters (Gillman *et al.*, 2009; Wright *et al.*, 2011). The general pattern of faster rates of evolution for warmer climate mammal species was also weaker amongst a data subset in which the cooler climate species was larger. Similarly, in a study involving birds, body size was only weakly related to rates of evolution, whereas latitude and elevation show a strong relationship (Gillman *et al.*, 2012). Furthermore, Cooper and Purvis (2009) found strong positive, rather than negative, correlations between body size and phenotypic evolution. Body size asymmetries cannot therefore explain associations between thermal environment and rates of genetic evolution.

3.   Generation time

Simpson (1953) suggested that, because the frequency of genetic replication error increases with the frequency of reproduction, species that have shorter generation times might be expected to have more mutations occurring over any given time period. More mutations occurring per unit time is then posited to result in faster rates of evolution. This hypothesis assumes that, for multicellular sexual organisms, mutations are either more likely during sexual reproduction, and in particular during meiosis, than during germ-line mitotic cell divisions, or that the total number of germ-line cell divisions does not increase commensurately with generation time. It has also been proposed that generation time correlates positively with latitude. Therefore, shorter generation times at lower latitudes might explain faster rates of genetic evolution and diversification in the tropics (Rohde, 1992).

Inverse correlations between genetic evolution and generation time have been found among some groups of invertebrates and angiosperms (Smith & Donoghue, 2008; Thomas *et al.*, 2010). This relationship is apparent for both synonymous and non-synonymous substitutions. Synonymous mutations are assumed to have little effect on phenotypic evolution and therefore are unlikely to be implicated in speciation and diversification, whereas non-synonymous mutations have a direct bearing on protein synthesis and phenotypic evolution. Thus, generation time provides a plausible explanation for the relationships found between latitude, genetic evolution, and diversification for some ectotherms.

However, the influence of generation time on diversification via genetic evolution among mammals is much less clear because, although negative relationships have been observed between generation time and synonymous substitutions within both nuclear and mitochondrial DNA, no correlations with non-synonymous substitutions have been found (Bromham *et al.*, 1996; Nikolaev *et al.*, 2007; Nabholz *et al.*, 2008; Welch *et al.*, 2008). If speciation is dependent on genetic evolution, it will largely involve non-synonymous protein-altering substitutions. Generation time is therefore unlikely to influence diversification in mammals and nor can it be invoked to account for the negative association between non-synonymous genetic change and latitude reported for mammals by Gillman *et al.* (2009). Therefore, generation time does not appear to be implicated in putative latitudinal effects on rates of protein evolution and diversification among mammals.

4.   Annual metabolic activity

Basal metabolic rate may not reflect long-term metabolic activity and yet if metabolic activity has an effect on mutation rates it will not be the basal or resting rate that will be important. Instead, maximum or long-term average metabolic rates are more likely to be causally linked to mutations. Therefore, total metabolic activity measured over a full 12-month cycle may yet prove to be associated with rates of evolution (Gillman et al., 2009). In cooler environments, periods of hibernation or torpor that conserve energy in response to low energetic supply (McKechnie & Lovegrove, 2002; Munro et al., 2005) may reduce total metabolic activity over full annual cycles. It is therefore possible that mutagenesis, and therefore rates of genetic evolution, among birds and mammals might reflect metabolic averages that are reduced in cooler less productive environments. Such a hypothesis may be testable in the future if real-time field monitoring of metabolic activity becomes practicable.

5.   UV radiation

Ultraviolet radiation is known to induce mutations and has been proposed as a possible mechanism inversely linking mutation rate with latitude (Rohde, 1992). Of all the environmental variables tested by Davies et al. (2004), UV radiation correlated with rates of genetic evolution among angiosperms most strongly. However, the latitudinal diversity gradient applies equally well to taxa largely shielded from UV radiation such as forest-floor plants and animals and marine animals as it does for those taxa exposed fully to UV radiation such as canopy trees (Hillebrand, 2004), and UV radiation generally increases with elevation and with aridity, whereas species richness tends to decline along these gradients. Furthermore, rates of genetic evolution have been shown to be slower among plants in more arid environments where UV radiation is higher (Goldie et al., 2010) and to be slower among mammals and amphibians at higher elevations where UV radiation can also be higher (Gillman et al., 2009; Wright et al., 2010).

6.   The Red Queen hypothesis

A third explanation for an association between climate and rates of genetic evolution among both ectotherms and endotherms is that metabolic rates may influence the rate of genetic evolution of ectotherms directly via a metabolic effect, whereas the rate of evolution among endotherms in the same community may depend on the rate of evolution among co-evolving ectotherms via a Red Queen effect (VanValen, 1973). That is, if the rate of evolution of a given species in a particular ecosystem is dependent on the rate of evolution among other species within that community with which it is co-evolving, the rate of genetic evolution among endotherms might be linked to the corresponding rate of evolution among ectotherms within the same community (Rohde, 1992; Gillman et al., 2009).

Given that rates of genetic evolution among ectotherms are more rapid in warmer locations it is possible that the biotic environment engineered by ectotherms may also be more dynamic in warmer locations. The probability of a mutation possessing a positive selection coefficient may be greater if it occurs within an endothermic population living among more rapidly changing ectotherms, than if that same mutation were to occur within a population of endotherms living in a cooler more static ectothermic milieu (Gillman *et al.*, 2009). For example, if a mutation in a warm climate amphibian produced a defense against a mammalian predator, a subsequent mutation within a mammalian predator species that overcame this defense would be rapidly selected and fixed within the mammal species population. However, if the same "anti-defense" mutation were to occur in a mammalian predator occupying a cooler environment, where the amphibians being predated had not produced the relevant "defensive" mutation – because they were evolving their defenses more slowly – the corresponding anti-defense mutation would not be selected and fixed in the predator population.

Therefore, endotherms living within a community in which the ectotherms with which they interact are evolving more rapidly might also be evolving more rapidly. The rate of fixation of novel alleles in endotherms may, on average, be faster in warmer climates due to a greater proportion of mutations possessing a selective advantage.

If it is assumed that the rate of synonymous fixation is dependent on the rate of mutation, then a Red Queen effect would be evidenced by an increase in non-synonymous evolution due to positive selection without a corresponding increase in the mutation rate, or rate of synonymous evolution. Therefore, the ratio of dN/dS should be elevated in warm climate species relative to cold climate species if there has been a Red Queen effect. However, there was no such elevation in dN/dS detected among the warm climate mammals (Gillman *et al.*, 2009).

A "Red Queen" dynamic equilibrium may exist within climate zones in terms of the pace of evolution. However, there is no empirical evidence to support the hypothesis that elevated selection due to a Red Queen effect occurs among mammals in warmer climates. By contrast, there is a body of empirical evidence strongly suggesting that the pace of evolution varies across thermal differentials. A balance in nature may exist with similar rates of evolution occurring among taxa within communities experiencing a uniform climatic environment, but the pace of evolution appears to vary among taxa in a systematic manner across climatic zones.

Cause or effect

The effective evolutionary time hypothesis predicts that faster rates of genetic evolution at lower, warmer, latitudes produces faster rates of diversification and ultimately higher diversity at these latitudes. This is perhaps a parsimonious concept given that adaptation appears to be limited by the supply of novel mutations and not just by standing genetic variation (Gossmann *et al.*, 2012).

Using a selected angiosperm species from each of 86 sister families, Davies *et al.* (2004) found relationships between environmental energy and species richness, energy and rates of genetic evolution, and genetic evolution and species richness. However, genetic evolution dropped out of the regression model as a significant predictor variable for species richness when the model was simplified. Davies *et al.* suggest this indicates that the influence of energy on species richness is independent of the rate of genetic evolution. However, the rate of genetic evolution within each sister family in that study was represented by one species and therefore it is unclear whether or not genetic evolution within a single species from each family is sufficiently free from error to adequately characterize the predictor variable.

The alternative hypothesis to the effective evolutionary time hypothesis, under nearly neutral theory, is that faster rates of diversification cause an increase in the rate of genetic evolution (Cardillo, 1999; Pagel *et al.*, 2006). Nearly neutral theory posits that small populations will accumulate a greater number of mildly deleterious mutations over a given time than a larger population (Ohta, 1992). This might occur because in small populations purifying selection that eliminates such mutations will be less efficient. Therefore, a greater number of mildly deleterious mutations will drift to fixation, resulting in an overall faster rate of genetic evolution, within smaller populations. It is assumed that during speciation population sizes are small, causing a spike in the rate of genetic evolution, and therefore where diversification rates have been high, such as at low latitudes, rates of genetic evolution are posited to be commensurately high.

As far as we are aware there have been no studies that have tested the assumption that population sizes are substantially reduced during speciation or that, if such reductions do occur, populations remain small for long enough on an evolutionary timescale for this to significantly affect rates of genetic evolution. There are also surprisingly few studies, using adequate sample replication and controls for confounding factors, which have tested for an influence of population size on genetic evolution. Most studies that have tested for population size effects have used comparisons that have suffered from a lack of phylogenetic independence and therefore have an effective sample size of one to three (Woolfit & Bromham, 2005). In many cases the compared lineages, such as rodents versus primates, differ substantially in their biology, thereby potentially confounding the influence due to population size on rates of genetic evolution with variables such as body weight, metabolic rate, temperature, generation time, and DNA repair mechanisms (Gillooly *et al.*, 2005; Wright *et al.*, 2009). Furthermore, justification of the assumptions about relative population sizes inherent in examinations of evolutionary rate and population size has usually been lacking.

The first well-replicated study of population size effects used 70 phylogenetically independent contrasts between island and mainland taxa (Woolfit & Bromham, 2005). However, this study did not find a statistically significant difference in rates of genetic evolution between large and small populations. This result may have been due to the many contrasts in the study that were only related at familial or ordinal levels. Such distantly related comparisons introduce the likelihood of ancestral population-size variation such that it is unclear whether the majority of genetic evolution that has been measured has occurred in large or small populations. Wright *et al.* (2009) addressed this issue by using 48 sister species comparisons of birds where one of each species occurred on a small island and the other on a landmass at least five times larger. However, this study reported that smaller populations had slower rates of genetic evolution, not faster rates as predicted by nearly neutral theory.

If greater rates of diversification at lower latitudes were responsible for the observed pattern of faster rates of genetic evolution at such low latitudes under the nearly neutral model, we would expect genera that have temperate diversity that is greater than tropical diversity to show a reverse pattern in rates of genetic evolution. Wright *et al.* (2006) tested this prediction using a data subset in which plant genera were more speciose at high latitudes. However, among this data subset rates of genetic evolution remained significantly faster in the tropical species. This finding therefore does not support the nearly neutral hypothesis; greater rates of cladogenesis are not the cause of accelerated genetic evolution in warmer climates.

A related hypothesis, again invoking nearly neutral theory, is that the greater number of species per unit area in warmer latitudes results in smaller average population sizes and again the smaller populations are posited to evolve faster (Stevens, 1989). This hypothesis might be able to potentially explain faster rates of genetic evolution in warmer latitudes, but faster rates of genetic evolution have also been found among mammals, amphibians and birds at lower, warmer, elevations and population sizes tend to be larger at lower elevations, not smaller (e.g., Patterson *et al.*, 1989).

Nearly neutral theory predicts that smaller populations will evolve faster than large ones due to an increase in the number of mildly deleterious mutations moving to fixation in the smaller populations (Ohta, 1992). Deleterious mutations will in most cases be limited to those mutations that affect protein synthesis (i.e., they are non-synonymous). By contrast, fixation of synonymous mutations is not predicted to be influenced by population size. Therefore, nearly neutral theory also predicts that smaller populations will have a greater non-synonymous to synonymous substitution ratio (dN/dS) than larger populations. If faster rates of genetic evolution in warmer lower latitude or elevation species are due to nearly neutral effects related to smaller populations, then we would expect the ratio of dN to dS to be elevated in the warmer climate species. This prediction was tested for mammals, amphibians and fish and in all cases, despite faster overall rates of genetic evolution, there was no evidence of elevated dN/dS in the warmer climate species (Gillman *et al.*, 2009; Wright *et al.*, 2010, 2011).

Similarly, if correlations between speciation and rates of genetic evolution, which have been found for a range of taxa (e.g., Webster *et al.*, 2003; Smith & Donoghue, 2008), are due to nearly neutral effects of small populations persisting following speciation, then faster-evolving lineages in putatively small populations will have a higher ratio of dN/dS than those evolving more slowly in larger populations. Lanfear *et al.* (2010) found a significant correlation between bird diversity and rates of genetic evolution within clades. However, they found no evidence of elevated dN/dS in faster-evolving lineages and therefore concluded that the correlation was not due to nearly neutral effects. By contrast, Goldie *et al.* (2011) failed to find a correlation between rates of mammal evolution and speciation.

A different approach to this question was taken by Lancaster (2010). She reasoned that if small population sizes during speciation cause an increase in rates of genetic evolution, then substitution rate heterogeneity should also be apparent among clades, with shorter branches (that have been in the process of speciation for a greater proportion of their span) exhibiting higher rates of genetic evolution than longer branches. Using aged lineages, Lancaster (2010) found a positive correlation between substitution rates and rates of diversification among 13 clades of angiosperms, but failed to find a correlation between substitution rate variation and diversification rate. She therefore rejected the hypothesis that speciation had caused an increase in substitution rate.

The balance of evidence therefore suggests that greater rates of speciation are unlikely to be the cause of the higher tempo of genetic evolution at lower latitudes. However, the Goldie *et al.* (2011) result for mammals is difficult to reconcile with the effective evolutionary speed hypothesis, unless higher rates of extinction have been occurring in more rapidly evolving and diversifying clades involved in their study. There is therefore a need for more testing of the hypothesis that rates of genetic evolution influence rates of diversification.

Influence of climate change on rates of evolution and diversification

If rates of genetic evolution and diversification are underpinned by available energy either via annual metabolic activity or via a direct metabolic influence on ectotherm evolution and an indirect "Red Queen" effect among endotherms, an increase in global temperatures would suggest faster rates of contemporary evolution and diversification. There is evidence of rapid evolution over contemporary timeframes (Stockwell *et al.*, 2003). However, counteracting any putative effect due to faster evolution will be the disruptive influence of climate change on species and their habitats and the consequential enhanced extinction rate. If we also consider elevated rates of extinction due to other anthropogenic influences such as overharvesting, pollution and habitat destruction (e.g., Brook *et al.*, 2003), it is hard to imagine how a process of diversification which operates over millions of years could possibly compensate for the contemporary extinction event that is occurring within a time frame of decades.

Conclusion

There is a substantial body of empirical evidence showing that rates of genetic evolution are elevated in species occurring in warmer environments at lower latitudes, lower elevations and in shallower waters. By contrast, attempts to understand what factors might be responsible for this relationship have not been able to identify a likely mechanism: there is either no evidence in support of putative mechanisms or contradictory evidence. Additionally, although the weight of evidence indicates that nearly neutral effects do not generate the association between rates of genetic evolution and rates of diversification, there remains a lack of unequivocal evidence demonstrating that rates of diversification are driven by genetic evolution.

If the effective evolutionary time hypothesis (or evolutionary speed hypothesis) is valid then we might expect enhanced rates of genetic evolution across the planet as global temperatures increase. Given enough time this might eventually lead to an increase in speciation. Counteracting and overwhelming this effect, however, will be the exceptionally high rates of human-driven contemporary extinctions.

References

Allen, A., Gillooly, J., Savage, V., & Brown, J. (2006). Kinetic effects of temperature on rates of genetic divergence and speciation. *Proceedings of the National Academy of Sciences of the USA*, **103**, 9130–9135.

Anderson, K. J., & Jetz, W. (2005). The broad-scale ecology of energy expenditure of endotherms. *Ecology Letters*, **8**, 310–318.

Ashton, K. G., Tracy, M. C., & Dequeiroz, A. (2000). Is Bergmann's rule valid for mammals. *The American Naturalist*, **156**, 390–415.

Bleiweiss, R. (1998). Relative-rate tests and biological causes of molecular evolution in humming-birds. *Molecular Biology and Evolution*, **15**, 481–491.

Bromham, L., & Cardillo, M. (2003). Testing the link between the latitudinal gradient in species richness and rates of molecular evolution. *Journal of Evolutionary Biology*, **16**, 200–207.

Bromham, L., Rambaut, A., & Harvey, P. H. (1996). Determinants of rate variation in mammalian DNA sequence evolution. *Journal of Molecular Evolution*, **43**, 610–621.

Brook, B. W., Sodhi, N. S., & Ng, P. K. L. (2003). Catastrophic extinctions follow deforestation in Singapore. *Nature*, **424**, 420–423.

Brown, J. H. (1981). Two decades of homage to Santa Rosalia: toward a general theory of diversity. *American Zoology*, **21**, 877–888.

Cardillo, M. (1999). Latitude and rates of diversification in birds and butterflies. *Proceedings of the Royal Society of London B*, **266**, 1221–1225.

Clarke, A., Rothery, P., & Isaac, N. J. (2010). Scaling of basal metabolic rate with body mass and temperature in mammals. *Journal of Animal Ecology*, **79**, 610–619.

Cooper, N., & Purvis, A. (2009). What factors shape rates of phenotypic evolution? A comparative study of cranial morphology of four mammalian clades. *Journal of Evolutionary Biology*, **22**, 1024–1035.

Davies, T., Savolainen, V., Chase, M., Moat, J., & Barraclough, T. (2004). Environmental energy and evolutionary rates in flowering plants. *Proceedings of the Royal Society of London B*, **271**, 2195–2200.

Gillman, L. N., Ross, H. A., Keeling, J. D., & Wright, S. D. (2009). Latitude, elevation and the tempo of molecular evolution in mammals. *Proceedings of the Royal Society of London B*, **276**, 3353–3359.

Gillman, L. N., Keeling, D. J., Gardner, R. C., & Wright, S. D. (2010). Faster evolution of highly conserved DNA in tropical plants. *Journal of Evolutionary Biology*, **23**, 1327–1330.

Gillman, L. N., McCowan, L., & Wright, S. D. (2012). The tempo of genetic evolution in birds: body mass, population size and climate effects. *Journal of Biogeography*, **39**, 1567–1572.

Gillooly, J. F., Allen, A. P., West, G. B., & Brown, J. H. (2005). The rate of DNA evolution: effects of body size and temperature on the molecular clock. *Proceedings of the National Academy of Sciences of the USA*, **102**, 140–145.

Goldie, X., Gillman, L. N., Crisp, M., & Wright, S. D. (2010). Evolutionary speed limited by water in arid Australia. *Proceedings of the Royal Society of London B*, **277**, 2645–2653.

Goldie, X., Lanfear, R., & Bromham, L. (2011). Diversification and the rate of molecular evolution: no evidence of a link in mammals. *BMC Evolutionary Biology*, **11**, 1471–2148.

Gossmann, T. I., Keightley, P. D., & Eyre-Walker, A. (2012). The effect of variation in the effective population size on the rate of adaptive molecular evolution in eukaryotes. *Genome Biology and Evolution*, **4**, 658–667.

Hillebrand, H. (2004). On the generality of the latitudinal diversity gradient. *The American Naturalist*, **163**, 192–211.

Hutchinson, G. E. (1959). Homage to Santa Rosalia or why are there so many kinds of animals? *American Naturalist*, **93**, 145–159.

Lancaster, L. T. (2010). Molecular evolutionary rates predict extinction and speciation in temperate angiosperm lineages. *BMC Evolutionary Biology*, **10**, 162.

Lanfear, R., Ho, S. Y. W., Love, D., & Bromham, L. (2010). Mutation rate is linked to diversification in birds. *Proceedings of the National Academy of Sciences of the USA*, **107**, 20423–20428.

Lanfear, R., Thomas, J. A., Welch, J. J., Brey, T., & Bromham, L. (2007). Metabolic rate does not calibrate the molecular clock. *Proceedings of the National Academy of Sciences of the USA*, **104**, 15388–15393.

Martin, A. P., & Palumbi, S. R. (1993). Body size, metabolic rate, generation time, and the molecular clock. *Proceedings of the National Academy of Sciences of the USA*, **90**, 4087–4091.

Mckechnie, A. E., & Lovegrove, B. G. (2002). Avian facultative hypothermic responses: a review. *Condor*, **104**, 705–724.

Munro, D., Thomas, D. W., & Humphries, M. M. (2005). Torpor patterns of hibernating eastern chipmunks *Tamias striatus* vary in response to the size and fatty acid composition of food hoards. *Journal of Animal Ecology*, **74**, 692–700.

Nabholz, B., Glemin, S., & Galtier, N. (2008). Strong variations of mitochondrial mutation rate across mammals – the longevity hypothesis. *Molecular Biology and Evolution*, **25**, 120–130.

Nikolaev, S. I., Montoya-Burgos, J. I., Popadin, K., Parand, L., & Margulies, E. H. (2007). Life-history traits drive the evolutionary rates of mammalian coding and noncoding genomic elements. *Proceedings of the National Academy of Sciences of the USA*, **104**, 20443–20448.

Nunn, G., & Stanley, S. (1998). Body size effects and rates of cytochrome b evolution in tube-nosed seabirds. *Molecular Biology and Evolution*, **15**, 1360–1371.

Ohta, T. (1992). The nearly neutral theory of molecular evolution. *Annual Review of Ecology and Systematics*, **23**, 263–286.

Pagel, M., Venditti, C., & Meade, A. (2006). Large punctuational contribution of speciation to evolutionary divergence at the molecular level. *Science* **314**, 119–121.

Patterson, B. D., Meserve, P. L., & Lang, B. K. (1989). Distribution and abundance of small mammals along an elevational transect in temperate rainforests of Chile. *Journal of Mammalogy*, **70**, 67–78.

Rensch, B. (1959). *Evolution Above the Species Level*. London: Methuen.

Rohde, K. (1978). Latitudinal gradients in species diversity and their causes. I. A review of the hypotheses explaining the gradients. *Biologisches Zentralblatt*, **97**, 393–403.

Rohde, K. (1992). Latitudinal gradients in species diversity: the search for the primary cause. *Oikos*, **65**, 514–527.

Simpson, G. G. (1953). *The Major Features of Evolution*. New York: Columbia University Press.

Smith, S. A., & Donoghue, M. J. (2008). Rates of molecular evolution are linked to life history in flowering plants. *Science,* **322**, 86–89

Stevens, G. C. (1989). The latitudinal gradient in geographical range: how so many species coexist in the tropics. *The American Naturalist*, **133**, 240–256.

Stockwell, C. A., Hendry, A. P., & Kinnison, M. T. (2003). Contemporary evolution meets conservation biology. *Trends in Ecology & Evolution*, **18**, 94–101.

Thomas, J. A., Welch, J. J., Lanfear, R., & Bromham, L. (2010). A generation time effect on the rate of molecular evolution in invertebrates. *Molecular Biology and Evolution*, **27**, 1173–1180.

VanValen, L. M. (1973). A new evolutionary law. *Evolutionary Theory*, **1**, 1–30.

Von Humboldt, A. (1808). *Ansichten der Natur mit wissenschaftlichen Erlauterungen*. Tübingen.

Wallace, A. R. (1878). *Tropical Nature and Other Essays*. London: Macmillan.

Webster, A. J., Payne, R. J. H., & Pagel, M. (2003). Molecular phylogenies link rates of evolution and speciation. *Science*, **301**, 478.

Weir, J. T., & Schluter, D. (2008). Calibrating the avian molecular clock. *Molecular Ecology*, **17**, 2321–2328.

Welch, J. J., Bininda-Emonds, O. R., & Bromham, L. (2008). Correlates of substitution rate variation in mammalian protein-coding sequences. *BMC Evolutionary Biology*, **8**, 1471–2148.

Woolfit, M., & Bromham, L. (2005). Population size and molecular evolution on islands. *Proceedings of the Royal Society of London B*, **272**, 2277–2282.

Wright, D. H. (1983). Species-energy theory: an extension of species-area theory. *Oikos*, **41**, 496–506.

Wright, S. D., Gray, R. D., & Gardner, R. C. (2003). Energy and the rate of evolution: inferences from plant rDNA substitution rates in the western Pacific. *Evolution*, **57**, 2893–2898.

Wright, S., Keeling, J., & Gillman, L. (2006). The road from Santa Rosalia: a faster tempo of evolution in tropical climates. *Proceedings of the National Academy of Sciences of the USA*, **103**, 7718–7722.

Wright, S. D., Gillman, L. N., Ross, H. A., & Keeling, J. D. (2009). Slower tempo of micro-evolution in island birds: implications for conservation biology. *Evolution*, **63**, 2276–2287.

Wright, S. D., Gillman, L. N., Ross, H. A., & Keeling, D. J. (2010). Energy and the tempo of evolution in amphibians. *Global Ecology and Biogeography*, **19**, 733–740.

Wright, S. D., Ross, H. A., Keeling, D. J., McBride, P., & Gillman, L. N. (2011). Thermal energy and the rate of genetic evolution in marine fishes. *Evolutionary Ecology*, **25**, 525–530.

Part V

Effects Due to Invading Species, Habitat Loss and Climate Change

13   The physics of climate: equilibrium, disequilibrium and chaos

Michael Box

Although our knowledge of the Earth's climate becomes less and less detailed the further back we try to probe, two facts are clear. Firstly, the climate has changed on a wide range of timescales, in response to natural process we only partially understand. And secondly, despite this, the Earth's globally averaged surface temperature has remained within a relatively narrow range for most of its history. Over the past 10 000, as human civilization has arisen, this has been no more than 1°C. On geological timescales the range is perhaps 10 to 20°C, or roughly 5% of the mean when measured on the Kelvin (absolute) temperature scale.

Thus the definition of equilibrium is not straightforward. Nevertheless, it will make sense to firstly examine the Earth's climate in an equilibrium state before we more closely examine its current disequilibrium. Changes on longer timescales will be briefly discussed in the final section of this chapter.

1.   The Earth system in equilibrium

The Earth's global climate is maintained by a balance between energy flows: incoming energy from the sun – solar radiation – and energy emitted to space from both the surface and the atmosphere – terrestrial radiation. The laws of physics cover such information as the amount of radiation absorbed, emitted and scattered, as well as the wavelength ranges involved. In principle at least, it is possible to perform highly accurate calculations of all these radiative interactions.

The situation is much simpler if we assume (initially) that the atmosphere does not contain any gases which absorb terrestrial radiation – "greenhouse", or radiatively active gases. At the top of the atmosphere, the arriving energy (per square metre, per second) is denoted by F, the solar constant. (The solar constant varies inversely as the square of the distance from the Sun.) The Earth intercepts this energy in cross-section, as a disc of radius R, so the actual intercepted energy is $E_{int} = \pi R^2 F$. However, clouds, ice sheets and other components of the atmosphere and surface reflect a fraction α of this (known as the planetary albedo), so the actual amount of energy absorbed by the Earth-atmosphere system is $E_{abs} = (1-\alpha)\pi R^2 F$.

The Balance of Nature and Human Impact, ed. Klaus Rohde. Published by Cambridge University Press.
© Cambridge University Press 2013.

The energy (per square metre, per second) emitted by a perfect radiator (known as a black body) is given by the Stefan-Boltzmann equation, σT^4, where σ is the Stefan-Boltzmann constant, and T is the absolute temperature. A solid or liquid surface is a reasonable approximation to a black body: a gas is anything but! Since the Earth will emit as a sphere of radius R, the total energy emitted will be $E_{emit} = 4\pi R^2 \sigma T^4$, where T may be regarded as some average (or effective) temperature.

When the planetary climate is in equilibrium, the emitted energy must exactly balance the absorbed energy (over an appropriate interval of time): hence

$$4\pi R^2 \sigma T^4 = (1-\alpha)\pi R^2 F$$

F and α are accurately measurable from satellites, and σ is derived from several of the fundamental constants of physics, so the only free parameter in this equation is the temperature, T, which must take the value which ensures this equilibrium:

$$T = \sqrt[4]{\frac{(1-\alpha)F}{4\sigma}} = 255\,K = -18°C$$

This, of course, is much lower than the observed (global average) temperature of the Earth's surface of around 15°C, and reflects the fact that we have ignored the effects of greenhouse gases.

1.1.   The greenhouse effect

Incoming solar radiation is mostly at wavelengths between 0.2 and 4.0 μm, with a peak around 0.5 μm (hence "shortwave radiation"), while the Earth's emitted radiation is mostly at wavelengths between 4.0 and 100.0 μm, with a peak around 10.0 μm (hence "longwave radiation") – this is determined mainly by the temperatures of the Sun and Earth, respectively. These wavelength ranges also correspond to significantly different photon energies, and hence how the radiation interacts with gases.

In the longwave, low-photon-energy region, a number of atmospheric gases have vibrational and rotational energy levels which can lead to absorption and emission of radiation. In the Earth's atmosphere the most important are water vapor (H_2O) and carbon dioxide (CO_2), but also methane (CH_4), ozone (O_3) and nitrous oxide (N_2O); and more recently the CFCs and related man-made gases. Note that nitrogen and oxygen, the dominant gases in the atmosphere, are not radiatively active.

When we add these gases to a more sophisticated model of radiation flows we find that, while much of the solar radiation still reaches the surface, most of the terrestrial radiation emitted by the Earth's surface is absorbed before it can escape to space. This absorbed energy warms the atmosphere. As a result of Kirchhoff's law (fractional absorptivity equals fractional emissivity, at all wavelengths), the atmosphere must also emit radiation, some to space, and some – in fact the majority – back to the surface, in the same wavelength bands.

The additional energy arriving at the surface helps raise its temperature – and hence radiate more energy to be absorbed by the atmosphere. This sequence continues until the atmosphere is warm enough that it radiates sufficient energy to space to balance the incoming solar energy. In the process, the surface temperature will have risen significantly from the value we calculated in the absence of an atmosphere.

To complete this picture we need to realize that a warm ocean surface and atmosphere will lead to the evaporation of water vapor – a strong greenhouse gas – also contributing to this warming process. (This is an example of a positive feedback.) The natural greenhouse effect is the product of these two key gases, and an equilibrium exists with the atmospheric fluxes of each gas in balance. On Earth, we have reached this equilibrium state with almost all of the free water still in liquid form (the oceans): on Venus this is not the case!

The inflow of solar energy is concentrated mostly in equatorial regions, which receive almost ten times the annual input of the poles. By contrast, the variation of emitted energy is much smaller. This implies a net flow of energy from tropics to poles, driven by the temperature gradient, and it is this flow of energy which drives much of the world's weather. (Ocean circulation also transports energy poleward.)

2.   Climate: current disequilibrium

Roughly two centuries ago, the rise of industrialization began to perturb a number of the planet's natural systems, including its atmosphere, producing what we may see as a chemical disequilibrium. (Some scientists identify the current era, with a debatable starting point, as the Anthropocene.) With these changes in atmospheric composition, the Earth's climate is being progressively "forced" from the radiative equilibrium it had experienced for roughly the previous 10 000 years – the Holocene.

2.1.   Chemical disequilibrium

Over the past couple of centuries, humanity has come to manipulate many facets of our planet, mostly to our advantage. However, these manipulations have caused many disruptions to natural systems, and in particular to some key biogeochemical cycles. Of these, the most important are the sulfur, nitrogen and carbon cycles.

Perturbation of the sulfur cycle is now severe, with the vast majority of atmospheric sulfur being of anthropogenic origin. Sulfur impurities in many fuels are converted to SO_2 in the combustion process, and then once in the atmosphere to sulfuric acid, a major component of acid rain. Some is also converted to droplets, causing a significant increase of sulfate aerosol particles.

Sulfate aerosols (along with most aerosols, natural or anthropogenic) scatter sunlight back to space, cooling the planet and partially countering global warming. They may also enhance cloud reflectivity, and even cloud lifetimes, again increasing the planetary albedo. Note, however, that aerosols (unless injected into the stratosphere) have atmospheric lifetimes of only about a week, in contrast to CO_2 with an effective lifetime of roughly a century. (Lifetime, or average residence time, may be defined as the concentration divided by the flux into or out of the system. Because of the large fluxes of CO_2 between atmosphere, upper ocean, soils and biosphere, the atmospheric lifetime of a CO_2

molecule is around 10 years. However, the time taken to remove such a molecule from this larger reservoir is around 100 years.)

Perturbation of the nitrogen cycle comes in two forms: fertilizer use (the dominant contributor) and combustion. The latter is central to photochemical smog and related urban air problems, and is also a component of acid rain. Triggered by photochemical reactions, and thus quite seasonal, nitric oxide (NO) generated in combustion processes may produce nitrogen dioxide (NO_2) and ozone, both key factors in the poor air quality of many large urban areas. This problem, and the acid rain problem, are (mostly) local to regional (say ~1000 km.).

However, it is the perturbations to the carbon cycle which are likely to have the most significant global effects. Coal and oil deposits are components of the global carbon cycle which have been sequestered for many millions of years – back to times when the output of solar energy was somewhat lower than today, and the world may have needed higher greenhouse gas concentrations in its atmosphere. Today, things are different. Today we are returning these deposits to the atmosphere, in the form of the key greenhouse gas carbon dioxide, at an unprecedented rate, with its concentration rising from 280 ppm to over 390 ppm in the past 200 years.

Of the carbon dioxide emitted to the atmosphere, roughly half is absorbed by the ocean, changing its pH balance and carbonate chemistry, with potential impacts on marine ecosystems, especially corals. There are also many open questions as to the impact of elevated CO_2 levels on plant life, especially when coupled with higher temperatures and other climatic changes (some of these are discussed below).

2.2.   Radiative disequilibrium

The increased atmospheric CO_2 concentration also leads to an immediate increase in the longwave radiation which is absorbed by the atmosphere (hence less is transmitted to space), causing it to warm. This results in an increase in the longwave radiation emitted by the greenhouse gases, much of which returns to the Earth's surface. Overall there is a net decrease of longwave radiation emitted to space, which no longer balances the incoming solar radiation. The net effect is known as a "radiative forcing" – a change in the net radiative energy at an atmospheric boundary, resulting in a perturbation to, and disequilibrium in, the energy balance of the planet.

The first consequence, just as with the natural greenhouse effect, is an increase in the evaporation of water vapor (followed later by increased precipitation), and an increase in the atmospheric concentration of this greenhouse gas, amplifying the initial forcing. Other feedbacks also occur, which will be considered shortly.

This disequilibrium in the energy flows has consequences for the thermodynamic state of the planet. While energy can manifest itself as either work or heat, any work component (a slight increase in average wind speeds) will be negligible compared to the heating component. The most obvious such heating effect is a temperature increase in the lower atmosphere, leading to increased radiation emission which will, in time, restore the radiative equilibrium, at least once the gas concentrations are stabilized (although not to the same equilibrium state as existed prior to the CO_2 emissions).

The second way in which this heat is used is to warm the ocean, or at least the top ~100 metres of it – down to the thermocline. This is a large mass of water, with an enormous heat capacity, so its temperature will always lag behind the forcing in times of varying radiative forcing, such as the past century. (The atmospheric temperature rise will also lag, but by a shorter period.) It is for this reason that the climate system is out of equilibrium, and would continue to warm, even if gas concentrations could be stabilized today.

Finally, some heat will be used to melt glaciers, ice caps, "permanent" sea ice, and permafrost. (We may ignore the annual cycles of ice/snow formation and melting.) Ice and snow are very reflective surfaces, and are major contributors to the planetary albedo. Reduced reflection resulting from this increased melting implies more solar energy absorbed, especially at high latitudes – another positive feedback. Thus the radiative disequilibrium is felt most strongly in polar regions, and especially so in the northern hemisphere, and explains why the most dramatic warming is seen there.

3. Modeling and prediction

Trying to predict the future of our planet's climate, or to reconstruct the past, requires a good understanding of all components of the "Earth system" – atmosphere, ocean, cryosphere, biosphere and econosphere. Then we need to combine all of these pieces, including all their interactions and feedbacks, into one consistent model. This sounds more like a task for Don Quixote than for hard-headed scientists. Nevertheless it is a challenge that must be addressed.

3.1. Weather and climate modeling

If we want to predict the evolution of a physical system we need to model it: this applies to an Atlas rocket launch, or the atmosphere. Modeling the atmosphere requires the equations of fluid dynamics, radiative transfer, and thermodynamics – all well under-stood. However, atmospheric processes take place over an enormous range of scales, from planetary-scale circulation (including wave motion and major pressure systems) down to the micro-scale for the formation of individual cloud droplets (see Figure 13.1). Given limitations on computing resources, this inevitably calls for the use of approx-imations – or parameterizations.

A numerical weather prediction model chops the globe up into "grid boxes" of about 10–20 km on a side, with 40 or more vertical layers. (Individual national weather services may, in fact, run "nested models", with smaller grid boxes down to 1 km over their area of primary forecast responsibility.) Such models are now able to provide valuable predic-tions – "forecasts" – for 3 or 4 days, and useful indicators – "outlooks" – for up to a week or more. Further improvements to models, and to the data supplied to them from satellite and ground observations around the world, are likely to extend this time frame even further.

While grid size and other computational constraints are partly responsible for these limits, more important is sensitivity to initial conditions – otherwise known as chaos. The dynamical equations are non-linear, so that their solutions over time can diverge

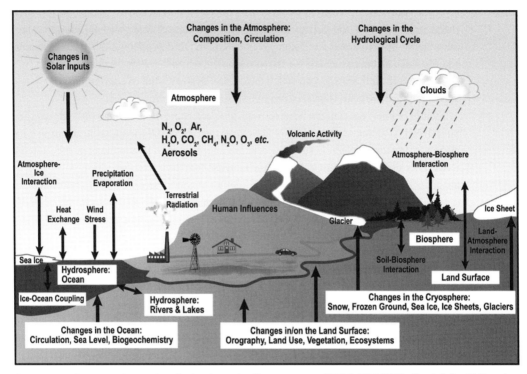

Figure 13.1.   Important interactions in the dynamic climate system. [IPCC AR4 WG1, faq-1-2-fig-1] See plate section for color version.

significantly due to tiny variations in starting values. Indeed, one of the early papers in the field that was to become known as Chaos Theory was by Lorenz, who observed the phenomenon while running one of the earliest versions of a weather model.

While chaos will always place strict limitations on weather forecasting (and seasonal climate prediction), the situation with longer-term climate prediction is different. Climate is fundamentally about statistics, such as average temperature and rainfall in a locality, plus appropriate measures of their variability. It doesn't really matter which day it rains, climate is mainly interested in (say) seasonal rainfalls – plus any changes in extreme events – and, especially in countries like Australia which are so dominated by the El Niño/La Niña cycle, the interannual variability. Long-term prediction is thus concerned with the envelope of such parameters, with that envelope representing the chaotic effects of natural variability.

Modeling the climate, past or future, requires simulation runs of decades to centuries, rather than just a week. Climate models have much in common with weather models, plus a number of key differences. The most obvious is grid size, currently ~100 km. This is a direct consequence of the limitations of computing resources, as such models need to be run for more than 1000 times longer simulation time.

Among the inputs to a numerical weather model are surface conditions, especially sea surface temperature. This can be assumed fixed over short periods, but for longer time frames must be allowed to vary. Hence the next step in climate modeling, whether seasonal or long term, is to couple an ocean model to an atmosphere model (and maybe also a cryosphere model). As well as the obvious increased demand on computing resources, there are additional numerical challenges due to the different densities of these two fluids.

Such coupled models are useful for climate prediction in the medium term (say a few decades), provided they are able to predict oceanic oscillations such as those driving El Niño/La Niña – still a significant challenge. Current coupled models are capable of providing the statistics of phenomena such as El Niño, although not of predicting an individual El Niño event some years into the future.

3.2. Twentieth-century climate

Throughout the last century or so, the atmospheric concentrations of the greenhouse gases carbon dioxide, methane and nitrous oxide have all risen (Figure 13.2). Globally averaged surface temperature has also increased, by around 0.7°C (Figure 13.3). This has been accompanied by ice cover reductions, as well as earlier flowering of plants and other ecological signatures. However, this rise has been far from steady, with regular fluctuations of 0.2°C, and more, from year to year. Are these phenomena connected, or are there other explanations? While a rise in temperature is to be expected, due to increasing greenhouse gas forcing (allowing for the oceanic thermal time lag), the details are less straightforward, and require careful analysis.

Firstly there have been other forcings which need to be considered. Solar output varies over the 11-year solar cycle, and may vary, though no doubt slowly, on longer timescales. Measurements using satellites over several decades, plus various proxy indicators, all suggest that solar forcing is much smaller than that due to increased greenhouse gases. Massive volcanic eruptions inject large amounts of sulfur gases (H_2S and SO_2) into the stratosphere, where they are converted into sulfuric acid droplets. Because of the greater stability of the stratosphere than the troposphere, the resulting sulfate aerosol layer may diffuse globally, and last a year or two, cooling the planet. The last such event, Mt Pinatubo in 1991, caused a temperature decrease of around 0.3°C (see Figure 13.3).

As well as these forcings, there is internal variability, on yearly to decadal timescales, mostly driven by changes in ocean circulation. Scientists have carefully studied ENSO – El Niño/La Niña and the southern oscillation – the best understood, and globally influential. On regional scales we also recognize NAO (the North Atlantic Oscillation), which has impacts for Europe, and IOD (the Indian Ocean Dipole) and SAM (the Southern Annular Mode), which both have impacts on Australian rainfall. The ocean component of a climate model can now simulate many of the important features of these phenomena, but not all the details.

During an El Niño event, dynamical changes in the equatorial Pacific Ocean lead to a warming in the surface layers, starting near Peru and moving westward. Contact with the atmosphere transfers some of this heat upwards, producing an atmospheric warming which might last for a year or so. This is not climate change, but a temporary transfer of heat from one part of the climate system – the ocean – to another – the atmosphere. The last major El Niños were 1998 and 2010, each producing a global average temperature well above the long-term trend.

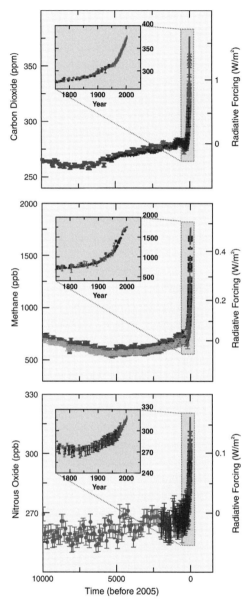

Figure 13.2.   Concentrations of key greenhouse gases over the past 10 000 years. [IPCC AR4 WG1, fig2–3] See plate section for color version.

In many countries, industrial development has been seen as a national imperative, overriding everything else. As a consequence, SO_2 emissions have risen over the past 60 years or more, with much of it ending up as aerosols, partially masking the warming effects of the CO_2 increase. This rise has been variable, as developed nations have put in place more stringent pollution control measures, while emerging nations have placed a higher priority on raising living standards. Should these nations make the shift to greater

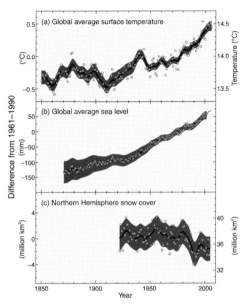

Figure 13.3.   Changes in temperature, sea level and snow cover since 1850. [IPCC AR4 WG1, fig1–1] See plate section for color version.

pollution control, SO_2 emissions could fall significantly, with corresponding falls in sulfate aerosols, and their cooling effects. This would expose the world to the "full force" of the warming of the greenhouse gas increases.

3.3.   Change and attribution

Climate scientists have run models to try to reconstruct the observed climate of the past century, employing some or all of the forcings mentioned above: increases in greenhouse gases, increases in aerosols, volcanic eruptions, solar output variation, and internal variability, particularly ENSO. Simulations which include only the natural forcings are unable to reproduce the temperature rise of the past 40 years, while those which include greenhouse gas increases do demonstrate warming. However, it is the model runs which include all forcings, including aerosol increases, which are closest to the observed temperature record (Figure 13.4).

There are a number of other indicators which give us confidence that we do understand climate change during the past century. Although climate models are at their most robust when providing outputs such as global average surface temperature, regional climatic results are now becoming more reliable. Figure 13.4 provides results not just for the world as a whole, but for a number of regions as well, again showing consistency between the data and model runs including all forcings. In fact it is this geographic pattern of consistency which is one of the key factors in giving the science community confidence in these results.

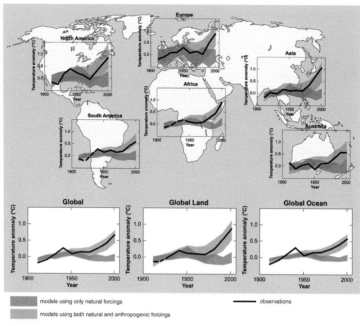

Figure 13.4.   Global and regional temperature changes in the twentieth century. [IPCC AR4 WG1, fig2–5] See
plate section for color version.

The strongest warming is observed at high northern latitudes, and is a consequence of
the strong ice-albedo feedback operating there. This mechanism is also contributing to
the observed decrease in Arctic sea ice, and the retreat of many glaciers. It is also one of
several factors at work in the observed loss of ice from Greenland. The situation in the
Antarctic is more complicated, mainly due to the effects of the circumpolar current. (Such
oceanic flows are not possible in the Arctic.)

As has been mentioned above, most of the energy trapped by the increase in greenhouse gases has actually gone into warming the top layers of the ocean. This ought to
have produced a measurable rise in sea level. A further contribution has also come from
the melting of ice on land – glaciers, and Greenland. Modeling the rise due to thermal
expansion is relatively straightforward, and the reduction of land ice is observable. While
the resulting sea level rise has been only a couple of millimetres per year, once again there
is good agreement between models and measurements.

One additional piece of evidence deserves a mention. Official temperature records
generally consist of the daily minimum and maximum at each station. When we study the
trends in each of these we note that the overnight minima have increased faster than the
daily maxima, although with some geographic variability. This is consistent with the two
anthropogenic forcings: greenhouse gas warming is able to act both day and night, while
aerosol cooling can only act during daylight hours. Simulations which assume a significant rise in solar output, while ignoring greenhouse gas increases, cannot reproduce this
important result.

4.   Predicting the future

4.1.   Extending the models

While coupled ocean-atmosphere models may provide useful guidance into the near future (several decades), decision-makers need guidance as to potential longer-term changes. To make such predictions, especially in an era of dynamic radiative forcing, two further factors are required. Firstly we need some idea of future atmospheric composition; and secondly we need to have a reasonable idea of how the biosphere will respond to forcings it may experience (including potential land-use changes).

Atmospheric composition, at least in terms of its carbon dioxide and aerosol concentrations, depends very much on human decisions – industrialization, standard of living, population, choice of energy sources and pollution controls, and energy efficiency measures. These are not subjects for discussion amongst physicists and other climate scientists/modelers, but are, instead, the province of economists and technologists. Since meaningful predictions decades into the future are impossible, they outline a number of "scenarios" for global economic/technological development (grouped into four families, or storylines), and the emissions which would ensue, and climate modelers run their models for one or more scenarios.

The A1 family assumes very rapid economic growth, population peaking mid century, and rapid introduction of efficient technologies. It is split into three groups: A1FI (fossil fuel intensive); A1T (non-fossil fuel sources); and A1B (a balance across fuel sources). The A2 family assumes a very heterogeneous world, with regionally focused economic growth, and slower technological change than other storylines. The B1 storyline is similar to A1 until mid century, followed by a rapid change to a service economy with reductions in material intensity. The B2 storyline emphasizes local solutions to sustainability, with intermediate levels of economic development.

Predicting how the biosphere may respond to multiple disequilibria of warming temperatures, elevated CO_2 levels, more intense droughts and floods, not to mention the effects of species migration where human constraints still permit it, is a task which science has only recently begun to address. Nevertheless these interactions, and their potential biogeophysical and biogeochemical feedbacks on the climate, for example via further changes in the carbon cycle, must be included in the models.

Land surfaces influence the climate through a number of biogeophysical effects. The nature of a surface – bare or vegetated, plus quantities such as leaf area – has a direct effect on surface albedo. A second direct effect is via surface roughness, which affects boundary layer turbulence and cloud formation. A less direct effect is through evapotranspiration, which again affects hydrology, and also how energy is partitioned between latent and sensible heat.

The biogeochemical effects of the land surface mainly comprise the exchange of various compounds with the atmosphere, particularly carbon dioxide. However, plants also exchange mineral compounds and organic compounds, for example isoprene, which is an important precursor to secondary organic aerosols. Finally, plants (and their associated root microbes) are also central to the nitrogen and phosphorus cycles.

These various effects are all required by an "Earth system" model. Historical data sets are available covering anthropogenic land use changes which may be applied to reconstructions of recent climate change. Simulations of future climate may be run with scenarios of future land use change, based on economic models and assumptions consistent with the emissions scenarios.

4.2.   Twenty-first century climate

Because of the need to employ emissions, and land use change scenarios, the scientific community does not make predictions of the climate of the twenty-first century, but rather "projections". Modeling groups around the world run their models for a selection of the scenarios supplied by the Intergovernmental Panel on Climate Change (IPCC). These results are subjected to peer review, as with all science, and then collated and assessed in the next IPCC Assessment Report.

The most widely reported output from these models is the global mean surface temperature: results from the Fourth Assessment Report for three key scenarios are shown in Figure 13.5 (plus ranges for three others). Of these, A1B is the most widely reported. Also shown in this figure are the patterns of warming for 2020–2029 and 2090–2099. As with the warming of the late twentieth century, the strongest warming is set to occur in high northern latitudes, due to ice-albedo feedback.

Of comparable importance to temperature is the expected change to rainfall patterns: this has proved challenging for modelers. The increased evaporation inherent in the water vapor feedback must lead to increased precipitation, including more intense precipitation events, although large regional variation is expected. Increased temperatures are likely to

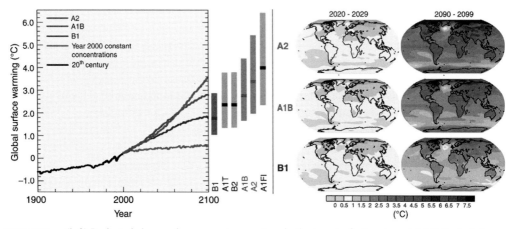

Figure 13.5.   (left) Projected changes in average temperature in the twenty-first century. (right) Projected warming patterns for 2020–2029 and 2090–2099. [IPCC AR4 WG1, fig3–2] See plate section for color version.

lead to increased re-evaporation of the rain that does fall, potentially amplifying drought conditions when they do develop.

There is, by now, broad agreement on some details. Poleward movement and intensification of the subtropical ridges (high pressure regions of dry, sinking air which largely define the world's deserts) will lead to a reduction of rainfall on the poleward edges of these ridges. This suggests that southern parts of Europe, USA, Africa and Australia can expect drier summers. The tropical Pacific may become more El Niño-like, with an increase in drought conditions over northeast Australia.

Thermal expansion of the ocean will continue, even if greenhouse gas concentrations were to stabilize. Projections for the end of the twenty-first century are for a rise of between 0.5 and 0.9 metres. However, this does not include contributions due to ice loss from Greenland or Antarctica. There are reasons to believe that these sources may make increasing contributions at some point in the future, although the physics is complex, making prediction challenging. (Some recent research suggests that sea level could rise by up to 1.5 metres, possibly even higher, by 2100.)

The possible breakup of sections of the Greenland or West Antarctic icecaps is an example of a major threat to the climate which it is very difficult to forecast: are there any others? High-latitude wetlands and tundra contain very large reservoirs of methane trapped in hydrate form – also known as clathrates. Should global warming continue unchecked for many decades, the overlying sediments and permafrost could become permeable, releasing large quantities of this potent greenhouse gas. We must learn the key lesson of the ozone hole, and be ever vigilant for the unexpected.

4.3.   Impacts of change

One direct consequence of warming is a significant increase in the occurrence of heat waves, often defined as three consecutive days above a threshold temperature. Temperatures for a given location and season tend to follow a "bell curve" type of distribution. A shift in the mean of this distribution would push many additional days into the high tail, while an increase in its variance would be even more damaging. A heat wave in Europe in 2003 is believed to have been responsible for at least 20 000 premature deaths: more such episodes can be expected this century.

Sea level rise will be felt most keenly by the over 300 million people who live in some of the world's largest river deltas, many of whom will also be at the mercy of the possible increase in intensity of tropical cyclones, and associated storm surge. (Since 1980, over a quarter of a million lives have been lost to such devastation.) Densely populated Bangladesh, with a population of 150 million, is particularly vulnerable, as 10% of its habitable land would be lost with a half metre sea level rise, and 20% lost to a 1 metre rise.

Temperature increases, along with a poleward movement in rainfall patterns, will have significant implications for forestry, agriculture and food supply, as will continuing population growth. However, managed agricultural ecosystems should have at least some advantages over natural systems. Continued endeavors in crop substitution, and the breeding of more drought-tolerant plant species (possibly also involving genetic modification), should allow agriculture to remain productive in most parts of the world.

Forests, vital ecosystems with large carbon stores, cover about 30% of the world's land area. Because of their long lifetimes, trees are very slow to migrate, yet in some situations a rise of 1°C in average temperature can affect productivity. Changes now expected in the twenty-first century could see much of our forests suffering from unsuitable climatic conditions. This is especially the case in the northern boreal forests, which will be prone to pests, dieback and forest fires. It is quite possible that the terrestrial biosphere, currently a net carbon sink, could become a carbon source.

While higher CO_2 levels may have some positive ecosystem effects, particularly on short (decadal) timescales, other potential effects – caused by floods, drought, fire, insect invasion – are mostly negative. This suggests significant threats to biodiversity, where it is not just species, but entire ecosystems, that may not be able to migrate at a sufficient pace. The combination of warmer temperatures and ocean acidification will have severe impacts on corals and their ecosystems.

5.   Climate change on longer timescales

Climate has changed in the past, quite independently of human activity, or even human existence. In the long sweep of Earth's history, many changes have taken place. The Sun's output has increased by around 25% since its birth, yet at no point does the Earth appear to have been fully frozen. This can only imply a much greater greenhouse effect in the distant past. In fact, the Earth's atmosphere has changed significantly over its history, in response to changing geology (e.g., plate tectonics, volcanism). For example, the clustering of the continents in Pangaea may well have allowed more efficient oceanic heat transport, resulting in warmer polar regions.

Biology has also played a major role. In particular, oxygen has only been abundant during the second half of Earth's existence, and this change was undoubtedly one of the greatest environmental impacts the planet has ever experienced, as it was toxic to almost all the then existing life forms. (Bolide impacts have also contributed to several mass extinctions.) However, it also allowed the creation of the ozone shield, which is essential for terrestrial life.

The Gaia hypothesis asserts (at least in its strongest version) that life forms interact with their environment in ways that are designed to ensure an environment suitable for the continued existence of life. While Lovelock's idea has sometimes been "over-interpreted", it remains a fascinating way to think about the contribution that life forms undoubtedly have on atmospheric composition, and especially on the concept of (bio-chemical) waste products, as oxygen was in the early Earth.

Studying past climates requires the use of proxy data, such as tree rings, isotope ratios in ice cores and sediments, micro-flora remains, etc., so that the picture we have becomes progressively murkier the further back we look. However, considerable efforts have been made to unravel the details of the quaternary – the past ~2.5 million years which have been ice-dominated. (An "ice age" consists of glacials and interglacials: we are currently in an interglacial.) In particular, studies of ice cores from both Greenland and Antarctica have provided data on temperature and atmospheric composition for around the past 600 000 years, and this time is being steadily pushed back.

5.1.   Biogeochemical feedbacks

Climate modeling studies of this glacial-interglacial cycle need additional science components. In the absence of measurable human impact, we must allow the climate and the land surface to mutually interact: to exert feedbacks on each other. For example, the replacement of taller trees and shrubs with greater leaf area by low tundra plants during a period of glaciation allows more sunlight to reach a snow-covered surface, and hence be reflected, strengthening the ice-albedo feedback.

Another important feedback concerns the hydrological cycle. Modeling has shown that reductions in rainfall can lead to changes in plant variety with reduction in transpiration, which then feeds back on rainfall. This process has been implicated in the abrupt drying of the Sahara in the mid-Holocene (~6000 years ago). Modeling has also suggested that Amazonian deforestation could become "permanent" via this process.

The driving force behind the glacial-interglacial cycle is the variation in the Earth's orbital parameters, and especially its ~100 000 year eccentricity cycle. While the effects on global total solar energy received are tiny, changes in its spatial and temporal distribution range up to ~10%, with potentially significant effects on snow melt. In effect, the climate is periodically forced at this frequency.

However, the climate may have its own resonant frequency (or frequencies), governed by its feedbacks, so that we may be looking at one of the classic examples of chaos: a "driven oscillator". (Another common characteristic of complex nonlinear systems is the existence of multiple equilibrium states, and the potential to jump from one to another in response to some form of excitation.) Some recent studies by Rial (2004) show how this might be modeled.

6.   Synopsis

Over geological time, the Earth's climate has mostly been in a state of equilibrium, or at least near equilibrium. However, this situation has been subject to change on a range of timescales, of which the 100 000 year glacial-interglacial cycling of the Pleistocene has been closely tied to human evolution. By contrast, the last ~10 000 years have been particularly stable, no doubt assisting the rise of civilization.

Over the past century or so, this situation has begun to change, due primarily to emissions from human activity. Today the climate system is in an increasing disequilibrium, and is likely to remain so for a number of decades at least, as the rise in atmospheric carbon dioxide seems unlikely to stop any time soon. Many of the consequences of this continuing rise are likely to be disruptive for the planet, from natural systems to human comfort.

Acknowledgments

All figures are from the IPCC Fourth Assessment Report, Climate Change 2007 (AR4) WG 1, as indicated in the captions. ipcc.ch/publications_and_data/publications_and_data_figures_and_tables.shtml.

Further reading

The Intergovernmental Panel on Climate Change. (A vast range of information, including the most recent Working Group Reports.). Available at: www.ipcc.ch.

Sir John Houghton (2009). *Global Warming, The Complete Briefing* (4th edn.) Cambridge: Cambridge University Press.

F. W. Taylor (2005). *Elementary Climate Physics*. Oxford: Oxford University Press.

T. E. Graedel & P. J. Crutzen (1993). *Atmospheric Change, An Earth System Perspective*. New York: W. H. Freeman.

James Lovelock (2000). *Gaia: A New Look at Life on Earth* (3rd edn.) Oxford: Oxford University Press.

S. Levis. Modeling vegetation and land use in models of the Earth System. wires.wiley.com/climatechange. doi: 10.1002/wcc.83.

J. A. Rial (2004). Abrupt climate change: chaos and order at orbital and millennial scales. *Global and Planetary Change*, **41**, 95–109.

14 Episodic processes, invasion and faunal mosaics in evolutionary and ecological time

Eric P. Hoberg and Daniel R. Brooks

Episodic processes and faunal structure

Episodes of ecological perturbation and faunal turnover represent crises for global biodiversity and have occurred periodically across Earth's history on a continuum linking deep evolutionary and shallow ecological time (Briggs, 1995; Hallam & Wignall, 1997; Hoberg & Brooks, 2008; Stigall, 2010). Major extinction events and biodiversity crises across the 540 million years of the Phanerozoic are equated with periods of maximum ecological disruption associated with geological, oceanographic and atmospheric (climatological) mechanisms which have influenced patterns and processes for diversification (dispersal and isolation), species diversity, community and faunal structure, turnover, and distribution on global to regional and landscape scales (Briggs, 1995; Stigall, 2012a, 2012b). Episodic or punctuated events set the stage for patterns of diversification and faunal associations downstream for extended periods of time (Eldredge & Gould, 1972; Eldredge *et al.*, 2005). In essence, the cascading effects of ecological disruption may canalize faunal structure, eliminating evolutionary potential through differential extinction events, but concurrently may heighten faunal mixing and interchange through breakdown in ecological isolation during biotic expansion and geographic colonization (Rode & Lieberman, 2005; Hoberg & Brooks, 2010). Paradoxically, ecological crises may also be precursors for subsequent radiation and diversification in taxa which have persisted through events of maximal ecological perturbations (e.g., Hoberg & Brooks, 2008), and elevated rates for speciation are often linked to periods of rapid climatological and environmental change (Vrba, 1996). These processes and their influence on faunal structure and diversity are equivalent in evolutionary and ecological time and thus can serve as analogs for understanding and predicting the general outcomes of invasion and range shifts in contemporary communities and faunas (Hoberg, 2010; Hoberg & Brooks, 2010; Peterson, 2011; Stigall, 2012b).

Episodes of perturbation result in extinction or biodiversity crises through both *in situ* and invasive processes. At regional and local scales within a circumscribed geographic arena, *in situ* events such as those driven by climate can result in fragmentation and may "reshuffle" faunal associations, leading to changes in range overlap, competition, trophic/predator-prey interactions, and connectivity among symbionts including hosts

The Balance of Nature and Human Impact, ed. Klaus Rohde. Published by Cambridge University Press.
© Cambridge University Press 2013.

and parasites. In contrast, events of invasion, geographic colonization and establishment introduce "new players" which may be accommodated (or not) into local and regional faunas, leading to competition, displacement, extinction, divergent adaptation (assuming directional selection and sufficient time) and overall novel faunal associations. Patterns of diversity then result from intricate interactions of regional history, changing environmental parameters over time, and the linkage between large-scale and local processes (Rickleffs, 2004).

Invasions may constitute isolated events such as a one-time (or protracted) expansion into new geographic and faunal space following the dissolution of a long permanent barrier, as exemplified by biotic interchange with the emergence of the Panamian Isthmus about 3 Mya (million years ago) (Webb & Marshall, 1981; Marshall *et al.*, 1982). Recurrent processes, particularly those driven by cyclic climate change, directly influence macro- and microevolutionary outcomes, faunal associations and biogeographic patterns (Jansson & Dynesius, 2002). As a consequence, dispersal mechanisms also can be recurrent in space and time, leading to episodic events of expansion/isolation involving phylogenetically disparate taxa as exemplified in both marine and terrestrial faunas in the Beringian region, or nexus for the Nearctic and Palearctic, during the Quaternary (Vermeij, 1991a, 1991b; Sher, 1999; Hewitt, 1996; Hoberg & Adams, 2000; Waltari *et al.*, 2007; Hoberg *et al.*, 2012). In either set of circumstances, the outcome of geographic colonization will also be directly influenced by the degree of ecological packing in ancestral and recipient regions, and whether expansion brings similar or dissimilar faunal elements together in space and time (e.g., Rickleffs, 2005).

Significantly, the effects of directional climate change and substantial invasive processes (facilitated by climate and with both anthropogenic and natural drivers) are in a regime of convergence in contemporary systems, and resulting synergy between these phenomena can have considerable implications for ecological continuity and faunal structure (Thuiller, 2007; Hoberg *et al.*, 2008; Lawler *et al.*, 2009; Stigall, 2012a, 2012b). An exploration of complex host-parasite systems provides a nuanced window into the role of episodic events and invasion as determinants of diversity at varying temporal and spatial scales across the biosphere (e.g., Hoberg *et al.*, 2012). Parasites are particularly revealing as indicators of historical biogeography, historical ecology and more broadly biotic structure and ecological connectivity as these diverse organisms track predictably within and across ecosystems in space and time (e.g., Brooks & McLennan, 1993; Hoberg, 1997; Marcogliese, 2005; Lafferty *et al.*, 2005; Nieberding & Olivieri, 2007; Hoberg & Brooks, 2008, 2010).

Some generalities of faunal structure and assembly

Diversification, largely by allopatric speciation, and faunal assembly are complex and constitute an interaction with varying contributions from vicariance and dispersal (e.g., Wiley, 1981; Brooks & McLennan, 2002). Vicariance results from the formation of permanent or relatively impermeable geographic barriers, usually as a large-scale process, leading to isolation and the origins of sister species, or more broadly faunal assemblages of equivalent age, and has been identified as the primary mechanism determining the structure of regional biotas at all timescales (e.g., Nelson & Platnick, 1981; Stigall, 2010). The overriding explanatory power for vicariance and its coevolutionary counterpart termed "maximum cospeciation" (e.g., Page, 2003) is increasingly challenged by recognition of a substantial role for dispersal (invasion) and processes for geographic colonization as determinants of biotic structure (e.g., Brooks & McLennan, 2002; Rode & Lieberman, 2005; Hoberg & Brooks, 2008). Invasion as a mechanism of diversification is embodied in the concept of the "taxon pulse" which contrasts with vicariance in some fundamental ways (Erwin, 1985; Halas *et al.*, 2005): (1) recurrent breakdown in ecological/physical barriers allowing dispersal, with resulting expansion, rather than isolation, being a driver for diversification; and (2) origin of complex or "mosaic faunas" exhibiting reticulate histories relative to sources, ages and phylogenetic composition resulting from recurrent expansion over time (Hoberg & Brooks, 2008, 2010; Hoberg *et al.*, 2012).

Although both vicariance and dispersal are important mediators of diversity, it is the recurrent episodes of expansion/isolation over extended time frames that may provide a more comprehensive model for historical biogeography and which has considerable explanatory power for understanding the structure of contemporary systems (Erwin, 1985; Halas *et al.*, 2005; Folinsbee & Brooks, 2007; Hoberg & Brooks, 2008, 2010).

Dynamic climate, biotic expansion and diversity

Environmental forcing, through shifts in climate regimes, is essential in creating new circumstances, and the timing and duration of events is a first-order determinant of spatial distribution and composition of faunal assemblages over evolutionary history (e.g., Vrba, 1996; Dynesius & Jansson, 2000; Jansson & Dynesius, 2002). Consequences of such environmental change are seen as faunal stasis or diversification.

Under stasis, geographic expansion or contraction may track suitable environments relative to characteristics for resilience and tolerance and specific thresholds for development that are the determinants of distribution in complex biotic assemblages including host-parasite systems (e.g., Lafferty, 2009). Essentially, stasis represents an outcome from a new set of "old" circumstances, resulting in no net change in diversity, but may for example lead to substantial changes (shifts) in spatial distribution and apparent host range for parasites (Hoberg & Brooks, 2008). This dynamic underexpansion and geographic colonization is described by "ecological fitting" (EF), which recognizes that species and their phylogenetically conservative traits may disperse through space and time, further explaining the extent of initial host range and potential for host colonization (Janzen, 1985; Agosta & Klemens, 2008; Hoberg & Brooks, 2010). Among members of a fauna, EF accommodates the recognition of a continuum from generalists to specialists, where for example phylogenetic conservatism in parasite biology, in concert with conservatism in host biology, creates a considerable arena for host switching even without the evolution of novel capacities for host utilization (Brooks & McLennan, 2002). In this arena, niche conservatism (= stasis) contrasts with niche evolution (= potential for diversification) through adaptation to changing conditions or new circumstances (e.g., Stigall, 2012a). Ecological fitting constitutes the null hypothesis against which to assess the downstream effects of episodic change and dispersal on faunal continuity and structure.

The continuity of faunal assemblages in space and time reflects or is a function of lineage persistence, and only secondarily diversification (e.g., Hoberg & Brooks, 2008). Persistence, rather than diversification, is the measure of "success" in a Darwinian context. In an arena of episodic perturbation that may also involve expansion and geographic colonization over time, persistence is linked to EF and phylogenetic conservatism, whereas diversification may be a secondary outcome. This idea of persistence contrasts with a neo-Darwinian expectation that success is equated with diversification, and diversification rates. Persistence, however, is the foundation, and even with a high generation of variation, or short generation time and turnover, these phenomena can only represent the potential for change. In the absence of persistence, there simply is no evolutionary potential for diversification and radiation (e.g., Hoberg & Brooks, 2008).

In lineages of symbionts or parasites, persistence emerges from coevolutionary mechanisms (cospeciation) and colonization processes, with the latter being particularly common in evolutionary and ecological time (Hoberg & Klassen, 2002). As host-parasite systems are components of complex biotas (parasites nested within hosts), such continuity linking evolutionary and ecological time, and across multiple events of ecological perturbation, serves to reveal considerable and general insights about the structure and maintenance of ecological connectivity across the biosphere (Hoberg & Brooks, 2008, 2010). The origin, radiation and persistence of complex host-parasite assemblages in terrestrial and aquatic systems extends well into deep time and transcends even the series of global-level mass extinctions and biodiversity crises that have characterized Earth's history. Consequently when we view contemporary processes, it is apparent that history constrains phenomena in ecological time (Rickleffs, 2004). History is the backbone from which contemporary systems are emergent (Hoberg & Brooks, 2008). Historical constraints (phylogenetic conservatism) are thus the most uniformitarian aspect of biology and are the basis for all truly predictive explanations about persistence in space and time.

We can predict in a general sense what happens during episodes of ecological perturbation, but not the details. Reconstruction of events and their outcomes occurs in retrospect: (1) they cannot predict diversity or diversification; but (2) they can relate persistence. Consequently, the general rules for faunal structure should be addressed in the context of persistence: (1) surviving the pulse or episode of perturbation (through geographic colonization, ecological fitting, and/or host switching); and (2) secondarily, between pulses the possibility of diversification emerges. Thus, it is not diversification, but lineage continuity in time that is critical, and this is the general dynamic recognized in the dichotomy between niche conservatism and niche evolution (Eldredge *et al.*, 2005; Stigall, 2012a).

Episodes, and especially climatological forcing influencing habitat and ecological continuity, are the external drivers (e.g., Vrba, 1996; Jansson & Dynesius, 2002). In this instance, pulses of perturbation set the stage sequentially for (1) geographic colonization,

(2) host switching and (3) ecological re-organization. Host colonization emerges through disruption of conservative trophic pathways, changing ecological associations, ecological release, and breakdown in mechanisms for ecological isolation (Hoberg & Brooks, 2008). Diversification (origin of new species or differentiated population structure) occurs between these events and is the dynamic specified in an integration of taxon pulses and ecological fitting (Hoberg & Brooks, 2008, 2010).

Factors influencing outcomes of expansion

Biotic expansion or invasion has been a common phenomenon in freshwater, marine, and terrestrial systems linking evolutionary and ecological time (Vermeij, 2005; Hoberg & Brooks, 2008; Stigall, 2010; Hoberg, 2010). Invasion can be multi-faceted and there are prominent examples: (1) single or isolated events of expansion following dissolution of a barrier (e.g., the Great American Interchange between North America and South America at the Pliocene-Pleistocene boundary) (Webb & Marshall, 1981; Marshall *et al.*, 1982); (2) hierarchical expansion/isolation processes linking independent events across extended time frames (e.g., invasion from the Pacific in the early Tertiary with origin of the Amazonian ray fauna and a complex assemblage of fishes, invertebrates and parasites; subsequent radiation in these aquatic freshwater faunas during recurrent marine transgression of Amazonia since the Miocene) (Brooks *et al.*, 1981; Webb, 1995); (3) episodic or recurrent expansion over time in terrestrial systems during the Tertiary (e.g., terrestrial faunal exchanges between Eurasia and Africa; Africa and Eurasia; Eurasia and North America) (Beard, 2002; Folinsbee & Brooks, 2007; Waltari et al., 2007); (4) recurrent events in marine systems driven directly by episodic climate shifts in the Pliocene and Quaternary (e.g., at Bering Straits involving marine invertebrates, mammals and seabirds, following prolonged hemispheral isolation of the North Pacific and Atlantic across the Arctic) (Vermeij, 1991a; Sher, 1999; Hoberg & Adams, 2000; Briggs, 2003); and (5) recurrent events in terrestrial systems of the Holarctic under glacial-interglacial cycles (e.g., intercontinental and intracontintential episodes of expansion/isolation with substantial refugial effects; largely asymmetrical expansion from Eurasia) (Zarlenga *et al.*, 2006; Waltari *et al.*, 2007; Shafer *et al.*, 2010; Hoberg *et al.*, 2012). To some extent these episodes represent a continuum where events may be protracted or relatively finite and ephemeral. Events of large scale or magnitude in deep time may have pervasive downstream effects on diversity (Vermeij, 2005).

Episodic invasion emerges as a model that contributes to a general understanding of processes that have structured diversity in space and time. Among the best exemplars are the assemblages of hosts and parasites that characterize high-latitude systems of the Holarctic where a history of periodic expansion and isolation has been defined by variable regimes of climate, most notably the alternation of glacial/interglacial cycles (Jansson & Dynesius, 2002; Shafer *et al.*, 2010; Hoberg *et al.*, 2012). These dominant shifts in climate and environment during the Quaternary over the past 2.6 Myr serve as an analog for the processes that are more generally involved in invasion (Hoberg 2010; Hoberg *et al.*, 2012). Recurrent expansion involves: (1) taxon pulse with diversification,

(2) development of mosaics with diversification, (3) mosaics with differential extinction, and (4) mosaics with stasis (failure to diversify).

Recurrent events constitute turnover often on short time frames (e.g., pulses of environmental perturbation in stadial-interstadial cycles of the Pleistocene). These are dominated by changes, or constant change, in the absence of steady states being attained across a continuum of events. Systems in transition result in unstable or unpredictable outcomes. A feature of these northern systems is multi-species expansion events over time, with recurrence of invasion involving different assemblages or homologous faunal assemblages and considerable linkage for geographic and host colonization. Overall, this results in an accumulation of species by invasion over time (e.g., Hoberg et al., 2012). Idiosyncrasies relate to gaps in distribution (that is, who didn't make the trip?) and relative age of assemblages may directly influence diversity.

Species are unlikely to adapt in a regime of episodic perturbation, but phylogenetic or niche conservatism may allow persistence (through tracking habitat or other requirements during range shifts) (e.g., Eldredge et al., 2005). This may explain why some events that otherwise drive isolation are insufficient as drivers of diversification. For example, in the Pliocene/Quaternary-age faunas, diversification among many mammalian species appears to be a cumulative process of genetic change that crosses multiple glacial/interglacial cycles (Jansson & Dynesius, 2002; Lister, 2004). This phenomenon has been demonstrated for ungulates and rodents, so does not appear to be taxon specific or linked to life history or generation time. Cumulative genetic change that is trans-stadial, however, would assume conservation of conditions which transcend multiple events (Lister, 2004). Species have apparently geographically tracked expanding and contracting environments but in the absence of substantial genetic change. In many instances, shifts were influenced by relative vagility, velocity, population representation and absence of substantial founder events that may drive divergence in peripheral populations (Hewitt, 1996; Sandel et al., 2011; Arenas et al., 2012).

Mosaics also result from processes unfolding in the absence of extensive perturbation, particularly where life history patterns for hosts involve long-distance dispersal. For example, among seabirds and marine mammals overall distributions for parasites may be driven to a great extent by focality associated with breeding sites often on oceanic islands (e.g., philopatry), modified by spatially extensive dispersal that may promote faunal mixing and host switching for pathogens (e.g., Hoberg, 1995, 1996; Hoberg & Adams, 2000). A remarkable demonstration of faunal mosaic structure was observed within the complex of argasid ticks, Ornithodoros capensis, which are ectoparasites on a diverse assemblage of tropical and temperate seabirds representing 12 families distributed across the oceans of the world (Gómez-Díaz et al., 2012).Tick diversity appears to have been structured by events of recurrent dispersal over extended time frames with trans-oceanic and host colonization among volant avian hosts linking geographically isolated breeding sites. These largely stochastic events were drivers for development of multi-species assemblages (mosaics) of phylogenetically disparate ticks associated with particular island/archipelago systems. It appears that philopatry for respective avian host species may be the initial driver for allopatric speciation for parasites on isolated insular habitats. Speciation generally appears to precede dispersal, and secondary oceanic expansion, establishment and host switching at novel localities leads to increases in diversity for parasites. Dispersal or invasion apparently does not often involve gene flow, which would dampen patterns of speciation, but involves episodes of range expansion for otherwise discrete and isolated insular species of parasites. Thus, post-speciation dispersal and colonization, rather than coevolutionary processes, were the drivers for assembly of temporally and spatially complex faunal mosaics in these systems.

Invasion and community assembly

Invasion is a prominent component of community assembly, and episodes of dispersal differentially involving diverse taxa will influence ultimate faunal structure in aquatic, marine and terrestrial habitats (Vermeij, 2005). The sequential or hierarchical order of arrival and establishment may influence the potential for subsequent waves of geographic colonization (Hoberg *et al.*, 2012). The degree of taxonomic and ecological saturation in the recipient region also serves as a determinant of structure during and following an expansion event. Relative vagility for a complex assemblage and the form of association between hosts and parasites or among hosts, herbivores and parasitoids (in the instance of arthropods) may strongly influence the process for community assembly under regimes for invasion. For example, community assembly among obligate endoparasites under invasion contrasts with the process of delayed assembly observed among some assemblages of plant/herbivores and parasitoids (Stone *et al.*, 2012). Essentially, obligate endoparasites are dependent and limited by host vagility, dispersion and density (either for intermediate or definitive hosts) during dispersal and the processes of establishment of a new population. Although delayed assembly over time may occur for endoparasites, natural dispersal necessarily requires sympatry and ecological continuity for entire assemblages of hosts and parasites involved in transmission (e.g., Hoberg *et al.*, 2012). In anthropogenic events characterized by jump or long-range dispersal, introductions may be facilitated by prior dissemination for either intermediate or definitive hosts (Miura *et al.*, 2006; Hoberg, 2010). Consequently, parasitoids are a contrast as there are generally greater capacities for expansion and "catching up" to host assemblages which have previously become established in space and time (Stone *et al.*, 2012). Volant parasitoids are not immediately limited by host vagility in dispersal, but successful establishment (as with obligate endoparasites) requires sympatry for plants and susceptible herbivore populations. Overall this process for assembly extended over time and involving a series of events emphasizes the importance of faunal mosaics.

Mosaic structure and invasion

Recurrent events, or a dynamic of taxon pulses leading to geographic expansion and isolation in conjunction with spatially and temporally discrete episodes of host colonization, result in a mosaic structure for parasite faunas, and more generally

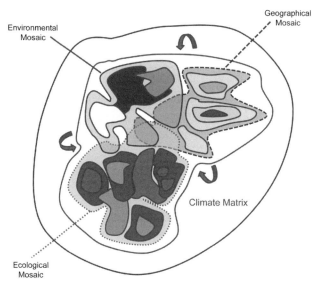

Figure 14.1.   Mosaic structure in space and time. Faunas are structured relative to mosaics resulting from an intricate interaction among ecological (= species richness, population structure, diversity, faunal associations and processes), environmental (= habitat) and geographical (= structural homogeneity and heterogeneity, barriers, invasion corridors) drivers unfolding within a climatological matrix. Autonomous, overlapping and interdependent mosaics emerge and interact within regimes of climate that contribute to periods or episodes of stasis, perturbation and transition. See plate section for color version.

are indicative of the processes involved in overall faunal assembly from landscape to regional scales (Thompson, 2005; Hoberg & Brooks, 2008; Gómez-Díaz *et al.*, 2012; Hoberg *et al.*, 2012). Faunas emerge within intricate mosaics resulting from environmental, geographical and ecological drivers that can be autonomous, but also interacting, interdependent, and correlated in space and time and which exist within a larger climatological matrix (Figure 14.1). Mosaics are described through temporal and spatial hierarchies that extend, and constitute the linkages, from populations to regional faunas and communities. Macroevolutionary mosaics were originally discussed in the context of invasion biology related to faunal (regional interspecific) or populational (landscape intraspecific) admixtures of endemic (native) and introduced species brought into contact and sympatry by anthropogenic translocation and introduction (e.g., Hoberg, 2005, 2010; Hoberg *et al.* 2008). Mosaic structure, however, is a fundamental phenomenon, an observation that emerges from the idea that processes serving as determinants of complex faunas, including those associated with parasites, are equivalent in evolutionary and ecological time (Hoberg & Brooks, 2008, 2010). A conceptual linkage to the original terminology for coevolutionary mosaics is established (Thompson, 2005). In this view, mosaic structure constitutes a continuum resulting from large-scale (regions and faunas) to fine-scale (landscapes and populations) processes acting on the distribution of diversity. Mosaics result from faunal interchange and invasion, both general phenomena at all temporal and spatial scales (e.g., Marshall *et al.*, 1982; Vermeij, 1991a, 2005; Hoberg & Brooks, 2008; Hoberg, 2010; Gómez-Díaz *et al.*, 2012; Hoberg *et al.*, 2012).

Outcomes for faunal mixing apply generally in evolutionary and ecological time: (1) faunas may remain discrete in respective endemic and newly arrived hosts; (2) parasite species may remain discrete, but faunal structure reflects reciprocal exchange (or unidirectional host colonization) often at borderlands defined by ecotones; (3) hybridization or introgression may occur among closely related genera and species; and (4) displacement of an endemic fauna may occur with differential or heterogeneous patterns of extinction/extirpation. Processes may be played out over extended spatiotemporal scales from primary intercontinental events to regions and landscapes. Our developing understanding of the historical origins and complex structure of mosaics is essential in formulating predictions about responses to environmental perturbation in contemporary ecosystems.

Invasion mechanisms leading to introductions of pathogens and parasites are of fundamental importance in contemporary ecosystems and emergent diseases are a critical cornerstone of the current biodiversity crisis (Daszak *et al.*, 2000; Brooks & Ferrao, 2005; Brooks & Hoberg, 2006; Hoberg, 2010). "Anthropogenic invasion mosaics" result directly from human-mediated translocation and introduction leading to admixtures of endemic and introduced species. These mechanisms, often involving long-range and jump dispersal, influence diversity at regional, landscape and populational scales and may have considerable implications in conservation biology (Hoberg 2010). "Contemporary invasion mosaics" represent outcomes of accelerated expansion resulting in rapid and changing patterns of faunal distribution played out in ecological time and often lead to diminished diversity downstream for invaders (and pathogens) (e.g., Torchin *et al.*, 2003); as demonstrated for seabirds, this may not always be a general phenomenon (Gómez-Díaz *et al.*, 2012). In these instances, which reflect ongoing processes across global ecosystems, invasion mosaics involving faunas, species or populations may result from shifts in distribution mediated by accelerated climate warming and habitat perturbation (Parmesan, 2006; Lawler *et al.*, 2009) and other mechanisms for dispersal. The development of new ecotones, patterns of infection in host-pathogen systems, and emergent disease may be one outcome (Brooks & Hoberg, 2007; Hoberg *et al.*, 2008; Kutz *et al.*, 2012).

Articulating a cohesive picture?

Irrespective of the recognized significance of geographic colonization, the process appears to be difficult and stochastic. Successful establishment downstream from source areas requires time, often episodic expansion events, and recurrent or conservative circumstances relative to habitat and ecological structure (e.g., Rickleffs, 2005; Hoberg 2010; Hoberg *et al.*, 2012; Stone *et al.*, 2012). For example, downstream gradients recognized for complex helminth faunas in terrestrial mammals linking Eurasia and North America demonstrated diminished diversity for most parasite groups (Waltari *et al.*, 2007; Hoberg *et al.*, 2012). Differential reductions associated with expansion may or may not be counter-balanced by secondary diversification in regions that have been colonized. Considering examples from Nearctic faunas: specialist helminths in pikas (species of *Ochotona*) include 17 genera in Eurasia and only six are known in the western Nearctic. Among these Holarctic parasite groups, species-richness is centered in Eurasia, but diversification among tapeworms and nematodes was relatively extensive following the initial occupation of North America during the early Pleistocene (Galbreath & Hoberg, 2012). In contrast, ungulate nematode faunas generally exhibit patterns of reduced diversity, with some components being established relatively late in the Quaternary coincidental with expansion by cervids (antlered ungulates) and caprine bovids (wild sheep) into North America from Eurasia (Hoberg *et al.*, 2012).

Perturbation is both isolated and recurrent, and in Earth's history these events have been to a large extent driven by climate (e.g., Vrba, 1996; Jansson & Dynesius, 2002). Events may be short-term (ephemeral) or protracted and may be local, regional or on global spatial scales (see Hoberg & Brooks 2008). Climate-mediated ecological perturbation is the driver, has considerable influence on faunal structure, and can result in substantial ecological re-organization or permanent turnover (e.g., Stigall, 2010, 2012b). Forcing through climate change is essential in creating new circumstances, but is counterbalanced in evolutionary time by phylogenetic conservatism.

Events of ecological disruption have had a pervasive effect on patterns of global diversity, and have in many instances been the precursors and irreversible determinants of faunal structure and history in evolutionary time (e.g., Briggs, 1995). For example, invasion and turnover at the Paleocene/Eocene boundary set the stage for the radiation of modern mammals across the Holarctic (Beard, 2002). Invasion and the resultant faunal mosaics may be a consequence of rapid climate change, setting new boundary conditions which influence tolerance, resilience and species persistence, and constitute new circumstances, particularly with respect to species-level interactions. Invasion through heightened gene flow and connectivity across regions and landscapes dampens or eliminates speciation by vicariance of ancestral ranges (Stigall, 2010).

There are substantial downstream effects linked to episodes of dispersal and invasion that may unfold or cascade over extended time periods. Assuming lineage persistence, these are embodied by (1) perturbation as a mediator of geographic and host colonization; (2) development of new mosaics, or in recurrent events, continual disruption of mosaics with additive processes for species interchange; (3) ecological restructuring; (4) new selection regimes; and (5) new coevolutionary regimes (e.g., Lafferty *et al.*, 2005; Hoberg & Brooks, 2008). Dampening of speciation rates is a further outcome of invasion (Stigall, 2010). In synergy with increasing rates for actual extinction during an ecological crisis, this may drive considerable reductions in species diversity over time. In these instances, there is also a shift from mechanisms of diversification/speciation driven by vicariance to those linked to dispersal. Thus, it is relevant to consider whether the contemporary crisis for biodiversity, which to some degree is dominated by invasion, is one where actual rates of extinction are elevated concurrently with a dampening of diversification rates, leading to a greater overall reduction in species richness and complexity (Stigall, 2010, 2012b).

Human history and invasion

The mode and pace of invasion have shifted over the course of human history such that few corners of the planet are now free of the impacts from exotic introduced species (e.g., Davis, 2009; Stigall, 2012b; Brooks & Hoberg, Chapter 15 of this volume). The character of human-mediated invasion differs to some degree relative to process in natural systems, due not only to the intentional or accidental aspect, but also the degree to which long-range introductions typify these events leading to colonization, establishment, amplification and dissemination (Rickleffs, 2005). Largely anthropogenic drivers increasingly influence invasion and the distribution of invasive species, with attendant threats across a matrix linking environments, economies and societies (e.g., Pimentel *et al.*, 2005).

Thresholds and tipping points in human history have directly influenced the character and evolution of geographic expansion for both free-living and parasitic species. The human imprint on the biosphere has accelerated, and has been directly linked to the distribution of pathogens and host-pathogen systems since initial expansion out of Africa near 40 to 60 kya. The advent of agriculture and animal husbandry near 10 to 11 kya, the age of European exploration ensuing around 1500, and the industrial revolution have represented irreversible points of change for people and our interface with the environment (e.g., Riccardi, 2007; Hoberg, 2010). Increasingly it has been argued that human influence now is a pervasive force in evolution with effects visible in natural systems, and diverse assemblages of pathogens in free-ranging and domesticated hosts (Palumbi, 2001). Tipping points in human history emerged from a burgeoning population and a transition from a slow and large world dominated by relative isolation and local effects, to a rapid and small world associated with extensive globalization, homogenization and integrated but often fragmented environmental networks. It is in this arena that we can see the continuing significance of invasive events that emerge from a deep temporal continuum established over a dynamic history of environmental and ecological perturbation that is episodic in evolutionary and ecological time.

Acknowledgments

Some research results discussed here are contributions of the Beringian Coevolution Project, a multi-national research collaboration to explore diversity across the roof of the world, supported by grants from the National Science Foundation (DEB 0196095 and 0415668) to J. A. Cook at the Museum of Southwestern Biology, University of New Mexico and EPH at the US National Parasite Collection.

References

Agosta, S. J., & Klemens, J. A. (2008). Ecological fitting by phenotypically flexible genotyopes: implications for species associations, community assembly and evolution. *Ecology Letters*, **11**, 1–12.

Arenas, M., Ray, N., Currat, M., & Excoffier, L. (2012). Consequences of range contractions and range shifts on molecular diversity. *Molecular Biology and Evolution*, **29**, 207–218.

Beard, C. (2002). East of Eden at the Paleocene/Eocene boundary. *Science*, **195**, 2028–2029.

Briggs, J. C. (1995). *Global Biogeography*. Amsterdam: Elsevier.

Briggs, J. C. (2003). Marine centres of origin as evolutionary engines. *Journal of Biogeography*, **30**, 1–18.

Brooks, D. R., & Ferrao, A. (2005). The historical biogeography of coevolution: emerging infectious diseases are evolutionary accidents waiting to happen. *Journal of Biogeography*, **32**, 1291–1299.

Brooks, D. R., & Hoberg, E. P. (2006). Systematics and emerging infectious diseases: from management to solution. *Journal of Parasitology*, **92**, 426–429.

Brooks, D. R., & Hoberg, E. P. (2007). How will global climate change affect parasite-host assemblages? *Trends in Parasitology* **23**, 571–574.

Brooks, D. R., & McLennan, D. A. (1993). *Parascript: Parasites and the Language of Evolution*. Washington DC: Smithsonian Institution Press.

Brooks, D. R., & McLennan, D. A. (2002). *The Nature of Diversity: An Evolutionary Voyage of Discovery*. Chicago, IL: University of Chicago Press.

Brooks, D. R., Thorson, T. B., & Mayes, M. A. (1981). Freshwater stingrays (Potamotrygonidae) and their helminth parasites: testing hypotheses of evolution and coevolution. In V. A. Funk & D. R. Brooks (Eds.), *Advances in Cladistics* (pp. 147–175). New York: New York Botanical Garden.

Daszak, P., Cunningham, A. A., & Hyatt, A. D. (2000). Emerging infectious diseases of wildlife: threats to biodiversity and human health. *Science*, **287**, 443–449.

Davis, M. A. (2009). *Invasion Biology*. Oxford: Oxford University Press.

Dynesius, M., & Jansson, R. (2000). Evolutionary consequences of changes in species' geographical distributions driven by Milankovitch climate oscillations. *Proceedings of the National Academy of Sciences of the USA*, **97**, 9115–9120.

Eldredge, N., & Gould, S. J. (1972). Punctuated equilibria: an alternative to phyletic gradualism. In T. J. M. Schopf (Ed.), *Models in Paleobiology* (pp. 82–115). San Francisco, CA: Freeman, Cooper.

Eldredge, N., Thompson, J. N., Brakefield, P. M., *et al.* (2005). The dynamics of evolutionary stasis. *Paleobiology*, **31**, 133–145.

Erwin, T. L. (1985). The taxon pulse: a general pattern of lineage radiation and extinction among carabid beetles. In G. E. Ball (Ed.), *Taxonomy, Phylogeny, and Biogeography of Beetles and Ants* (pp. 437–472). Dordrecht: W. Junk.

Folinsbee, K. E., & Brooks, D. R. (2007). Early hominid biogeography: pulses of dispersal and differentiation. *Journal of Biogeography*, **43**, 383–397.

Galbreath, K. E., & Hoberg E P. (2012). Return to Beringia: parasites reveal cryptic biogeographic history of North American pikas. *Proceedings of the Royal Society of London B*, **279**, 371–378.

Gómez-Díaz, E., Morris-Pococik, J. A., González-Solis, J., & McCoy, K. D. (2012). Trans-oceanic host dispersal explains high seabird tick diversity on Cape Verde Islands. *Biology Letters*. doi:10.1098/rsbl.2012.0179.

Halas, D., Zamparo, D., & Brooks, D. R. (2005). A historical biogeographical protocol for studying diversification by taxon pulses. *Journal of Biogeography*, **32**, 249–260.

Hallam, A., & Wignall, P. B. (1997). *Mass Extinctions and Their Aftermath*. Oxford: Oxford University Press.

Hewitt, G. M. (1996). Some genetic consequences of ice ages and their role in divergence and speciation. *Biological Journal of the Linnaean Society*, **58**, 247–276.

Hoberg, E. P. (1995). Historical biogeography and modes of speciation across high-latitude seas of the Holarctic: concepts for host-parasite coevolution among the Phocini (Phocidae) and Tetrabothriidae (Eucestoda). *Canadian Journal of Zoology*, **73**, 45–57.

Hoberg, E. P. (1996). Faunal diversity among avian parasite assemblages: the interaction of history, ecology and biogeography. *Bulletin of the Scandinavian Society of Parasitology*, **6**, 65–89.

Hoberg E. P. (1997). Phylogeny and historical reconstruction: host parasite systems as keystones in biogeography and ecology. In M. Reaka-Kudla, D. E. Wilson & E. O. Wilson (Eds.), *Biodiversity II: Understanding and Protecting Our Resources* (pp. 243–261). Washington DC: Joseph Henry Press, National Academy of Sciences.

Hoberg, E. P. (2005). Coevolution and biogeography among Nematodirinae (Nematoda: Trichostrongylina) Lagomorpha and Artiodactyla (Mammalia): exploring determinants of history and structure for the northern fauna across the Holarctic. *Journal of Parasitology*, **91**, 358–369.

Hoberg, E. P. (2010). Invasive processes, mosaics and the structure of helminth parasite faunas. *Revue Scientifique et Technique Office International des Épizooties*, **29**, 255–272.

Hoberg, E. P., & Adams, A. M. (2000). Phylogeny, history and biodiversity: understanding faunal structure and biogeography in the marine realm. *Bulletin Scandinavian Society of Parasitology*, **10**, 19–37.

Hoberg, E. P., & Brooks. D. R. (2008). A macroevolutionary mosaic: episodic host-switching, geographic colonization, and diversification in complex host-parasite systems. *Journal of Biogeography*, **35**, 1533–1550.

Hoberg E. P., & Brooks, D. R. (2010). Beyond vicariance: integrating taxon pulses, ecological fitting and oscillation in historical biogeography and evolution. In S. Morand & B. Krasnov (Eds.), *The Geography of Host-Parasite Interactions* (pp. 7–20). Oxford: Oxford University Press.

Hoberg, E. P., & Klassen, G. J. (2002). Revealing the faunal tapestry: co-evolution and historical biogeography of hosts and parasites in marine systems. *Parasitology*, **124**, S3–S22.

Hoberg, E. P., Polley, L., Jenkins, E. J., & Kutz, S. J. (2008). Pathogens of domestic and free ranging ungulates: global climate change in temperate to boreal latitudes across North America. *Office International des Épizooties Revue Scientifique et Technique*, **27**, 511–528.

Hoberg, E. P., Galbreath, K. E., Cook, J. A., Kutz, S. J., & Polley L. (2012). Northern host-parasite assemblages: history and biogeography on the borderlands of episodic climate and environmental transition. *Advances in Parasitology*, **79**, 1–97.

Jansson, R., & Dynesius, M. (2002). The fate of clades in a world of recurrent climatic change: Milankovitch oscillations and evolution. *Annual Reviews of Ecology and Systematics*, **33**, 741–777.

Janzen, D. (1985). On ecological fitting. *Oikos*, **45**, 308–310.

Kutz, S. J., Ducrocq, J., Verocai, G., *et al.* (2012). Parasites in ungulates of Arctic North America and Greenland: a view of contemporary diversity, ecology and impact in a world under change. *Advances in Parasitology*, **79**, 99–252.

Lafferty, K. D. (2009). The ecology of climate change and infectious diseases. *Ecology*, **90**, 888–900.

Lafferty, K. D., Smith, K. F., Torchin, M. E., Dobson, A. P., & Kuris, A. M. (2005). The role of infectious diseases in natural communities. In D. F. Sax, J. Stachowicz & S. D. Gaines (Eds.), *Species Invasions: Insights into Ecology, Evolution and Biogeography* (pp. 111–134). Sunderland, MA: Sinauer Associates.

Lawler, J. J., Shafer, S. L., White, D., *et al.*, 2009. Projected climate-induced faunal change in the Western Hemisphere. *Ecology*, **90**, 588–597.

Lister, A. (2004). The impact of Quaternary ice ages on mammalian evolution. *Philosophical Transactions of the Royal Society B: Biological Sciences*, **359**, 221–241.

Marcogliese, D. J. (2005). Parasites of the superorganism: are they indicators of ecosystem health. *International Journal for Parasitology*, **35**, 705–716.

Marshall, L. G., Webb, S. D., Sepkoski, J. J., & Raup, D. M. (1982). Mammalian evolution and the Great American Interchange. *Science*, **215**, 1351–1357.

Miura, O., Torchin, M. E., Kuris, A. M., Hechinger, R. F., & Chiba, S. (2006). Introduced cryptic species of parasites exhibit different invasion pathways. *Proceedings of the National Academy of Sciences of the USA*, **103**, 19818–19823.

Nelson, G., & Platnick, N. I. (Eds.) (1981). *Systematics and Biogeography: Cladistics and Vicariance*. New York: Columbia University Press.

Nieberding, C., & Olivieri, I. (2007). Parasites: proxies for host genealogy or ecology? *Trends in Ecology & Evolution*, **22**, 156–165.

Page, R. D. M., (Ed.). (2003). *Tangled Trees: Phylogeny, Cospeciation and Coevolution*. Chicago, IL: University of Chicago Press.

Palumbi, S. (2001). Humans as the world's greatest evolutionary force. *Science*, **293**, 1786–1790.

Parmesan, C. (2006). Ecological and evolutionary responses to recent climate change. *Annual Review of Ecology and Systematics*, **37**, 637–669.

Peterson, A. T. (2011). Ecological niche conservatism: a time-structured view of evidence. *Journal of Biogeography*, **28**, 817–827.

Pimentel, D., Zuniga, R., & Morrison, D. (2005). Update on the environmental and economic costs associated with alien-invasive species in the United States. *Ecological Economics*, **52**, 273–288.

Riccardi, A. (2007). Are modern biological invasions an unprecedented form of global change? *Conservation Biology*, **21**, 239–336.

Rickleffs, R. E. (2004). A comprehensive framework for global patterns in biodiversity. *Ecological Letters*, **7**, 1–5.

Rickleffs, R. E. (2005). Taxon cycles: insights from invasive species. In D. F. Sax, J. Stachowicz & S. D. Gaines (Eds.), *Species Invasions: Insights into Ecology, Evolution and Biogeography* (pp. 165–193). Sunderland, MA: Sinauer Associates.

Rode, A. L., & Lieberman, B. S. (2005). Integrating evolution and biogeography: a case study involving Devonian crustaceans. *Journal of Paleontology*, **79**, 267–276.

Sandel, B., Arge, L., Dalsgaard, B., *et al.* (2011). The influence of Late Quaternary climate-change velocity on species endemism. *Science*, **334**, 660–664.

Shafer, A. B. A., Cullingham, C. I., Côté, S. D., & Coltman, D. W. (2010). Of glaciers and refugia: a decade of study sheds new light on the phylogeography of northwestern North America. *Molecular Ecology*, **19**, 4589–4621.

Sher, A. (1999). Traffic lights at the Beringian crossroads. *Nature*, **397**, 103–104.

Stigall, A. L. (2010). Invasive species and biodiversity crises: testing the link in the late Devonian. *PLoS One*, **5** (12): e15584. doi:10.1371/journal.pone.0015584.

Stigall, A. L. (2012a). Using ecological niche modeling to evaluate niche stability in deep time. *Journal of Biogeography*, **39**, 772–781.

Stigall, A. L. (2012b). Invasive species and evolution. *Evolutionary Education Outreach*. doi: 10.1007/s12052-012-0410-5.

Stone, G. N., Lohse, K., Nicholls, J. A., *et al.* (2012). Reconstructing community assembly in time and space reveals enemy escape in a western Palearctic insect community. *Current Biology*, **22**, 532–537.

Thompson, J. N. (2005). *The Geographic Mosaic of Coevolution*. Chicago, IL: University of Chicago Press.

Thuiller, W. (2007). Biodiversity: climate and the ecologist. *Nature*, **448**, 550–552.

Torchin, M. E., Lafferty, K. D., Dobson, A. P., Mackenzie, V. J., & Kuris, A. N. (2003). Introduced species and their missing parasites. *Nature*, **412**, 628–629.

Vrba, E. S. (1996). On the connections between paleoclimate and evolution. In E. S. Vrba, G. H. Denton, T. C. Partridge & L. H. Burkle (Eds.), *Paleoclimate and Evolution with Emphasis on Human Origins* (pp. 24–48). New Haven: Yale University Press.

Waltari, E., Hoberg, E. P., Lessa, E. P., & Cook, J. A. (2007). Eastward Ho: phylogeographic perspectives on colonization of hosts and parasites across the Beringian nexus. *Journal of Biogeography*, **34**, 561–574.

Webb, S. D. (1995). Biological implications of the Middle Miocene Amazon seaway. *Science*, **269**, 361–362.

Webb, S. D., & Marshall, L. G. (1981). Historical biogeography of recent South American land mammals. *Pymatuning Symposia in Ecology*, **6**, 39–52.

Wiley, E. O. (1981). *Phylogenetics: The Theory and Practice of Phylogenetic Systematics*. New York: John Wiley.

Vermeij, G. (1991a). When biotas meet: understanding biotic interchange. *Science*, **253**, 1099–1104.

Vermeij, G. J. (1991b) Anatomy of an invasion: the trans-Arctic interchange. *Paleobiology*, **17**, 281–307.

Vermeij, G. J. (2005). Invasion as expectation: a historical fact of life. In D. F. Sax, J. Stachowicz & S. D. Gaines (Eds.), *Species Invasions: Insights into Ecology, Evolution and Biogeography* (pp. 315–339). Sunderland, MA: Sinauer Associates.

Zarlenga, D. S., Rosenthal, B. M., La Rosa, G., Pozio, E., & Hoberg, E. P. (2006). Post Miocene expansion, colonization, and host switching drove speciation among extant nematodes of the archaic genus *Trichinella*. *Proceedings of the National Academy of Sciences of the USA*, **103**, 7354–7359.

15 The emerging infectious diseases crisis and pathogen pollution

Daniel R. Brooks and Eric P. Hoberg

The human population grows daily, it's on the move and it's carving a deep technological footprint on this planet. We alter landscapes and perturb ecosystems, inserting ourselves and other species into novel regions of the world, leading to potentially irreversible changes in the biosphere. This is not news. Half a century ago, Charles Elton (1958), a founder of modern ecology, wrote, "We must make no mistake; we are seeing one of the greatest historical convulsions in the world's fauna and flora". This is the biodiversity crisis.

Even as our species engineers this planet, the planet itself is changing with perturbations emerging from overall warming which is accelerating over time (e.g., Parry *et al.*, 2007). Some areas are getting wetter, some drier. Some areas are warmer, others cooler. Weather patterns are becoming more extreme – more droughts, more floods – and our ability to predict the weather as it emerges from a deeper climatological background seems to have taken a step backwards in recent years. The growing body of empirical evidence accords with predictions made by most models of climate change, which has led to even more dire predictions about the short-term future and major shifts in the structure of ecosystems and the distribution of biodiversity (Parmesan, 2006; Lawler *et al.*, 2009; Dawson *et al.*, 2011). This is the global climate change crisis.

We are also in the midst of an epidemiological crisis. Climate change alters movements and geographic distributions for myriad species and their associated pathogens (e.g., Dobson & Carper, 1992; Daszak *et al.*, 2000; Hoberg *et al.*, 2008; Patz *et al.*, 2008). Transporting people and goods carries countless pathogens around the globe, bringing isolated species into sudden contact (Kilpatrick, 2011). Pathogens encounter hosts with no resistance and no time to evolve resistance. This is also not news – maladies rare or unknown two or three decades ago, like HIV and Ebola, West Nile virus and avian influenza, have become commonplace. In such a world – this world – events like these are ongoing. Scarcely a week passes without news of some freshly discovered strain of pathogen trading up to a human host. This is the emerging infectious disease (EID) crisis.

In the past 10 000–15 000 years agriculture, domestication and urbanization disseminated EID risk on a global scale as people and their interfaces with the environment were altered over time (e.g., Daszak *et al.*, 2000; Rosenthal, 2008). If doctors had existed in those times, they would have remarked on a worrisome surge in the number of EIDs,

The Balance of Nature and Human Impact, ed. Klaus Rohde. Published by Cambridge University Press.
© Cambridge University Press 2013.

responding to each new crisis as best they could, while waiting for events to "burn out" in a vacuum of limited knowledge but considerable superstition about the biosphere. In the past 50 years, exploding human population, rapid and global transportation networks (connectivity), and accelerating climate change have produced the real-time crisis we are experiencing (e.g., Hoberg, 2010). Popular opinion, including the world of biotechnology and genetic engineering, treats the EID crisis as something novel in the history of this planet and our species. We think of EIDs as isolated events, and react only after the fact. We allocate massive resources to pathogens that have already made themselves known, while ignoring those waiting in the wings. They're discovering us easily enough – weekly outbreaks and endlessly mutating strains of recent years are ample evidence. This succession of crises is the new status quo.

How do emerging diseases happen?

We believe the EID crisis is a medical issue in only a superficial sense. It's more fundamentally an evolutionary and ecological issue, a predictable consequence of species that evolved in isolation being brought into close contact with breakdown in mechanisms for biogeographic and ecological isolation. The difference today is that human activity accelerates the rate of introductions, so outbreaks occur more frequently and over a wider geographic range than ever before.

One reason for a general belief that emerging diseases will be rare is the recognition that emerging diseases are the result of pathogens switching hosts, and the conventional wisdom in evolutionary biology has been that host switches are difficult to achieve. One of the most obvious and intriguing features of parasitism is pronounced conservatism in the range of hosts used, both on ecological (Thompson, 1994, 2005) and evolutionary time-scales (Ehrlich & Raven, 1964; Brooks & McLennan, 1991, 1993, 2002; Thompson, 1994, 2005; Futuyma & Mitter, 1996; Janz & Nylin, 1998). Most parasites appear to be resource specialists. The overwhelming majority of parasites use only a tiny fraction of the available host species in the habitat (best documented for phytophagous insects and their host plants, e.g., Futuyma & Moreno, 1988; Bernays & Chapman, 1994; Thompson, 1994). Not surprisingly, this pattern has given rise to the long-standing idea that specialization is a one-way street, an evolutionary dead-end where parasites become increasingly well-adapted to their hosts at the expense of the ability to perform on alternative hosts. The idea of specialization as a dead-end originated in the nineteenth century (Thompson, 1994) and some recent studies support the idea (Moran, 1988; Wiegmann et al., 1993; Kelley & Farrell, 1998). Other studies, however, conclude that generalized lineages are often derived from specialists (Scheffer & Wiegmann, 2000; Janz et al., 2001; Termonia et al., 2001; Radtke et al., 2002; Kergoat et al., 2005; Yotoko et al., 2005). Many groups exhibit higher transition rates from generalization to specialization than vice versa, but host specialization appears to be a dynamic trait with no inherent necessary directionality (Janz et al., 2001; Nosil, 2002; Nosil & Mooers, 2005). If this is the case, why are wide host ranges so rare? Opportunities to broaden a species' niche should always exist, so something must consistently select against this

(Futuyma & Moreno, 1988). Many studies have also shown that selection should favor increased host specificity over time for a variety of reasons (Futuyma & Moreno, 1988; Bernays, 1989; Via, 1991; Agrawal, 2000; Bernays, 2001; Janz, 2002; Janz *et al.*, 2005). So many factors favor host specialization that it has become a challenge to find any good scenario that would reverse the process (Johansson *et al.*, 2007; Singer, 2008).

Hence, the solution for the original problem of widespread specialization has in many ways led to the new problem of understanding why there are exceptions at all. And yet, without such exceptions there would be no host shifts, and thus no emerging diseases. Every host shift must begin with colonization, implying a capacity to use both the ancestral and novel host. Multiple host use following such a colonization may be brief or it may be prolonged (Janz *et al.*, 2006; Janz & Nylin, 2008), but host shifts must begin with a host range expansion. Additional hosts should be inferior alternatives to the original host, to which the parasite is specifically adapted, and special circumstances should be needed to incorporate such a host into the repertoire. And yet, host shifts and host range expansions occur often, and can happen rapidly (Tabashnik, 1983; Singer *et al.*, 1993; Carroll *et al.*, 1997; Fox *et al.*, 1997; Thompson, 1998; van Klinken & Edwards, 2002; Agosta, 2006). How parasites can be highly specialized and shift to novel hosts often is the parasite paradox (Agosta *et al.*, 2010).

To understand and resolve this paradox, it is important to understand how to complete a shift to a novel host if specialization results only from coevolution that constrains parasites to their current hosts. In this case, a full host shift will require more or less simultaneous correlated evolution across a number of traits. In order to successfully colonize a novel host, a parasite will need to modify traits that enable it to locate the new resource, identify it as a possible host, and ensure reproductive continuity in association with the new host. In addition, offspring finding themselves on this novel resource will need to be able to sustain themselves nutritionally, and their metabolic system will have to be able to digest the new resource and overcome its chemical defense (or immune system). Each new host may also come with a different set of external enemies requiring new methods of defense or evasion and a different micro-habitat requiring novel physiological adaptations.

Such correlated changes occurring simultaneously across all these sets of characters ought to be so unlikely that host shifts ought to border on the impossible. Yet they are common. Phylogenetic comparative studies of hosts and parasites demonstrate two macroevolutionary patterns: (1) host range is narrow for most parasite species, but (2) there is substantial evidence of host switching and in some cases host switching seems to have been the primary driver of diversification (Brooks & McLennan, 1991, 1993, 2002; Hoberg & Klassen, 2002; Janz *et al.*, 2006; Agosta, 2006; Brooks *et al.*, 2006a; Janz & Nylin, 2008; Brooks & Van Veller, 2008; Hoberg & Brooks, 2008, 2010; Agosta *et al.*, 2010).

The resolution to the parasite paradox assumes that host shifts comprise two different phases: (1) colonization of the novel host (host range expansion) followed by (2) loss of the ancestral host (host specialization). Hence, for a host shift to be completed, there must first be a mechanism for generalization (increased host breadth) and then a mechanism for specialization (decreased host breadth). Furthermore, in order for specialization not to be an evolutionary "dead-end", these mechanisms must be at least partly independent, so specialists maintain the potential to become generalists and generalists maintain the potential to become specialists. This links host shifts to the microevolutionary processes that determine host specificity.

Resolving the paradox

It is hard to escape the conclusion that the capacity to utilize the novel resource must have existed before a successful shift was initiated. Some might object that this is not possible, because evolution cannot pre-assemble the necessary features involved in completing a shift to a new resource. However, mechanisms allowing organisms to colonize and persist in novel environments exist. Phenotypic accommodation (West-Eberhard, 2003) stemming from phenotypic plasticity, in addition to factors discussed below, provides a mechanistic basis for an ecological concept called ecological fitting (Janzen, 1985). Janzen suggested that species often enter into new communities and form new interactions without any prior evolutionary changes, and claimed that "ecological fitting" has played a major role in shaping ecological communities. When a species encounters a new set of circumstances – a new habitat, a changed climate, a change in resource availability – it will persist where (and if) it "fits" into them by means of characters it already possesses. In other words, successful establishment in a novel environment requires that the species have a reaction norm including the novel resource. Ecological and macroevolutionary evidence for ecological fitting among hosts and parasites is abundant in natural and human-manipulated systems (Agosta, 2006; Hoberg & Brooks, 2008; Agosta et al., 2010). This suggests that host shifts are often initiated because the insect or parasite is "pre-adapted" or "exapted" (Gould & Vrba, 1982) to the novel resource. The novel host might share important characteristics with the current host or might have been used in the past (Futuyma et al., 1995; Futuyma & Mitter, 1996; Janz et al., 2001; Wahlberg, 2001; Brooks & McLennan, 2002; Radtke et al., 2002), or the parasite might fortuitously possess capabilities to use a novel resource (Agosta & Klemens, 2008).

Agosta and Klemens (2008) proposed a general framework for ecological fitting as a mechanism behind the assembly of ecological communities and the formation of novel interactions between species within communities. They posited three factors giving rise to ecological fitting and the ability of organisms (genotypes) to achieve realized fitness under novel conditions (e.g., a novel host). First, phenotypic plasticity can allow organisms to mount a response to novel conditions (West-Eberhard, 2003). Second, correlated trait evolution (Lande & Arnold, 1983) can produce phenotypes that are "preadapted" to some future novel condition. Third, phylogenetic conservatism in traits related to resource use provide the latent ability to perform under apparently novel conditions, such as a herbivore encountering a new host that is actually the same or sufficiently similar to some ancestral host (Brooks & McLennan, 2002; Hoberg & Brooks, Chapter 14 of this volume). These capacities produce organisms possessing potential fitness outside the range of conditions in which the species evolved. Agosta and

Klemens (2008) termed this region of fitness space "sloppy fitness space", a by-product of direct selection under some other set of conditions (the ancestral operative environment: Agosta & Klemens, 2008). The operative environment comprises all components defining a host as a resource and therefore the range of host-related variables affecting parasite evolution. Each parasite evolves in response to the ancestral host operative environment, but as a consequence also has potential fitness (sloppy fitness space) beyond the range of conditions encountered with the ancestral host. Thus, the parasite has some ability to perform and persist on other hosts that represent a novel operative environment, in addition to the ability to add new hosts representing the same or highly similar operative environments.

Armed with adaptations to their ancestral hosts, and the sloppy fitness space that results, parasites can ecologically fit with new hosts in at least two ways (Agosta & Klemens, 2008). First, due to phylogenetic conservatism or convergence of host resources, parasites may shift to a new host species because the new host possesses the same (or highly similar) resources as the old host: ecological fitting via resource tracking. Second, parasites may use sloppy fitness space to shift onto new hosts representing a novel resource: ecological fitting via sloppy fitness space. In either case, observed interactions between hosts and parasites will appear to be newly evolved, but are products of ecological fitting. Ecological fitting via resource tracking and via sloppy fitness space are not mutually exclusive and may represent two ends of a continuum (Agosta & Klemens, 2008). A typical host shift may involve both tracking pleisomorphic resources and use of sloppy fitness space. In the case of ecological fitting via resource tracking, the new host may be novel only in the sense that it is a different species – from the perspective of the parasite it may not be novel at all. This situation illustrates how parasites can shift to relatively unrelated hosts if the operative environments (e.g., chemicals recognized as oviposition stimulants: Murphy & Feeny, 2006) are plesiomorphic or convergent; it also explains the ecological context such as guild structure that can define the potential for host switching in space and time (Hoberg & Brooks, 2008; Hoberg & Brooks, Chapter 14 of this volume). It also illustrates conceptual problems with our understanding of what constitutes specialists and generalists.

The terms specialist and generalist are vaguely defined, and used variously by researchers. Typically, the terms refer to the dimension of niche width that reflects diet, and are then often simply measured as the number of hosts in a parasite's repertoire. As host species are themselves hierarchically related, counting host species may be misleading, and some authors have adopted host use indices that to various extents take host relatedness into account (Symons & Beccaloni, 1999; Janz *et al.*, 2001, 2006). More important is that species with virtually identical present-day host ranges can achieve them in fundamentally different ways, and as a consequence have divergent evolutionary potentials. Some species will be restricted to one host due to a narrow niche width, others will be restricted to one host because it is the only host locally available.

A true generalist parasite has evolved a more general host recognition and tolerance system that allows it to infect any host that falls within a broad tolerance range. By contrast, a polyspecialist parasite infects different hosts because it has adapted separately to each. Brooks and McLennan (2002) suggested that hidden among the "true" specialists and generalists are "faux specialists" and "faux generalists". Faux specialists are generalists restricted to a few or a single resource by ecological factors, such as competition, local micro-climate or non-concordant distributional ranges. Faux generalists are specialists specialized on a resource that is phylogenetically widespread. Parasites are not specialized on particular host species; they are specialized on resources that may or may not be shared among several species. In some cases, a plesiomorphic resource can be shared among a set of hosts to the extent that, from the parasite's perspective, they are identical and interchangeable. In other cases there will be a quantitative difference between host species in the amount of the resource(s) they possess. For faux generalists and faux specialists, host shifts can be initiated simply by a change in ecological circumstances, e.g., a shift in local host availability or local extinction of a competitor.

Host shifts are difficult to observe. We must often be content with scattered "snapshots" from different stages of the process to draw a composite picture of the whole process. Researchers have occasionally observed a host shift from the initial colonization to the final (local) loss of the ancestral host. The butterfly *Euphydryas editha* occurs in a fragmented population structure across the western USA and Canada. Host use has been studied extensively in several populations over many years (Singer, 2003; Singer *et al.*, 2008). Singer and colleagues have observed two instances of anthropogenically induced shifts in host use. In one case, the local flora was altered by the introduction of an exotic plant, and in the other logging removed the original host (Singer *et al.*, 2008). In both cases, at least some of local *E. editha* accepted a novel host for oviposition and survived on it at first encounter (Singer *et al.*, 2008).

Transfer to a novel environment may remove the ancestral host and offer alternatives, but more often a parasite population comes into contact with a novel host in sympatry with the ancestral host. Antonovics *et al.* (2002) showed that the recent shift by the anther-smut disease (*Microbotryum violaceum*) from the ancestral host *Silene alba* to *S. vulgaris* was contingent on local-scale co-occurrence of both species. The pathogen was imperfectly adapted to its new host and susceptibility to the pathogen varied considerably. Similarly, some populations of the prodoxid moth *Greya politella* in central Idaho have shifted from the ancestral host *Lithophragma parviflorum* to the related *Heuchera grossulariifolia*. In this case, populations feeding on novel hosts are locally adapted to them, and preference for tetraploid variants seems to have evolved independently in several populations (Segraves *et al.*, 1999; Nuismer & Thompson, 2001). *Lithophragma*-feeding populations exposed to the novel host oviposited in it to a low degree, but did not differentiate between plants of different ploidies (Janz & Thompson, 2002). Hence, the shift was probably initiated by local contingency, and local preference for tetraploid variants of the novel host evolved independently after the initial colonization (Thompson *et al.*, 2004). These examples suggest shifts initiated without any evolutionary change, through ecological fitting by phenotypic accommodation, followed by rapid evolution of traits associated with host use (Antonovics *et al.*, 2002; Thompson *et al.*, 2004; Singer *et al.*, 2008).

As mentioned above, host use involves a number of different processes that must all function in concert. A potential host may possess some but not all of the required resources. The first step towards a host shift involves a failure to fully discriminate against a host sympatric with the ancestral host, at least in part of the parasite's geographic distribution (Larsson & Ekbom, 1995). If offspring have non-zero fitness on the new host, natural selection can then begin playing a role in modifying traits involved in host use on this plant, through genetic accommodation and later possibly through character release, allowing utilization of the novel plant to evolve more independently (West-Eberhard, 2003; Nylin & Janz, 2009). Host shifts with loss of the ancestral host are likely to require evolutionary modification of host utilization traits after the initial colonization. The butterfly *Pieris napi* regularly oviposits on the introduced *Thlaspi arvense*, although the plant is lethal for the larvae (Chew, 1977). Presumably, the introduced plant contains an oviposition "resource" similar to that of local hosts, but a different larval feeding "resource". Mortality is 100%, leaving no opportunity for evolutionary modification enabling a host shift. Initial colonization of *Chloroleucon ebano* by the seed beetle *Stator limbatus* depended on pre-existing variance in the capacity to accept and utilize this novel host, and populations expanding their host range to include this species had not locally adapted to it. One reason for the lack of local adaptation was a significant non-genetic effect of maternal host plant on offspring survival on the novel host (Fox *et al*., 1997). Hence, successful establishment on *C. ebano* depends on local co-occurrence of one of the other hosts in the repertoire of *S. limbatus*. These examples show that successful host shifts depend on the history of association, as well on life history, abundance and distribution of the species involved. Understanding and potentially predicting host shifts emerges from recognition of three interrelated processes: (1) ecological mechanisms (ecological fitting, local contingency); (2) evolutionary mechanisms (evolutionary past, degree of plasticity, phylogenetic conservatism, genetic accommodation); and (3) physical processes (e.g., climate, episodes of environmental perturbation at varying scales) (Brooks & McLennan 2002; Hoberg & Brooks, 2008, Chapter 14 of this volume; Agosta *et al*., 2010; Hoberg, 2010).

Complete understanding of the complex history of host-parasite associations requires consideration of multiple, non-mutually exclusive evolutionary and ecological mechanisms and phenomena, but we believe that host shifts via ecological fitting provide a missing link in our general understanding of the evolution and diversification of host-parasite interactions (Brooks & McLennan, 2002; Brooks *et al*., 2006b; Hoberg & Brooks, 2008, 2010; Agosta *et al*., 2010). Ecological fitting not only provides the necessary first step in the colonization process by initiating a host shift, it also provides essential raw material for coevolutionary interactions. The geographic mosaic theory of coevolution (Thompson, 1994, 2005) emphasizes the interplay between local adaptation and gene flow in geographically structured populations. Species can interact in different ways, and with different species in different parts of its geographic range, leading to the ephemeral or transient build-up and breakdown of locally adapted coevolutionary hot spots depending on gene flow and the local presence or absence of other interacting species (Thompson & Fernandez, 2006). This perspective provides an appreciation of the observation that parasite species infecting multiple hosts often comprise several locally specialized populations (Fox & Morrow, 1981). It further explains the observation that the distribution of a parasite is considerably greater than the often focal distribution of disease within a complex assemblage on regional to local scales (Audy, 1958). Raw material for coevolution, however, must be constantly regenerated through colonization of novel resources in parts of a species' geographic distribution (Janz & Thompson, 2002; Singer et al., 2008), made possible by ecological fitting and the exploration of sloppy fitness space. Coevolutionary processes can buffer a species against fragmentation or promote diversification, depending on the nature and strength of processes acting on local and regional scales (Benkman, 1999; Thompson, 1994, 2005; Godsoe et al., 2008). Host colonization by ecological fitting promotes diversification only to the extent that the geographic mosaic allows local adaptation to newly formed hosts and sufficient isolation from the ancestral population (Agosta & Klemens, 2008; Hoberg & Brooks, Chapter 14 of this volume).

Host shifts initiated by ecological fitting are one step in the process that fuels biological expansion and generates novel combinations of interacting species. From these novel interactions, ecological fitting may promote evolutionary stasis (e.g., if there are many ecologically fit populations connected by sufficient gene flow), or may facilitate evolutionary diversity (Agosta, 2006; Agosta & Klemens, 2008; Agosta et al., 2010). Janz and colleagues recently proposed that diversification in these systems is driven by repeated "oscillations" in host range (Janz & Nylin, 1998; Janz et al., 2006; Nylin & Janz, 2009). Phases of host expansions, coupled with geographic range expansions, are followed by phases of host specialization and geographic isolation (or, alternatively, sympatric speciation by host race formation). As with the geographic mosaic theory (Thompson, 2005), the oscillation hypothesis relies on constant regeneration of variation in host use; it is the diversification of host use that drives diversification of species regardless of speciation mode (Janz & Nylin, 2008; Agosta et al., 2010). Conceptually these ideas are central to a synergy for episodic geographic expansion and invasion, taxon pulses, and host colonization as principal mechanisms involved in structuring complex assemblages of hosts and parasites in evolutionary and ecological time (Hoberg & Brooks, 2008, 2010, Chapter 14 of this volume).

This paradigm allows us to see that the current EID crisis is a manifestation of an old and repeating phenomenon. The rules have not changed. Every episode of global climate change and ecological perturbation throughout Earth's history has produced new pathogens or new associations (Brooks & Hoberg, 2006; Hoberg & Brooks, 2008, 2010, Chapter 14 of this volume). For example, more than a million years ago, our African ancestors moved from forest to savannah. Adopting a predatory lifestyle, sharing prey with grassland carnivores, early humans acquired tapeworm parasites previously found in circulation among hyenas, large cats and African hunting dogs (e.g., Hoberg et al., 2001); this scenario has played out for a considerable array of pathogens including viruses derived from free-ranging wildlife species during human evolution (Wolfe et al., 2007). Biotic expansion beyond Africa and geographic colonization have contributed to the interchange of pathogens among people, wildlife and domesticated animals (e.g., Folinsbee & Brooks, 2007). Agriculture and urbanization brought people and animals into even closer contact, making infection and transmission easier than ever. Further, human-mediated introductions and invasion through these tipping points in population size and mobility have resulted in global dissemination for a complex array of parasites in both domestic animals and wildlife species (Rosenthal, 2008; Hoberg,

2010; Hoberg & Brooks, Chapter 14 of this volume). The potential for EID is thus a "built-in feature" of evolution. Research shows that those species best at surviving climate change will be the primary sources of EID. Parasites are not only good at finding us, they are really good at surviving, and persistence rather than diversification is often of critical importance (Hoberg & Brooks, Chapter 14 of this volume). There are many, not a few, evolutionary "accidents waiting to happen" out there, requiring only the catalyst of climate change, invasion with species introductions, and the intrusion of humans and their livestock and crops into areas they have never inhabited before (Hoberg *et al.*, 2008; Hoberg, 2010; Kilpatrick, 2011).

If we are correct, the current EID crisis arises precisely because we exist today in the intersection of the three main drivers that restructure the biosphere: (1) human activities/population overgrowth and consequences for social infrastructure and behavior; (2) technology (we can move ourselves and other species around the planet rapidly, we can produce drug-resistant strains; agricultural and land use changes bring potential pathogens into contact with humans and their livestock and crops); and (3) ecological perturbation and invasion, representing the long-term cumulative outcomes and ephemeral but extreme impacts of a rapidly changing environmental arena under natural and anthropogenic forcing (e.g., Daszak *et al.*, 2000; Hoberg, 2010; Weaver *et al.*, 2010; Kilpatrick, 2011) (Figure 15.1). These drivers exist and interact through feedback loops

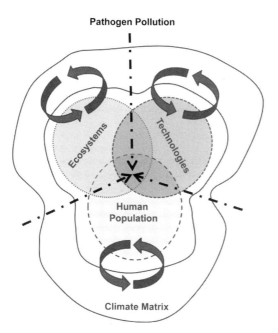

Figure 15.1.   Drivers for emerging infectious diseases and pathogen pollution. A relationship for pathogen pollution and processes for emerging infectious diseases arising from interactions among human populations, ecosystem structure (and biodiversity), and the development of technologies which influence globalization. Feedback loops within a scale-dependent matrix for climate and accelerated warming influence patterns of invasion and emergence for pathogens. See plate section for color version.

within a larger abiotic matrix defined by climate, and accelerated warming (along with a cascade of extreme events for temperature, precipitation and drought), that now represents the most pervasive impact for the continuity of biodiversity on landscape to global scales (Hannah *et al.*, 2005).

Periods of human population growth and movement, periods of human technological advancement and periods of regional to global climate change have each led to emerging diseases (e.g., Daszak *et al.*, 2000; Hoberg, 2010; Kilpatrick, 2011). Throughout Earth's history invasion has been a pervasive phenomenon and our current understanding of these episodes indicates the substantial effects that cascade through ecosystems and regions (e.g., Hoberg & Brooks,Chapter 14 of this volume). Historical studies inform us that as the effects of climate warming continue to emerge, range shifts and development of novel ecotones can drive host switching and dissemination of pathogens (Brooks & Hoberg, 2007; Hoberg & Brooks, 2008; Lafferty, 2009). As the biosphere is restructured, each EID will exact an economic cost, and even when host immune systems catch up to a particular "new" pathogen, it will not go away; it will persist as pathogen pollution, which exists at the intersection of human populations, ecosystems, and globalization (Daszak *et al.*, 2000) (Figure 15.1). For example, West Nile virus is no longer an acute invader in North America, but now that it is established across the continent, it will always be a chronic problem (Kilpatrick, 2011). Other pathogens can be predicted to follow in our current regime of globalization interacting with climate, each repeating that pattern.

Conclusions and an action plan

Today's EID crisis stems directly from fundamental ignorance about the biosphere: we simply don't know what's out there. And what you don't know can hurt you. Undiscovered pathogens and their vectors lurk everywhere as we move into novel habitats, trans-locate species and alter ecosystems. Emerging parasitic disease threatens human health, agriculture, natural systems, conservation practices and the global economy (e.g., Daszak *et al.*, 2000; Hoberg, 2010; Kilpatrick, 2011).

Information about the diversity and distribution of known and potential pathogens is critical for limiting their socio-economic impacts (Brooks & Hoberg, 2006; Weaver *et al.*, 2010). Yet our knowledge of the identities, geographic locations and threat potential for the world's pathogens can only be called fragmentary. At most, 10% of the world's pathogens have been documented – the rest remain utterly unknown. This massive ignorance is reason to be concerned about our preparedness to handle the crisis. It's impossible to prepare for a threat whose very existence is unknown. You can't monitor, much less seek cures or develop vaccines for, undiscovered maladies. We act as if EID were a rare phenomenon and engage in crisis response mode. The evidence is that the potential for EID is large, and in many cases climate change will make more of the world accessible to more pathogens (e.g., Hoberg *et al.*, 2008; Patz *et al.*, 2008). This makes the planet an evolutionary minefield into which millions of people, not to mention their crops, livestock and pets, wander daily (Brooks & Ferrao, 2005).

In the near future – if it's not already true – our crisis response abilities will be overwhelmed. We need strategic planning based on solving the problem rather than managing it. Prevention and anticipation is always cheaper than crisis response – game planning for contingencies that never happen is cheaper than not planning at all, then being overwhelmed by unanticipated crises. We cannot ignore the accumulating costs of pathogen pollution while focusing only on the crisis of the moment.

We must maintain and upgrade our ability to deal with the crises of the moment – the diseases that have emerged. But we must also be able to anticipate potential EID in the short term of the next one to three generations. If we are to cope with the future in a timely and economical manner, we must learn the lessons of the past. We need to monitor pathogens so we can assess EID risk before medical or veterinary clinicians see their symptoms. This information needs to be placed in a long-term evolutionary context (thousands to millions of years) connecting previous cycles of climate change, biodiversity and EID in a way that allows for adaptive planning for the future (e.g., Dawson *et al.*, 2011).

The information from that inventory must be freely and widely available online. This will require massive societal support for natural history collections and the taxonomic specialists who can identify each species by name, using those names as indices of information. No name, no information, wrong name, wrong information. In the context of the EID crisis, an incorrect diagnosis by a clinician is a taxonomic mistake that may have devastating consequences. It is both simple and critical.

The database has to grow fast, because nothing substantial can be accomplished until we know what we're up against. It's essential that we complete a global inventory of species rapidly. It sounds daunting. But 500 000 years of experience in hunting and gathering, cheaper, faster DNA analysis, faster, cheaper computers – all this makes the task feasible. It's a massive undertaking, but it must be done. It'll be costly, but it will never be cheaper, and the alternative is to let this chance slip away, passively accepting the consequences, knowing that we could have done better. As Voltaire wrote," There is a certain inevitability about inaction".

References

Agosta, S. J. (2006). On ecological fitting, plant-insect associations, herbivore host shifts, and host plant selection. *Oikos*, **114**, 556–565.

Agosta, S. J., & Klemens, J. A. (2008). Ecological fitting by phenotypically flexible genotypes: implications for species associations, community assembly and evolution. *Ecology Letters*, **11**, 1123–1134.

Agosta, S. J., Janz, N., & D. R. Brooks. (2010). How generalists can be specialists: resolving the "parasite paradox" and implications for emerging disease. *Zoologia,* **27**, 151–162.

Agrawal, A. A. (2000). Host-range evolution: adaptation and trade-offs in fitness of mites on alternative hosts. *Ecology,* **81**, 500–508.

Antonovics, J., Hood, M., & Partain, J. (2002). The ecology and genetics of a host shift: *Microbotryum* as a model system. *The American Naturalist*, **160**, S40–S53.

Audy, J. R. (1958). The localization of disease with special reference to the zoonoses. *Transactions of the Royal Society of Tropical Medicine and Hygiene*, **52**, 309–328.

Benkman, C. W. (1999). The selection mosaic and diversifying coevolution between crossbills and lodgepole pine. *The American Naturalist*, **153**, S75–S91.

Bernays, E. A. (1989). Host range in phytophagous insects: the potential role of generalist predators. *Evolutionary Ecology*, **3**, 299–311.

Bernays, E. A. (2001). Neural limitations in phytophagous insects: implications for diet breadth and evolution of host affiliation. *Annual Review of Entomology*, **46**, 703–727.

Bernays, E. A., & Chapman, R. F. (1994). *Host-Plant Selection by Phytophagous Insects*. London: Chapman & Hall.

Brooks D. R., & Ferrao, A. L. (2005). The historical biogeography of coevolution: emerging infectious diseases are evolutionary accidents waiting to happen. *Journal of Biogeography*, **32**, 1291–1299.

Brooks, D. R., & Hoberg, E. P. (2006). Systematics and emerging infectious diseases: from management to solution. *Journal of Parasitology*, **92**, 426–429.

Brooks, D. R., & Hoberg, E. P. (2007). How will global climate change affect parasites? *Trends in Parasitology*, **23**, 571–574.

Brooks, D. R., & McLennan, D. A. (1991). *Phylogeny, Ecology, and Behavior*. Chicago, IL: University of Chicago Press.

Brooks, D. R & McLennan, D. A. (1993). *Parascript: Parasites and the Language of Evolution*. Washington DC: Smithsonian Institution Press.

Brooks, D. R., & McLennan, D. A. (2002). *The Nature of Diversity: An Evolutionary Voyage of Discovery*. Chicago, IL: University of Chicago Press.

Brooks, D. R., & Van Veller, M. G. P. (2008). Assumption 0 analysis: comparative evolutionary biology in the age of complexity. *Annals of the Missouri Botanical Garden*, **95**, 201–223.

Brooks, D. R., McLennan, D. A., León-Règagnon, V., & Hoberg, E. P. (2006a). Phylogeny, ecological fitting and lung flukes: helping solve the problem of emerging infectious diseases. *Revista Mexicana de Biodiversidad*, **77**, 225–234.

Brooks, D. R., McLennan, D. A., León-Règagnon, V., & Zelmer, D. (2006b). Ecological fitting as a determinant of parasite community structure. *Ecology*, **87**, S76–S85.

Carroll, S. P., Dingle, H., & Klassen, S. P. (1997). Genetic differentiation of fitness-associated traits among rapidly evolving populations of the soapberry bug. *Evolution*, **51**, 1182–1188.

Chew, F. S. (1977). Coevolution of pierid butterflies and their cruciferous food plants. II. The distribution of eggs on potential food plants. *Evolution*, **31**, 568–579.

Daszak, P., Cunningham, A. A., & Hyatt, A. D. (2000). Emerging infectious diseases of wildlife: threats to biodiversity and human health. *Science*, **287**, 443–449.

Dawson, T. P., Jackson, S. T., House, J. I., Prentice, I. C., & Mace, G. M. (2011). Beyond predictions: biodiversity conservation in a changing climate. *Science*, **332**, 53–58.

Dobson, A., & Carper, R. (1992). Global warming and potential changes in host-parasite and disease vector relationships. In R. L. Peters & T. E. Lovejoy (Eds.), *Global Warming and Biological Diversity* (pp. 201–220). New Haven, CT: Yale University Press.

Ehrlich, P. R., & Raven, P. H. (1964). Butterflies and plants: a study in coevolution. *Evolution*, **18**, 586–608.

Elton, C. (1958). *The Ecology of Invasions by Animals and Plants*. London: Methuen.

Folinsbee, K., & Brooks, D. R. (2007). Early hominoid biogeography: pulses of dispersal and differentiation *Journal of Biogeography*, **34**, 383–397.

Fox, C. W., Nilsson, J. A., & Mousseau, T. A. (1997). The ecology of diet expansion in a seed-feeding beetle: pre-existing variation, rapid adaptation and maternal effects? *Evolutionary Ecology*, **11**, 183–194.

Fox, L. R., & Morrow, P. A. (1981). Specialization: species property or local phenomenon? *Science*, **211**, 887–893.

Futuyma, D. J., & Mitter, C. (1996). Insect-plant interactions: the evolution of component communities. *Philosophical Transactions of the Royal Society of London, Series B*, **351**, 1361–1366.

Futuyma, D. J., & Moreno, G. (1988). The evolution of ecological specialization. *Annual Review of Ecology and Systematics*, **19**, 207–233.

Futuyma, D. J., Keese, M. C., & Funk, D. J. (1995). Genetic constraints on macroevolution: the evolution of host affiliation in the leaf beetle genus *Ophraella*. *Evolution*, **49**, 797–809.

Godsoe, W., Yoder, J. B., Smith, C. I., & Pellmyr, O. (2008). Coevolution and divergence in the Joshua tree/yucca moth mutualism. *The American Naturalist*, **171**, 816–823.

Gould, S. J., & Vrba, E. S. (1982). Exaptation – a missing term in the science of form. *Paleobiology*, **8**, 4–15.

Hannah, L., Lovejoy, T. E., & Schneider, S. H. (2005). Biodiversity and climate change in context. In T. E. Lovejoy & L. Hannah (Eds.), *Climate Change and Biodiversity* (pp. 3–13). New Haven, CT: Princeton University Press.

Hoberg, E. P. (2010). Invasive processes, mosaics and the structure of helminth parasite faunas. *Revue Scientifique et Technique Office International des Épizooties*, **29**, 255–272.

Hoberg, E. P., & Brooks. D. R. (2008). A macroevolutionary mosaic: episodic host-switching, geographic colonization, and diversification in complex host-parasite systems. *Journal of Biogeography*, **35**, 1533–1550.

Hoberg, E. P., & Brooks, D. R. (2010). Beyond vicariance: integrating taxon pulses, ecological fitting and oscillation in historical biogeography and evolution. In S. Morand & B. Krasnov (Eds.), *The Geography of Host-Parasite Interactions* (pp. 7–20). Oxford: Oxford University Press.

Hoberg, E. P., & Klassen, G. J. (2002). Revealing the faunal tapestry: co-evolution and historical biogeography of hosts and parasites in marine systems. *Parasitology*, **124**, S3–S22.

Hoberg, E. P., Alkire, N. L., de Queiroz, A., & Jones, A. (2001). Out of Africa: origins of *Taenia* tapeworms in humans. *Proceedings of the Royal Society of London B*, **268**, 781–787.

Hoberg, E. P., Polley, L., Jenkins, E. J., & Kutz, S. J. (2008). Pathogens of domestic and free ranging ungulates: global climate change in temperate to boreal latitudes across North America. *Office International des Épizooties Revue Scientifique et Technique*, **27**, 511–528.

Janz, N. (2002). Evolutionary ecology of oviposition strategies. In M. Hilker & T. Meiners (Eds.), *Chemoecology of Insect Eggs and Egg Deposition* (pp. 349–376). Berlin: Blackwell.

Janz, N., & Nylin, S. (1998). Butterflies and plants: a phylogenetic study. *Evolution*, **52**, 486–502.

Janz, N., & Nylin, S. (2008). The oscillation hypothesis of host plant-range and speciation. In K. J. Tilmon (Ed.), *Specialization, Speciation, and Radiation: the Evolutionary Biology of Herbivorous Insects* (pp. 203–215). Berkeley, CA: University of California Press.

Janz, N., & Thompson, J. N. (2002). Plant polyploidy and host expansion in an insect herbivore. *Oecologia*, **130**, 570–575.

Janz, N., Nylin, S., & Nyblom, K. (2001). Evolutionary dynamics of host plant specialization: a case study of the tribe Nymphalini. *Evolution*, **55**, 783–796.

Janz, N., Bergström, A., & Sjögren, A. (2005). The role of nectar sources for oviposition decisions of the common blue butterfly *Polyommatus icarus*. *Oikos*, **109**, 535–538.

Janz, N., Nylin, S., & Wahlberg, N. (2006). Diversity begets diversity: host expansions and the diversification of plant-feeding insects. *BMC Evolutionary Biology*, **6**, 4.

Janzen, D. H. (1985). On ecological fitting. *Oikos*, **45**, 308–310.

Johansson, J., Bergstrom, A., & Janz, N. (2007). Search efficiency and host range expansion in a polyphagous butterfly; the benefit of additional oviposition targets. *Journal of Insect Science*, **7**, 3.

Kelley, S. T., & Farrell, D. B. (1998). Is specialization a dead end? The phylogeny of host use in *Dendroctonus* bark beetles (Scolytidae). *Evolution*, **52**, 1731–1743.

Kergoat, G. J., Delobel, A., Fediere, G., Le Ru, B., & Silvain, J. F. (2005). Both host-plant phylogeny and chemistry have shaped the African seed-beetle radiation. *Molecular Phylogenetics and Evolution*, **35**, 602–611.

Kilpatrick, A. M. (2011). Globalization, land use, and the invasion of West Nile Virus. *Science*, **334**, 323–327.

Lafferty, K. (2009). The ecology of climate change and infectious diseases. *Ecology*, **90**, 888–900.

Lande, R., & Arnold, S. J. (1983). The measurement of selection on correlated characters. *Evolution*, **37**, 1210–1226.

Larsson, S., & Ekbom, B. (1995). Oviposition mistakes in herbivorous insects: confusion or a step towards a new host plant? *Oikos*, **72**, 155–160.

Lawler, J. J., Shafer, S. L., White, D., *et al.* (2009). Projected climate-induced faunal change in the Western Hemisphere. *Ecology*, **90**, 588–597.

Moran, N. A. (1988). The evolution of host-plant alternation in aphids: evidence for specialization as a dead end. *The American Naturalist*, **132**, 681–706.

Murphy, S. M., & Feeny, P. (2006). Chemical facilitation of a naturally occurring host shift by *Papilio machaon* butterflies (Papilionidae). *Ecological Monographs*, **76**, 399–414.

Nosil, P. (2002). Transition rates between specialization and generalization in phytophagous insects. *Evolution*, **56**, 1701–1706.

Nosil, P., & Mooers, A. Ø. (2005). Testing hypotheses about ecological specialization using phylogenetic trees. *Evolution*, **59**, 2256–2263.

Nuismer, S. L., & Thompson, J. N. (2001). Plant polyploidy and non-uniform effects on insect herbivores. *Proceedings of the Royal Society of London B*, **268**, 1937–1940.

Nylin, S., & Janz. N. (2009). Butterfly host plant range: an example of plasticity as a promoter of speciation? *Evolutionary Ecology*, **23**, 137–146.

Parmesan, C. (2006). Ecological and evolutionary responses to recent climate change. *Annual Reviews of Ecology, Evolution and Systematics*, **37**, 637–669.

Parry, M. L., Canziani, O. F., Palutikof, J. P., van der Linden, P. J., & Hansen, C. E. (Eds.) (2007). *Climate Change 2007: Impacts, Adaptation and Vulnerability*. Contribution of Working Group II to the Fourth Assessment Report of the Intergovernmental Panel on Climate Change. Cambridge: Cambridge University Press.

Patz, J. A., Olson, S. H., Uejio, C. K., & Gibbs, H. K. (2008). Disease emergence from global climate and land use change. *Medical Clinics of North America*, **92**, 1473–1491.

Radtke, A., McLennan, D. A., & Brooks, D. R. (2002). Evolution of host specificity in *Telorchis* spp. (Digenea: Plagiorchiformes: Telorchiidae). *Journal of Parasitology*, **88**, 874–879.

Rosenthal, B. M. (2008). How has agriculture influenced the geography and genetics of animal parasites? *Trends in Parasitology*, **25**, 67–70.

Scheffer, S. J., & Wiegmann, B. M. (2000). Molecular phylogenetics of the holly leaf miners (Diptera: Agromyzidae: *Phytomyza*): species limits, speciation, and dietary specialization. *Molecular Phylogenetics and Evolution*, **1**, 244–255.

Segraves, K. A., Thompson, J. N., Soltis, P. S., & Soltis, D. E. (1999). Multiple origins of polyploidy and the geographic structure of *Heuchera grossulariifolia*. *Molecular Ecology*, **8**, 253–262.

Singer, M. C. (2003). Spatial and temporal patterns of checkerspot butterfly-hostplant association: the diverse roles of oviposition preference. In C. L. Boggs, W. B. Watt & P. R. Ehrlich (Eds.), *Ecology and Evolution Taking Flight: Butterflies as Model Study Systems* (pp. 207–208). Chicago, IL: University of Chicago Press.

Singer, M. S. (2008). Evolutionary ecology of polyphagy. In K. J. Tilmon (Ed.), *Specialization, Speciation, and Radiation: The Evolutionary Biology of Herbivorous Insects* (pp. 29–42). Berkeley, CA: University of California Press.

Singer, M. C., Thomas, C. D., & Parmesan, C. (1993). Rapid human-induced evolution of insect-host associations. *Nature*, **366**, 681–683.

Singer, M. C., Wee, B., & Hawkins, S. (2008). Rapid anthropogenic and natural diet evolution: three examples from checkerspot butterflies. In K. J. Tilmon (Ed.), *Specialization, Speciation, and Radiation: The Evolutionary Biology of Herbivorous Insects* (pp. 311–324). Berkeley, CA: University of California Press.

Symons, F. B., & Beccaloni, G. W. (1999). Phylogenetic indices for measuring the diet breadths of phytophagous insects. *Oecologia*, **119**, 427–434.

Tabashnik, B. E. (1983). Host range evolution: the shift from native legume hosts to alfalfa by the butterfly, *Colias philodice eriphyle*. *Evolution*, **37**, 150–162.

Termonia, A., Hsiao, T. H., Pasteels, J. M., & Milinkovitch, M. C. (2001). Feeding specialization and host-derived chemical defense in Chrysomeline leaf beetles did not lead to an evolutionary dead end. *Proceedings of the National Academy of Sciences of the USA*, **98**, 3909–3914.

Thompson, J. N. (1994). *The Coevolutionary Process*. Chicago, IL: University of Chicago Press.

Thompson, J. N. (1998). Rapid evolution as an ecological process. *Trends in Ecology & Evolution*, **13**, 329–332.

Thompson, J. N. (2005). *The Geographic Mosaic of Coevolution*. Chicago, IL: University of Chicago Press.

Thompson, J. N., & Fernandez, C. C. (2006). Temporal dynamics of antagonism and mutualism in a geographically variable plant-insect interaction. *Ecology*, **87**, 103–112.

Thompson, J. N., Nuismer, S. L., & Merg, K. (2004). Plant polyploidy and the evolutionary ecology of plant/animal interactions. *Biological Journal of the Linnaean Society*, **82**, 511–519.

Van Klinken, R. D., & Edwards, O. R. (2002). Is host-specificity of weed biological control agents likely to evolve rapidly following establishment? *Ecology Letters*, **5**, 590–596.

Via, S. (1991). The population structure of fitness in a spatial network: demography of pea aphid clones from two crops in a reciprocal transplant. *Evolution*, **45**, 827–852.

Wahlberg, N. (2001). The phylogenetics and biochemistry of host plant specialization in melitaeine butterflies (Lepidoptera: Nymphalidae). *Evolution*, **55**, 522–537.

Weaver, H. J., Hawdon, J. M., & Hoberg, E. P. (2010). Soil-transmitted helminthiases: implications of climate change and human behavior. *Trends in Parasitology*, **26**, 574–581.

West-Eberhard, M. J. (2003). *Developmental Plasticity and Evolution*. New York: Oxford University Press.

Wiegmann, B. M., Mitter, C., & Farrell, B. (1993). Diversification of carnivorous parasitic insects – extraordinary radiation or specialized dead-end. *The American Naturalist*, **142**, 737–754.

Wolfe, N. D., Panosian Dunavan, C., & Diamond, J. (2007). Origins of major human infectious diseases. *Nature*, **447**, 279–283.

Yotoko, K. S. C., Prado, P. I., Russo, C. A. M., & Solferini, V. N. (2005). Testing the trend towards specialization in herbivore–host plant associations using a molecular phylogeny of *Tomoplagia* (Diptera: Tephritidae). *Molecular Phylogenetics and Evolution*, **35**, 701–711.

16 Establishment or vanishing: fate of an invasive species based on mathematical models

Yihong Du

Understanding the nature of spreading of invasive species is a central problem in invasion ecology. This is a problem of nonequilibrium. If we represent the population distribution of an invasive species as a function of time t and space location x, written as $u(t,x)$, then it is possible to establish mathematical models that govern the evolution of u. We will look at several such mathematical models in this article, and discuss the predictions they offer for the invasion problem.

1.   The logistic model

Although ecology was traditionally a largely observational and descriptive subject, mathematical models have been proposed and developed to help with the understanding of ecological problems for over a century. The most basic mathematical model in ecology was proposed in 1836 by the Belgian mathematician Pierre F. Verhilst. The model is now known as the logistic model, which has the form

$$u' = au\left(1-\frac{u}{K}\right),$$

where $u = u(t)$ is a function of time, representing the total population of a certain species at time t, and u' is the derivative of u, representing the rate of change of the population. The parameters a and K are positive constants measuring the growth rate and crowding effect of the population, respectively. The model predicts a self-regulation property of population growth: if the initial population is low, the population will increase towards the number K, while starting with a large population, it will decrease towards K. The stabilizing number K is called the carrying capacity of the habitat, which is the maximum population size of the species that the environment can sustain indefinitely, given the food, habitat, water and other necessities available in the environment.

   This kind of model is clearly insufficient for understanding the invasion problem mentioned above, since the variation in space is missing.

The Balance of Nature and Human Impact, ed. Klaus Rohde. Published by Cambridge University Press.

2. Fisher's model

In order to understand the spreading of certain advantageous genes among members of a population, Fisher (1937) considered an extension of the above logistic equation, where he used a two-variable function $u(t,x)$ to represent the population (in one space dimension) that carries a certain advantageous gene, and assumes that u satisfies the "diffusive" logistic equation

$$u_t = du_{xx} + au\left(1-\frac{u}{K}\right), \tag{1}$$

where u_t stands for the derivative of u with respect to time, representing the rate of change of the population at time t and location x, and u_{xx} denotes the second derivative of u with respect to the space variable x, representing "diffusion", and the parameter d is the so-called diffusion rate. The diffusion term du_{xx} is added here to describe the movement of the species in space, assuming that the movement is random and has the tendency of moving from locations of high density to locations of low density, like the way heat is diffused. So in this diffusive logistic model, the rate of change of the population "u_t" is determined by a combination of diffusion "du_{xx}" and reaction "$au\left(1-\frac{u}{K}\right)$".

Fisher made a remarkable observation; he found a class of solutions to the above diffusive logistic equation, for which the graph of u as a function of x moves with constant speed without changing its shape, like a water wave traveling in a canal. Different solutions in the class may move with different speeds, and Fisher demonstrated that for every number $c \geq 2\sqrt{ad}$, there exists a solution moving with speed c. Naturally, $c_* = 2\sqrt{ad}$ is called the minimal speed of these "traveling wave" solutions.

If a traveling wave moves in the increasing x direction, then its profile is a decreasing function of x, which approaches 0 as x goes to positive infinity, and it approaches K as x goes to negative infinity. Therefore to anyone moving with the same speed and direction as the traveling wave, the population curve represented by the traveling wave appears unchanged, but to anyone who does not move, the population curve is observed to increase to the horizontal line representing the carrying capacity.

Fisher proposes that the minimal speed $c_* = 2\sqrt{ad}$ should be the spreading speed of the advantageous gene. Independently, Kolmogorov *et al.* (1937) made the same discovery. However, these pioneering works did not receive the deserved attention until the 1950s.

Skellam (1951) investigated the spreading of muskrat in Europe in the early 1900s based on existing empirical data: he calculated the area of the muskrat range from a map obtained from field data over several decades, took the square root and plotted it against years, and found that the data points lay on a straight line. As the square root of the area is roughly the average distance from the boundary to the center of the range (multiplied by $\sqrt{\pi}$), this indicates that the spreading speed is roughly a constant. Other examples revealing the same phenomenon were subsequently observed following the same method; see Shigesada and Kawasaki (1997).

Skellam demonstrated through a simplified mathematical model that this constant speed can be calculated by the formula proposed by Fisher. Since the intrinsic growth rate a and diffusion rate d in this formula can often be calculated from field data for specific animal species, it is possible to compare the theoretical spreading speed c_* obtained from the formula $c_* = 2\sqrt{ad}$ with the observed spreading rate, and this was done in a number of works; on page 55 of Shigesada and Kawasaki (1997) one may find a comparison table. For most species listed in this table the theoretical rate c_* agrees reasonably well with the observed rate, with one exception where the theoretical rate is one magnitude smaller than the observed one. Such exceptional cases may be caused by "long-distance dispersal", such as small insects getting carried a long distance away during a storm, well beyond normal spreading via diffusion (random walk).

These early works have inspired extensive mathematical research on traveling wave solutions and their applications, ranging from species invasion in ecology to signal and flame propagations in neural science and combustion theory. For example, Fisher's proposal and many other properties of traveling wave solutions have been proved by a rigorous mathematical theory in Aronson and Weinberger (1978). We will not discuss these developments here, except to mention that this is still a very active and important area of current research in mathematics and its applications, and Fisher's model (also called the KPP equation due to the important contribution of Kolmogorov et al., 1937) has been recognized as one of the most important models in reaction-diffusion theory.

As a model for the spreading of species, the Fisher equation carries some shortcomings. For example, by the theory in Aronson and Weinberger (1978), it predicts successful spreading and establishment of a new species regardless of its initial size and range. This is not consistent with numerous pieces of empirical evidence; for instance, the introduction of several bird species from Europe to North America in the 1900s was successful only after many initial attempts. Secondly, although this model gives a convenient and often effective formula for predicting the spreading speed, it does not provide precise information on the location of the spreading front.

3.   Allee effect

The persistent spreading drawback mentioned above can be eliminated if the reaction term $au\left(1-\frac{u}{K}\right)$ is suitably modified to reflect the "Allee effect" on the growth rate of the species. One key feature of the Allee effect is that populations shrink at very low densities because reproduction of the species becomes difficult in such a situation. This can be taken into account in the model by replacing the logistic reaction term $au\left(1-\frac{u}{K}\right)$, for example, by a function of the form

$$f(u) = au(1-u)(u-b), \quad (0<\mathrm{b}<1/2).$$

Since $f(u) < 0$ for $0 < u < b$, this reaction term contributes negatively to the growth of the population when u is below the level b.

With this new reaction function, it is known that every traveling wave solution has the speed $\bar{c}_* = \left(\frac{1}{2}-b\right)\sqrt{2ad}$ (Hadeler & Rothe, 1975), and the model predicts spreading of the species with asymptotic speed $\tilde{c}_*$ if the initial population size is big enough, and the population vanishes in the long run if its initial size is small enough (Aronson & Weinberger, 1978). Thus the persistent spreading problem is rectified.

The Allee effect is generally believed to play a crucial role at the early stage of establishment of a spreading species, and once establishment is guaranteed the spreading process of the species is not expected to be greatly influenced by the Allee effect. Therefore one may think that the asymptotic spreading speed for an establishing invading species should not be affected in a significant manner by the Allee effect. However, this may not be the case. Lewis and Kareiva (1993) demonstrate, with the help of the above reaction function $f(u)$, that the Allee effect may significantly reduce the spreading speed. On the other hand, no Allee effect has been observed in some invading species (Fauvergue *et al.*, 2008). Further discussions on Allee effects can be found in the review article by Kramer *et al.* (2009) and references therein.

We observe that, since $\tilde{c}_* = \left(\frac{1}{2} - b\right)c_*$, which is less than half the value of c_* for any b taken in the range $0 < b < \frac{1}{2}$, the theoretical speed $\tilde{c}_*$ would not work as well whenever c_* is a good approximation of the real spreading speed, and vice versa.

4.   Spreading front as a free boundary

Recently in Du and Lin (2010) a new variation of the Fisher equation (1) has been proposed to model the spreading of species in one space dimension. The variation is that, while Fisher's model allows the space variable to run over all real numbers without restriction, the new model assumes that x varies in a finite range $g(t) < x < h(t)$ at time t, with the two ends representing the spreading fronts of the species. Thus now equation (1) is satisfied for positive time t and for x in the range $g(t) < x < h(t)$, where the population function u is positive, and u is zero at the spreading fronts $x = g(t)$ and $x = h(t)$. To determine $g(t)$ and $h(t)$, we need to know the equations they satisfy. These are supplied by

$$g'(t) = -\mu u_x(t, g(t)), \quad h'(t) = -\mu u_x(t, h(t)),$$

meaning that the rate of change of the spreading front is proportional to the spatial population gradient at the fronts. Here μ is a positive parameter, whose meaning is explained by Bunting *et al.* (2011).

This is known as a free boundary problem in mathematics, since apart from $u(t,x)$, the functions $g(t)$ and $h(t)$ determining the boundary of the range of the space variable x are also unknowns.

The new model has the following properties:

(a) **Spreading-vanishing dichotomy.**
     One of the following alternatives happens:
     - (Spreading) The population range $g(t) < x < h(t)$ increases to $-\infty < x < \infty$ as t increases to infinity, and at the same time the population u converges to K.
     - (Vanishing) The population range $g(t) < x < h(t)$ stays inside a bounded interval of length no bigger than $\pi\sqrt{\frac{d}{a}}$ for all time and the population u goes to 0 as time goes to infinity.

(b) **Spreading-vanishing criteria.**
   - If the size of the initial population range (namely $h(0) - g(0)$) exceeds $\pi\sqrt{\frac{d}{a}}$, then spreading will happen.
   - If the size of the initial population range is below $\pi\sqrt{\frac{d}{a}}$, then one can find a critical value $\mu^* > 0$ depending on the initial population $u(0,x)$ such that vanishing happens if $\mu \leq \mu^*$, and spreading happens if $\mu > \mu^*$.

(c) **Spreading speed.**
   - When spreading happens, the two fronts spread at the same speed k_0 for large time, that is, for all large t, the function $h(t)$ behaves like $k_0 t$, and $g(t)$ is like $-k_0 t$.

In ecological terms, when spreading happens, the invasive species eventually invades into all the available space, and establishes itself towards the carrying capacity of the environment, while vanishing represents a situation where the invasion is confined in a finite region all the time and the invasive species dies out in the long run, so failure of establishment occurs. Thus the persistent spreading drawback of the Fisher model is now eliminated. Let us also point out that the new model gives a precise prediction of the location of the spreading fronts: $x = g(t)$ and $x = h(t)$. Furthermore, when spreading occurs, the spreading speed will be determined through the introduction of a "semi-wave", which is a variation of the classic traveling wave solution first found by Fisher, and the population curve near the spreading front is like a semi-wave moving with a constant speed (the spreading speed); this will be discussed in further detail later.

We note that the number $\pi\sqrt{\frac{d}{a}}$ serves as a barrier for the spreading of the species: the size of population range, $h(t) - g(t)$, either never breaks this barrier, and the population vanishes in the long run, or it breaks this barrier at some time, and the population eventually spreads to the entire available space and establishes itself towards the carrying capacity.

What is the relationship between the spreading speeds k_0 and c_*? This has been examined in Bunting, Du and Krakowski (2011). It is shown that there is an increasing function $\lambda_0(s)$ defined for $0 < s < \infty$, such that

$$k_0 = \lambda_0\left(\frac{K\mu}{d}\right)\sqrt{ad}, \text{ and hence } k_0 = \frac{1}{2}\lambda_0\left(\frac{K\mu}{d}\right)c_*.$$

The function $\lambda_0(s)$ goes to 2 as s increases to infinity, and it goes to 0 as s decreases to 0. Moreover $\lambda_0(s)/s$ goes to $1/\sqrt{3}$ as s decreases to 0.

These properties of $\lambda_0(s)$ indicate that k_0 is always smaller than c_*, and when $\frac{K\mu}{d}$ is large, k_0 can be approximated by the formula

$$k_0 \approx 2\sqrt{ad} = c_*, \tag{2}$$

while when $\frac{K\mu}{d}$ is small, k_0 can be approximated by the formula

$$k_0 \approx \frac{1}{\sqrt{3}}\frac{K\mu}{d}\sqrt{ad} = \frac{1}{\sqrt{3}}\frac{K\mu}{2d}c_*. \tag{3}$$

Table 16.1 shows how $\lambda_0(s)$ varies as s varies in $[1, \infty)$. The changes of $\lambda_0(s)$ and $\frac{\lambda_0(s)}{s}\sqrt{3}$ for $s \in (0,1]$ are described in Table 16.2.

Table 16.1. $\lambda_0(s)$ for $s \geq 1$.

s	1	10	10^2	10^3	10^4	10^5	10^6	10^7	10^8
$\lambda_0(s)$	0.36	1.01	1.49	1.72	1.84	1.90	1.93	1.95	1.96

s	10^9	10^{10}	10^{11}	10^{12}	10^{13}	10^{14}	10^{15}	10^{16}	∞
$\lambda_0(s)$	1.97	1.98	1.98	1.99	1.99	1.99	1.99	1.99	2.00

Table 16.2. $\lambda_0(s)$ for $s \leq 1$.

s	0	0.01	0.1	0.2	0.3	0.4	0.5	0.6	0.7	0.8	0.9	1.0
$\lambda_0(s)$	0	0.006	0.05	0.10	0.15	0.19	0.22	0.25	0.28	0.31	0.34	0.36
$\frac{\lambda_0(s)}{s}\sqrt{3}$	1	0.99	0.94	0.89	0.84	0.80	0.77	0.73	0.70	0.68	0.65	0.63

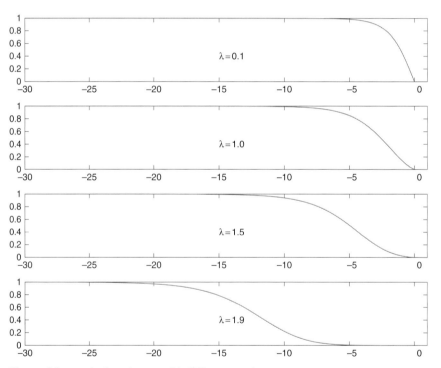

Figure 16.1. Shape of the standard semi-wave with different speed.

As mentioned earlier, when spreading occurs, near the spreading front, the population curve u (as a function of x) at a fixed large time looks the same as the semi-wave, and as time increases the front looks like a moving curve (with the shape of the semi-wave) with speed k_0. The shape of the semi-wave changes when the parameter μ and hence the traveling speed varies.

Figure 16.1 shows the graph of the standard semi-wave (obtained by letting $a = K = d = 1$) for various values of $\lambda = \lambda_0(\mu)$. Let us note that when $a = K = d = 1$,

$k_0 = \lambda_0 \left(\frac{K\mu}{d} \right) \sqrt{ad} = \lambda_0(\mu)$. So λ represents the spreading speed. Since $0 < \lambda_0(\mu) < 2$ for all $\mu > 0$, the allowed range for λ is $0 < \lambda < 2$. The profile of each standard semi-wave is a decreasing function defined for $x \leq 0$, which takes the value 0 at $x = 0$, and converges to 1 as x goes to negative infinity. So the front of the semi-wave is at $x = 0$, and when spreading happens, the population curve near the front $x = h(t)$ behaves like a semi-wave with its front shifted from $x = 0$ to $x = h(t)$, while at the front $x = g(t)$, the population curve behaves like a semi-wave with its direction reversed and its front put at $x = g(t)$.

These graphs reveal that the gradient of the front of the semi-wave gets smaller and smaller as its traveling speed is increased closer and closer towards the limiting value 2. We remark that as λ goes to 2, the semi-wave converges to the traveling wave with speed $c_*(= 2)$ determined by the Fisher model with $a = K = d = 1$. Indeed, it has been shown that the Fisher model is the limiting case of the free boundary model as μ goes to infinity (Du & Guo, 2012). (Let us note that by the property of the function $\lambda_0(s)$, μ goes to infinity is the same as $\lambda = \lambda_0(\mu)$ goes to 2.)

It was demonstrated in Bunting *et al.* (2011) that the critical value μ^* can be numerically estimated. The impact of the initial distribution on the chance of successful spreading of the species was also examined in that paper through numerical simulation, and it was demonstrated that the chance of successful spreading is increased if the distribution of the initial population concentrates more near the edge of the population range.

Similar to the case of the Fisher model, where the phenomena shown in one space dimension continue to hold in higher space dimensions (Aronson & Weinberger, 1978), the same is true for the free boundary model (see Du & Guo, 2011, and Du *et al.*, 2012).

5. Concluding remarks

Are mathematical models useful in ecology? The answer is probably yes, and increasingly more people are taking this view. However, due to the complexity and lack of first principles in most ecological problems, at least in the traditional sense, it is unlikely for mathematical models to be found as precise and powerful as in classic physics, for example.

The free boundary model discussed here has not been tested by ecologists, though it appears to possess several advantages over the classic Fisher model, for which some tests have been done, especially about the speed c_* as mentioned earlier. Since the new model behaves similarly to Fisher's model for large μ, it will not be worse than Fisher's for the use in invasion problems. It would be interesting to see some tests of the new model against field data, to determine whether it indeed gives better predictions for the spreading of invasive species, and whether the mathematical criteria governing the spreading-vanishing dichotomy are reflected in real-world invasion processes.

The theoretical research on the free boundary model is mostly for the ideal situation where the environment is homogeneous. It is possible to use numerical simulation to analyze variations of the model with heterogeneous environments, including the effect of climate changes. Also it is more realistic to consider models that involve more than one species. Work in these directions is continuing.

Finally we refer the reader to Shigesada and Kawasaki (1997) for a systematic discussion of mathematical modeling in invasion ecology.

References

Aronson, D. G., & Weinberger, H. F. (1978). Multidimensional nonlinear diffusions arising in population genetics. *Advances in Mathematics*, **30**, 33–76.

Bunting, G., Du, Y., & Krakowski, K. (2011). Spreading speed revisited: analysis of a free boundary model. *Networks and Heterogeneous Media*, in press.

Du, Y., & Guo, Z. M. (2011). Spreading-vanishing dichotomy in the diffusive logistic model with a free boundary, II. *Journal of Differential Equations*, **250**, 4336–4366.

Du, Y., & Guo, Z. M. (2012). The Stefan problem for the Fisher-KPP equation. *Journal of Differential Equations*, **253**, 996–1035.

Du, Y., & Lin, Z. G. (2010). Spreading-vanishing dichotomy in the diffusive logistic model with a free boundary. *SIAM Journal of Mathematical Analysis*, **42**, 1305–1333.

Du, Y., Matano, H., & Wang, K. (2012). Regularity and asymptotic behavior of nonlinear Stefan problems. preprint.

Fauvergue, X., Malausa, J. C., Giuge, L., & Courchamp, F. (2008). Invading parasitoids suffer no Allee effect: a manipulative field experiment. *Ecology*, **88**, 2392–2403.

Fisher, R. A. (1937). The wave of advance of advantageous genes. *Annals of Eugenics*, **7**, 335–369.

Hadeler, K. P., & Rothe, F. (1975). Travelling fronts in nonlinear diffusion equations. *Journal of Mathematical Biology*, **2**, 251–263.

Kolmogorov, A. N., Petrovsky, I. G., & Piskunov, N. S. (1937). Ètude de l'équation de la diffusion avec croissance de la quantité de matière et son application à un problème biologique. Bull. Univ. Moscou Sér. Internat. A1, 1–26; English translation in P. Pelcé (Ed.), *Dynamics of Curved Fronts* (pp. 105–130). New York: Academic Press, 1988.

Kramer, A. M., Dennis, B., Liebhold, A. M., & Drake, J. M. (2009). The evidence for Allee effects. *Population Ecology*, **51**, 341–354.

Lewis, M. A., & Kareiva, P. (1993). Allee dynamics and the spreading of invasive organisms. *Theoretical Population Biology*, **43**, 141–158.

Shigesada, N., & Kawasaki, K. (1997). Biological Invasions: theory and Practice*; Oxford Series in Ecology and Evolution*. Oxford: Oxford University Press.

Skellam, J. G. (1951). Random dispersal in theoretical populations. *Biometrika*, **38**, 196–218.

17 Anthropogenic footprints on biodiversity

Camilo Mora and Fernando A. Zapata

Introduction

One of the most concerning issues to modern ecology and society is the ongoing loss of biodiversity. Ecosystems are now losing species at rates only seen in previous mass extinction events (Hails, 2008; Barnosky *et al.*, 2011) with rates of extinction between 100 and 1000 times higher than pre-human levels (Pimm *et al.*, 1995). This loss, in turn, is impairing the functioning of ecosystems (Worm *et al.*, 2006; Mora *et al.*, 2011a) and their capacity to deliver goods and services to mankind (Díaz *et al.*, 2006). The sharp contrast between the declining "supply" of the Earth's services and the rising "demand" from a growing human population indicates that such services will increasingly fall short, leading to the exacerbation of hunger, poverty and human suffering (Campbell *et al.*, 2007; Mora & Sale, 2011).

There is relatively good consensus that biodiversity loss is being driven directly or indirectly by human stressors such as overexploitation, habitat loss, invasive species and climate change (Myers, 1995; Sala *et al.*, 2000; Novacek & Cleland, 2001; Gaston *et al.*, 2003; Jackson, 2008; Weidenhamer & Callaway, 2010). The relative role of such stressors, however, has been a focus of controversy as all threats do provide rational mechanisms to explain biodiversity loss and unfortunately most threats co-occur in natural conditions, making it difficult to isolate their individual effects (Myers, 1995; Sala *et al.*, 2000; Novacek & Cleland, 2001; Mora *et al.*, 2007). Since the cost of mitigating specific stressors could be considerable but disproportionate among different sectors of the economy (e.g., industries vs. fishers, fishermen vs. tourism developers, etc.), this uncertainty over the relative effect of anthropogenic stressors is often used as an argument to prevent the implementation of mitigation policies (e.g., Schiermeier, 2004; Worm & Myers, 2004; Grigg & Dollar, 2005). A counter-argument, however, is that any stressor at play, if proven to have a considerable effect on biodiversity, should be mitigated regardless of its effect relative to other stressors. This would, of course, require demonstrating the significance of the stressor(s) at play. In this review, we provide an overview of the current biodiversity crisis and the role of anthropogenic stressors. The evidence is considerable and although some uncertainties remain and will probably never be answered, there is considerable knowledge to suggest that a lack of policy action

The Balance of Nature and Human Impact, ed. Klaus Rohde. Published by Cambridge University Press.

should not be further justified on the basis of a lack of knowledge (see similar pledge by Worm & Myers, 2004; Jackson, 2008; Knowlton & Jackson, 2008). Our assessment is focused on the effect of anthropogenic factors on interrelated patterns of abundance, evolution, distribution and extinction. Note that a deterring stressor could trigger a cascade of responses starting with changes in abundance (e.g., new colonization, population increase or decline) and adaptation, and this in turn can lead to changes in geographical distribution ranging from expansion to extinction.

To obtain some insight into the comparative effect of anthropogenic drivers, we complemented our literature review with an analysis of the factors that have been associated with the extinction and critical endangerment of species in recent times. For this analysis, we collected the list of species classified as "extinct" and "critically endangered" in the IUCN Red List of Species and the threats leading to that classification (IUCN, 2012). Information on threats leading to the extinction of species was available for 266 species of animals and 27 species of plants; information on the threats leading to the critical endangering of species was available for 1743 species of animals and 877 species of plants. Threats were broadly categorized as habitat loss (i.e., the alteration of habitats into ones no longer suitable for species, including deforestation, pollution, infrastructure development, etc.), invasive species, overharvesting and climate change.

Technical considerations

It will be almost impossible to determine the geographical position of all individuals of a given species, even those of decimated species because the rarer they become the harder they are to detect (Dulvy *et al.*, 2004). Consequently, assessing changes in the distribution of species relies on multiple extrapolation and simplifying sources of data. Some common sources of information include the geographical range or extent of occurrence (i.e., the outermost boundary within which individuals of a particular species have been recorded), the area of occupancy (i.e., the inhabited parts of the extent of occurrence), and spatial patterns of abundance throughout the geographical range of species (Gaston, 2003). These types of data provide different strengths and weaknesses. The extent of occurrence, for instance, is easy to quantify but overestimates the actual geographical distribution because individuals are not found everywhere within such an area (Gaston, 2003; Jetz *et al.*, 2008); in turn, this can cause underestimation of extinction risk (Jetz *et al.*, 2008), flawed conservation decisions about the size and position of protected areas (Rondinini *et al.*, 2006), and failure to identify actual changes in species' distributions (Burgman & Fox, 2003). Similarly, the area of occupancy fails to capture variations in abundance within occupied areas, which may lead to inappropriate decisions about the most effective placement of protected areas (Gaston, 2003) and to inappropriate conservation action for species undergoing range contractions (Channell & Lomolino, 2000). Finally, spatial patterns in abundance could be more precise to identify distributional changes but they require substantially more sampling (Gaston, 2003).

In addition to extrapolation limitations, identifying changes in the distribution of species can also be affected by the resolution of data. As an example, the use of areas of occupancy with geographical coordinates at a precision of 0.1° can lead to errors of ±11 km (i.e., at the tropics, 1° equals ~111 km, thus a precision of 0.1° will equal ±11 km; the error at other latitudes should be within that range). As a result of data precision, range shifts will remain undetected if the precision of available data is smaller than the shift itself (Gaston, 2003; Thomas *et al.*, 2006). An analogous bias is related to sampling effort. Because of the strong relationship between range size and abundance (Gaston, 2003), similar sampling efforts could considerably underestimate the distribution of range-restricted species. Consequently, it is common to see records of range expansions which are difficult to differentiate between prior lack of sampling and a causal mechanism (e.g., Mora *et al.*, 2000; Botts *et al.*, 2012); similarly, many species are difficult to accurately define as extinct due to the lack of sampling (Dulvy *et al.*, 2004).

The issue of species classifications is also important. On the one hand, synonyms (i.e., multiple names for the same species) will underestimate the distribution of species whose ranges are fractioned under different names for the same species. On the other hand, homonyms (i.e., the same name for different species) will overestimate the distribution of species whose ranges are actually the concatenation of the ranges of several species. Unfortunately, both of these issues are considerable: within reviewed taxonomic groups synonyms can account for up to 50% of named species (Alroy, 2002); in turn, new genetic developments in the identification of species are ramping up the number of new species due to the discovery of cryptic species (i.e., individuals reproductively isolated from each other, but whose morphology is virtually identical) (e.g., Fouquet *et al.*, 2007; Vieites *et al.*, 2009).

Magnitude of the biological crisis

In spite of limitations in the quality of data and accuracy in the metrics that describe the distribution of species, there is great certainty that species have undergone considerable changes in their geographical distribution due to human activities. Evidence of such an effect is clear even in the fossil record, which has revealed the extinction of multiple species in the wake of human arrival in previously uninhabited regions of the world. Burney and Flannery (2005), for instance, tracked human migrations from the Australian continent some 50 000 years, to the New World by the end of the Pleistocene, to remote islands in recent centuries and show how each of those new arrivals were followed by the collapse of several species in relation to factors such as overkilling, biological invasions, habitat transformation, disease and their aggravation with climatic change. Evidence of the human footprint on biodiversity is also evident and more severe in recent times. According to the IUCN Red List of Species (IUCN, 2012), 712 species of animals and 89 species of plants have been documented to have gone extinct in modern times and for evaluated groups such as scleractinian corals, amphibians, birds and mammals between 20% and 43% of their species are currently threatened with extinction. On average, monitored populations of some 1700 vertebrate species across all regions of the world have declined by nearly 30% over the past 35 years (Hails, 2008). These numbers, however, may be serious underestimates because we lack data on the status of most species and because most species have not yet been formally described. Of the ~1.2 million species that have been described and compiled in a central database (Mora *et al.*, 2011b), only ~25 000 have complete information of their distributions, population trends and threats (IUCN, 2012). The extent to which local and global extinctions are probably passing unnoticed is further illustrated by the fact that the ~1.2 million species currently named and catalogued represent only ~14% of the species estimated to exist on Earth (Mora *et al.*, 2011b). Modeling the number of species as a function of area with a power-law relationship (i.e., $S = cA^z$, where S is the number of species, A is area, and c and z are constants) it has been possible to quantify that some 27 000 species could go extinct each year due to deforestation alone (Sax & Gaines, 2008, but see He & Hubbell, 2011). According to Barnosky *et al.* (2011), if the ongoing loss of species is to continue, the current rate of extinction could resemble levels seen in prior mass extinctions in which over 75% of all species went extinct.

Human footprints on the biodiversity crisis

Introduced species

The globalization of the economy and trade, combined with an increasing trade in wildlife, is directly and indirectly leading to the introduction of species into foreign areas. While most species introductions are unsuccessful or benign, some are devastating. The probability of successfully establishing alien populations is low (in general about 1 in 10; Williamson & Fitter, 1996) depending on the number of individuals released, the number of release events (Kolar & Lodge, 2001), the time of such events (Gertzen *et al.*, 2011), reproductive strategy (Kolar & Lodge, 2001), history of prior colonizations (Kolar & Lodge, 2001), speed of genetic adaptation to novel environments (Prentis *et al.*, 2008), and avoidance of genetic bottlenecks (e.g., purging deleterious alleles) associated with initial low genetic diversity (Frankham, 2004), etc. Of the species that make the cut of becoming successful residents, about 9 in 10 will have negligible effects (Williamson & Fitter, 1996); the remaining few, however, often have devastating ecological consequences. Invasive species have been responsible for 32% of extinctions of animal species, principally on islands, although only 5% of currently critically endangered species are threatened by invasive species (Figure 17.1). Among plants, no extinction has been related to the unique effect of invasive species and less than 2% are currently endangered for this reason alone (Figure 17.1).

   The importance of invasive species for extinction has been a topic of debate. For instance, among oceanic islands there is a strong correlation between the number of exotic predatory mammal species established after European colonization and the magnitude of bird extinctions (Blackburn *et al.*, 2004). Yet the rate of mammal invasions was also correlated with the magnitude of habitat loss (Didham *et al.*, 2005), and this collinearity complicates determining whether bird extinctions were caused by the loss of their habitat or by mammal invasions (Didham *et al.*, 2005). Another debate has been related to the role of invasive species on species currently threatened (Gurevitch &

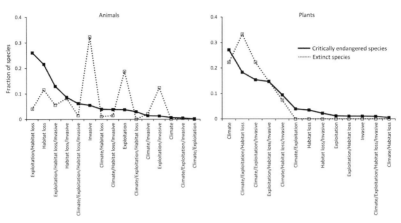

Figure 17.1. Frequency distribution of extinct and critically endangered animal and plant species according to their most likely threats. Data from IUCN (2012).

Padilla, 2004; vs. Ricciardi, 2004; Clavero & García-Berthou, 2005); this debate clarified that while documented extinctions were greatly linked to invasive species, such effect is minor among species currently at risk of extinction: species most vulnerable to invasive species have already gone extinct (this is confirmed in Figure 17.1).

Among animal species, the mechanism through which invasive species drive other species to extinction or affect their geographical distribution is most likely related to predation, often of eggs or young (Sax & Gaines, 2008) and indirectly through the loss of ecosystem functions (e.g., the loss of birds can lead to the decline of plant populations in relation to the reduction of pollination, seed dispersal and increases in insect populations that feed on plants; Pearson & Callaway, 2003). Among invasive plants, the process of impact is mediated by competitive exclusion through allelopathy or "novel weapons", and indirectly by altering nutrient cycles through litter and root exudates; at times the use of herbicides to control invasive plants may also affect ecosystems overall (Weidenhamer & Callaway, 2010). Of the documented extinctions caused by invasive species, predation alone (i.e., in the absence of other factors) is listed as being responsible for 30% of extinct animal species (Sax & Gaines, 2008), whereas competition has never been listed as the sole factor responsible for species extinctions (Sax & Gaines, 2008). Predation by invasive species, in concert with other factors, is believed to account for 98% of all animal extinctions (Sax & Gaines, 2008).

Overexploitation

Overexploitation can be defined as a human-induced source of mortality beyond natural levels of replenishment. Since the loss of individuals is larger than the gain, populations decline. The reasons for human exploitation of certain species are multiple and include supplying an increasing demand for food (e.g., ~15% of the animal protein consumed by humans is directly or indirectly derived from fisheries; FAO, 2011), cultural reasons (e.g., ornaments and jewels derived from animal parts such as rhino horns, fur, etc; Loveridge *et al.*,

2012), medicines and remedies (e.g., penises of tigers, shark fins, etc.; Loveridge *et al.*, 2012), recreational purposes (e.g., hunting and fishing; Loveridge *et al.*, 2012), and limiting human fatalities and livestock losses (e.g., Michalski *et al.*, 2006; Loveridge *et al.*, 2012; Marchini & Macdonald, 2012) to name a few. As a single threat, overexploitation has been the second leading cause of extinction among animals, accounting alone for 18% of extinct animal species; in contrast, no plant species has gone extinct due to the unique effects of overexploitation; only 4% and 1% of animal and plant species, respectively, currently at risk of extinction are so by the unique effect of overexploitation (Figure 17.1).

The mechanism through which overexploitation affects the distribution of species is directly through mortality (e.g., killing of adults, collection of eggs). In principle, over-exploitation should be self-regulated because declining populations will increase the cost of harvesting beyond profitability, at which point harvesting pressure should decrease. Unfortunately, there are multiple reasons why this is not the case and harvesting is continued despite the ongoing decline of exploited species. First, in some instances the declining supply of overexploited species can also lead to an increase in their market price (Courchamp *et al.*, 2006). This process is known to trigger exploitation vortices, in which smaller populations enhance further exploitation as rarer individuals become increasingly more valuable (Courchamp *et al.*, 2006). A similar mechanism results from access roads which open new markets for trade and add value to the exploitation of certain species (e.g., Macdonald *et al.*, 2012). A second reason promoting exploitation of declining species is economic subsidies. Here governments grant different types of aids to compensate the monetary loss associated with declining stocks, thus preventing social turmoil but further intensifying exploitation (Sumaila *et al.*, 2008; Mora *et al.*, 2009; Mora & Sale, 2011).

Habitat loss

Habitat loss could be defined as the process by which the area inhabited by a species is rendered functionally unsuitable to further sustain individuals of the species. The causes of habitat loss are diverse and include the expansion for agriculture and urban areas (Short & Burdick, 1996; Gaston *et al.*, 2003), climate change (Thomas *et al.*, 2004), sea-level rise and sedimentation (Valiela *et al.*, 2001), pollution (Short & Burdick, 1996), roads (Rytwinski & Fahrig, 2012), and excessive tourism (Schlacher & Thompson, 2012), among others. As a reference, the expansion of agricultural land has been associated with a 25% decline of the world's bird numbers since pre-agricultural times (Gaston *et al.*, 2003), whereas ongoing climate change is expected to make certain areas inhospitable "committing to extinction" between 15% and 37% of the world's birds by 2050 (Thomas *et al.*, 2004). Individually, habitat loss is the third leading cause of extinction in animals, being the sole factor in 11% of extinctions and accounting for the current vulnerability of 21% of threatened animal species. Yet of the documented plant extinctions not one occurred for this reason alone and only ~4% of currently threatened plant species are so by the sole effect of habitat loss (Figure 17.1). Habitat loss is considered the most pervasive driver of current biodiversity change (Sala *et al.*, 2000) with an estimated 27 000 species extinct each year due to deforestation alone

(Sax & Gaines, 2008, but see He & Hubbell, 2011). Many of these extinct species are likely not to have been described yet (Mora *et al*., 2011b).

The mechanism through which habitat loss increases extinction risk is probably related to the maintenance of populations in metapopulation systems (i.e., a group of spatially separated populations interlinked by dispersal). Three characteristics of habitats are critical to the maintenance of such systems: size, isolation and habitat quality (Hill *et al*., 1996; Griffen & Drake, 2008). Size will influence directly the number of individuals that can be supported and the probability of immigration (Hill *et al*., 1996; Griffen & Drake, 2008); additionally, depending on the area/perimeter ratio, it could also trigger edge effects, which may cause changes in populations due to drastic environmental shifts and susceptibility to negative interactions with other species (Murcia, 1995). Isolation can create genetic bottlenecks and increase the probability of extinction by haphazard phenomena (Hill *et al*., 1996) whereas habitat quality can influence the amount and resilience of individuals in the population and reduce the chance of recolonization of previously occupied patches (Griffen & Drake, 2008). An important aspect of metapopulation dynamics and habitat loss is the existence of extinction debts (i.e., delayed extinction triggered by the loss of habitats; Tilman *et al*., 1994). According to this idea, isolation, reduced amount and quality of habitats and accentuated ecological interactions arising from habitat loss can lead to transient/unsustainable populations; because extinction may occur long after habitat loss, such extinctions represent a debt of current habitat loss (Tilman *et al*., 1994).

Climate change

Increasing greenhouse gas emissions are changing the Earth's climate and biodiversity. CO_2 concentrations have increased from 280 ppm in pre-industrial times to 380 by 2000 and could reach between 550 and 800 ppm by 2100 depending on the emissions scenario (Solomon *et al*., 2007). Since pre-industrial times and as a response to the greenhouse effect of CO_2 and other human-generated gases, average global temperature has increased by 0.74°C, whereas the mixture of CO_2 with water has acidified the world's oceans by 0.1 pH units (Solomon *et al*., 2007). Future projections indicate that temperature could rise by up to 5°C and pH decrease by 0.3 units before 2100 (Solomon *et al*., 2007). In addition to temperature and pH, climate change may also trigger changes in rainfall, extreme weather events and sea level rise. The magnitude of these changes would be unprecedented in the Earth's history during the last 20 million years (RSL, 2005). As a single factor, climate change is the leading factor in extinction (~22%) and imperilment (27%) of plants but has played only a minor role in the extinction and endangerment of animals (Figure 17.1); one would expect, however, an indirect effect of plant loss on animals, as the former provide critical resources for the latter. Climate change has also been linked to substantial pole-ward shifts, and tropical and low-elevation retreats in multiple species (Parmesan & Yohe, 2003; Wilson *et al*., 2005; Franco *et al*., 2006; Devictor *et al*., 2012). For example, range shifts have been consistent with climate change in over 80% of monitored species, averaging 6.1 km per decade (Parmesan & Yohe, 2003), ranging as high as 114 km northward over 18 years in Europe

(Devictor *et al.*, 2012). The best-documented cases of uphill displacement and tropical retreat are available for butterflies, which are moving uphill at 70 m per decade in Spain (Wilson *et al.*, 2005), ~50 m per decade in Britain (Franco *et al.*, 2006) and retreating from lower towards higher latitudes at ~44 km per decade in Britain (Franco *et al.*, 2006). In fishes, climate change is expected to cause changes in community composition in over 60% of the world oceans by 2050 due to local extinctions and facilitation of invasions (Cheung *et al.*, 2009).

Linking changes in the distribution of species to climate change, however, has been challenging. In some instances, for example, patterns that could be explained by temperature could alternatively be explained by a covariant of temperature. Gaylord and Gaines (2000), for example, documented that while many boundaries in the distribution of marine organisms with pelagic larvae are coincident with abrupt changes in temperature, these same places coincide with changes in the directions of currents, which prevent the dispersal and settlement of populations up-current. Likewise, transplant experiments have indicated that while in some cases species fail to survive in more poleward locations, in other cases they thrive, highlighting the complexity of generalizing patterns of expansion solely from temperature changes (Gaston, 2003). The existence of source-sink populations has also been implied to yield errors on projection of climate change effects as species could persist in some places only because individuals disperse into them from elsewhere (Davis *et al.*, 1998). Additionally, climate change may unbalance ecological interactions, and thus local extinctions or population increases could be caused by the latter but not the former (Davis *et al.*, 1998). Conversely, there may also be a lag in response to the effects of climate change, in which species track changes in temperature at a rate slower than the rate of change in temperature; this may lead to underestimate actual responses to climate change (Devictor *et al.*, 2012).

The mechanisms through which climate change affects the distribution of species are diverse, including changes in the suitability of habitats, influencing physiological processes, increasing exposure and susceptibility to pathogens, affecting the timing, extent, direction and even propensity to migrate, etc. (Parmesan & Yohe, 2003; Fabry, 2008; Hoegh-Guldberg & Bruno, 2010; Devictor *et al.*, 2012). Climate change has also been related to changes in other geophysical processes and biological responses such as ocean acidification impacting calcification (Fabry, 2008), oxygen depletion and the growing expansion of dead zones in the world oceans (Stramma *et al.*, 2010) and iron enrichment and ocean productivity (Shi *et al.*, 2010). The great variety of physiological processes that are mediated by temperature, pH, oxygen and nutrients highlights the broad spectrum of processes potentially influenced by climate change. Among coral and reef fish species, it is often argued that these changes are happening too fast for species to adapt, although this has been a topic of debate (e.g., Mora & Ospina, 2001; Baird & Maynard, 2008; Hoegh-Guldberg & Bruno, 2010; Pandolfi *et al.*, 2011).

Interaction among drivers

A major unknown about projections of biodiversity change is the extent to which anthropogenic threats interact and potentially accelerate the risk of extinction (Myers,

1995; Sala *et al.*, 2000; Mora *et al.*, 2007; Brook *et al.*, 2008; Mora, 2008). Around 70% of the animal and plant species currently critically endangered are so by the effects of two or more anthropogenic factors (Figure 17.1). Although it is often said that anthropogenic stressors interact in a synergistic manner (i.e., causing a larger effect than the individual effects combined), a recent meta-analysis indicated that cases of synergies are rarer than simple additive or antagonistic effects (Darling & Côté, 2008). Examination of 57 experimental studies combining the effects of multiple stressors indicated that synergistic interactions occurred in 36% of the cases, whereas non-synergistic responses were reported in the remaining 64% of the cases (i.e., 42% and 22% reported antagonistic and additive effects, respectively). These results, however, need to be considered within context. For instance, Mora *et al.* (2007) showed that the interacting effects of habitat loss and overharvesting were additive whereas the interaction of either habitat loss or over-harvesting with warming led to synergistic declines in population size in either case.

Although the effect of interacting factors is likely to be case specific, empirical studies highlight multiple mechanisms through which co-occurring factors interact and add to the decline of wild populations (Figure 17.2). In the marine realm, for instance, climate change can accelerate the development of pelagic larvae, which in turn reduces the potential for dispersal. Simultaneously, climate change is reducing the suitability of habitats, therefore increasing their isolation. This combination of reduced dispersal capabilities and increasing isolation of habitats adds to the species' risks of extinction (Mora & Sale, 2011). Another example is the ongoing decline in wild frog populations, which appears to be linked to an interaction between global warming and disease. Some suggested mechanisms include an upward extension of fungal pathogens to encompass virtually all high-elevation anuran

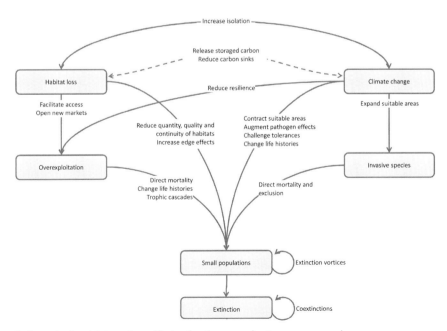

Figure 17.2.   Independent and interacting effects of anthropogenic stressors on species.

habitats (Seimon *et al.*, 2007) and a shift in environmental conditions that are optimum for the growth of pathogens, thus encouraging outbreaks (Pounds *et al.*, 2006). A similar situation has been reported for coral species, where ongoing exposure to stressors such as warming and pollution appear to have undermined their resilience, making them particularly vulnerable to mortality by diseases (Bruno *et al.*, 2007; Mora, 2009; Rogers, 2009). In general, climate change might facilitate invasions by non-indigenous species that act as novel competitors, predators or pathogens (Brook *et al.*, 2008) and such effects may be exacerbated if invasions occur simultaneously with the loss and deterioration of habitats (Didham *et al.*, 2005). Another interesting mechanism involving the interaction of stressors is how roads and associated habitat loss grant access to new grounds for exploitation and trade of species (Macdonald *et al.*, 2012).

Intrinsic characteristics and resilience to stressors

The extent to which anthropogenic factors will result in changes in the distribution of species depends to a large degree upon life history characteristics or intrinsic attributes of species. For example, the contrasting effect of anthropogenic stressors in animals and plants (Figure 17.1) is very likely to be due to different life histories offering different levels of resilience. More specifically, for instance, the expansion and/or collapse of species subjected to excessive fishing appears to be strongly mediated by behavioral aggregations which increase their rate of harvest, body size which makes them particularly targeted, and longevity and fecundity which affect their rate of recovery (Pauly *et al.*, 1998; Worm & Tittensor, 2011). Likewise the resilience to habitat loss appears to be undermined in species that are rare (in abundance), specialized (in habitat and food), have little dispersing capabilities, high population variability and high trophic position (Davies *et al.*, 2000; Tscharntke *et al.*, 2002; Öckinger *et al.*, 2010). Resilience to climate change may also vary among species depending upon their abilities to disperse (Dolman & Sutherland, 1995; Warren *et al.*, 2001; Parmesan & Yohe, 2003; Franco *et al.*, 2006; Lawler *et al.*, 2010), ability to alter migration routes (Dolman & Sutherland, 1995), physiological tolerance (Mora & Ospina, 2001; Jiguet *et al.*, 2007) and the timing and frequency of reproduction to maximize recovery and chances of adaptation through mutations (Mora *et al.*, 2007) to name some. It should be noted, however, that some traits of resilience may not be sufficient to withstand widespread and intensifying anthropogenic stressors. For instance, abundant and widespread species are markedly underrepresented in the extinction record, indicating their resilience to extinction, perhaps due to possessing these traits (McKinney, 1997; Şekercioğlu *et al.*, 2012), yet a number of species that are currently highly threatened or have recently become extinct due to anthropogenic factors were at some point considered common and widespread (Gaston & Fuller, 2007).

Alternative equilibrium states

One of the most worrisome pieces of knowledge about the current loss of biodiversity is the possibility that such changes could be irreversible if species get locked into alternative and less desirable states. This may occur if the processes of adaptation to ongoing and intensifying anthropogenic stressors leads to a loss of genetic diversity and the fixation of "maladaptive" traits (Walsh *et al.*, 2006; Swain *et al.*, 2007; Allendorf & Hard, 2009; Conover *et al.*, 2009; Darimont *et al.*, 2009). It should also be noted that "maladaptations" that impair demographic processes can potentially cause extinction in what has been defined as "evolutionary suicide" (Rankin & Lopez-Sepulcre, 2005). Evidence of changes in life history traits induced by human stressors is diverse. For instance, overexploitation, by taking large proportions of the populations and targeting large, reproductive-aged adults, has been shown to induce particularly rapid and dramatic changes in life history traits of fishes, mammals and plants (Allendorf & Hard, 2009; Darimont *et al.*, 2009). Experimental evidence suggests that intensified size-selective fishing induces substantial declines in mean body size, fecundity and larval viability, which in turn detrimentally affect recruitment, yield and biomass (Conover & Munch, 2002; Walsh *et al.*, 2006). Likewise, increasing warming has been related to reductions in body size (Sheridan & Bickford, 2011), clutch size (Winkler *et al.*, 2002), accelerated early development among marine organisms (Hoegh-Guldberg & Bruno, 2010; Mora & Sale, 2011), and mismatches between key reproductive and migratory cycles (e.g., matching reproduction to the production of early stages of zooplankton is essential to ensure sufficient food supply and larval survival in cod (Bollens *et al.*, 1992); some birds also link their reproduction to coincide with peaks in food availability, which maximize offspring survival; Visser *et al.*, 1998). Since temperature is fundamental to the biochemistry of most organisms, some physiological responses may be "inescapable" in the face of changing climate (Sheridan & Bickford, 2011). Induced evolution may also occur in response to habitat loss. For example, differential sex-mortality combined with an isolation of habitats by fences and roads has led to an increase in clutch size but a reduction in the years nesting in American prairie-chickens (*Tympanuchus pallidicinctus*); this trade-off increased the species' susceptibility to year-to-year environmental variations and appears to explain observed population declines in the species (Patten *et al.*, 2005). Habitat loss and subsequent fragmentation may also induce genetic bottlenecks since isolation may lead to inbreeding, reduced reproductive output and genetic drift; these responses are often associated with extinction debts in which genetic diversity is lost slowly over multiple generations (Lowe *et al.*, 2005). Even invasive species have been documented to induce morphological, behavioral, physiological and life history changes in native species (Carroll, 2007); as a case example, some Australian snakes have evolved a reduced size of their mouth as a way to escape the lethal effect of eating poisonous invasive cane toads (Phillips & Shine, 2004).

The potential to reverse "maladaptations" depends on the degree to which such changes are genetically based and the severity of the loss of genetic diversity. Recent advances on genome-wide scanning are likely to improve our understanding of the genetic basis of phenetic changes induced by anthropogenic stressors (Dettman *et al.*, 2012). However, empirical studies have shown that many collapsed fish populations have failed to recover after a decade or more with little fishing (Hutchings, 2000; Swain *et al.*, 2007), supporting the hypothesis that induced changes are affecting the genetic makeup of species and that such changes may be irreversible (Swain *et al.*, 2007) or very hard to reverse (Allendorf & Hard, 2009). The challenge of reversing genetic changes is supported by some models (e.g., de Roos *et al.*, 2006). If ultimately such changes lead to extinction, it is worth noting that the replacement of extinct species (i.e., through origination of new species) in prior mass extinction events has been of the order of tens of millions of years (note that a large number of niches were probably vacant after those mass extinction events) (Kirchner & Weil, 2000).

Extinction vortices and chains of extinction

Studies into patterns of extinction have revealed that declines in abundance and distri-bution imposed by anthropogenic stressors can be self-accelerated (i.e., the so-called extinction vortices). In turn, the decimation and/or extinction of species may lead to the extinction of other species (i.e., the so-called "chain of extinction"; Brook *et al.*, 2008) and disruption of ecological processes (i.e., the so-called "phase shifts", "trophic cas-cades" or "alternative states"; Bellwood *et al.*, 2004; Myers *et al.*, 2007; Heithaus *et al.*, 2008; Jackson, 2008). Patterns of extinction in monitored vertebrate species confirmed, for instance, that time to extinction scales to the logarithm of population size (i.e., the decline of populations is accelerated as the time-to-extinction is approached; Fagan & Holmes, 2006). It is likely that a combination of loss of genetic variation (which decreases population adaptive potential; Elam *et al.*, 2007), inbreeding depression (due to the limited number of mates to choose from, which in turn increases the expression of recessive deleterious genes and over-dominant genes; Soulé & Mills, 1998; Tanaka, 1998), Allee effects (Myers *et al.*, 1995), and environmental variability (Soulé & Mills, 1998; Tanaka, 1998) may all contribute to a general corrosion of population dynamics, causing a negative per-capita replacement rate as extinction is approached.

The consequences of a particular anthropogenic stressor may not end with the extinc-tion of the species directly under threat, if such a species is critical to the viability of other species or ecosystem processes. In one of the most comprehensive assessments of "co-extinctions" or "chains of extinction", Koh *et al.* (2004) analyzed coevolved interspecific systems (e.g., predator/prey, herbivore/plant and parasite/host) among species currently listed as extinct or endangered in the IUCN Red List. Their results indicate that 204 affiliate species may have become extinct historically due to the extinction of 399 host species. Likewise 9491 species currently endangered may be host to 6088 affiliate species, which thus are currently "coendangered" and likely to go extinct if their hosts become extinct. As a result of tight ecological interactions, induced changes on specific species can transcend to broader ecosystem consequences in what is named phase shifts, trophic cascades or alternative states (Bellwood *et al.*, 2004; Myers *et al.*, 2007; Heithaus *et al.*, 2008; Jackson, 2008).

Concluding remarks

Quantifying the relative role of anthropogenic drivers in the loss of biodiversity is likely to remain challenging and, if shown, such studies are likely to demonstrate that the relative role of such factors may not be generalized among and even within species. For instance, the response of primates to forestry, agriculture and hunting showed that the relative vulnerability of particular species to one threat does not predict its response to others, which is expected considering that different biological traits will determine the response to each type of threat (Isaac & Cowlishaw, 2004). Likewise, Cardillo *et al.* (2008) showed that a global model of different types of predictors related to human and environmental impacts, species' life history traits and ecology did have a low explanatory power in determining extinction risk among mammals. These studies suggest that species responses to anthropogenic stressors are likely to be highly individualistic. However, despite this lack of generality and difficulty in quantifying the relative effect of individual drivers and mechanisms, there is overwhelming evidence to suggest that, from a cautionary and ethical perspective, threatened species should be managed as if all stressors at play were responsible for their decline. Arguably, the focus could also be shifted to what drives overexploitation, habitat loss, climate change and invasive species in the first place. This focus will make clear that at the core of these stressors are our patterns of consumption and ongoing population growth (Mora & Sale, 2011). This focus will require a shift in conservation perspectives but could provide more definitive solutions to a broad range of stressors. The relentless loss of biodiversity and associated loss of goods and services suggest that we cannot afford much delay before choosing the right response to these stressors.

Synopsis

Humanity has taken a heavy toll on the Earth's biodiversity. Clues for such a human footprint are found ever since prehistoric times but have become considerably more evident and severe in recent times. We are now on the verge of a sixth mass extinction event whose causes are well connected to stressors such as habitat loss, overexploitation, invasive species and climate change. While the relative effect of such stressors will remain challenging to quantify and will most likely be individualistic, there is considerable evidence to suggest that from a cautionary and ethical perspective, threatened species should be managed as if all stressors at play were responsible for their decline. Arguably, a less contentious conservation strategy could focus on what drives such stressors in the first place, which will probably reveal the role of our patterns of consumption and ongoing population growth. This new focus will require a shift in conservation perspectives but should deliver more definitive solutions to a broad range of issues. Regardless of the solution, the rapid loss of biodiversity, goods and services suggests that we cannot afford much delay before choosing the right response to these stressors.

References

Allendorf, F. W., & Hard, J. J. (2009). Human-induced evolution caused by unnatural selection through harvest of wild animals. *Proceedings of the National Academy of Sciences of the USA*, **106**, 9987–9994.

Alroy, J. (2002). How many named species are valid? *Proceedings of the National Academy of Sciences of the USA*, **99**, 3706–3711.

Baird, A., & Maynard, J. A. (2008). Coral adaptation in the face of climate change. *Science*, **320**, 315–316.

Barnosky, A. D., Matzke, N., Tomiya, S., *et al.* (2011). Has the Earth's sixth mass extinction already arrived? *Nature*, **471**, 51–57.

Bellwood, D. R., Hughes, T. P., Folke, C., & Nystrom, M. (2004). Confronting the coral reef crisis. *Nature*, **429**, 827–833.

Blackburn, T. M., Cassey, P., Duncan, R. P., Evans, K. L., & Gaston, K. J. (2004). Avian extinction and mammalian introductions on oceanic islands. *Science*, **305**, 1955–1958.

Bollens, S. M., Frost, B. W., Schwaninger, H. R., *et al.* (1992). Seasonal plankton cycles in a temperate fjord and comments on the match-mismatch hypothesis. *Journal of Plankton Research*, **14**, 1279–1305.

Botts, E. A., Erasmus, B. F. N., & Alexander, G. J. (2012). Methods to detect species range size change from biological atlas data: a comparison using the South African Frog Atlas Project. *Biological Conservation*, **146**, 72–80.

Brook, B. W., Sodhi, N. S., & Bradshaw, C. J. A. (2008). Synergies among extinction drivers under global change. *Trends in Ecology & Evolution*, **23**, 453–460.

Bruno, J. F., Selig, E. R., Casey, K. S., Page, C. A., Willis, B. L., Harvell, C. D., Sweatman, H., & Melendy, A. M. (2007). Thermal stress and coral cover as drivers of coral disease outbreaks. *PLoS Biology*, **5**, 1220–1227.

Burgman, M. A., & Fox, J. C. (2003). Bias in species range estimates from minimum convex polygons: implications for conservation and options for improved planning. *Animal Conservation*, **6**, 19–28.

Burney, D. A., & Flannery, T. F. (2005). Fifty millennia of catastrophic extinctions after human contact. *Trends in Ecology & Evolution (Personal edition)*, **20**, 395–401.

Campbell, M., Cleland, J., Ezeh, A., & Prata, N. (2007). Public health – return of the population growth factor. *Science*, **315**, 1501–1502.

Cardillo, M., Mace, G. M., Gittleman, J. L., *et al.* (2008). The predictability of extinction: biological and external correlates of decline in mammals. *Proceedings of the Royal Society of London B*, **275**, 1441–1448.

Carroll, S. (2007). Brave New World: the epistatic foundations of natives adapting to invaders. *Genetica*, **129**, 193–204.

Channell, R., & Lomolino, M. V. (2000). Dynamic biogeography and conservation of endangered species. *Nature*, **403**, 84–86.

Cheung, W. W. L., Lam, V. W. Y., Sarmiento, J. L., *et al.* (2009). Projecting global marine biodiversity impacts under climate change scenarios. *Fish and Fisheries*, **10**, 235–251.

Clavero, M., & García-Berthou, E. (2005). Invasive species are a leading cause of animal extinctions. *Trends in Ecology & Evolution*, **20**, 110.

Conover, D. O., & Munch, S. B. (2002). Sustaining fisheries yields over evolutionary time scales. *Science*, **297**, 94–96.

Conover, D. O., Munch, S. B., & Arnott, S. A. (2009). Reversal of evolutionary downsizing caused by selective harvest of large fish. *Proceedings of the Royal Society of London B*, **276**, 2015–2020.

Courchamp, F., Angulo, E., Rivalan, P., *et al.* (2006). Rarity value and species extinction: the anthropogenic Allee effect. *PLoS Biology*, **4**, e415.

Darimont, C. T., Carlson, S. M., Kinnison, M. T., *et al.* (2009). Human predators outpace other agents of trait change in the wild. *Proceedings of the National Academy of Sciences of the USA*, **106**, 952–954.

Darling, E. S., & Côté, I. M. (2008). Quantifying the evidence for ecological synergies. *Ecology Letters*, **11**, 1278–1286.

Davies, K. F., Margules, C. R., & Lawrence, J. F. (2000). Which traits of species predict population declines in experimental forest fragments? *Ecology*, **81**, 1450–1461.

Davis, A. J., Jenkinson, L. S., Lawton, J. H., Shorrocks, B., & Wood, S. (1998). Making mistakes when predicting shifts in species range in response to global warming. *Nature*, **391**, 783–786.

De Roos, A. M., Boukal, D. S., & Persson, L. (2006). Evolutionary regime shifts in age and size at maturation of exploited fish stocks. *Proceedings of the Royal Society of London B*, **273**, 1873–1880.

Dettman, J. R., Rodrigue, N., Melnyk, A. H., *et al*. (2012). Evolutionary insight from whole-genome sequencing of experimentally evolved microbes. *Molecular Ecology*, **21**, 2058–2077.

Devictor, V., Van Swaay, C., Brereton, T., *et al*. (2012). Differences in the climatic debts of birds and butterflies at a continental scale. *Nature Climate Change*, **2**, 121–124.

Díaz, S., Fargione, J., Chapin, F. S., Iii & Tilman, D. (2006). Biodiversity loss threatens human well-being. *PLoS Biology*, **4**, e277.

Didham, R. K., Ewers, R. M., & Gemmell, N. J. (2005). Comment on "avian extinction and mammalian introductions on oceanic islands". *Science*, **307**, 1412.

Dolman, P. M., & Sutherland, W. J. (1995). The response of bird populations to habitat loss. *Ibis*, **137**, S38–S46.

Dulvy, N. K., Ellis, J. R., Goodwin, N. B., *et al*. (2004). Methods of assessing extinction risk in marine fishes. *Fish and Fisheries*, **5**, 255–276.

Elam, D. R., Ridley, C. E., Goodell, K., & Ellstrand, N. C. (2007). Population size and relatedness affect fitness of a self-incompatible invasive plant. *Proceedings of the National Academy of Sciences of the USA*, **104**, 549–552.

Fabry, V. J. (2008). Marine calcifiers in a high-CO_2 ocean. *Science*, **320**, 1020–1022.

Fagan, W. F., & Holmes, E. E. (2006). Quantifying the extinction vortex. *Ecology Letters*, **9**, 51–60.

FAO. (2011). State of the world's land and water resources for food and agriculture. Food and Agriculture Organization of the United Nations, Rome and Earthscan, London. http://www.fao.org/nr/solaw/the-book/en/.

Fouquet, A., Gilles, A., Vences, M., *et al*. (2007). Underestimation of species richness in neo-tropical frogs revealed by mtDNA analyses. *PLoS ONE*, **2**, e1109.

Franco, A. M. A., Hill, J. K., Kitschke, C., *et al*. (2006). Impacts of climate warming and habitat loss on extinctions at species' low-latitude range boundaries. *Global Change Biology*, **12**, 1545–1553.

Frankham, R. (2004). Resolving the genetic paradox in invasive species. *Heredity*, **94**, 385–385.

Gaston, K. (2003). *The Structure and Dynamics of Geographic Ranges*. Oxford: Oxford University Press.

Gaston, K. J., Blackburn, T. M., & Goldewijk, K. (2003). Habitat conversion and global avian biodiversity loss. *Proceedings of the Royal Society of London B*, **270**, 1293–1300.

Gaston, K. J., & Fuller, R. A. (2007). Biodiversity and extinction. *Progress in Physical Geography*, **31**, 213–225.

Gaylord, B., & Gaines, S. D. (2000). Temperature or transport? Range limits in marine species mediated solely by flow. *The American Naturalist*, **155**, 769–789.

Gertzen, E. L., Leung, B., & Yan, N. D. (2011). Propagule pressure, Allee effects and the probability of establishment of an invasive species (*Bythotrephes longimanus*). *Ecosphere*, **2**, a30.

Griffen, B. D., & Drake, J. M. (2008). Effects of habitat quality and size on extinction in experimental populations. *Proceedings of the Royal Society of London B*, **275**, 2251–2256.

Grigg, R. W., & Dollar, S. J. (2005). Reassessing US coral reefs. *Science*, **308**, 1740–1742.

Gurevitch, J., & Padilla, D. K. (2004). Are invasive species a major cause of extinctions? *Trends in Ecology & Evolution*, **19**, 470–474.

Hails, C. (2008). *Living Planet Report 2008*. Gland, Switzerland: WWF International.

He, F., & Hubbell, S. P. (2011). Species-area relationships always overestimate extinction rates from habitat loss. *Nature*, **473**, 368–371.

Heithaus, M. R., Frid, A., Wirsing, A. J., & Worm, B. (2008). Predicting ecological consequences of marine top predator declines. *Trends in Ecology & Evolution*, **23**, 202–210.

Hill, J. K., Thomas, C. D., & Lewis, O. T. (1996). Effects of habitat patch size and isolation on dispersal by Hesperia comma butterflies: implications for metapopulation structure. *Journal of Animal Ecology*, **65**, 725–735.

Hoegh-Guldberg, O., & Bruno, J. F. (2010). The impact of climate change on the world's marine ecosystems. *Science*, **328**, 1523–1528.

Hutchings, J. A. (2000). Collapse and recovery of marine fishes. *Nature*, **406**, 882–885.

Isaac, N. J. B., & Cowlishaw, G. (2004). How species respond to multiple extinction threats. *Proceedings of the Royal Society of London B*, **271**, 1135–1141.

IUCN (2012). IUCN Red List of Species.

Jackson, J. B. C. (2008). Ecological extinction and evolution in the brave new ocean. *Proceedings of the National Academy of Sciences of the USA*, **105**, 11458–11465.

Jetz, W., Sekercioglu, C. H., & Watson, J. E. M. (2008). Ecological correlates and conservation implications of overestimating species geographic ranges. *Conservation Biology*, **22**, 110–119.

Jiguet, F., Gadot, A.-S., Julliard, R., Newson, S. E., & Couvet, D. (2007). Climate envelope, life history traits and the resilience of birds facing global change. *Global Change Biology*, **13**, 1672–1684.

Kirchner, J. W., & Weil, A. (2000). Delayed biological recovery from extinctions throughout the fossil record. *Nature*, **404**, 177–180.

Knowlton, N., & Jackson, J. B. C. (2008). Shifting baselines, local impacts, and global change on coral reefs. *PLoS Biology*, **6**, 215–220.

Koh, L. P., Dunn, R. R., Sodhi, N. S., *et al.* (2004). Species coextinctions and the biodiversity crisis. *Science*, **305**, 1632–1634.

Kolar, C. S., & Lodge, D. M. (2001). Progress in invasion biology: predicting invaders. *Trends in Ecology & Evolution*, **16**, 199–204.

Lawler, J. J., Shafer, S. L., Bancroft, B. A., & Blaustein, A. R. (2010). Projected climate impacts for the amphibians of the western hemisphere. *Conservation Biology*, **24**, 38–50.

Loveridge, A. J., Wang, S. W., Frank, L. G., & Seidensticker, J. (2012). People and wild felids: conservation of cats and management of conflicts. In D. Macdonald & A. Loveridge (Eds.), *The Biology and Conservation of Wild Felids* (pp. 161–195). Oxford: Oxford University Press.

Lowe, A. J., Boshier, D., Ward, M., Bacles, C. F. E., & Navarro, C. (2005). Genetic resource impacts of habitat loss and degradation; reconciling empirical evidence and predicted theory for neotropical trees. *Heredity*, **95**, 255–273.

Macdonald, D. W., Johnson, P. J., Albrechtsen, L., *et al.* (2012). Bushmeat trade in the Cross–Sanaga rivers region: evidence for the importance of protected areas. *Biological Conservation*, **147**, 107–114.

Marchini, S., & Macdonald, D. W. (2012). Predicting ranchers' intention to kill jaguars: case studies in Amazonia and Pantanal. *Biological Conservation*, **147**, 213–221.

Mckinney, M. L. (1997). Extinction vulnerability and selectivity: combining ecological and pale-ontological views. *Annual Review of Ecology and Systematics*, **28**, 495–516.

Michalski, F., Boulhosa, R. L. P., Faria, A., & Peres, C. A. (2006). Human–wildlife conflicts in a fragmented Amazonian forest landscape: determinants of large felid depredation on livestock. *Animal Conservation*, **9**, 179–188.

Mora, C. (2008). A clear human footprint in the coral reefs of the Caribbean. *Proceedings of the Royal Society of London B*, **275**, 767–773.

Mora, C. (2009). Degradation of Caribbean coral reefs: focusing on proximal rather than ultimate drivers. Reply to Rogers. *Proceedings of the Royal Society of London B*, **276**, 199–200.

Mora, C., & Ospina, A. F. (2001). Tolerance to high temperatures and potential impact of sea warming on reef fishes of Gorgona Island (tropical eastern Pacific). *Marine Biology*, **139**, 765–769.

Mora, C., & Sale, P. (2011). Ongoing global biodiversity loss and the need to move beyond protected areas: a review of the technical and practical shortcomings of protected areas on land and sea. *Marine Ecology Progress Series*, **434**, 251–266.

Mora, C., Jiménez, J. M., & Zapata, F. A. (2000). *Pontinus clemensi* (Pisces: Scorpaenidae) at Malpelo Island, Colombia: new specimen and geographic range extension. *Boletín de investigaciones marinas y costeras*, **29**, 85–88.

Mora, C., Metzger, R., Rollo, A., & Myers, R. A. (2007). Experimental simulations about the effects of overexploitation and habitat fragmentation on populations facing environmental warming. *Proceedings of the Royal Society of London B*, **274**, 1023–1028.

Mora, C., Myers, R. A., Coll, M., *et al.* (2009). Management effectiveness of the world's marine fisheries. *PLoS Biology*, **7**, e1000131.

Mora, C., Aburto-Oropeza, O., Ayala Bocos, A., *et al.* (2011a). Global human footprint on the linkage between biodiversity and ecosystem functioning in reef fishes. *PLoS Biology*, **9**, e1000606.

Mora, C., Tittensor, D. P., Adl, S., Simpson, A. G. B., & Worm, B. (2011b). How many species are there on Earth and in the ocean? *PLoS Biology*, **9**, e1001127.

Murcia, C. (1995). Edge effects in fragmented forests: implications for conservation. *Trends in Ecology & Evolution*, **10**, 58–62.

Myers, N. (1995). Environmental unknowns. *Science*, **269**, 358–360.

Myers, R. A., Barrowman, N. J., Hutchings, J. A., & Rosenberg, A. A. (1995). Population dynamics of exploited fish stocks at low population levels. *Science*, **269**, 1106–1108.

Myers, R. A., Baum, J. K., Shepherd, T. D., Powers, S. P., & Peterson, C. H. (2007). Cascading effects of the loss of apex predatory sharks from a coastal ocean. *Science*, **315**, 1846–1850.

Novacek, M. J., & Cleland, E. E. (2001). The current biodiversity extinction event: scenarios for mitigation and recovery. *Proceedings of the National Academy of Sciences of the USA*, **98**, 5466–5470.

Öckinger, E., Schweiger, O., Crist, T. O., *et al.* (2010). Life-history traits predict species responses to habitat area and isolation: a cross-continental synthesis. *Ecology Letters*, **13**, 969–979.

Pandolfi, J. M., Connolly, S. R., Marshall, D. J., & Cohen, A. L. (2011). Projecting coral reef futures under global warming and ocean acidification. *Science*, **333**, 418–422.

Parmesan, C., & Yohe, G. (2003). A globally coherent fingerprint of climate change impacts across natural systems. *Nature*, **421**, 37–42.

Patten, M. A., Wolfe, D. H., Shochat, E., & Sherrod, S. K. (2005). Habitat fragmentation, rapid evolution and population persistence. *Evolutionary Ecology Research*, **7**, 235–249.

Pauly, D., Christensen, V., Dalsgaard, J., Froese, R., & Torres, F. (1998). Fishing down marine food webs. *Science*, **279**, 860–863.

Pearson, D. E., & Callaway, R. M. (2003). Indirect effects of host-specific biological control agents. *Trends in Ecology & Evolution*, **18**, 456–461.

Phillips, B. L., & Shine, R. (2004). Adapting to an invasive species: toxic cane toads induce morphological change in Australian snakes. *Proceedings of the National Academy of Sciences of the USA*, **101**, 17150–17155.

Pimm, S. L., Russell, G. J., Gittleman, J. L., & Brooks, T. M. (1995). The future of biodiversity. *Science*, **269**, 347–350.

Pounds, J. A., Bustamante, M. R., Coloma, L. A., *et al*. (2006). Widespread amphibian extinctions from epidemic disease driven by global warming. *Nature*, **439**, 161–167.

Prentis, P. J., Wilson, J. R. U., Dormontt, E. E., Richardson, D. M., & Lowe, A. J. (2008). Adaptive evolution in invasive species. *Trends in Plant Science*, **13**, 288–294.

Rankin, D. J., & Lopez-Sepulcre, A. (2005). Can adaptation lead to extinction? *Oikos*, **111**, 616–619.

Ricciardi, A. (2004). Assessing species invasions as a cause of extinction. *Trends in Ecology & Evolution*, **19**, 619.

Rogers, C. (2009). Coral bleaching and disease should not be underestimated as causes of Caribbean coral reef decline. *Proceedings of the Royal Society of London B*, **276**, 197–198.

Rondinini, C., Wilson, K. A., Boitani, L., Grantham, H., & Possingham, H. P. (2006). Tradeoffs of different types of species occurrence data for use in systematic conservation planning. *Ecology Letters*, **9**, 1136–1145.

RSL (2005). Ocean acidification due to increasing atmospheric carbon dioxide. In Policy Document 12/05, Royal Society of London.

Rytwinski, T., & Fahrig, L. (2012). Do species life history traits explain population responses to roads? A meta-analysis. *Biological Conservation*, **147**, 87–98.

Sala, O. E., Chapin, F. S., Armesto, J. J., *et al*. (2000). Global biodiversity scenarios for the year 2100. *Science*, **287**, 1770–1774.

Sax, D. F., & Gaines, S. D. (2008). Species invasions and extinction: the future of native biodiversity on islands. *Proceedings of the National Academy of Sciences of the USA*, **105**, 11490–11497.

Schiermeier, Q. (2004). Climate findings let fishermen off the hook. *Nature*, **428**, 4–4.

Schlacher, T. A., & Thompson, L. (2012). Beach recreation impacts benthic invertebrates on ocean-exposed sandy shores. *Biological Conservation*, **147**, 123–132.

Seimon, T. A., Seimon, A., Daszak, P., *et al*. (2007). Upward range extension of Andean anurans and chytridiomycosis to extreme elevations in response to tropical deglaciation. *Global Change Biology*, **13**, 288–299.

Şekercioğlu, Ç. H., Primack, R. B., & Wormworth, J. (2012). The effects of climate change on tropical birds. *Biological Conservation*, **148**, 1–18.

Sheridan, J. A., & Bickford, D. (2011). Shrinking body size as an ecological response to climate change. *Nature Climate Change*, **1**, 401–406.

Shi, D., Xu, Y., Hopkinson, B. M., & Morel, F. M. M. (2010). Effect of ocean acidification on iron availability to marine phytoplankton. *Science*, **327**, 676–679.

Short, F., & Burdick, D. (1996). Quantifying eelgrass habitat loss in relation to housing development and nitrogen loading in Waquoit Bay, Massachusetts. *Estuaries and Coasts*, **19**, 730–739.

Solomon, S., Qin, D., & Manning, M. (2007). The Physical Science Basis. Contribution of Working Group I to the Fourth Assessment Report of the Intergovernmental Panel on Climate Change, pp. 104, Geneva.

Soulé, M. E., & Mills, L. S. (1998). No need to isolate genetics. *Science*, **282**, 1658–1659.

Stramma, L., Schmidtko, S., Levin, L. A., & Johnson, G. C. (2010). Ocean oxygen minima expansions and their biological impacts. *Deep Sea Research Part I: Oceanographic Research Papers*, **57**, 587–595.

Sumaila, U. R., Teh, L., Watson, R., Tyedmers, P., & Pauly, D. (2008). Fuel price increase, subsidies, overcapacity, and resource sustainability. *ICES Journal of Marine Science*, **65**, 832–840.

Swain, D. P., Sinclair, A. F., & Mark Hanson, J. (2007). Evolutionary response to size-selective mortality in an exploited fish population. *Proceedings of the Royal Society of London B*, **274**, 1015–1022.

Tanaka, Y. (1998). Theoretical aspects of extinction by inbreeding depression. *Researches on Population Ecology*, **40**, 279–286.

Thomas, C. D., Cameron, A., Green, R. E., *et al.* (2004). Extinction risk from climate change. *Nature*, **427**, 145–148.

Thomas, C. D., Franco, A. M. A., & Hill, J. K. (2006). Range retractions and extinction in the face of climate warming. *Trends in Ecology & Evolution*, **21**, 415–416.

Tilman, D., May, R. M., Lehman, C. L., & Nowak, M. A. (1994). Habitat destruction and the extinction debt. *Nature*, **371**, 65–66.

Tscharntke, T., Steffan-Dewenter, I., Kruess, A., & Thies, C. (2002). Characteristics of insect populations on habitat fragments: a mini review. *Ecological Research*, **17**, 229–239.

Valiela, I., Bowen, J. L., & York, J. K. (2001). Mangrove forests: one of the world's threatened major tropical environments. *BioScience*, **51**, 807–815.

Vieites, D. R., Wollenberg, K. C., Andreone, F., *et al.* (2009). Vast underestimation of Madagascar's biodiversity evidenced by an integrative amphibian inventory. *Proceedings of the National Academy of Sciences of the USA*, **106**, 8267–8272.

Visser, M., Van Noordwijk, A., Tinbergen, J., & Lessells, C. (1998). Warmer springs lead to mistimed reproduction in great tits (*Parus major*). *Proceedings of the Royal Society of London B*, **265**, 1867–1870.

Walsh, M. R., Munch, S. B., Chiba, S., & Conover, D. O. (2006). Maladaptive changes in multiple traits caused by fishing: impediments to population recovery. *Ecology Letters*, **9**, 142–148.

Warren, M. S., Hill, J. K., Thomas, J. A., *et al.* (2001). Rapid responses of British butterflies to opposing forces of climate and habitat change. *Nature*, **414**, 65–69.

Weidenhamer, J., & Callaway, R. (2010). Direct and indirect effects of invasive plants on soil chemistry and ecosystem function. *Journal of Chemical Ecology*, **36**, 59–69.

Williamson, M., & Fitter, A. (1996). The varying success of invaders. *Ecology*, **77**, 1661–1666.

Wilson, R. J., Gutiérrez, D., Gutiérrez, J., *et al.* (2005). Changes to the elevational limits and extent of species ranges associated with climate change. *Ecology Letters*, **8**, 1138–1146.

Winkler, D. W., Dunn, P. O., & Mcculloch, C. E. (2002). Predicting the effects of climate change on avian life-history traits. *Proceedings of the National Academy of Sciences of the USA*, **99**, 13595–13599.

Worm, B., & Myers, R. A. (2004). Managing fisheries in a changing climate – no need to wait for more information: industrialized fishing is already wiping out stocks. *Nature*, **429**, 15–15.

Worm, B., & Tittensor, D. P. (2011). Range contraction in large pelagic predators. *Proceedings of the National Academy of Sciences of the USA*, **108**, 11942–11947.

Worm, B., Barbier, E. B., Beaumont, N., *et al.* (2006). Impacts of biodiversity loss on ocean ecosystem services. *Science*, **314**, 787–790.

18   Worldwide decline and extinction of amphibians

Harold Heatwole

Amphibians constitute the most threatened major taxon on Earth today. Their dependence on cutaneous respiration necessitates a thin, moist, permeable skin that makes them vulnerable to desiccation, toxic chemicals, endocrine disruptors and changes in their physical environment. The seasonal migration of many species between terrestrial habitats and aquatic breeding sites exposes them to hazards such as increased risk of predation, traversing of inhospitable habitats and automobile traffic. Invasive species and destruction and fragmentation of habitat are implicated in some declines and humans collect amphibians for food, pets, research and medicines. Although amphibians cutaneously secrete a wide variety of antibiotics (Erspamer, 1994), they are susceptible to some viral, bacterial, parasitic and fungal infections. Thus, the alarming rate of decline and extinction of amphibians globally is not caused by a single agent (Halliday, 2005), but by a suite of them that vary geographically (Stuart *et al.*, 2010), and interact with each other. Changes in global climate have exacted a toll on amphibians already and are projected to be increasingly severe in the future. The present chapter reviews the causes of global decline and extinction of amphibians around the world. The reasons vary from one place to another.

History of amphibian decline

Following a growing, but mainly anecdotally based, awareness among herpetologists during the 1970s and 1980s that many amphibian populations were declining, special sessions to discuss the phenomenon began to be held at meetings of scientific societies, notably one by the United States Academy of Sciences in February 1990 that concluded that declines should be treated as a possible emergency and that recommended a body be established to determine the extent of the problem. The resulting agency was the Declining Amphibian Populations Task Force (DAPTF), set up in December 1990 under the umbrella of the Species Survival Commission (SSC) of the International Union for the Conservation of Nature (IUCN). It is now the IUCN/SSC Amphibian Specialist Group (ASG). It has fostered research on amphibian decline, set up protocols, collated databases, organized workshops, set up working groups to study diseases, chemical pollutants, and climatic influences, instituted local groups around the world to monitor conditions and collect and report data, and established an ongoing newsletter,

The Balance of Nature and Human Impact, ed. Klaus Rohde. Published by Cambridge University Press.
© Cambridge University Press 2013.

Froglog (Heyer & Murphy, 2005). With this impetus there was a burgeoning of research and literature on the topic. Because of the enormous volume of this literature and the limitations of space in the present volume, the references herein mainly are to reviews in which more extensive literature is cited.

Perhaps the most important breakthrough in research on decline and extinction of amphibians was the discovery by Lee Berger of chytridiomycosis (Skerratt, 2009), a cause of amphibian declines second in importance only to destruction of habitat. Scientists had been unable to explain a number of mysterious massive declines and extinctions of amphibians in protected, undisturbed habitats. In 1996, Dr. Berger was examining samples of sick and dead frogs from a massive mortality event at Big Tableland, Queensland, Australia, for potential viral agents. The frogs tested negative for viruses but she did find that their skins were infected by a previously unknown species of chytrid fungus, subsequently described by Joyce Longcore *et al.* (1999) as a new genus and species, *Batrachochytrium dendrobatidis*, now commonly abbreviated as *Bd*. Lee Berger's discovery opened the floodgates for research on this pathogen, and a plethora of publications around the world. The nature of the disease is now well known (Berger *et al.*, 2009), and its presence has been documented from every continent (Figure 18.1), except Antarctica, and has been implicated in numerous declines and extinctions (Heatwole *et al.*, 2010, 2011, 2012, in press).

The general topic of amphibian conservation and decline was reviewed by Beebee (1996), Semlitsch & Wake (2003), and Stuart *et al.* (2004, 2008) and in two volumes in the series *Amphibian Biology* (Heatwole & Wilkinson, 2009, 2012). One (volume 8) treated diseases, parasites, maladies, and pollution, whereas another (volume 10) dealt with the more ecological aspects (habitat loss, invasive species), anthropogenic effects (roadkills, humans' use of amphibians for food, pets, and medicine), phylogenetic, geographic, and life-history correlates of risk of extinction, methods of monitoring, and conservation practices (captive breeding, refuges and reserves). Another major impetus has been reviews of conservation and decline for particular regions: Lesser Antilles (Kaiser & Henderson, 1994), the former Soviet Union (Kuzmin *et al.*, 1995), Canada (Green, 1997), USA (Lannoo, 2005), Latin America (multiple papers in a special section of Biotropica, synthesized by Lips *et al.*, 2005), and Central America (Wilson *et al.*, 2010). Two volumes of *Amphibian Biology* treat the status of conservation and decline of amphibians on a country-by-country basis, including review of local literature and governmental reports that otherwise might escape notice by the broader scientific community. These volumes are published as individual parts, much as the individual issues of a scientific journal. One volume (9, edited by Heatwole, Barrio-Amorós, & Wilkinson) covers the Western Hemisphere and is now complete with parts 1 (2010), 2 (2011) and 3 (2012) already published and reviews for the remaining countries in press. The other volume (11, edited by Heatwole *et al.*, in press) covers the Eastern Hemisphere, with most countries of Asia, Europe and North Africa either in press or as completed manuscripts, but with sub-Saharan Africa, Madagascar, Australia and New Zealand still pending. The above regional treatments provide a historical baseline, a time-capsule, against which future changes can be evaluated. In one sense they are already going out of date at the time of publication, but in another sense they are timeless as one can

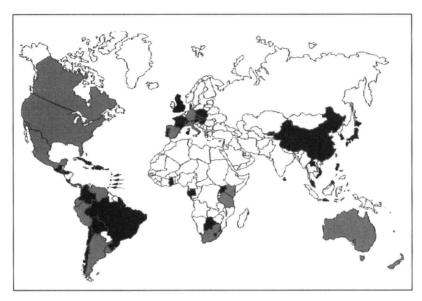

Figure 18.1.   Map of the world showing the countries in which *Batrachochytrium dendrobatidis* (*Bd*) was reported by Berger *et al.* (2009) (gray) supplemented by other reports (black). New data from volumes 9 and 11 of *Amphibian Biology* (Heatwole *et al.*, 2010, 2011, 2012, in press and in preparation) and from Weldon *et al.* (2004), Weldon (2005), Goldberg *et al.* (2007), Alemu *et al.* (2008), Bovero *et al.* (2008), Bourke *et al.* (2010), Wilson *et al.* (2010), Bell *et al.* (2011), Swei *et al.* (2011) and Civis and Vojar (2012). Illnesses have occurred in amphibians in Belgium and Serbia (Heatwole *et al.*, in press and in preparation) for which the cause was undetermined. Based on what is known of its ecology Ron (2005) predicted that *Bd* will eventually spread to those areas in the Caribbean, Central America, and South America (in white) in which it has not yet been reported. Arrows (from top to bottom respectively) indicate Montserrat, Dominica, Grenada, and Tobago. Not shown on map is presence of *Bd* in an introduced frog in Hawaii (Beard & O'Neill, 2005).

return to them for information on status at a particular time as a means of assessing subsequent extents and rates of change. There are a number of websites that keep abreast of current knowledge, and can be used to compare to that baseline. The IUCN devised a classification of the degree of endangerment and keep it up to date as the statuses of species change (http://www.iucnredlist.org/initiatives/amphibians), and several other major websites provide periodic assessments of amphibian decline (*Froglog*) and disease (*Disease*).

Major threats to amphibians

The sources of threat vary from one location to another, and their relative degrees of danger change in space and time. There are many, however, that are pervasive and recur in country after country. Most are anthropogenic, i.e., caused by humans, or at least abetted by humans, and are escalating at unprecedented magnitude and rapidity. Figure 18.2 provides a flow chart of the way major threats interact with each other in contributing to amphibian decline.

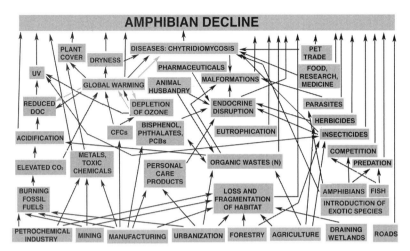

Figure 18.2.   Simplified diagram of the interaction of factors influencing amphibian decline. Arrowheads point from the influencing factor to the affected attribute. Double-headed arrows indicate that both factors interact, each having an effect on the opposite one. Arrows of different shades indicate that the influencing factor can either enhance (black arrows) or inhibit (gray arrows) the effect, depending on circumstances.

The diagram is simplified in several ways. Some boxes list only general categories of influences, e.g., groups of chemicals, rather than individual chemicals. A more detailed flow chart would list the individual chemicals separately. Also, such items as herbicides and insecticides are manufactured and hence could be linked to "manufacturing", but rather are linked to "agriculture", which is the activity that applies the chemicals. Similarly pharmaceuticals are manufactured but are not linked to manufacturing, as the diagram would become unduly cluttered. Note that the effect of "food, research, medicine" on chytridiomycosis is via trade in infected animals that serve as vectors. Some of the arrows are of direct links; others subsume a cascade of links. For example, the single arrow linking herbicides with amphibian decline is a composite of direct and indirect links. The direct one is the effect of some herbicides on the development and survival of amphibian larvae; the indirect one is the effect of herbicides on algae that form the diet of amphibian larvae. Many of these links could be expanded in this way, each forming an entire new flowchart. Perhaps a useful exercise would be to number each arrow in this figure and then compile a series of sub-flowcharts, each relating to a numbered arrow. An atlas of such diagrams would allow sequential assessment of multifaceted interactions. More detailed flowcharts of some particular segments have already been published for amphibians, e.g., global climatic change and amphibian decline (Burrowes, 2009), relationships among stressors, organismal responses, and population performance (Rohr *et al.*, 2009), effects of pesticides on amphibian declines (Hayes, 2005; Boone *et al.*, 2009) and on the mechanisms of endocrine disruption (McCoy & Guillette, 2009), and the negative impacts of contaminants on larvae (Bridges & Semlitsch, 2005). Blaustein *et al.* (2011) presented flowcharts showing the relationships between (1) multiple stressors and population declines and extinctions of amphibians, (2) climatic change, UV radiation, and pathogens of amphibians, (3) effects of insecticides on tadpoles, and (4) role of evolutionary constraints in molding behavior and life histories of amphibians. CFCs, chlorofluorocarbons (used as refrigerants, aerosol propellants and solvents); DOC, dissolved organic compounds; PCBs, polychlorinated biphenyls (used as coolants, insulating fluids and flame retardants); UV, ultraviolet radiation. The diagram is based on the reviews cited in this chapter, and on the literature cited therein.

1.   Pollution

Pollution is a major threat to amphibians globally and the sources are myriad. Industry is the most conspicuous source, as wastes from mining, manufacturing and petrochemical processing are often discharged into streams and rivers where they leave visible residues and sediments, and demonstrate obvious harm to wildlife in the form of dead or dying animals. Experimental studies have confirmed the toxicity of these pollutants to amphibians (Heatwole & Wilkinson, 2009).

The airborne residues from industry and exhaust fumes from vehicles are less directly apparent, but are insidious for that reason and because their toxins, carried on the wind, may exert their effects far from the source of pollution and in otherwise pristine areas, even in reserves set aside for the protection of wildlife. Indeed, sulfur dioxide and nitrogen oxides produced by combustion of fossil fuels may have their greatest effect not at the site of emission but at distant places where they are delivered as acid rain (Räsänen & Green, 2009).

Natural environments vary greatly in pH, with naturally acidic ones having resulted from high concentrations of organic acids forming gradually over hundreds or thousands of years; now, however, large parts of the world have become quickly acidified as a result of industrialization, with a time frame of decades or even years, insufficient time for evolutionary adaptation to occur (Räsänen & Green, 2009). The most-affected ecosystem is freshwater, the habitat of many amphibians.

Carbon compounds ("greenhouse gases") released by burning hydrocarbon fuels trap heat within the Earth's atmosphere. These have increased in recent decades and have contributed to global warming (Kemp, 1990). Chlorofluorocarbons (CFCs) and their replacement chemicals used in refrigerants, aerosol propellants, solvents, fire extinguishers and other industrial applications not only are greenhouse gases (Ko *et al.* 1993), but they also contribute to depletion of stratospheric ozone (IPCC, 2007) that in turn allows greater penetration of deleterious levels of ultraviolet radiation (Marco *et al.*, 2009). CFCs are being phased out but it will be a long time before the ozone layer repairs itself.

The demands of a burgeoning human population have resulted in elevated levels of nitrogen in freshwater systems in recent years. Nitrogenous pollutants (ammonia, urea, nitrate and nitrite) are contained in domestic wastewater effluents, fertilizers and manure in runoff from fields and farmyards, and wastes from the paper and munitions industries. These wastes can cause eutrophication of bodies of water with consequent degradation of the habitat; they also have direct toxic effects (e.g., mortality, malformations) (Marco & Ortiz-Santaliestra, 2009). Sediments from such runoff may have an independent effect, especially if additionally carrying residues of toxic pesticides. Nitrogen is also emitted as trace gases, such as nitrous oxides, that serve as greenhouse gases, contribute to acidification, and catalyze destruction of the ozone layer.

Pesticides, in addition to directly being deleterious to amphibians, may also influence amphibians indirectly and insidiously (Boone *et al.*, 2009). Herbicides may reduce the abundance of algae that forms the basis of aquatic food webs in which amphibians participate, and negatively affect survival, biomass and time to metamorphosis of amphibians. Insecticides with only sublethal direct effects on amphibians can launch trophic cascades that can have either indirect negative effects on amphibians by decreasing abundances of their invertebrate prey, or indirect positive ones by reducing abundances of their invertebrate competitors. The complexity of such community interactions makes complete assessment of the overall effect of pesticides difficult.

Endocrine-disrupting chemicals (EDCs) are ubiquitous pollutants to which virtually any physiological, behavioral, or developmental process that is controlled by the endocrine system is susceptible (McCoy & Guillette, 2009). Some are natural biological products such as phytoestrogens, estrogenic hormones excreted by women, and steroids in the wastes of domestic animals. Others are artificially synthesized or deliberately produced, such as 17α-ethinyl estradiol (the synthetic estrogen in contraceptives), pharmaceutical agents, surfactants, pesticides (organochlorines like DDE and DDT), herbicides (e.g., atrazine), industrial and household chemicals such as coolants and insulating fluids (polychlorinated biphenyls [PCBs]), flame retardants (polybrominated biphenyl ethers [PBDEs]), detergents, plastics containing bisphenol A or phthalates, additives to stock feed, and personal care and beauty products (Snyder *et al.*, 2003). These chemicals enter aquatic ecosystems and water resources mainly through runoff from feedlots and fields treated with manure, effluents from paper-mills, and disposal of domestic wastes; they have even penetrated aquifers in which they can be dispersed long distances in the ground water. Some can be dispersed aerially.

EDCs mimic or block transcriptional activation elicited by endogenous hormones such that they resemble or alter natural biological signals and are thereby misconstrued by the target organism, with the result that gene expression may be altered and abnormal phenotypes occur (McCoy & Guillette, 2009). They alter various endocrine signaling pathways via multiple mechanisms. Even the evolutionary trajectory of a species can be altered because of induction of persistent heritable phenotypes independent of mutagenesis. Specific effects common to many taxa, including amphibians, are the feminization of males, development of intersexed gonads, reduced fertility, and altered sex ratios. In amphibians EDCs can also cause hypothyroidism, alter the developmental rate of tadpoles, and the rate and timing of metamorphic changes, influence behaviors such as swimming, avoidance of predators, and foraging in tadpoles, and affect chemical communication and learning in adults. EDCs are especially insidious because they can reach distant areas through runoff, fallout, and via ground water, have long half-lives, and can magnify in the food chain, with repercussions throughout the entire ecosystem.

2.   Destruction, modification and fragmentation of habitats

Destruction of habitat is perhaps the major present cause of decline in amphibians. Almost all aspects of human endeavor involve alteration or removal of natural habitats: clearing land for agriculture, timber production, mining, residential areas, shopping malls, roadways, golf courses and other recreational areas, airfields, military bases and industrial complexes; salinization can occur from inappropriate irrigation or use of arid lands; aquatic habitats are destroyed by pollution from mining, agricultural and industrial wastes (see above). Especially in tropical areas land is being cleared at an unprecedented rate for agricultural use and for timber. Many amphibians are restricted to forests.

As amphibian habitats are destroyed, suitable habitat becomes fragmented into pro-gressively smaller units, thereby leading to random extirpation in the small, isolated fragments (Lemckert *et al.*, 2012), exacerbated by greater "edge effects" as the perimeter of ecotones between forest and open areas increases relative to the area of primary habitat.

Although anthropogenic effects are predominantly negative for amphibians, a few species may be favored. Urbanization removes the habitats of many amphibians, especially those requiring forest, but may expand the available habitat for those adapted to open, disturbed areas (Kusrini *et al.*, 2012). Even activities as destructive as war can benefit some species; Stuart and Davidson (1999) found that water-filled craters originating from bombing of Laos during the Vietnam War provided additional breeding sites for some frogs.

3.   Diseases, parasites and maladies

Amphibians are subject to a number of viral and bacterial diseases that cause high levels of mortality (Hemingway *et al.*, 2009). By far the most important amphibian disease is the fungal infection chytridiomycosis. Sporangia initially form deep inside living epidermal cells of all stages of amphibians and in the keratinized "teeth" of larvae where they develop discharge tubes that push through the cellular membranes and release zoospores as the epidermal cells move outward in the skin, die, become keratinized, and are shed from the surface (Berger *et al.*, 2009). *Bd* grows optimally at temperatures between 17° and 25°C and more slowly at lower temperatures down to 5°C. At 28°C it survives but cannot grow and at 30°C survival is only 50% after 8 days, and death occurs within 4 hours at 30°C, 30 minutes at 47°C and within 5 minutes at 60°C. The fungus is sensitive to dry conditions and zoospores are killed by an hour of desiccation. Incubation time of the disease ranges from 9 to 83 days, with most frogs dying between 18 and 70 days. Susceptibility, mortality rate, and time until death vary among species. In some species infection is not fatal. Young frogs are more susceptible than older ones.

Detection of the fungus in preserved museum specimens has allowed retrospective assessment of the presence of this pathogen in wild populations before it was discovered in living animals. In Australia it first appeared in the late 1970s in southeastern Queensland and probably caused the extinction of two species there (Berger *et al.*, 2009). It then spread northward and southward along the eastern coast. Its first known occurrence in Western Australia was just south of Perth in 1985, from whence it spread in all directions. It is now known also from the vicinity of Adelaide and from Tasmania. It may have reached its geographic limits in Australia as its present distribution accords with a model predicting the limits of its spread, based on temperature and precipitation (Retallick, 2003); the disease has not moved into the hotter, drier parts of the continent. It did spread, however, from Australia to New Zealand in about 1999.

The spread of chytridiomycosis has been documented for a number of areas, especially in the Western Hemisphere, where it is now widespread (Figure 18.1) (Berger *et al.*, 2009; Heatwole *et al.* 2010, 2011, 2012; Wilson *et al.*, 2010; Hedges, in press; Whitfield *et al.*, in press).

In 2005, Henle reviewed the causes of amphibian decline in Europe and didn't include diseases as one of them. Now the disease is found in a number of European countries (Figure 18.1).

The origin of *Bd* may well have been Africa (see 5. Introduced invasive species). Hence, lack of records for much of Africa (Figure 18.1) may represent lack of sampling for the fungus, rather than a real absence.

Asia, by contrast, seems to have been largely free of the disease until later (Figure 18.1). Swei *et al.* (2011) suggested three possible hypotheses: (1) *Bd* is newly invading Asia and hasn't become established in many places yet, (2) *Bd* is an old Asian disease but has little effect because native species have evolved immunity to it, (3) there are conditions in Asia that are unique to the region and that inhibit the emergence of *Bd*. There is some evidence that the Japanese giant salamander (*Andrias japonicus*) may harbor ancient haplotypes of the *Bd* ITS region and may have established a commensal relationship with this fungus (Goka *et al.*, 2009); more research is needed to assess the relative merits of these three hypotheses.

Infections by parasites have also been important. One of the first indications that something was wrong with amphibian assemblages was the sudden occurrence of high incidences of grotesquely malformed frogs, mostly with supernumerary limbs (Lannoo, 2009). Initially, it was believed that the malformations were caused by UV radiation, pollution, or a combination of environmental factors. Later it was discovered that an important proximal cause was trematodiasis (infection by digenetic trematodes) (Rohr *et al.*, 2009). The cercariae mostly attach to tadpoles in the crease between body and tail, where they form metacercarial cysts that sink into the skin and become embedded in and around the hind limb-buds. Disruption of the tissues of the limb-bud by the parasite results in malformation. Insecticides and other agents have also been implicated in malformations of amphibians (Lannoo, 2009).

4.   Ultraviolet radiation

With depletion of the ozone layer (see above), the amount of UV radiation reaching the Earth's surface has increased and has adverse effects on amphibians, such as greater mortality among embryos, larvae and adults; higher incidence of deformities; delayed metamorphosis and smaller size at metamorphosis (Marco *et al.*, 2009).

5.   Introduced invasive species

Predatory mammals, when introduced into new areas, can reduce amphibian populations; if the mammals are removed before extirpation of the amphibians occurs, amphibian populations may recover (e.g., Aloha *et al.*, 2006). Introduction of fish into otherwise fish-free habitats often results in extirpation of native amphibians, either because of interspecific competition or because of predation by the fish upon the amphibians (Piliod *et al.*, 2012). Introduced amphibians can affect native amphibians in the same ways as exotic fish, but they also can serve as vectors of disease. For example it is likely that infected Australian *Litoria raniformis* introduced into New Zealand spread chytridiomycosis to amphibians native to the latter country (Berger *et al.*, 2009). Two species of amphibians that are relatively resistant to chytridiomycosis and can thus be undetected carriers of the disease are the African clawed frog (*Xenopus laevis*) and the American bullfrog (*Lithobates catesbeiana*; formerly *Rana catesbeiana*) (Weldon *et al.*, 2004; Piliod *et al.*, 2012). Both have been distributed around the world, the former for use in pregnancy tests and the latter for farming for human consumption. Their inadvertent or purposive release into the wild in their new localities has carried chytridiomycosis to many parts of the world; more recently the pet trade in amphibians has contributed to the spread of this disease (Heatwole, *et al.*, 2010, 2011, 2012; also see section 6 below).

6.   Roadkills

Roads transect amphibian habitats and cross amphibians' migration routes to breeding sites. Crossing of roads during ingress and egress of adults at breeding ponds or the dispersal of young can lead to high mortality, and is a problem in many countries (Puky, 2012; Heatwole *et al.*, in press).

7.   Exploitation of amphibians by humans for food, medicine and pets

Traditionally, amphibians have been the subjects of superstitions or symbols in many human cultures, and/or have been used for food and medicine (Das, 2012), and today there is a brisk trade both locally and internationally in amphibians for food, scientific research, and, increasingly, for pets (Kusrini, 2012; Kusrini *et al.*, 2012; Heatwole *et al.*, 2010, 2011, in press). The volume of international trade is great; for example, in one year Hong Kong alone imported for food, pets and scientific research about 4.3 million amphibians of 45 species from 11 different countries (Rowley *et al.*, 2007).

Global warming/climatic change

Weather is one of the most common topics of conservation, probably because of its importance in people's everyday lives and because it varies daily, if not hourly, and continuously provides new material for discussion. Climate, the prevailing pattern of weather for a given place, also varies, but over annual, decadal and geological time-scales or with the El Niño southern oscillation. The fact that long-term trends may be obscured by shorter-term fluctuations makes assessment of the causes of change and predictions of future climates difficult. Nevertheless, intense research has now made it abundantly clear that there is an overall trend of increasing global temperature and that humans are in part responsible for it (Chapter 13 of this volume). The human-induced rate of change is rapid, compared to geological scales, and outstrips the ability of many amphibians to make adaptive evolutionary adjustments and, accordingly, many species have declined in abundance, have been locally extirpated, have had their geographic ranges truncated, or have been driven into extinction (Lannoo, 2005; Wake, 2007; Burrowes, 2009).

There are several examples that clearly illustrate the vulnerability of amphibians to climatic change (Reaser & Blaustein, 2005). The golden toads of Monteverde, Costa Rica (*Bufo periglenes*) declined from 1500 known breeding adults to 11 in 1988. In 1989, the last live individual ever to be seen was a single adult male (Crump, 2005). Forty percent of the amphibians in the Monteverde region underwent synchronous rapid declines during peaks of warm, dry conditions during the following decade, perhaps because of stress from desiccation (Reaser & Blaustein, 2005).

Species of *Atelopus*, a genus especially the subject of recent declines and extinctions, are more likely to be seen for the last time following an unusually warm year (Wells, 2007).

Based on the prediction from various sources of an increase in temperature of 0.4°C per decade in northern Argentina and of 0.25°C per decade in the south while precipitation will decrease in the west and increase in the east, Lavilla and Heatwole (2010) predicted a diminution in the number of sites suitable for reproduction of amphibians in the arid western part of the country north of the 45th parallel, an area in which there are a high number of endemic species of amphibians. Ron (2005) modeled the potential spread of chytridiomycosis in Latin America and predicted that it would be widely distributed, including the countries from which it had not been reported then, a prediction that has subsequently been largely realized (Figure 18.1). The validity of such predictions can be tested by monitoring both climatic change and populations of amphibians of particular areas and should become a priority in research. If found to be valid, such predictions could be used for establishing conservation practices. In the present example, it is likely that some of the endemics will become extinct unless remedial measures are taken.

Climatic change may not affect all species of amphibians in the same ways. For example, infections by ranaviruses are more prevalent in spring and summer, perhaps in response to the higher temperatures then (Hemingway *et al.*, 2009). By contrast, *Bd* does much better under cooler conditions (see section on diseases, parasites, and maladies). Accordingly, at a particular location, global warming could increase the incidence of ranaviral diseases but inhibit chytridiomycosis. Amphibians can take advantage of the sensitivity of *Bd* to high temperature and select microenvironments that are warmer than the favorable limits of the fungal pathogen, and even above the normal body temperatures that the amphibians would otherwise maintain (behavioral fever). In some cases this eliminates the infection, but in others is not completely effective (Rowley & Alford, 2009). Burrowes (2009) suggested that the higher temperatures and drier conditions of climatic change cause greater clumping of drought-stressed individuals in small, moist patches, thereby facilitating transmission of such diseases as chytridiomycosis.

Synergistic effects

Many factors impinge on amphibian assemblages at the same time and few factors operate alone. Indeed, Blaustein (2011) suggested that single-factor explanations for amphibian population declines are likely to be the exception rather than the rule. Synergisms are known to occur among some of the agents of amphibian decline, but only a few have been studied. Following are only a few of the known examples; many more synergisms, still unstudied, also probably occur.

Carbaryl, an insecticide, is ten-fold more toxic to amphibians when combined with sublethal doses of UV radiation than it is at the same concentration alone (Zaga *et al.*, 1998).

Endocrine disruptors can contribute to outbreaks of infectious disease in amphibians. Hayes *et al.* (2006) exposed frogs to nine different agricultural pesticides and found adverse effects on growth and development. When the pesticides were administered in combination the effect was more severe. The mixtures damaged the thymus and the immune system, thereby leading to greater susceptibility to the disease of flavobacterial meningitis. Control animals tested positive for the bacterial agent but did not contract the disease.

Wells (2007) reviewed the interactive effects of various pesticides and natural environmental factors on amphibians. In some cases exposure to multiple pesticides resulted in higher mortality than did exposure to either singly; in other cases the effects were not additive. Carbaryl (an insecticide) and atrazine (a herbicide) both caused mortality in some species. Larvae of *Ambystoma maculatum*, however, did better when exposed to both than it did when exposed only to carbaryl. Tadpoles of *Rana sphenocephala* increased in size in response to carbaryl because of an increase in algal food, but decreased in size and in rate of development in response to atrazine because that chemical reduced the biomass of algae on which the tadpoles preyed. At low doses carbaryl also had a positive effect on bullfrog tadpoles in two ways: it stimulated growth of the algal food of the tadpoles, and killed two of their predators, crayfish and sunfish. At high doses, however, it also killed the tadpoles. Carbaryl has a number of other effects on amphibians that affect their survival, including reduction of activity, inhibition of anti-predator behavior, and causing stress.

Carbaryl also interacts with the fertilizer ammonium nitrate. Each chemical acting alone stimulates growth of algae and thereby benefits tadpoles. When tadpoles were exposed to both chemicals together, however, they did not show the enhancement of growth and development seen in those exposed to either chemical alone. It is likely that the metabolic costs and stress of detoxification of multiple chemicals outweigh the benefit of greater availability of food.

Temperature influences the susceptibility of amphibians to chytridiomycosis, with outbreaks being more prevalent when conditions are cooler, e.g., in winter and at higher elevations (Berger *et al.*, 2009). Thus, global warming may make populations in some areas less prone to catastrophic declines.

Synergisms may have unexpected effects. For example, in coastal New South Wales, Australia, remnant populations of *Litorea aurea* persist in areas of high salinity and at sites contaminated by heavy metals from gold mines, copper smelters and tanneries. It is likely that salt and metals act as fungicides against *Batrachochytrium dendrobatidis* in this species (Berger *et al.*, 2009).

Some of the examples above were of factors operating indirectly upon the amphibian through intermediary factors. Because of such indirect effects, one needs to distinguish between proximal (immediate) and distal (ultimate) causes of decline.

For example, a species may have declined in response to an increase in ultraviolet (UV) radiation (proximal cause); in turn, the increase in UV may have occurred because of a depletion in stratospheric ozone that allowed more UV to penetrate to ground level.

This is still not the ultimate cause because depletion of the ozone layer occurred in response to humans' release of pollutants. Thus, release of ozone-depleting chemicals by humans could be considered as the ultimate cause. Even this conclusion is simplistic, however, as the deleterious effects of UV are linked to another series of events. Dissolved organic carbons (DOC) inhibit the penetration of UV into bodies of water, and thus bestow some protection from UV. That protection, in turn, is reduced by acidification that lowers the level of DOCs in the water (Marco *et al.* 2009).

A lowered pH (acid rain) derives from combustion of fossil fuels (Box, Chapter 13 of this volume). Those gases also are partly responsible for global warming. Thus, anthropogenic pollution can affect amphibians via three different routes: (1) via release of chemicals that produce acid rain, which causes acidification of aquatic habitats, which reduces the amount of dissolved organic carbon, which results in greater penetration of UV, which adversely affects amphibians in various ways; (2) via release of CFCs, which deplete the ozone layer, which results in increased levels of UV, which adversely affects amphibians; and (3) via greenhouse gases that raise global temperature. Clearly there are three proximal causes of amphibian decline in this example, as well as several intermediate ones, and at least one modifying one, but the ultimate cause along all three trajectories is chemical pollution of the air by humans. A further complication is that there is a complex interaction of UV radiation, climatic change and outbreaks of disease (Marco *et al.*, 2009). Warmer conditions lower pond levels; eggs in shallower water are exposed to higher levels of radiation, which makes them more susceptible to a pathogenic oomycete, *Saprolegnia*, and hence they have higher mortality. This example illustrates the difficulties in teasing out the causes of amphibian declines. Conservation measures that counteract the ultimate causes will be more effective that those that merely treat symptoms arising somewhere else along the chain.

Stress, in general, seems capable of making amphibians susceptible to other agents, and in that sense all stressors can be viewed as acting synergistically with all others. Stressors of all kinds, both natural and anthropogenic, stimulate the release of glucocorticosteroid hormones that can be immunosuppressive and elevate the risk of disease (Rohr *et al.*, 2009). Thus, even if some stressor is at sublethal levels per se, it can have a great effect on population declines of amphibians through its facilitation of other deleterious influences.

Understanding the ultimate causes and adopting appropriate conservation strategies constitute one of the main challenges in counteracting amphibian decline. It is clear that the complexity of these interactions makes it virtually impossible to manage amphibian decline merely by focusing on only one, or a few, aspects. It is imperative to take a holistic approach and begin to quantify the various links in the interactive web for particular habitats so as to identify where control measures should be targeted, and to devise conservation strategies most appropriate for given regions.

There are multiple causes of decline and extinction that may vary with size of range, diversity within taxa, geographic locality, elevation, climate, life history and taxonomic position, and these may be masked by differences among taxa in the extent of available knowledge (see Hero & Morrison, 2012; Hero *et al.*, 2012; Kuzmin, 2012; Morrison & Hero, 2012). Consequently, it is difficult to tease apart the various ecological and phylogenetic correlates of decline and extinction. Some broad correlates have emerged, however. Certain genera seem especially vulnerable (Crump, 2005) and, in various parts of the world, species that breed in streams at high elevations seem to have a greater risk of decline than do most other species (Wells, 2007; Hero & Morrison, 2012). This is not universal, however, as Green (2005) concluded that in North America some declines were identified among highly fecund pond-breeding species.

Cascading effects of loss of amphibians

Loss of a taxon in any biotic community is likely to have repercussions throughout the entire system. Lips *et al.* (2005) pointed out that loss of amphibians would produce cascading effects throughout aquatic and terrestrial food webs because of the sheer biomass of amphibians, and the linkages they form between aquatic and terrestrial habitats. In a review of the effect of amphibian decline on Neotropical ecosystems, Whiles *et al.* (2006) concluded that evidence suggested that amphibian declines will have large-scale and lasting ecosystem-level effects, including changes in algal assemblage structure and primary production, altered dynamics of organic matter, changes in other consumers such as aquatic insects and riparian predators, and reduced energy transfers between streams and riparian habitats. Furthermore, because of habitat and functional differences between larvae and adults in most amphibians, they considered the loss of a single species is akin to losing two "species" (i.e., two basic functional units of the biotic community). Heatwole (1989) termed such infraspecific functional units "econes".

Accelerated rate of change

In the face of environmental change, animals are faced with the alternatives of adapting to the new situation, migrating to areas where conditions are still suitable for them, or becoming extinct. Most environmental changes, such as warming climates, occur relatively gradually over geological spans of time, and many lineages of animals are able to survive through evolutionary adaptation in response to natural selection, including the capacity to accommodate small departures from usual conditions through seasonal acclimation of physiological processes or of tolerance limits. When climatic changes are more extreme over shorter periods, e.g., oscillations between the cold glacials and warmer interglacials during the late Pliocene and Pleistocene, extinction rates may increase and various elements of the biota be lost, while other species adjust their geographical ranges. When conditions change drastically and rapidly, major extinctions occur because species are unable to adapt quickly enough to keep pace with the changing environment.

   There have been five, relatively rapid, massive extinctions in the past 545 million years, the End-Ordovician (444 million years ago (Mya)), Late-Devonian (375 Mya), End-Permian (251 Mya), End-Triassic (200 Mya), and End-Cretaceous (or K/T) (66 Mya), which were correlated with periods of environmental instability, such as climatic changes, marine regressions, loss of carbonate sediments and extensive vulcanism. The last of these (K/T) ushered out the dinosaurs and other forms of life. Although the impact of a large extra-terrestrial object (e.g., meteor or asteroid) with the Earth, which caused global changes in temperature, acid rain emanating from vaporized rock, and elevated infrared radiation, is credited as the principal cause, it was likely to be merely the sudden critical event that triggered the extinction of many species already stressed by a deteriorating environment (e.g., marine regression, vulcanism) (Archibald, 2011). Thus, these cataclysmic extinctions probably had multiple causes operating over relatively short periods (geologically speaking).

A sixth massive extinction, now in progress, is being caused largely by anthropogenic changes such as destruction of habitat, pollution of water and air, global warming, destruction of the ozone layer, and introduction of invasive species and the vectors of emergent diseases – in short most of the sources of amphibian decline mentioned above. The rapidity of these changes outstrips the capacity of many animals to evolutionarily adapt in response to selection of the new environments and consequently those not vagile enough to migrate to suitable habitats are declining toward extinction, or already have become extinct. Of the world's known species of amphibians, 168 species are believed to have gone extinct in the past two decades and at least 2469 (43%) more are now in decline. About 32% are currently listed as threatened (AmphibiaWeb, 2012). However, some may persist, as phenotypic plasticity, physiological acclimation, and/or behavioral adjustment may be avenues of more immediate adaptive response. For example, Beebee (1995) showed that in Britain over a 17-year period, amphibians adjusted their repro-ductive cycles in concert with rising environmental temperatures. They spawned 2 to 3 weeks earlier in the last 5-year period (1990–1994) than they did in the first 5-year period (1978–1982), for an advance in the reproductive cycle of 9 to 10 days per 1°C rise in maximum environmental temperature.

What can be done?

As discussed above, amphibian decline has numerous causes that act not only individually but in various interactive ways, and consequently there is no simple formula for arresting or reversing it, or preventing its spread. Since many of the causes are related to increasing deleterious effects of a burgeoning human population, the prospectus for improvement is not optimistic. Nevertheless, some partial remedies have been implemented or are being planned. These include programs for reduction in pollution of water and air, building safe passages under roads for migrating amphibians, restriction of export and/or import of amphibians and other taxa, eliminating introduced invasive species, gathering baseline data and subsequently monitoring the status of populations, promulgation of refuges and reserves, leaving connecting corridors between fragments of favorable habitat, making the public and political leaders cognizant of the issues, applying mitigative measures, and breeding endangered species in captivity with a view to re-introduction into restored habitats (Aloha et al., 2006; Dodd *et al.*, 2012; Griffiths & Kuzmin, 2012; Hecknar & Lemckert, 2012; Lemckert *et al.*, 2012; Morrison *et al.*, 2012; Piliod *et al.*, 2012; Soorae & Launay, 2012). Some of these are meeting with success; others have proven less propitious. Hayes (2005) appealed for greater integration of data from disparate fields of study and for merging relevant disciplines into a new field, much in the manner of cooperative trends in evolution and development ("evo-devo") and suggested adding endocrinology to the mix to produce "evo-devo-endo". Carey (2005) similarly called for greater participation by physiologists in the field of conservation biology and pointed out that they could contribute in important ways, including providing proof of cause-and-effect relationships, showing how environmental change affects organismal energetics, host-pathogen relationships, and immune defenses.

References

Alemu, I., Cazabon, M. N. E., Dempewolk, L., *et al.* (2008). Presence of the chytrid fungus *Batrachochytrium dendrobatidis* in populations of the critically endangered frog, *Mannophryne olmonae* in Tobago, West Indies. *EcoHealth*, **5**, 34–39.

Aloha, M., Nordström, M., Banks, P. B., Laanetu, N., & Korpimäki, E. (2006). Alien mink predation induces prolonged declines in archipelago amphibians. *Proceedings of the Royal Society of London B*, **275**, 1261–1265.

AmphibiaWeb. (2012). Worldwide amphibian declines: how big is the problem, what are the causes and what can be done? http://amphibiaweb.org/ (accessed 22 April 2012).

Archibald, J. D. (2011). *Extinction and Radiation. How the Fall of Dinosaurs Led to the Rise of Mammals*. Baltimore, MD: The Johns Hopkins University Press.

Beard, K. H., & O'Neill, E. M. (2005). Infection of an invasive frog *Eleutherodactylus coqui* by the chytrid fungus *Batrachochytrium dendrobatidis* in Hawaii. *Biological Conservation*, **126**, 591–595.

Beebee, T. J. C. (1995). Amphibian breeding and climate. *Nature*, **374**, 219–220.

Beebee, T. J. C. (1996). *Ecology and Conservation of Amphibians*. New York: Chapman & Hall.

Bell, R. C., Gata Garcia, A. V., Stuart, B. L., & Zamudio, K. R. (2011). High prevalence of the amphibian chytrid pathogen in Gabon. *EcoHealth*, **8**, 116–120.

Berger, L., Longcore, J. E., Speare, R., Hyatt, A., & Skerratt, L. F. (2009). Fungal diseases of amphibians. In H. Heatwole & J. W. Wilkinson (Eds.), *Amphibian Decline: Diseases, Parasites, Maladies and Pollution* (Ch. 2, pp. 2986–3052). Vol. 8 in *Amphibian Biology*. Baulkham Hills, Australia: Surrey Beatty & Sons.

Blaustein, A. R., Han, B. A., Relyea, R. A., Johnson, P. T. J., Buck, J. C., Gervas, S. S., & Kats, L. B. (2011). The complexity of amphibian population declines: understanding the role of cofactors in driving amphibian losses. *Annals of the New York Academy of Sciences*, **1223**, 108–119.

Boone, M. D., Davidson, C., & Bridges-Britton, C. (2009). Evaluating the impact of pesticides in amphibian declines. In H. Heatwole & J. W. Wilkinson (Eds.), *Amphibian Decline: Diseases, Parasites, Maladies and Pollution* (Ch. 8, pp. 3186–3207). Vol. 8 in *Amphibian Biology*. Baulkham Hills, Australia: Surrey Beatty & Sons.

Bourke, J., Mutschmann, F., Ohst, T., *et al.* (2010). *Batrachochytrium dendrobatidis* in Darwin's frog *Rhinoderma* spp. in Chile. *Diseases of Aquatic Organisms*, **92**, 217–221.

Bovero, S., Angelini, G., Doglio, S., Gazzaniga, E., & Cunningham, A. A. (2008). Detection of chytridiomycosis caused by *Batrachochytrium dendrobatidis* in the endangered Sardinian newt (*Euproctus platycephalus*) in southern Sardinia, Italy. *Journal of Wildlife Diseases*, **44**, 712–715.

Bridges, C. M., & Semlitsch, R. D. (2005). Xenobiotics. In M. Lannoo (Ed.), *Amphibian Declines, the Conservation Status of United States Species* (Ch. 15, pp. 89–92). Berkeley, CA: University of California Press.

Burrowes, P. A. (2009). Climatic change and amphibian declines. In H. Heatwole & J. W. Wilkinson (Eds.), *Amphibian Decline: Diseases, Parasites, Maladies and Pollution* (Ch. 12, pp. 3268–3279). Vol. 8 in *Amphibian Biology*. Baulkham Hills, Australia: Surrey Beatty & Sons.

Carey, C. (2005). How physiological methods and concepts can be useful in conservation biology. *Integrative and Comparative Biology*, **45**, 4–11.

Civis, P., & Vojar, J. (2012). Current state of *Bd* occurrence in the Czech Republic. *Herpetological Review*, **43**, 75–78.

Crump, M. L. (2005). Why are some species in decline but others not? In M. Lannoo (Ed.), *Amphibian Declines, the Conservation Status of United States Species* (Ch. 2, pp. 7–9). Berkeley, CA: University of California Press.

Das, I. (2012). Man meets frog: perceptions, use and conservation of amphibians by indigenous people. In H. Heatwole & J. W. Wilkinson (Eds.), *Conservation and Decline of Amphibians: Ecological Aspects, Effect of Humans, and Management* (Ch. 3, pp. 3383–3468). Vol. 10 in *Amphibian Biology*. Baulkham Hills, Australia: Surrey Beatty & Sons.

Disease. available at: http://www.jcu.edu.au/school/phtm/PHTM/frogs/ampdis.html.

Dodd, C. K. Jr., Loman, J., Cogălniceanu, D., & Puky, M. (2012). Monitoring amphibian populations. In H. Heatwole & J. W. Wilkinson (Eds.), *Conservation and Decline of Amphibians: Ecological Aspects, Effect of Humans, and Management* (Ch. 11, pp. 3577–3635). Vol. 10 in *Amphibian Biology*. Baulkham Hills, Australia: Surrey Beatty & Sons.

Erspamer, V. (1994). Bioactive secretions of the amphibian integument. In H. Heatwole, G. T. Barthalmus & A. Y. Heatwole (Eds.), *The Integument* (Ch. 9, pp. 178–350). Vol. 1 in *Amphibian Biology*. Chipping Norton: Surrey Beatty & Sons.

Froglog. http://www.amphibianark.org/the-crisis/; availaible at: http://amphibiaweb.org/declines/declines.html [31.5.2012].

Goka, K., Yokoyama, J., Une, Y., *et al.* (2009). Amphibian chytridiomycosis in Japan: distribution, haplotypes and possible route of entry into Japan. *Molecular Ecology*, **18**, 4757–4774.

Goldberg, T. L., Readel, A. M., & Lee, M. H. (2007). Chytrid fungus in frogs from an equatorial African montane forest in Western Uganda. *Journal of Wildlife Diseases*, **43**, 521–524.

Green, D. M. (Ed.) (1997). Amphibians in decline: Canadian studies of a global problem. *Herpetological Conservation*, **1**, 1–138.

Green, D M. (2005). Biology of amphibian declines. In M. Lannoo (Ed.), *Amphibian Declines, the Conservation Status of United States Species* (Ch. 7, pp. 28–33). Berkeley, CA: University of California Press.

Griffiths, R. A., & Kuzmin, S. L. (2012). In H. Heatwole & J. W. Wilkinson (Eds.), *Conservation and Decline of Amphibians: Ecological Aspects, Effect of Humans, and Management* (Ch. 14, pp. 3687–3703). Vol. 10 in *Amphibian Biology*. Baulkham Hills, Australia: Surrey Beatty & Sons.

Halliday, T. (2005). Diverse phenomena influencing amphibian population declines. In M. Lannoo (Ed.), *Amphibian Declines, the Conservation Status of United States Species* (Ch. 1, pp. 3–6). Berkeley, CA: University of California Press.

Hayes, T. B. (2005). Welcome to the revolution: integrative biology and assessing the impact of endocrine disruptors on environmental and public health. *Integrative and Comparative Biology*, **45**, 321–329.

Hayes, T. B., Case, O., Chui, S., *et al.* (2006). Pesticide mixtures, endocrine disruption, and amphibian declines: are we underestimating the impact? *Environmental Health Perspectives*, **114**, 40–50.

Heatwole, H. (1989). The concept of the econe, a fundamental ecological unit. *Tropical Ecology*, **30**, 13–19.

Heatwole, H., & Wilkinson, J. W. (Eds.) (2009). *Amphibian Decline: Diseases, Parasites, Maladies and Pollution.* Vol. 8 in *Amphibian Biology.* Baulkham Hills, Australia: Surrey Beatty & Sons.

Heatwole, H., & Wilkinson, J. W. (Eds.) (2012). *Conservation and Decline of Amphibians: Ecological Aspects, Effect of Humans, and Management.* Vol. 10 in *Amphibian Biology.* Baulkham Hills, Australia: Surrey Beatty & Sons.

Heatwole, H., Barrio-Amorós, C. L., & Wilkinson, J. W. (Eds.) (2010). Paraguay, Chile and Argentina, *Part 1* of Status of Decline of Amphibians: Western Hemisphere. Vol. 9 in *Amphibian Biology.* Baulkham Hills, Australia: Surrey Beatty & Sons.

Heatwole, H., Barrio-Amorós, C. L., & Wilkinson, J. W. (Eds.) (2011). Uruguay, Brazil, Ecuador and Colombia, *Part 2 of* Status of Decline of Amphibians: Western Hemisphere. Vol. 9 in *Amphibian Biology.* Baulkham Hills, Australia: Surrey Beatty & Sons.

Heatwole, H., Barrio-Amorós, C. L., & Wilkinson, J. W. (Eds.) (2012). Venezuela, Guyana, Suriname, French Guiana, *Part 3* of Status of Decline of Amphibians: Western Hemisphere. Vol. 9 in *Amphibian Biology.* Baulkham Hills, Australia: Surrey Beatty & Sons.

Heatwole, H., Das, I., Busack, S., & Wilkinson, J. W. (Eds.) (in press). *Status of Decline of Amphibians: Eastern Hemisphere.* Vol. 11 in *Amphibian Biology.* Baulkham Hills, Australia: Surrey Beatty & Sons.

Hecnar, S. J., & Lemckert, F. (2012). Habitat protection: refuges and reserves. In H. Heatwole, & J. W. Wilkinson. *Conservation and Decline of Amphibians: Ecological Aspects, Effect of Humans, and Management* (Ch. 12, pp. 3636–3676). Vol. 10 in *Amphibian Biology.* Baulkham Hills, Australia: Surrey Beatty & Sons.

Hedges, S. B. (in press). Amphibian conservation in the West Indies. In H. Heatwole, C. L. Barrio-Amorós (Eds.), Central America, the Caribbean and North America, *Part 4* in Status of Decline of Amphibians: Western Hemisphere (Ch. 13). Vol. 9 in *Amphibian Biology.* Baulkham Hills, Australia: Surrey Beatty & Sons.

Hemingway, H., Brunner, J., Speare, R., & Berger, L. (2009). Viral and bacterial diseases of amphibians. In H. Heatwole & J. W. Wilkinson (Eds.), *Amphibian Decline: Diseases, Parasites, Maladies and Pollution* (Ch. 1, pp. 2963–2985). Vol. 8 in *Amphibian Biology.* Baulkham Hills, Australia: Surrey Beatty & Sons.

Henle, K. (2005). Lessons from Europe. In M. Lannoo (Ed.), *Amphibian Declines, the Conservation Status of United States Species* (Ch. 12, pp. 64–74). Berkeley, CA: University of California Press.

Hero, J.-M., & Morrison, C. (2012). Life history correlates of extinction and risk in amphibians. In H. Heatwole & J. W. Wilkinson (Eds.), *Conservation and Decline of Amphibians: Ecological Aspects, Effect of Humans, and Management* (Ch. 10, pp. 3567–3576). Vol. 10 in *Amphibian Biology.* Baulkham Hills, Australia: Surrey Beatty & Sons.

Hero, J.-M., Morrison, C., Chanson, J., Stuart, S., & Cox, N. A. (2012). In H. Heatwole & J. W. Wilkinson (Eds.), *Conservation and Decline of Amphibians: Ecological Aspects, Effect of Humans, and Management* (Ch. 8, pp. 3539–3551). Vol. 10 in *Amphibian Biology.* Baulkham Hills, Australia: Surrey Beatty & Sons.

Heyer, W. R., & Murphy, B. (2005). Declining Amphibian Populations Task Force. In M. Lannoo (Ed.), *Amphibian Declines, the Conservation Status of United States Species* (Ch. 5, pp. 17–21). Berkeley, CA: University of California Press.

IPCC (2007). Changes in atmospheric halocarbons, stratospheric ozone, tropospheric ozone and other gases. Intergovernmental Panel on Climate Change Fourth Assessment Report: Climate Change 2007. AR4 WGI Technical summary. File://VolumesKINGSTON?TO%20PRINT/TS.2. 1.3%20Changes%20in%20A. . .Other%20Gases%20AR4%20WGI%20Technical%20Summary. webarchive.

Kaiser, H., & Henderson, R. W. (1994). The conservation status of Lesser Antillean frogs. *Herpetological Natural History*, **2**, 41–56.

Kemp, D. D. (1990). *Global Environmental Issues, a Climatological Approach*. London: Routledge.

Ko, M. K. W., Sze, N. D., Molnar, G., & Prather, M. J. (1993). Global warming [*sic*] from chloro-fluorocarbons and their alternatives: time scales of chemistry and climate. *Atmospheric Environment. Part A. General Topics*, **27**, 581–587.

Kusrini, M. D. (2012). International trade in amphibians. In H. Heatwole & J. W. Wilkinson (Eds.), *Conservation and Decline of Amphibians: Ecological Aspects, Effect of Humans, and Management* (Ch. 5, pp. 3494–3504). Vol. 10 in *Amphibian Biology*. Baulkham Hills, Australia: Surrey Beatty & Sons.

Kusrini, M. D., Heatwole, H., & Davenport, D. (2012). Harvesting of amphibians for food. In H. Heatwole & J. W. Wilkinson (Eds.), *Conservation and Decline of Amphibians: Ecological Aspects, Effect of Humans, and Management* (Ch. 4, pp. 3469–3493). Vol. 10 in *Amphibian Biology*. Baulkham Hills, Australia: Surrey Beatty & Sons.

Kuzmin, S. K., Dodd, C. K. Jr., & Pikulik, M. M. (Eds.) (1995). *Amphibian Populations in the Commonwealth of Independent States: Current Status and Declines*. Moscow: Pensoft.

Kuzmin, S. L. (2012). Declines and extinctions in amphibians: an evolutionary and ecological perspective. In H. Heatwole & J. W. Wilkinson (Eds.), *Conservation and Decline of Amphibians: Ecological Aspects, Effect of Humans, and Management* (Ch. 7, pp. 3522–3538). Vol. 10 in *Amphibian Biology*. Baulkham Hills, Australia: Surrey Beatty & Sons.

Lannoo, M. (Ed.) (2005). *Amphibian Declines, the Conservation Status of United States Species*. Berkeley, CA: University of California Press.

Lannoo, M. J. (2009). Amphibian malformations. In H. Heatwole & J. W. Wilkinson (Eds.), *Amphibian Decline: Diseases, Parasites, Maladies and Pollution* (Ch. 5, pp. 3089–3111). Vol. 8 in *Amphibian Biology*. Baulkham Hills, Australia: Surrey Beatty & Sons.

Lavilla, E. O., & Heatwole, H. (2010). Status of amphibian conservation and decline in Argentina. In H. Heatwole, C. L. Barrio-Amorós & J. W. Wilkinson (Eds.), Status of decline of amphibians: Western Hemisphere. *Part 1:* Paraguay, Chile and Argentina (Ch. 3, pp. 30–78). Vol. 9 in *Amphibian Biology*. Baulkham Hills, Australia: Surrey Beatty & Sons.

Lemckert, F., Hecnar, S. J., & Piliod, D. S. (2012). Loss and modification of habitat. In H. Heatwole & J. W. Wilkinson (Eds.), *Conservation and Decline of Amphibians: Ecological Aspects, Effect of Humans, and Management* (Ch. 1, pp. 3291–3342). Vol. 10 in *Amphibian Biology*. Baulkham Hills, Australia: Surrey Beatty & Sons.

Lips, K. R., Burrowes, P. A., Mendelson, J. R. III., & Parra-Olea, G. (2005). Amphibian population declines in Latin America: a synthesis. *Biotropica*, **37**, 222–226.

Longcore, J. E., Pessier, A. P., & Nichols, D. K. (1999). *Batrachochytrium dendrobatidis gen. et sp. nov.*, a chytrid pathogenic to amphibians. *Mycologia*, **91**, 219–227.

Marco, A., & Ortiz-Santaliestra, M. (2009). Pollution: impact of reactive nitrogen on amphibians (nitrogen pollution). In H. Heatwole & J. W. Wilkinson (Eds.), *Amphibian Decline: Diseases, Parasites, Maladies and Pollution* (Ch. 7, pp. 3145–3185). Vol. 8 in *Amphibian Biology*. Baulkham Hills, Australia: Surrey Beatty & Sons.

Marco, A., Bancroft, B. A., Lizana, M., & Blaustein, R. (2009). Ultraviolet-B radiation and amphibians. In H. Heatwole & J. W. Wilkinson (Eds.), *Amphibian Decline: Diseases, Parasites, Maladies and Pollution* (Ch. 6, pp. 3112–3144). Vol. 8 in *Amphibian Biology*. Baulkham Hills, Australia: Surrey Beatty & Sons.

McCoy, K. A., & Guillette, L. J. Jr. (2009). Endocrine disrupting chemicals. In H. Heatwole & J. W. Wilkinson (Eds.), *Amphibian Decline: Diseases, Parasites, Maladies and Pollution* (Ch. 9, pp. 3208–3238). Vol. 8 in *Amphibian Biology*. Baulkham Hills, Australia: Surrey Beatty & Sons.

Morrison, C., & Hero, J.-M. (2012). Geographic correlates of extinction risk in amphibians. In H. Heatwole & J. W. Wilkinson (Eds.), *Conservation and Decline of Amphibians: Ecological Aspects, Effect of Humans, and Management* (Ch. 9, pp. 3552–3566). Vol. 10 in *Amphibian Biology*. Baulkham Hills, Australia: Surrey Beatty & Sons.

Morrison, C., Hero, J. M., & Van Sluys, M. (2012). Integrated procedures: where do we go from here? In H. Heatwole & J. W. Wilkinson (Eds.), *Conservation and Decline of Amphibians: Ecological Aspects, Effect of Humans, and Management* (Ch. 15, pp. 3704–3707). Vol. 10 in *Amphibian Biology*. Baulkham Hills, Australia: Surrey Beatty & Sons.

Piliod, D. S., Griffiths, R. A., & Kuzmin, S. L. (2012). Ecological impacts of non-native species. In H. Heatwole & J. W. Wilkinson (Eds.), *Conservation and Decline of Amphibians: Ecological Aspects, Effect of Humans, and Management* (Ch. 2, pp. 3343–3382). Vol. 10 in *Amphibian Biology*. Baulkham Hills, Australia: Surrey Beatty & Sons.

Puky, M. (2012). Road kills. In H. Heatwole & J. W. Wilkinson (Eds.), *Conservation and Decline of Amphibians: Ecological Aspects, Effect of Humans, and Management* (Ch. 6, pp. 3505–3521). Vol. 10 in *Amphibian Biology*. Baulkham Hills, Australia: Surrey Beatty & Sons.

Räsänen, K., & Green, D. M. (2009). Acidification and its effects on amphibian populations. In H. Heatwole & J. W. Wilkinson (Eds.), *Amphibian Decline: Diseases, Parasites, Maladies and Pollution* (Ch. 11, pp. 3244–3267). Vol. 8 in *Amphibian Biology*. Baulkham Hills, Australia: Surrey Beatty & Sons.

Reaser, J. K., & Blaustein, A. (2005). Repercussions of global change. In M. Lannoo (Ed.), *Amphibian Declines, the Conservation Status of United States Species* (Ch. 11, pp. 60–63). Berkeley, CA: University of California Press.

Retallick, R. W. (2003). Bioclimatic investigations in the potential distribution of *Batrachochytrium dendrobatidis in Australia*. Draft report, University of Queensland [cited from Berger *et al.* 2009; not seen in original].

Rohr, J., Raffel, T., & Sessions, S. K. (2009). Digenetic trematodes and their relationship to amphibian declines and deformities. In H. Heatwole & J. W. Wilkinson (Eds.), *Amphibian Decline: Diseases, Parasites, Maladies and Pollution* (Ch. 4, pp. 3067–3088). Vol. 8 in *Amphibian Biology*. Baulkham Hills, Australia: Surrey Beatty & Sons.

Ron, D. (2005). Predicting the distribution of the amphibian pathogen *Batrachochytrium dendrobatidis* in the New World. *Biotropica*, **37**, 209–221.

Rowley, J. J. L., & Alford, R. A. (2009). Factors affecting interspecific variation in susceptibility to disease in amphibians. In H. Heatwole & J. W. Wilkinson (Eds.), *Amphibian Decline: Diseases, Parasites, Maladies and Pollution* (Ch. 3, pp. 3053–3066). Vol. 8 in *Amphibian Biology*. Baulkham Hills, Australia: Surrey Beatty & Sons.

Rowley, J. L., Chan, S. K. F., Tang, W. S., *et al.* (2007). Survey for the amphibian chytrid *Batrachochytrium dendrobatidis* in Hong Kong in native amphibians and in the international amphibian trade. *Diseases of Aquatic Organisms*, **78**, 87–95.

Semlitsch, R. D., & Wake, D. (2003). *Amphibian Conservation*. Washington DC: Smithsonian Institution.

Skerratt, L. F. (2009). Dedication. In H. Heatwole & J. W. Wilkinson (Eds.), *Amphibian Decline: Diseases, Parasites, Maladies and Pollution* (p. ix). Vol. 8 in *Amphibian Biology*. Baulkham Hills, Australia: Surrey Beatty & Sons.

Snyder, S. A., Westerhoff, P., Yoon, Y., & Sedlak, D. L. (2003). Pharmaceuticals, personal care products, and endocrine disruptors in water: implications for the water industry. *Environmental Engineering Science*, **20**, 449–469.

Sooray, P. S., & Launay, F. J. (2012). *Guidelines of the International Union for the Conservation of Nature (IUCN) for Re-Introductions and their Application to Amphibians* (Ch. 13, pp. 3677–3686). Gland, Switzerland: IUCN.

Stuart, B. L., & Davidson, P. (1999). Use of bomb crater ponds by frogs in Laos. *Herpetological Review*, **30**, 72–73.

Stuart, S. N., Chanson, J. S., Cox, N. A., *et al.* (2004). Status and trends of amphibian declines and extinctions worldwide. *Science* **306**, 1783–1786.

Stuart, S. N., Hoffman, M., Chanson, J. S., *et al.* (2008). *Threatened Amphibians of the World*. Arlington, VA, USA: IUCN, Gland, Switzerland & Conservation International.

Stuart, S. N., Chanson, J. S., Cox, N. A., & Young, B. E. (2010). The global decline of amphibians: current trends and future prospects. In L. D. Wilson, J. H. Townsend & J. D. Johnson (Eds.), *Conservation of Mesoamerican Amphibians and Reptiles* (pp. 2–15). Eagle Mountain, UT: Eagle Mountain Publishing.

Swei, A., Rowley, J. L., Rödder, D., *et al.* (2011). Prevalence and distribution of chytridiomycosis throughout Asia. *Froglog*, **98**, 33–34.

Wake, D. (2007). Climate change implicated in amphibian and lizard declines. *Proceedings of the National Academy of Sciences of the USA*, **104**, 8201–8202.

Weldon, C. (2005). *Chytridiomycosis, an emerging infectious disease of amphibians in South Africa*. PhD dissertation in Zoology, North-West University, Potchefstroom, South Africa. 213 pp.

Weldon, C., du Preez, L. H., Hyatt, A. D., Muller, R., & Speare, R. (2004). Origin of the amphibian chytrid fungus. *Emerging Infectious Disease Journal*, **10**, 1–11.

Wells, K. (2007). *The Ecology & Behavior of Amphibians*. Chicago, IL: The University of Chicago Press.

Whiles, M. R., Lips, K. R., Pringle, C. M., *et al.* (2006). The effects of amphibian population declines on the structure and function of Neotropical stream ecosystems. *Frontiers in Ecology and the Environment*, **4**, 27–35.

Whitfield, S. M., Lips, K. R., & Donnelly, M. A. (in press). Decline and conservation of amphibians in Central America. In H. Heatwole & C. L. Barrio-Amorós (Eds.), Status of Decline of Amphibians: Western Hemisphere. *Part 4:* Central America, the Caribbean and North America (Ch. 12). Vol. 9 of *Amphibian Biology*. Baulkham Hills, Australia: Surrey Beatty & Sons.

Wilson, L. D., Townsend, J. H., & Johnson, J. D. (Eds.) (2010). *Conservation of Mesoamerican Amphibians and Reptiles*. Eagle Mountain, UT: Eagle Mountain Publishing.

Zaga, A., Little, E. E., Rabeni, C. F., & Ellersieck, M. R. (1998). Photoenhanced toxicity of a carbamate insecticide to early life stage anuran amphibians. *Environmental Toxicology and Chemistry*, **17**, 2543–2553.

19 Climatic change and reptiles

Harvey B. Lillywhite

Life on Earth has experienced dramatic alterations of diversity and distribution as conditions of environment have changed over millennia of evolutionary history. The Earth's climate has cooled during the past 100 million years, and events such as volcanic activity and meteorite strikes have resulted in extinctions and collapses of ecosystems. There is now overwhelming evidence that the Earth's temperatures are warming and that human activities are driving this aspect of climatic change. Global average temperatures have increased by ~ 0.2°C per decade during the past 30 years due to increasing atmospheric concentrations of greenhouse gases (Hansen et al., 2006). Much of the added energy contributing to global warming is absorbed by the world's oceans, in which the upper layers have increased in temperature by 0.6°C during the past 100 years (IPCC, 2007). Further warming will produce thermal expansion which could result in an 18–30 cm rise in sea level by 2100 (Meehl *et al.*, 2005). Warming is larger in the Western Equatorial Pacific than in the Eastern Equatorial Pacific over the past century, probably due to strong El Niños, and comparisons with paleoclimatic data suggest that this critical region and possibly the planet as a whole is as warm as the Holocene maximum and the maximum temperature during the past million years (Hansen *et al.*, 2006).

Accepting the reality of climatic change in its current status, non-avian reptiles are an important and informative group of vertebrates to examine in related contexts. The utility of reptilian data for assessing the effects of climatic changes on organisms and communities relates to characteristic features of this group, including ectothermy, geographic distributions, reproductive modes (especially oviparity), breadth of diversity, biogeographical histories, local abundance, keystone status in some communities, dispersal capabilities, trophic relationships, and energetics. With regard to global warming, it is significant to note that two-thirds of the global variation in the species richness of reptiles can be explained by temperature alone (Qian, 2010). The diversity of reptiles is greater in the tropics and decreases with increasing latitude. Only three species of squamate reach the Arctic Circle, and no reptile ranges into the Antarctic Circle. However, rising temperatures will have multiple and complex effects even if considered the sole driver in climatic change. The responses of reptiles to climatic change are likely to be influenced by environmental factors related to thermal microenvironment, precipitation, phenology of reproduction, availability of trophic resources, interactions with pathogens and

The Balance of Nature and Human Impact, ed. Klaus Rohde. Published by Cambridge University Press.
© Cambridge University Press 2013.

invasive species, and synergistic interactions with other environmental stressors. Key issues to be investigated are: (1) can species respond to climatic change with sufficient plasticity or adaptation to track changes in the environment with sufficiently rapid adjustment for survival in specific regions? (2) Will appropriate habitat be available to accommodate changes in range if climatic change or anthropogenic activities eliminate the original habitat from occupation? (3) How will range extensions driven by climatic warming interact with other elements of new communities? (4) To what extent will diversity be affected by distributional changes that are climate-driven?

Paleoclimatic changes

The consequences of climatic changes for reptiles have been substantial during their evolutionary history. However, interpretation of relationships between past faunal changes and climate are complicated by the temporal and spatial heterogeneity of climatic changes combined with deficits in the fossil record and biases in the locations of interpretive studies. These considerations are too complex to be examined fully here.

Past climatic changes are represented by extreme events, gradual long-term changes, and local anomalies. With respect to general long-running trends during the past 100 million years, the Earth's climate changed from largely ice-free conditions of the Cretaceous and early Paleogene to the glacial conditions of the late Cenozoic (Crowley & North, 1991). During the mid Pliocene, about three million years ago, the climate was similar or somewhat warmer than present-day conditions, especially at high latitudes, and also wetter (Jansen et al., 2007). Climate has cooled since that time. Cooling through the Eocene was probably gradual or stepwise, with the greatest expansions of ice taking place during the Oligocene. Temperatures were about 5°C cooler globally, especially at higher latitudes, during the last glacial maximum about 21 000 years ago. From the middle Eocene through part of the Oligocene, evidence suggests there was a general increase in aridity, including the disappearance of aquatic reptiles and amphibians (Hutchison, 1992; Stucky, 1992). However, events in North America have biased much of the record, and the exact timing and nature of climatic transitions is equivocal.

Perhaps the most dramatic event related to climatic effects on reptiles, at least in part, was the mass Cretaceous-Tertiary (K-T) (also called Cretaceous-Paleogene, or K-Pg) extinction that occurred approximately 65.5 million years ago (MacLeod et al., 1997). The K-T extinctions are thought to be related to one or more catastrophic events such as volcanic activity and collision with an asteroid, the aftermath of which altered the Earth's ecology and gradually changed sea levels and climate (MacLeod et al., 1997). Although the geological disruptions and ecological changes were severe, the effects on organisms varied greatly among taxa with differing requirements and adaptability. Trophic or other linkages among the taxa were also important. For example, ammonites suffered heavy losses, and these were probably the principal prey of mosasaurs. Both mosasaurs and plesiosaurs – giant top marine predators – became extinct by the end of the Cretaceous. Photosynthetic organisms suffered, and the herbivores that depended on them also diminished or died out when the plants on which they depended became scarce. As a further consequence, top predators such as *Tyrannosaurus rex* also became extinct.

Archosaurs

Pterosaurs and probably all non-avian dinosaurs became extinct at the K-T boundary. Whether the extinction of dinosaurs was gradual or abrupt and the nature of the decline in diversity of dinosaurs are unclear due to limitations and differences of interpretation of the fossil record. All of the non-neornithine birds became extinct, whereas other birds diversified and replaced the archaic birds and pterosaur groups, due either to direct competition or the occupation of ecological niches that were available following the extinction of non-avian dinosaurs (Robertson *et al.*, 2004).

Approximately 50% of representative crocodilians survived the K-T extinctions, but none of the larger crocodiles did (MacLeod *et al.*, 1997). Extinction of the non-avian dinosaurs in contrast with survival of crocodilians is conceivably related to disruptions of the food chain and the greater energy demands of the former, assuming these were endothermic (Wilf & Johnson, 2004; Eagle *et al.*, 2011). Survival of crocodilians might also have been favored by ectothermy and the ability to feed on detritus and carrion (Sheehan & Hansen, 1986). Thus, survival in the face of climatic disturbances to ecosystems probably depends on physiological, life history, and other characteristics of crocodilians. Moreover, these reptiles may react differently to environmental changes depending on the structure of responses related to phylogenetic and ecological components (Piras *et al.*, 2009). Examination of crown-group crocodilian diversity suggests that the K-T event had little long-term impact, and whatever caused the mass extinctions in other groups of organisms, it was not a long-term change of climate (Markwick, 1998a).

The spatial distribution of fossil crocodilians has often been used to construct paleoclimates, or to interpret the consequences of climatic change (e.g., Lyell, 1830; Berg, 1965; Donn, 1987; Markwick, 1998a, 1998b). The distribution of crocodilians changed in concert with reasonably well-documented climatic changes. These were widespread throughout middle latitudes during non-glacial periods, suggesting that conditions were relatively equable and that bodies of water were available, whereas crocodilians were restricted to lower latitudes and more maritime environments during glacial intervals, in particular the Oligocene and Pliocene-Recent. The latter correlation suggests there was increasing aridity and thermal seasonality in mid-latitude continental interiors (Estes, 1970; Hutchison, 1992).

The data for crocodilians support the hypothesis that global climatic cooling largely affects biodiversity at higher latitudes. Patterns of crocodilian extinctions appear to reflect a greater sensitivity of high-latitude regions to cycling of ice sheets during the Pliocene to Recent (Markwick, 1998a). However, interpretations are subject to sampling effects and also non-climatic factors such as regional uplift or other geological changes. An analysis by Markwick (1998a) strongly suggests that there was an early exponential diversification of crown-group crocodilians during the late Cretaceous and Paleocene, followed by a major extinction in the Pliocene-Pleistocene. The diversification of crown-group crocodilians is a Northern Hemisphere phenomenon, and expansion into the Southern Hemisphere occurred after the K-T boundary and the extinction of mesosuchians. There is a strong link between the Pliocene-Pleistocene extinctions and climatic deterioration related to expansion of the Antarctic ice sheet and the formation of Arctic ice. Further details related to extinction gradients and speculations about causation are beyond the scope of this chapter.

The majority of crocodilian taxa surviving the K-T boundary inhabited freshwater and terrestrial environments. Survival might have been related to freshwater habitats and the ability to burrow for protection from deteriorating climate. Freshwater environments were less affected by K-T events than were marine environments, and the former might have been less stressful and therefore favorable for survival (Jouve *et al.*, 2008). Intolerance of marine environments might explain the absence of both gavialids and alligatorids from Australia and Africa (Markwick, 1998b).

Non-archosaurian reptiles

Turtles and lepidosaurs both survived through the K-T extinctions (MacLeod *et al.*, 1997). Six families of turtles that existed at the end of the Cretaceous (and about 80% of species) survived into the Tertiary and are represented by living species (Novacek, 1999). Rhynchocephalians were a relatively successful group in the Mesozoic but began to decline by the middle of the Cretaceous and are represented by only a single species today, the surviving tuatara of New Zealand. Squamates (lizards, snakes and amphisbaenians) were successful at evolutionary radiations during the Jurassic and survived through the Cretaceous to become the most successful group of reptiles, currently represented by more than 6000 extant species (MacLeod *et al.*, 1997; Novacek, 1999). The fossil evidence suggests that squamates did not decline significantly in numbers, and there is no family of squamates known to have become extinct during the K-T extinctions. Survival was probably favored by relatively small body size and versatility of adaptation related to low metabolic requirements in a diversity of niches. These opinions are necessarily speculative.

Marine environments

The surface temperatures of oceans varied, and changes in sea levels occurred during past ages, with implications for numerous ecological changes at shorelines. During the middle Pliocene sea levels were 15–25 m higher than at present, and maximum continental submergence occurred from 30 to 125 million years ago (Abbott, 1984). Sea levels were as much as 120 m lower during the last glacial maximum and have been rising since that time (Jansen *et al.*, 2007).

Mosasaurs evolved from shore-dwelling lizards and have been considered as possible candidates for the evolutionary ancestors of snakes. Mosasaurs are thought to have been highly adaptable, and representative of a rapid pre-Tertiary adaptive radiation. Two genera had grown to lengths of more than 15 m just before the K-T extinctions. These represented successful top predators that must have had considerable dominance in oceans. Marine mosasaurs and plesiosaurs became extinct by the end of the

Cretaceous. The extinction of these animals was probably due to trophic disruptions and the collapse of marine ecosystems.

Sea-level changes displaced sea turtles' nesting sites. These marine reptiles depend on beaches with appropriate features for successful incubation of eggs, and the sex determination of embryos is temperature-dependent (Janzen & Paukstis, 1991). There is evidence that Green turtle (*Cheloniamydas*) populations have adapted to past climatic changes by altering nesting beaches, migration routes, and foraging areas (Kennett et al., 2004; Dethmers *et al.*, 2006; Bourjea *et al.*, 2007; Limpus, 2008). Turtles nesting in the Gulf of Carpentaria appear to have invaded from western populations, but have since altered the timing of breeding from austral summer to the austral winter, presumably as adaptation to local temperature regimes. Genetic analysis of Green turtles has indicated displacement of nesting rookeries from the Atlantic Ocean to the southwestern Indian Ocean during periods of flows of warmer water around the tip of South Africa, although there is no evidence for gene flow from the Atlantic to the Indian Ocean during the past 150 million years. Such colonization is unlikely to be an ongoing process due to maintenance of the present pattern of oceanic currents (Bourjea *et al.*, 2007).

Variations of sea level are a major influence on sedimentation, changes in coastal habitats and biogeographic patterns. It is important to recognize that such changes produce not only threats of extinctions and losses of biodiversity, but also opportunities for adaptive radiations into newly opened niches. The rapid disappearance or formation of various habitat zones, such as mangrove fringes and mud flats, led to the establishment or truncation of species' distributions. Additionally, the opening or closure of ocean passages and the creation of narrow bridges or fully colonized corridors severely fragmented habitats and impacted the expansion or contraction of species. Numerous factors related to physical, demographic and ecological processes interact to influence the nature and rates of ecological succession in terrestrial, coastal, and marine communities. Such processes have produced an emerging perspective of a highly dynamic Pleistocene at the Sunda Shelf of southeast Asia, where palaeogeographic changes have strongly driven plant and animal diversification in relation to an ever-changing landscape and climate (Hanebuth *et al.*, 2011).

Genetic and other data strongly suggest that the Sunda Shelf and the so-called "Coral Triangle" region of southeastern Asia was a center of evolution and radiation of marine reptiles. The most speciose clade of sea snakes – the viviparous Hydrophiini with more than 60 extant species representing 19 genera – radiated rapidly from this center between 5 and 10 Mya (Sanders *et al.*, 2008). These are the most diverse and widely distributed lineages of marine snakes. The most widely distributed species, *Pelamis platurus*, is the only pelagic sea snake and has the widest distribution of any snake species. It ranges from the southeastern coast of Africa across the Indo-Pacific to the western shores of Central America (Heatwole, 1999). This marine snake originated within the past 5 million years, and genetic data suggest that it underwent explosive population growth within the recent past and associated with its westward and eastward expansion (Sheehy *et al.*, 2012). Its unusually broad distribution is attributable to passive drifting with oceanic currents, and genetic variation within this species is extremely low (Sheehy *et al.*, 2012). Individual snakes occasionally drift into colder water (< 19–20°C) where they do not reproduce or establish permanent range extensions (e.g., New Zealand, Ecuador and California in the USA). However, warming of oceanic waters might allow eventual polar range extensions of this species in the future.

Global trends in present and future abundance and diversity of reptiles

Probably every major taxon of reptile is threatened with ongoing diminishment of abundance and diversity, largely a result of anthropogenic causes (Gibbons *et al.*, 2000). Because of the scale of changes to habitat, and the rapidity with which new threats to communities of organisms appear, it is difficult to separate causation related to climate from a myriad other factors. Indeed, the various threats to wildlife are often interrelated and synergistic. Quantitative scenarios of future extinction rates consistently indicate that biodiversity will continue to decline during the twenty-first century, although various estimates vary in predictions, underlying hypotheses and complexity of changes (Pereira *et al.*, 2010). It seems clear that climatic changes will cause shifts in species' ranges as well as produce more dramatic or rapid changes at local scales (Hughes, 2000; Hansen *et al.*, 2006; Rosenzweig *et al.*, 2007). There is also strong evidence that the timing of life-cycle events has been altered by species largely in response to changes in temperature.

While causation remains a controversial issue, increasing numbers of studies suggest that numerous reptilian declines and extinctions have causal links with climatic change. Recent historical surveys of 48 lizard species in Mexico, comparing 300 field sites, indicated that 12% of local populations have gone extinct since 1975 (Sinervo *et al.*, 2010). Links to macroclimatic change and physiological models of extinction risk (see below) have implicated climatic causation in these extinctions, while global projections for extinction risks suggest that local extinctions will reach 39% and that species' extinctions could be 20% by 2080.

Many studies that document species' declines do not provide as rigorous a linkage to climatic change as the one just mentioned, but the implications are compelling. Investigation of 17 snake populations from the UK, France, Italy, Nigeria and Australia indicates that 11 have declined sharply during the past 30 years, and the synchrony of multiple factors likely to be causal again implicates the involvement of global climatic change (Reading *et al.*, 2010).

The geographic distributions of numerous terrestrial organisms have shifted toward higher latitudes and elevations during recent decades. Meta-analyses have estimated, for example, that species' distributions have shifted to higher elevations at a median rate of 11.0 meters per decade, and to higher latitudes at a median rate of 16.9 kilometers per decade. The distances moved are greater in areas where there were greater changes in temperature (Chen *et al.*, 2011). However, range shifts of species depend on multiple factors related to inherent characteristics of physiology and behavior as well as to the external drivers. In the tropics, montane faunas frequently include endemics that reside close to summits, and these are likely to be particularly vulnerable to extinction as the climate becomes warmer. Upward migration is considered to be an important biogeographical factor that is driven by past and future changes of climate in tropical mountains (Bush, 2002; Rull & Vegas-Vilarrúbia, 2006).

Herpetological assemblages have been investigated at the Tsaratanana Massif in northern Madagascar (the island's highest massif), and these included 30 species of reptiles and amphibians (Raxworthy *et al.*, 2008). Meteorological data and standard moist adiabatic lapse rates were used to predict upslope species displacements of 17–74 m per decade between 1993 and 2003, and these were supported by ground data which quantified mean shifts in the elevation midpoints of 19–51 m upslope. Recent warming in Madagascar is consistent with model simulations of recent climate. Local meteorological conditions and habitat changes suggest that many of Madagascar's montane species are potentially vulnerable to habitat loss and upslope extinction. More generally, biogeographical changes and species' extinctions might be comparatively high in the tropics due to relatively high species richness and a high incidence of endemism. Extinctions of high-altitude viviparous lizards in Mexico are expected to exceed those of more lowland species of lizards (Sinervo *et al.*, 2010).

On the other hand, there is also increasing evidence that reptiles living in cool climates can benefit from increasing temperatures by having activity and energetic advantages related to more thermally advantageous habitats and extended activity times (Kearney *et al.*, 2009). Empirical studies have reported that such positive outcomes are associated with increasing warmth during the past decade. Increased body size, female fecundity, and survival rates through longer active seasons have been documented for the high-altitude populations of the European lizard *Lacerta vivipara* (Chamaillé-Jammes *et al.*, 2006). Similarly, high-altitude reptiles (*Trimeresurus gracilis* in Taiwan) are predicted to have energetic advantages associated with climatic change, and biophysical modeling of niches demonstrates an increased digestive capacity related to extended time of activity, and to energetic benefits arising from warming of habitats (Huang *et al.*, in press).

In contrast to high-altitude reptiles, tropical species at lower elevations might be threatened by decreased activity time and by compromise of their physiological functions as temperatures increase (Tewksbury *et al.*, 2008; Huey *et al.*, 2009; Sinervo *et al.*, 2010). The difficulty in predicting responses to climatic warming, however, is to accurately forecast how habitats will change and to what extent such changes will affect biotic interactions. Biophysical modeling will provide a very useful tool for predicting changes in niche requirements and, to some extent, biotic interactions related to overlaps of energy demands and timing of activity (Porter & James, 1979; Porter *et al.*, 2002; Porter & Mitchell, 2006).

Impacts of climate, vulnerability and resilience

Increasing temperatures

Increasing environmental temperatures are of profound importance to all reptiles, affecting features of life history, behavioral activity patterns, interactions with resources and distribution of adults. For the numerous species of reptiles that are oviparous, nest temperatures can affect embryonic development and hatchling phenotypes such as sex, size, color, behavior, and aspects of physiology such as locomotor performance. Assessing the impacts of climatic changes will require consideration of the ecology of eggs as well as the responses of adults.

One important concern is the potential disturbance to sex ratios in species that exhibit temperature-dependent sex determination (TSD). These include all crocodilians, the majority of turtles, some species of lizards, and the tuatara. However, numerous species with TSD have persisted during cycles of ice ages during the past 100 000 years when significant changes occurred in temperature. Species can avoid deleterious effects of temperature change on TSD if they are able to shift the timing or location of their nesting, evolve compensatory changes in pivotal sex-determining temperature, or can select appropriate nest microenvironments at a given locality. Most reptiles lay eggs in nests that experience diurnal temperature fluctuations, and these fluctuations could be altered in such a manner as to avoid changes in sex ratios by global warming.

Studies of tuataras demonstrate the complexity of factors that can be involved in responses of reptiles to climatic change. Tuataras (*Sphenodon punctatus, S. guntheri*) are primitive cold-climate reptiles with relict populations surviving only on several offshore islands of New Zealand. Contemporary rates of climatic change threaten the survival of tuataras due to their isolation on islands that prevents a southward migration in response to warming climate. Tuataras are threatened by increasing temperature because of long generation time, TSD, and low genetic diversity of the surviving populations. Physiological and developmental data have been used in conjunction with microclimatic models for assessing the suitability of nesting sites under climatic change scenarios (Mitchell *et al.*, 2008). Soil temperatures were estimated using microclimatic models and climatic change scenarios projected to the mid-2080s.

Tuataras will be challenged by warming climate that imposes an increasing sex bias for males, earlier development, and increasing energy expenditure. Tuataras could compensate for the male-biasing effects of warmer temperatures by nesting later in the season or selecting shaded nest sites. Later nesting seems unlikely, however, as many species of oviparous reptiles are nesting earlier rather than later in response to warming climate (e.g., Parmesan, 2007). Warming climate could lead to earlier vitellogenesis and calcification of eggs and thus to earlier, rather than later, nesting. Another consequence of climatic warming might be that tuataras could complete development early in autumn rather than in spring or summer (Mitchell *et al.*, 2008). In such instances, hatchlings could emerge early from the nest during a time when resources are increasingly unavailable, or they could overwinter in the nest but consume increasing fractions of yolk as a consequence of warming nest temperatures. Smaller yolk reserves have been correlated with warmer overwintering temperatures in slider turtles (*Trachemys scripta elegans*) emerging from the nest (Willette *et al.*, 2005). This observation demonstrates a physiological change that is linked with climate. Thus, male-biased changes in emergence time and energy consumption could have important consequences for juveniles' activity, growth and mortality. The survival of tuataras will depend on some form of adaptation to counteract the warming trend to all-male clutches with increasing male bias of operational sex ratios, or on translocation of animals.

What about sea turtles which nest on open beaches that are exposed to changing coastal conditions, including temperature? Sea turtles lay eggs at depths in the soil that do not experience marked daily fluctuation, but long-term temperature changes at nesting beaches remain a special concern for these reptiles. Some nesting beaches have persisted with strong bias toward females for several decades, and there is no evidence to date that low production of male turtles has altered reproductive success (Hawkes *et al.*, 2009; Poloczanska *et al.*, 2009). Sex ratios can also change within and across seasons (Hawkes *et al.*, 2007, 2009). Other factors are also important for nest quality such as soil albedo, grain size, and precipitation. Temperature additionally drives the duration of incubation. The behavior and abundance of predators can also importantly affect the survival of hatchlings, and these can change due to climatically induced shifts in phenology and range of the predatory species (Root *et al.*, 2003).

Storms and cyclones

Destructive storms such as tropical cyclones could increase in intensity and shift pole-ward as global temperatures rise (e.g., Saunders & Lea, 2008). Regional threats to biota include rough surf conditions, damaging water surge, intense and prolonged rainfall, high-velocity winds, flooding, salt spray and other secondary impacts. Large storm and tidal surges can be extremely destructive and intensify the effects of rises in sea level (Zhang *et al*, 2004). Flooding can impact nesting success as well as displace adults, thereby affecting nesting sea turtles as well as the local abundance, demography, and the forced dispersal of various terrestrial reptiles (Ross, 2005). Storms generally do not result in long-term losses in either abundance or diversity of reptiles, but long-term quantitative studies are not available. There is some evidence that the approach of tropical cyclones can be anticipated, and that sea kraits living at coastal intertidal zones can avoid their destructive impacts (Liu *et al.*, 2010). Changes in coastal processes can also affect crocodilians, sea turtles, and numerous species of squamates.

Precipitation and moisture

Global warming and associated climatic changes are projected to include long-term changes and regional anomalies with respect to patterns of precipitation. Wet seasons are likely to become wetter, and dry seasons will intensify in the tropics (Neelin *et al.*, 2006; Chou *et al.*, 2007), with general drying of the Northern Hemisphere tropics and moistening of the Southern Hemisphere tropics (Zhang *et al.*, 2007). Changes in precipitation will produce flooding, alter vegetation, shift habitats, and affect the water budgets and nesting sites of reptiles. Increasing frequency and intensity of fires will also affect reptiles living in arid or semi-arid zones, both directly and indirectly. Losses of amphibian breeding sites can eliminate food resources for snakes that are dependent on them. The intensification of tropical drought can affect marine as well as terrestrial and freshwater aquatic reptiles, for example sea kraits (*Laticauda* spp.) that are dependent on freshwater sources of drinking water (Bonnet and Brischoux, 2008; Lillywhite *et al.*, 2008). Sea snake populations at Ashmore Reef, Australia, have been in serious decline during the past decade (Anonymous, 2012), and this phenomenon is possibly related (in part) to declines in precipitation. It may be predicted that sea snake populations will decline to extinction or migrate from areas that experience intensifying drought in future years.

Rise in sea level and attendant geographical disturbance

As described above, rises in sea level will cause numerous significant changes in coastal ecology and will impact importantly oviparous species that depend on nesting sites in the coastal zone (e.g., sea turtles, sea kraits). There is expected to be regional heterogeneity of changes in sea level in relation to a global mean rise of sea level due to local geology and hydrodynamics. Sea levels in the Indo-Pacific, where numerous tropical islands occur, are expected to change in accordance with mean global rise in sea level (Church *et al.*, 2006). Low-lying islands such as coral atolls are particularly vulnerable to rise in sea level, especially in the Pacific (Woodroffe, 2008), and unique populations and endemic species of reptiles are likely to be affected. More generally, it remains uncertain whether beach formation processes will keep up with rising sea levels, but in many regions this will be constrained by the "squeeze" of human development in coastal regions.

Wind and oceanic currents

Global warming will produce changes in circulation patterns of the atmosphere and the oceans, with likely regional trends in location, amplitude and intensity (IPCC 2007). Shifts in wind belts (e.g., Cai *et al.*, 2005) and changes in the El Niño southern oscillation will conceivably impact drift patterns of sea snakes (Sheehy *et al.*, 2012), feeding grounds of marine reptiles (especially sea snakes and sea turtles), and the health and persistence of marine ecosystems.

Oceanic acidification

Oceanic acidification is a secondary consequence of global warming, largely related to oceanic buffering of anthropogenic CO_2 emissions (Raven *et al.*, 2005). Complex and multifaceted effects that result from changes in oceanic acidification have great potential for impacting populations of marine reptiles through widespread disruption of marine food chains and ensuing negative consequences for marine ecosystems (Fabry *et al.*, 2008).

Challenges

Despite a growing body of research attempting to demonstrate or predict the influence of climatic change on reptiles, the base of evidence required to predict the impacts of climatic change remains relatively poor. It is also difficult to predict how plasticity in physiological and morphological features of reptiles might mitigate some of the effects of climatic changes. Reptiles are known for their ability to alter multiple aspects of phenotype in impressively short periods of time, and they can achieve rapid adaptive changes in morphology or performance on ecological timescales in response to changes in the environment (e.g., Losos *et al.*, 1997; Herrel *et al.*, 2008; Vervust *et al.*, 2010). Changing thermal conditions that are experienced early in life can have a positive influence on subsequent thermoregulatory tactics, and behavioral plasticity can allow snakes to adjust to suboptimal thermal conditions (Aubret & Shine, 2010). Also, some reptiles can adjust their foraging behaviors and feeding activities to changes in availability of prey attributable to seasonal changes in climate (e.g., Kadota, 2011). However, behavioral plasticity is limited, and predicting the future plasticity to anticipated changes of climate is likely to be tenuous. For example, there is considerable controversy related to attempts at predicting the future range expansions of Burmese pythons, which are recently invasive and evidently thriving in the Everglades of south Florida. Difficulty arises with respect to how these snakes might respond to climatic features of new environments that are encountered by dispersing individuals. Thus, different models have been used to infer the potential expansion and limits of distribution of Burmese pythons within the continental USA, and the results are variable (Pyron *et al.*, 2008; Rodda *et al.*, 2009; Van Wilgen *et al.*, 2009; Jacobson *et al.*, 2012).

Although the causal forces and details remain largely unknown, climatic changes are clearly altering ecosystems in fundamental ways. Emphasis of changes that are perceived by the research community is largely directed to the effects of warming, but cooling in certain regions also will have an important influence on biota and especially ectotherms. The impacts of climatic change will be numerous with complex and myriad interactions, including physiological stress, altered productivity and food web dynamics, shifts in species distributions and interactions, changes in biophysical interactions affecting activity and uses of energy, and increasing incidence of disease (IPCC, 2007; Hoegh-Guldberg & Bruno, 2010; Pereira *et al.*, 2010; Harley, 2011).

The future of research related to climatic effects on biota is likely to see an integrative merger of niche modeling, biophysical ecology and case studies of specific populations or species that are facing significant environmental changes. Reptiles will continue to provide model vertebrates for understanding the responses of organisms to climatic change. Therefore, it is important that the research community somehow provide quantitative baseline monitoring of abundance and habitat characteristics so that predicted changes or outcomes might actually be documented. Finally, all of this information will also be helpful in constructing and implementing plans for the conservation of these species.

Acknowledgments

I am grateful to Klaus Rohde for organizing this volume, and I thank Harold Heatwole and Coleman Sheehy III for editorial emendations that improved the manuscript. This article was written while the author's research was supported by the National Science Foundation [IOS-0926802 to HBL].

References

Abbott, D. H. (1984). Archaean plate tectonics revisited 2. Paleo-sea level changes, continental area, oceanic heat loss and the area-age distribution of the ocean basins. *Tectonics*, **3**, 709–722.

Anonymous. (2012). Dwindling sea snakes. (News of the week). *Science*, **335**, 150.

Aubret, F., & Shine, R. (2010). Thermal plasticity in young snakes: how will climate change affect the thermoregulatory tactics of ectotherms? *Journal of Experimental Biology*, **213**, 242–248.

Berg, E. E. (1965). Krokodileale Klimazeugen. *Geologische Rundschau*, **54**, 328–333.

Bonnet, X., & Brischoux, F. (2008). Thirsty sea snakes forsake refuge during rainfall. *Austral Ecology*, **33**, 911–21.

Bourjea, J., Lapegue, S., Gagnevin, L., *et al.* (2007). Phylogeography of the green turtle, *Chelonia mydas*, in the Southwest Indian Ocean. *Molecular Ecology*, **16**, 175–186.

Bush, M. B. (2002). Distributional change and conservation on the Andean flank, a palaeoecological perspective. *Global Ecology and Biogeography*, **11**, 463–473.

Cai, W., Shi, G., Cowan, T., Bi, D., & Ribbe, J. (2005). The response of the Southern Annular Mode, the East Australian Current, and the southern mid-latitude ocean circulation to global warming. *Geophysical Research Letters*, **32**, L23706.

Chamaillé-Jammes, S., Massott, M., Aragon, P., & Clobert, J. (2006). Global warming and positive fitness response in mountain populations of common lizards, *Lacerta vivipara*. *Global Change Biology*, **12**, 392–402.

Chen, I.-J, Hill, J. K., Ohlemüller, R., Roy, D. B., & Thomas, C. D. (2011). Rapid range shifts of species associated with high levels of climate warming. *Science*, **333**, 1024–1026.

Chou, C., Tu, J.-Y., & Tan, P. -H. (2007). Asymmetry of tropical precipitation change under global warming. *Geophysical Research Letters*, **34**, L17708, 1–5.

Church, J. A., White, N. J., & Hunter, J. R. (2006). Sea-level rise at tropical Pacific and Indian Ocean islands. *Global Planetary Change*, **53**, 155–168.

Crowley, T. J., & North, G. R. (1991). *Paleoclimatology*. New York: Oxford University Press.

Dethmers, K. E. M., Broderick, D., Moritz, C., *et al.* (2006). The genetic structure of Australasian green turtles (*Chelonia mydas*): exploring the geographical scale of genetic exchange. *Molecular Ecology*, **15**, 3931–3946.

Donn, W. L. (1987). Terrestrial climate change from the Triassic to Recent. In M. R. Rampino, J. E. Sanders, W. S. Newman & L. K. Königsson (Eds.), *Climate: History, Periodicity and Predictability* (pp. 343–352). New York: Van Nostrand Reinhold.

Eagle, R. A., Tütken, T., Martin, T. S., *et al.* (2011). Dinosaur body temperatures determined from isotopic (^{13}C-^{18}O) ordering in fossil biominerals. *Science*, **333**, 443–445.

Estes, R. (1970). Origin of the Recent North American lower vertebrate fauna: an inquiry into the fossil record. *Forma et Functio*, **3**, 139–163.

Fabry, V. J., Seibel, B. A., Feely, R. A., & Orr, J. C. (2008). Impacts of ocean acidification on marine fauna and ecosystem processes. *ICES Journal of Marine Science*, **65**, 414–432.

Gibbons, J. W., Scott, D. E., Ryan, T. J., *et al.* (2000). The global decline of reptiles, déjà vu amphibians. *BioScience*, **50**, 653–666.

Hanebuth, T. J. J., Voris, H. K., Yokoyama, Y., Saito, Y., & Okuno, J. (2011). Formation and fate of sedimentary depocentres on Southeast Asia's Sunda Shelf over the past sea–level cycle and biogeographic implications. *Earth-Science Reviews*, **104**, 92–110.

Hansen, J., Sato, M., Ruedy, R., *et al.* (2006). Global temperature change. *Proceedings of the National Academy of Sciences of the USA*, **103**, 14288–14293.

Harley, C. D. G. (2011). Climate change, keystone predation, and biodiversity loss. *Science*, **334**, 1124–1127.

Hawkes, L. A., Broderick, A. C., Godfrey, M. H., & Godley., B. J. (2007). Investigating the potential impacts of climate change on a marine turtle population. *Global Change Biology*, **13**, 923–932.

Hawkes, L. A., Broderick, A. C., Godfrey M. H., & Godley, B. J. (2009). Climate change and marine turtles. *Endangered Species Research*, **7**, 137–154.

Heatwole, H. (1999). *Sea Snakes*. Sydney: University of New South Wales Press.

Herrel, A., Huyghe, K., Vanhooydonck, K. B., *et al.* (2008). Rapid large-scale evolutionary divergence in morphology and performance associated with exploitation of a different dietary resource. *Proceedings of the National Academy of Sciences of the USA*, **105**, 4792–4795.

Hoegh–Guldberg, O., & Bruno, J. F. (2010). The impact of climate change on the world's marine ecosystems. *Science*, **328**, 1523–1528.

Huang, S.-P., Chiou, C.-R., Lin, T.-E., *et al.* (in press). Temperature-dependent digestive function and vegetation shade: critical elements affecting high–altitude snake's energetics and thermally suitable habitat with climate change. *Functional Ecology*.

Huey, R. B., Deutsch, C. A., Tewksbury, J. J., *et al.* (2009). Why tropical forest lizards are vulnerable to climate warming. *Proceedings of the Royal Society of London B*, **276**, 1939–1948.

Hughes, L. (2000). Biological consequences of global warming: is the signal already apparent? *Trends in Ecology & Evolution*, **15**, 56–61.

Hutchison, J. H. (1992). Western North American reptile and amphibian record across the Eocene/Oligocene boundary and its climatic implications. In D. R. Prothero & W. A. Berggren (Eds.), *Eocene-Oligocene Climatic and Biotic Evolution* (pp. 451–463). Princeton, NJ: Princeton University Press.

IPCC (2007). S. Solomon, D. Qin, M. Manning, *et al.* (Eds.), *Climate Change 2007: The Physical Science Basis. Contribution of Working Group I to the Fourth Assessment Report of the Intergovernmental Panel on Climate Change*. Cambridge: Cambridge University Press, 996 pp.

Jacobson, E. R., Barker, D. G., Barker, T., *et al.* (2012). Environmental temperatures, physiology, and behavior limit the range expansion of invasive Burmese pythons in the US. *Integrative Zoology*, **7**, 271–285.

Jansen, E., Overpeck, J., Briffa, K. R., *et al.* (2007). Palaeoclimate. In S. Solomon, D. Qin, M. Manning, *et al.* (Eds.), *Climate Change 2007: The Physical Science Basis. Contribution of Working Group I to the Fourth Assessment Report of the Intergovernmental Panel on Climate Change* (pp. 433–497). Cambridge: Cambridge University Press.

Janzen, F. J., & Paukstis, G. L. (1991). Environmental sex determination in reptiles: ecology, evolution and experimental design. *Quarterly Review of Biology*, **66**, 149–179.

Jouve, S., Bardet, N., Jalil, N.-E., *et al.* (2008).The oldest African crocodylian: phylogeny, paleobiogeography, and differential survivorship of marine reptiles through the Cretaceous-Tertiary Boundary. *Journal of Vertebrate Paleontology*, **28**, 409–421.

Kadota, Y. (2011). Is *Ovophiso kinavensis* active only in the cool season? Temporal foraging pattern of a subtropical pit viper in Okinawa, Japan. *Zoological Studies*, **50**, 269–275.

Kearney, M. R., Shine, R., & Porter, W. P. (2009). The potential for behavioral thermoregulation to buffer "cold-blooded" animals against climate warming. *Proceedings of the National Academy of Sciences of the USA*, **10**, 3835–3840.

Kennett, R., Munungurritj, N., & Yunupingu, D. (2004). Migration patterns of marine turtles in the Gulf of Carpentaria, northern Australia: implications for Aboriginal management. *Wildlife Research*, **31**, 241–248.

Lillywhite, H. B., Babonis, L. S., Sheehy III, C. M., & Tu, M.-C. (2008). Sea snakes (*Laticauda* spp.) require fresh drinking water: implications for the distribution and persistence of populations. *Physiological Biochemical Zoology*, **81**, 785–796.

Limpus, C. J. (2008). Adapting to climate change: a case study of the flatback turtle, *Natator depressus*. In E. S. Poloczanska, A. J. Hobday & A. J. Richardson (Eds.), *Hot Water: Preparing for Climate Change in Australia's Coastal and Marine Systems*. Proceedings of conference held in Brisbane, 12–14 November 2007. Hobart: CSIRO Marine and Atmospheric Research.

Liu, Y.-L., Lillywhite, H. B., & Tu, M.-C. (2010). Sea snakes anticipate tropical cyclone. *Marine Biology*, **157**, 2369–2373.

Losos, J. B., Warheit, K. I., & Schoener, T. W. (1997). Adaptive differentiation following experimental island colonization in *Anolis* lizards. *Nature*, **387**, 70–73.

Lyell, C. (1830). *Principles of Geology, Being an Attempt to Explain the Former Changes of the Earth's Surface, by Reference to Causes Now in Operation. 1*. London: John Murray.

MacLeod, N., Rawson, P. F., Forey, P. L., *et al.* (1997). The Cretaceous-Tertiary biotic transition. *Journal of the Geological Society*, **154**, 265–292.

Markwick, P. J. (1998a). Crocodilian diversity in space and time: the role of climate in paleoecology and its implications for understanding K/T extinctions. *Paleobiology*, **24**, 470–497.

Markwick, P. J. (1998b). Fossil crocodilians as indicators of late Cretaceous and Cenozoic climates: implications for using paleontological data in reconstructing palaeoclimate. *Palaeography, Palaeoclimatology, Palaeoecology*, **137**, 205–271.

Meehl, G. A., Washington, W. M., Collins, W. D., *et al.* (2005). How much more global warming and sea level rise? *Science*, **307**, 1769–1772.

Mitchell, N. J., Kearney, M. R., Nelson, N. J., & Porter, W. P. (2008). Predicting the fate of a living fossil: how will global warming affect sex determination and hatching phenology in tuatara? *Proceedings of the Royal Society of London B*, **275**, 2185–2193.

Neelin, J. D., Münnich, M., Su, H., Meyerson, J. E., & Holloway, C. E. (2006). Tropical drying trends in global warming models and observations. *Proceedings of the National Academy of Sciences of the USA*, **103**, 6110–6115.

Novacek, M. J. (1999). 100 million years of land vertebrate evolution: the Cretaceous-Early Tertiary transition. *Annals of the Missouri Botanical Garden*, **86**, 230–258.

Parmesan, C. (2007). Influences of species, latitudes and methodologies on estimates of phenological response to global warming. *Global Change Biology*, **13**, 1860–1872.

Pereira, H. M., Leadley, P. W., Proença, V., *et al.* (2010). Scenarios for global biodiversity in the 21st century. *Science*, **330**, 1496–1501.

Piras, P., Teresi, L., Buscalioni, A. D., & Cubo, J. (2009). The shadow of forgotten ancestors differently constrains the fate of Alligatoroidea and Crocodyloidea. *Global Ecology and Biogeography*, **18**, 30–40.

Poloczanska, E. S., Limpus, C. J., & Hays, G. C. (2009). Vulnerability of marine turtles to climate change. In D. W. Simms (Ed.), *Advances in Marine Biology*, Vol. 56 (pp. 151–211). Burlington, MA: Academic Press.

Porter, W. P., & James, F. C. (1979). Behavioral implications of mechanistic ecology II: the African rainbow lizard, *Agama agama*. *Copeia*, 594–619.

Porter, W. P., & Mitchell, J. W. (2006). Method and system for calculating the spatial-temporal effects of climate and other environmental conditions on animals. [U.S. Patent 7,155,377 in December 2006] http://www.warf.org/technologies.jsp?ipnumber=P01251US.

Porter, W. P., Sabo, J. L., Tracy, C. R., Reichman, O. J., & Ramankutty, N. (2002). Physiology on a landscape scale: plant-animal interactions. *Integrative and Comparative Biology*, **42**, 431–453.

Pyron, R. A., Burbrink, F. T., & Guiher, T. J. (2008). Claims of potential expansion throughout the US by invasive python species are contradicted by ecological niche models. *PLoS ONE*, **3**, e2931.

Qian, H. (2010). Environment-richness relationships for mammals, birds, reptiles, and amphibians at global and regional scales. *Ecological Research*, **25**, 629–637.

Raven, J., Caldeira, K., Elderfield, H., *et al.* (2005). *Ocean Acidification due to Increasing Atmospheric Carbon Dioxide*. Policy document. London: The Royal Society, 60 pp.

Raxworthy, C. J., Pearson, R. G., Rabibisoa, N., *et al.* (2008). Extinction vulnerability of tropical montane endemism from warming and upslope displacement: a preliminary appraisal for the highest massif in Madagascar. *Global Change Biology*, **14**, 1703–1720.

Reading, C. J., Luiselli, L. M., Akani, G. C., *et al.* (2010). Are snake populations in widespread decline? *Biology Letters*, **6**, 777–780.

Robertson, D. S., McKenna, M. C., Toon, O. B., Hope, S., & Lillegraven, J. A. (2004). Survival in the first hours of the Cenozoic. *GSA Bulletin*, **116**, 760–768.

Rodda, G. H., Jarnevich, C. H., & Reed, R. N. (2009). What parts of the US mainland are climatically suitable for invasive alien pythons spreading from Everglades National Park? *Biological Invasions*, **11**, 241–251.

Root, T. L., Price, J. T., Hall, K. R., *et al.* (2003). Fingerprints of global warming on wild animals and plants. *Nature*, **421**, 57–60

Rosenzweig, C., Casassa, G., Karoly, D. J., *et al.* (2007). Assessment of observed changes and responses in natural and managed systems. In M. L. Parry, O. F. Canziani, J. P. Palutikof, P. J. van der Linden & C. E. Hanson (Eds.), *Climate Change 2007: Impacts, Adaptation and Vulnerability. Contribution of Working Group II to the Fourth Assessment Report of the Interngovernmental Panel on Climate Change* (pp. 79–131). Cambridge: Cambridge University Press.

Ross, J. P. (2005). Hurricane effects on nesting *Caretta caretta*. *Marine Turtle Newsletter*, **108**, 13–14.

Rull, V., & Vegas-Vilarrúbia, T. (2006). Unexpected biodiversity loss under global warming in the neotropical Guayana Highlands: a preliminary appraisal. *Global Change Biology*, **12**, 1–9.

Sanders, K. L, Lee, M. S. Y., Leys, R., Foster, R., & Keogh, J. S. (2008). Molecular phylogeny and divergence dates for Australasian elapids and sea snakes (hydrophiinae): evidence from seven genes for rapid evolutionary radiations. *Journal of Evolutionary Biology* **31**, 682–95.

Saunders, M. A., & Lea, A. S. (2008). Large contribution of sea surface warming to recent increase in Atlantic hurricane activity. *Nature*, **451**, 557–561.

Sheehan, P. M, & Hansen, T. A. (1986). Detritus feeding as a buffer to extinction at the end of the Cretaceous. *Geology*, **14**, 868–870.

Sheehy III, C. M., Solórzano, A., Pfaller, J. B., & Lillywhite, H. B. (2012). Multilocus phylogeography of the Yellow-bellied sea snake, *Pelamis platurus* (Hydrophiini) across the Pacific Ocean. *Integrative and Comparative Biology*, in press.

Sinervo, B., Méndez-de-la-Cruz, F., Miles, D. B., *et al.* (2010). Erosion of lizard diversity by climate. *Science*, **328**, 894–899.

Stucky, R. K. (1992). Mammalian faunas in North America of Bridgerian to early Arikareean "ages" (Eocent and Oligocene). In D. R. Prothero & W. A. Berggren (Eds.), *Eocene-Oligocene Climatic and Biotic Evolution* (pp. 465–493). Princeton, NJ: Princeton University Press.

Tewksbury, J. J., Huey, R. B., & Deutsch, C. A. (2008). Putting the heat on tropical animals. *Science*, **320**, 1296–1297.

Van Wilgen, N. J., Roura-Pascual, N., & Richardson, D. M. (2009). A quantitative climate-match score for risk-assessment screening of reptile and amphibian introductions. *Environmental Management*, **44**, 590–697.

Vervust, B., Pafilis, P., Valakos, E. D., & Van Damme, R. (2010). Anatomical and physiological changes associated with a recent dietary shift in the lizard *Podarcis sicula*. *Physiological and Biochemical Zoology*, **83**, 632–642.

Wilf, P., & Johnson, K. R. (2004). Land plant extinction at the end of the Cretaceous: a quantitative analysis of the North Dakota megafloral record. *Paleobiology*, **30**, 347–368.

Willette, D. A., Tucker, J. K., & Janzen, F. J. (2005). Linking climate and physiology at the population level for a key life-history stage of turtles. *Canadian Journal of Zoology*, **83**, 845–850.

Woodroffe, C. D. (2008). Reef-island topography and the vulnerability of atolls to sea-level rise. *Global Planetary Change*, **62**, 77–96.

Zhang, K. Q., Douglas, B. C., & Leatherman, S. P. (2004). Global warming and coastal erosion. *Climate Change*, **64**, 41–58.

Zhang, X. B., Zweirs, F. W., Hegerl, G. C., *et al.* (2007). Detection of human influence on twentieth-century precipitation trends. *Nature*, **448**, 461–465.

20 Equilibrium and nonequilibrium in Australian bird communities – the impact of natural and anthropogenic effects

Hugh A. Ford

Introduction

The concept of nonequilibrium has been applied to populations that do not trend towards an equilibrium point, and in which the direction of changes in population size appear not to be dependent on density (Rohde, 2005). Furthermore, communities, such as those on islands or in nature reserves, would be considered to be in nonequilibrium if their species richness or diversity changes progressively over time. Many studies on the concept of equilibrium have been on birds.

In this chapter I identify several examples from Australian birds that I believe support the concept of nonequilibrium at both population and community levels. First, the majority of Australia is arid or semi-arid, with low and unpredictable rainfall, meaning that conditions are usually difficult for birds, but occasionally there are times of relative plenty, after heavy rain, which allow population increases. Next, I examine the concept of species equilibrium on Australian islands, and extend this to the declining woodland birds of the fertile grassy woodlands of southern Australia, which have become fragmented and degraded by human activity. Finally, Australia is experiencing climate change, which is likely to intensify in the future, so I shall consider the probable impacts of this on Australia's birds.

Birds in arid Australia

Two-thirds of Australia is regarded as arid or semi-arid, where shortage of water limits productivity most of the time (Kingsford & Norman 2002). Rainfall in arid Australia is also highly variable, with many dry years interspersed with one or two wetter years (Morton *et al.*, 2011). Many birds experience a "boom and bust" environment (Robin *et al.*, 2009), due to dramatic resource pulses following rain (Letnic & Dickman, 2010).

The Balance of Nature and Human Impact, ed. Klaus Rohde. Published by Cambridge University Press.
© Cambridge University Press 2013.

There are long periods of low reproductive success, high mortality and population decline, with heavy rain leading to opportunistic breeding, high productivity and population increase until the next dry spell. Rainfall patterns differ across the continent, meaning that usually somewhere has received adequate rain. Birds, unlike most animals, can respond to regional patterns of production by moving nomadically in pursuit of locally abundant resources (Schodde, 2006). When much of the arid zone is dry, some birds irrupt into more coastal parts of Australia. Possibly they survive there until the inland receives rain.

Waterbirds in the arid center

Many waterbirds respond to periodic heavy rainfall in inland Australia. Pelicans, cormorants, ibis, egrets and various waterfowl, such as grey teal *Anas gracilis*, head inland when the rivers flow and inundate the floodplains (Briggs, 1992; Kingsford & Norman, 2002; Reid, 2009; Roshier, 2009). Pink-eared ducks *Malacorhynchus membranaceus* depend entirely on good rains to breed; males show continuous spermogenesis and respond rapidly when they reach flooded sites (Briggs, 1992). Because most of the rivers in central Australia flow south-westwards from central Queensland, some water flows in them most years after the wet season. This may allow limited breeding by waterbirds. Less often, local and more northerly heavy rainfall combine to produce huge floods, and about one year in 12 salt lakes may fill. The Cooper Creek system, which drains from south-west Queensland to Lake Eyre, may hold as many as a million waterbirds during major floods (Kingsford & Norman, 2002), although Lake Eyre may only fill completely once every 30 years.

During extensive and prolonged floods, pelicans, ibis and egrets establish large breeding colonies and other species, such as the ducks, breed in a more dispersed fashion. Whereas the omnivorous ducks breed almost straight away, fish-eaters, such as the Australian pelican *Pelecanus conspicillatus*, wait several months until food is sufficiently abundant to feed their nestlings (Reid, 2009). Breeding continues as long as water levels are high, especially after repeated flood pulses, and successive broods are reared. Sometimes pelicans and ibis desert their colonies when water levels fall, and many young die.

How waterbirds detect inland floods and know when to move is poorly understood. However, satellite tracking has shown that grey teal make opportunistic flights to inland wetlands (Roshier, 2009). Pelicans, being long-lived, may rely on older birds, which have experienced previous floods, to lead the way to breeding sites. After the floods subside, the birds return to more coastal areas; some species, such as pelicans and ducks, reach New Guinea and New Zealand (Kingsford & Norman, 2002). Some grey teal also perish in the desert, whereas others find nearby wetlands (Roshier, 2009).

Ducks in arid Australia have large clutches of eight or nine eggs, similar to migratory ducks in the Northern Hemisphere, though arid-zone ducks breed for longer during suitable conditions. So, many waterbird populations show massive increases during flood years. During dry times, some, but not all, may breed in coastal areas. One notable exception is the banded stilt *Cladorhynchus leucocephalus*, which only breeds in flooded salt lakes. Its breeding was first recorded in 1930, but was suspected to be inland as juvenile birds turned up in coastal sites after heavy inland rain (Robin & Joseph, 2009). There may be little breeding inland during the longer dry periods, so populations of waterbirds progressively decline. Ducks may be shot by hunters or culled to protect rice crops, and occasionally disease kills many birds (Kingsford & Norman, 2002). However, food shortage and lack of suitable habitat probably contribute most to the declines. Ducks tend to be short-lived so several years without breeding may lead to substantial population declines, even without exceptional mortality.

Desert nomads

A number of land-birds that live in the arid center of Australia show similar high mobility and opportunistic breeding to the waterbirds (Schodde, 1982). These include species that feed on grass seeds, such as the zebra finch *Taeniopygia guttata*, budgerigar *Melopsittacus undulatus* and diamond dove *Geopelia cuneata* (Wyndham, 1982; Zann *et al.*, 1995; Morton, 2009). Their movements may be more local than those of the waterbirds; mostly they stay within the arid zone, but occasionally they may irrupt towards the coast. There is often a seasonal north-south pattern to their movements. For instance, Wyndham (1982) proposed that budgerigars move northwards in autumn and winter and southwards in spring and summer. This allows these birds to exploit seeds from grasses that flower in response to winter rains in the south and summer rains in the north. Numbers of budgerigars, zebra finches and other seed-eaters can change locally from scarce to abundant within weeks after good rain and between successive years in the same season. Zebra finches are opportunistic breeders in central Australia, responding to rain (Zann *et al.*, 1995), but may be regular seasonal breeders in more predictable locations, such as near irrigated land (Morton, 2009).

Overall, birds dependent on grass seeds show remarkable flexibility. They maintain a spatial map so that they can track food availability locally or even further afield. They also need to remember the location of water – essential for most seed-eaters (Morton, 2009). Rather than moving randomly, budgerigars may follow experienced birds to traditional sites when food becomes scarce locally (Wyndham, 1982).

Another group of land birds that respond nomadically to rainfall is the nectar-feeders. These are the "blossom nomads" noted by Keast (1968), including the pied (*Certhionyx variegatus*), black (*Sugomel niger*) and white-fronted honeyeaters (*Purnella albifrons*). Crimson chats (*Ephthianura tricolor*), which are ground-feeding honeyeaters that sometimes feed on nectar, are also irruptive. These honeyeaters are absent or scarce at a locality during the dry years, and arrive, sometimes abundantly, after heavy rain leads to shrubs, such as *Eremophila* and *Grevillea*, flowering (Paltridge & Southgate, 2001; Burbidge & Fuller, 2007).

Stability in the desert?

About half of the bird species in the arid and semi-arid zones of Australia are nomadic and breed in response to rain; the rest are sedentary or move only locally (Schodde, 1982).

These have been less well studied than the nomads. Considering the low relief and lack of substantial variation in average rainfall, Australian deserts possess an amazing variety of vegetation types, due to differences in the soil and how much water runs off or onto a site. In turn, the composition of bird communities is greatly influenced by the habitat (Pavey & Nano, 2009). Spinifex (*Triodia* spp.) hummock grassland, with scattered shrubs and small trees, is the most widespread vegetation type, and attracts many nomads. As well as responding rapidly to rain the hummock grasslands show great temporal and spatial patchiness due to frequent wildfires. Some bird species may be lost locally after fire and only return after the spinifex matures; other species exploit recently burnt areas.

The other widespread habitat is mulga, a name referring to both the vegetation and the dominant plant, *Acacia aneura*. This consists of low woodland, and has a different bird community from the spinifex (Burbidge & Fuller, 2007; Pavey & Nano, 2009). Cody (1994) identified about 18–20 core species in mulga across Australia, two-thirds of them insectivorous and sedentary. Thornbills, fairy-wrens, babblers and the crested bellbird (*Oreoica gutturalis*) forage on or just above the ground and remain even during the driest times. Whereas grasses, herbs and shrubs respond rapidly to rain, mulga appears superficially unchanged. In fact, Burbidge and Fuller (2007) found that the normalized difference vegetation index, detected by Landsat satellite imagery, did not alter in mulga following a dry spell and after heavy rain.

As well as low and unpredictable rainfall, the other universal characteristic of Australian deserts is the very low soil fertility, as it lacks phosphorus and numerous trace elements (Orians & Milewski, 2007). Plants adapt in various ways to these deficiencies; one is to produce copious nectar and other carbohydrate-rich resources, typically after rain. Many shrubs and trees produce strong chemical defenses against herbivory. Mulga leaves (phyllodes) last for 2 years, are rarely defoliated by insects and typically fall, along with branches, to the ground, where they are broken down slowly. Hence, many birds forage on fallen branches or the ground (Cody, 1994; Recher & Davis, 1997). The leaf litter and fallen timber accumulate until consumed by fire at approximately 100-year intervals (Orians & Milewski, 2007). A few bird species normally feed on insects in the canopy of mulga, such as western warbler (*Gerygone fusca*) (Cody, 1994) and rufous whistler (*Pachycephala rufiventris*) (Recher & Davis, 1997). Recher and Davis (1997) recorded an outbreak of geometrid moth caterpillars on mulga foliage. Other birds, such as thornbills and grey shrike-thrush (*Colluricincla harmonica*), which typically forage low, moved into the mulga to take these caterpillars.

How does the concept of equilibrium apply to bird communities in Australian deserts?

Greenslade (1982) discussed selection processes in arid Australia, by focusing on the dimensions of habitat favorability (average rainfall) and habitat predictability (variation in rainfall). As well as the familiar concepts of r and K selection she considered that adversity selection was an important process. This is where conditions are almost always tough; food is scarce or marginally palatable. Greenslade's account primarily discusses terrestrial arthropods, but she hoped that the same process would be found in vertebrates. As terrestrial arthropods are the main food of many birds in arid Australia, this seems highly likely.

The saw-tooth pattern of short bursts of steep population increase, followed by steady decline over longer periods, as shown by the nomadic waterbirds, does not fit comfortably with the concept of equilibrium. Added to this, breeding sites may be far from non-breeding sites and individuals may use several different breeding and non-breeding sites during their lifetimes. The absence of water will mean no breeding populations and very few individuals. Ultimately, drought refuges in mesic regions may determine the carrying capacity for most waterbirds. Even these may be subject to drought and, increasingly, a variety of human uses from irrigation to recreation have an impact on them.

The pattern is similar for the nomadic seed-eaters and nectarivores, though they tend to remain within the arid zone, and only occasionally irrupt towards the coast. They survive by moving among the rich patches. Despite the superficial appearance of homogeneity and monotony in arid Australia, as viewed from the top of a sand dune or while driving through endless mulga, there is remarkable patchiness. This occurs at numerous scales. At the smallest, there are run-off and run-on areas due to minor differences in topography, with the latter receiving more effective rainfall. On a grander scale rain that falls on the central ranges runs off into the plain, for instance from the Macdonell Ranges into the Finke River system (Pavey & Nano, 2009). Differences in soil type give rise to different vegetation types; saline clay soils are dominated by chenopods and deep sands by spinifex. Fire also leads to patchy mosaics of bare ground, regrowing, mature and senescing vegetation. Particularly in the northern inland, summer rainfall mostly comes from thunderstorms, which can be local, drenching one area whereas 10 km away it remains dry.

The desert nomads can thrive and breed in the best patches, survive in a range of others, but need to leave when resources decline. The pattern of discovery, exploitation and desertion of patches means that there is only a short period when competition may be intense. Dispersal to the next rich patch is probably the most hazardous stage of the life cycle of these nomads, with some failing to find favorable patches. Chance, and previous experience, may lead them to other rich patches or to their demise. They fit well with the concept of r selection, where the habitat is unpredictable, though at times and in places favorable. For most of the time, populations are in nonequilibrium.

In contrast, insectivorous birds in the mulga live in an almost perpetually poor environment, their food being suppressed by the indigestibility of the plant matter. The presence of flowers, fruit or seeds and occasional insect outbreaks give them periods when breeding can be prolonged and successful. But for most of the time, survival is a struggle. Mulga is perhaps a place where adversity selection operates on many bird species. Here, low mobility, more moderate reproductive potential and weak density dependence are relevant characteristics (Greenslade, 1982).

Are there any bird species in deserts nearer the K selected end of the spectrum? Birds in the more mesic woodlands of Australia are subject to reasonably aseasonal conditions, i.e., a lack of scarcity or superabundance (Ford et al. 1988; Ford, 1989). This has led to long breeding seasons, small clutch sizes, high longevity and a high incidence of cooperative breeding. The environments in arid Australia that most approach these are the woodlands that fringe the more permanent creeks and billabongs, for instance those where river red gum *Eucalyptus camaldulensis* and coolabah *E. coolibah* dominate. Here some species can be sedentary, breed regularly, and are probably more subject to intra-specific competition. In fact, Pavey and Nano (2009) found more nomadic species and individuals in the riverine woodland than in their other main vegetation types. This is possibly because the large eucalypts provide the nest sites for many parrots, which mostly feed in the surrounding grasslands.

Although we know something of the biology of the nomads, there have been few studies on the sedentary birds of arid Australia. The life histories of the thornbills have been well studied in mesic Australia, so comparative studies of the arid zone species would be interesting. The rufous whistler, a common bird in mulga, has been well studied in more coastal woodlands, allowing comparisons of the species' life history in the arid zone. Given the future changes in climate that are predicted, it is important that we find out more about the seasonal and year to year patterns of food abundance, breeding effort and success, longevity and survival of these species.

Letnic and Dickman (2010) proposed a state-and-transition model to describe the response of small mammals in Australian deserts to the resource pulses that follow rain. This is a nonequilibrium model, which recognizes three or four states. Regular or moderate rain may move a location to a higher state, whereas a dry period may drop it to a lower state. Exceptional heavy rain and floods move any states into the highest one. Predation and grazing by native and exotic species, and fire, may also force sites into a lower state. For instance, in spinifex hummock grasslands, in state 0, after prolonged drought, there are few small mammals. State 1, following average rainfall, has mostly insectivorous marsupials, whereas the progressively wetter states 2 and 3 are dominated by omnivorous and herbivorous rodents, respectively. Given that states may only last a year or less, there is probably insufficient time for equilibrium in populations and community structure to be established at any site. The state-and-transition model could probably also be applied to bird communities.

Equilibrium on islands

One of the frequently cited examples of equilibrium in ecology is the number of species on islands. The theory of island biogeography states that the number of species on any island reaches equilibrium when the rate of colonization by new species is balanced by the rate of extinction of existing species (MacArthur & Wilson, 1967). This is a dynamic equilibrium, with turnover as new species arrive and others go extinct. Extinction rate is related to population size, which in turn depends on area, whereas colonization rate depends on distance from a continent or larger island. Numerous studies in island biogeography involved birds, which are highly mobile, and have reached even the most remote islands. Birds are visible and so fairly complete species lists are available for most islands.

Proving that the number of species on an island is in equilibrium is more difficult than simply observing a relationship between species number and island area and isolation.

Abbott and Grant (1976) found that the number of land-bird species on islands around Australia and New Zealand had changed over periods of 59 to 184 years. On all but two of 15 islands the number of passerine birds had increased, on about half by 100% or more. This is especially surprising given that most islands had received a variety of introduced mammals, which had made a dramatic impact on the island ecosystems. Abbott and Grant (1976) excluded species directly introduced by humans, though they included bird species introduced from Europe to Australia or New Zealand which subsequently colonized islands. However, 10 islands received new species that were native to Australia or New Zealand. Abbott and Grant (1976) concluded that the numbers of passerine bird species on these islands were not in equilibrium.

Occasionally, events may dramatically shift the number of species on an island. The most extreme are major volcanic eruptions, as occurred on Krakatau Island in Indonesia in 1883. The bird fauna was well known before the eruption, which probably eliminated all bird species. As the vegetation recovered, birds recolonized. MacArthur and Wilson (1967) suggested that the number of species of land and freshwater birds reached an apparent equilibrium about 30 years after the eruption. This number was about what was predicted for an island the size of Krakatau. However, further surveys have indicated that new species are still colonizing more rapidly than species are going extinct, partly because the forest on the island is maturing (Thornton et al., 1993). Ongoing eruptions maintain secondary habitat, and the birds associated with it, which would be lost if volcanic activity ceased (Thornton et al., 1990). Equilibrium had apparently not been reached over 100 years after the destruction of Krakatau.

Less dramatic is the formation of new islands by rise in sea level. These "land-bridge" islands originally contained a large number of species from the mainland from which the island was excised. In time, their reduced area causes extinction rates to rise and, due to isolation, colonization rates to decline, leading to "species relaxation". The number of species should drop to a new equilibrium. Diamond (1972) identified many islands formed by rises in sea level some 10 000 years ago after the last glacial period. They lack many species found in New Guinea, or in the larger islands to which they had been joined. However, such islands appeared to have more species than predicted for their area – falling above the regression line between species number and island area (Diamond, 1972). Hence, there was evidence for species relaxation towards a new equilibrium, which had not yet been reached; the islands were supersaturated. Diamond (1972) estimated that large islands (3000–7500 km^2), separated from New Guinea, might reach a new equilibrium after about 7000–8000 years.

Australia, too, has land-bridge islands, and Abbott (1973, 1974) showed that Kangaroo Island, Tasmania and the Bass Strait islands are depauperate relative to sections of the neighboring mainland of comparable area and climate. For instance, Kangaroo Island (4500 km^2 in area) has only 61 species of land-birds, compared with 129 species on the Fleurieu Peninsula, from which it was separated after the end of the last glacial period and is only 14 km distant (Table 20.1). Kangaroo Island also has fewer bird species than the Yorke and Eyre Peninsulas, which have a comparable climate and vegetation, but which are more distant than the Fleurieu Peninsula. Abbott (1974) suggested that less than one-third of the 85 missing species had been present when

Table 20.1. The number of land-bird species on Kangaroo Island and three mainland peninsulas in South Australia, plus the number of species restricted to one region and numbers from each peninsula shared with or absent from Kangaroo Island (from Abbott, 1974).

Region	Number of species	Single site species	Shared with Kangaroo Is.	Absent from Kangaroo Is.
Kangaroo Is.	61	2	–	–
Fleurieu Pen.	129	35	57	72
Yorke Pen.	88	6	46	42
Eyre Pen.	84	5	47	37

Kangaroo Island was formed but had subsequently gone extinct. The remainder had probably failed to colonize, despite the short distance from the mainland.

Part of the reason for some species being absent is that suitable habitat may be missing or highly localized for some species (Ford & Paton, 1975). Whereas Abbott (1974) rejected this likelihood, he did admit that the "savanna forests" (or grassy woodlands) of the Fleurieu Peninsula and Adelaide Plains are absent from Kangaroo Island. Half of the missing species prefer this habitat or occur in drier or more open habitat than currently occurs on Kangaroo Island (Ford & Paton, 1975). Surprisingly, Aborigines were absent when Matthew Flinders visited Kangaroo Island over 200 years ago and he commented on the absence of smoke (Lampert, 1979). Perhaps the absence of Aboriginal burning contributed to the scarcity of savanna and grassland.

In fact, we need to consider the changes in climate, vegetation and area of land, as well as the impact of people, that have occurred since Kangaroo Island became isolated from the mainland. Over 10 000 years ago at the end of the last glacial period, low sea levels meant that a large land mass connected the mainland peninsulas and Kangaroo Island (Bye 1976). However, it would have been drier, therefore there would have been less sclerophyll forest than now occurs. Moister habitats would probably have occurred as patches on the higher ground, such as Mount Lofty, and in gullies along the coast, which was some 60 km south of the current Kangaroo Island coastline, from Eyre Peninsula to southeastern South Australia and western Victoria. There was also an Aboriginal population in the area at that time (Lampert, 1979). So probably extensive grassland and open woodlands would have occurred on what later became Kangaroo Island, with the bird species associated with these habitats. As the climate became warmer and wetter over the following few thousand years, the forest expanded while the land area declined as sea levels rose. Kangaroo Island separated from the mainland about 9500 years ago, before the warmest and wettest period 7000 to 4000 years ago (Lampert, 1979), when the sea reached its current level. The number of forest-dependent birds may have increased before isolation of the island. Other species may have subsequently colonized the mainland forests from eastern Australia, with some reaching Kangaroo Island, and others failing to, as proposed by Abbott. After 4000 years ago it became drier, probably followed by the extinction of the Aboriginal population around 2000 years ago (Lampert, 1979). The forests may have contracted, as may have the more open, fire-dependent habitats.

Overall then, the idea of the bird species of Kangaroo Island relaxing towards a new equilibrium after isolation from the mainland is simplistic. A more cautious hypothesis is that over the last 10 000 years the equilibrium number of bird species and mix of mesic and drier country species on Kangaroo Island fluctuated as the climate and vegetation changed. From Diamond's (1972) figures for land-bridge islands around New Guinea of comparable size to Kangaroo Island, the number of bird species would have taken thousands of years to reach each new equilibrium. Climate change and habitat loss would have caused some extinctions, first of savanna and grassland species, then of forest species. The pool of forest species on the neighboring mainland may also have changed, due to climate change, altering the colonization rate to Kangaroo Island. So, the time taken to reach each new equilibrium on land-bridge islands of this size would be long relative to changes in the climate and vegetation and in the pool of potential colonizers. Therefore, the current number of species may still be pursuing an elusive equilibrium number that is itself changing.

Mainland islands

MacArthur and Wilson (1967) also briefly mentioned the fact that isolated patches of habitat within continents, such as mountain-tops, can act in the same way as islands. Subsequently, the theory of island biogeography has been applied to nature reserves in urban and rural landscapes (Diamond, 1975; Wilson & Willis, 1975), albeit with modification to allow for the more hospitable matrix. The open forests and woodlands of the Fleurieu Peninsula and Mount Lofty Ranges behave as an island of mesic habitat isolated from the extensive forests of eastern Australia. Just as the fauna of Kangaroo Island is depauperate, the Mount Lofty Ranges also lack many of the bird species that occur in the eastern forests and woodlands.

The alternating glacial and interglacial periods, with drier and wetter climates respectively, allowed the mesic habitats around the coast of Australia to contract into and expand out of refuges (Schodde, 2006). As mentioned above for Kangaroo Island, the forests probably increased as temperature and rainfall increased from 10 000 to 7000 years ago, though total land area declined after sea levels rose. After 4000 years ago savanna woodland probably expanded at the expense of forest. So, any equilibrium in species number within individual forest patches or in mesic habitat collectively would have fluctuated. Hence, the number of bird species in the Mount Lofty Ranges could have been increasing or decreasing when Europeans arrived nearly 200 years ago.

European settlers extensively cleared the forests and open woodlands of the Mount Lofty Ranges, so that only about 10% remain, in a highly fragmented state. This has led to predictions that the number of bird species will decline (Ford & Howe, 1980), with the loss of perhaps 65 species present at European settlement. Already, some species have gone, with others precariously close to extinction (Table 20.2; Possingham & Field, 2001), and even common species are declining (Szabo *et al.*, 2011). Certainly, the number of species is currently far from equilibrium. It could take thousands of years for full species relaxation, so there is sufficient opportunity to rescue most bird species,

Table 20.2. Bird species that have gone extinct (Garnett & Crowley, 2000) in the Mount Lofty Ranges since European settlement or are now endangered and close to extinction (Possingham & Field, 2001).

Species	Status
King quail	Extinct
Glossy black cockatoo	Extinct
Swift parrot	Extinct
Ground parrot	Extinct
Barking owl	Extinct
Azure kingfisher	Extinct
Rufous fieldwren	Extinct
Regent honeyeater	Extinct as breeder (vagrant only)
Spotted quail-thrush	Probably extinct (Szabo *et al.*, 2011)
Brown quail	Endangered
Square-tailed kite	Endangered
Bush stone-curlew	Endangered
Little lorikeet	Endangered
White-throated gerygone	Endangered
Olive-backed oriole	Endangered
Flame robin	Endangered – non-breeding migrant

providing that there is the will to do so. There is now a major effort to expand and connect vegetation remnants, and manage the most threatened species. However, the scale of activity needs to be greatly increased (Paton *et al.*, 2010). Szabo *et al.* (2011) suggested that the scenario in the Mount Lofty Ranges is a "canary landscape" warning of what is likely to happen in the woodlands and open forests throughout eastern Australia.

A similar pattern of habitat loss, fragmentation and degradation has been experienced in the forests, and especially the more fertile grassy woodlands, of eastern Australia. Many bird species have declined or experienced range contractions (Ford, 2011). The reasons for these declines are many and complex. However, the loss of species from individual remnants suggests that the process of species relaxation is occurring, following reduction of patch size and increasing isolation from sources. It is interesting that many of the bird species that are absent from or only vagrant to Kangaroo Island and Tasmania are also declining in eastern Australia (Table 20.3).

Climate change and equilibrium in ecology

There is no doubt that average temperatures worldwide have risen over at least the last 50 years, and it is highly likely that human activity has contributed to this climate change. Animals and plants around the world have responded to such global warming (Rosenzweig *et al.*, 2008). Many birds have increased their ranges polewards, or shifted their migration and breeding earlier.

There is increasing evidence that birds in Australia have responded to climate change (Chambers *et al.*, 2005; Olsen, 2007). White-headed pigeons (*Columba leucomela*) and

Table 20.3. Examples of species of eucalypt woodlands that are absent from or vagrant in Kangaroo Island and/or Tasmania (from Abbott, 1973, 1974) and are declining in eastern Australia (Watson, 2011). Where a species is threatened in NSW or in Australia generally, this status is given rather than Declining. (I have omitted spotted nightjar, white-browed treecreeper, chestnut-rumped thornbill and crested bellbird, which were listed by Watson, but are found in drier woodland and forest.)

Species	Tasmania	Kangaroo Is.	Eastern Australia
Emu	Absent[a]	Absent[a]	Declining
Square-tailed kite	Absent	Absent	Vulnerable in NSW
Bush stone-curlew	Absent	Present	Endangered in NSW
Painted button-quail	Present	Present	Declining
Swift parrot	Endangered breeding migrant	Absent	Endangered winter migrant
Brown treecreeper	Absent	Absent	Vulnerable in NSW
Speckled warbler	Absent	Absent	Vulnerable in NSW
Southern whiteface	Absent	Absent	Declining
Regent honeyeater	Absent	Vagrant (formerly)	Critically endangered nationally
Grey-crowned babbler	Absent	Absent	Vulnerable in NSW
White-browed babbler	Absent	Absent	Declining
Varied sittella	Absent	Absent	Declining
Crested shrike-tit	Absent	Absent	Declining
Rufous whistler	Vagrant	Vagrant?	Declining
White-browed woodswallow	Uncommon visitor	Vagrant	Declining
Dusky woodswallow	Present	Present	Declining
Restless flycatcher	Absent	Vagrant	Declining
Jacky winter	Absent	Vagrant	Declining
Red-capped robin	Absent	Absent	Declining
Hooded robin	Absent[b]	Vagrant?	Vulnerable in NSW
Eastern yellow robin	Absent	Absent	Declining
Diamond firetail	Absent	Vagrant?	Vulnerable in NSW

[a] Endemic species of emu occurred on King and Kangaroo Island, but are now extinct. Emus have been introduced to Kangaroo Island.
[b] Dusky robin in Tasmania may be derived from hooded robin.

figbirds (*Sphecotheres vieilloti*) have spread southwards (Silcocks & Sanderson, 2007; McAllan *et al.*, 2007). Both are frugivores and it is possible that the planting of figs and other fruit trees has encouraged their spread. Pied butcherbirds (*Cracticus nigrogularis*) have invaded higher latitudes and altitudes (Silcocks & Sanderson, 2007) and Pacific bazas (*Aviceda subcristata*) have spread south and inland (McAllan *et al.*, 2007). Both feed on large invertebrates and small vertebrates, which may have increased in abundance due to rising temperatures. As well as changes in population distribution and abundance, the arrival of these herbivores and predators may have an impact on other species in their communities, potentially disrupting population and community equilibria.

Already, climate change has reduced the ranges of bird species, and in the future these may decline precipitously. Brereton *et al.* (1995) modeled the ranges of 17 rare and threatened bird species in southeastern Australia, following increases in mean temperatures of 1°C, 2°C and 3°C. Even with the most modest temperature increase the core ranges of five species disappeared, particularly affecting species in inland areas, such as the red-tailed black cockatoo (*Calyptorhynchus banksii graptogyne*) and Mallee emu-wren (*Stipiturus mallee*). Three more species lost their core ranges after a 3°C increase. Species at high altitude, such as the golden bowerbird (*Prionodura newtonia*) in the mountains of north Queensland, are predicted to contract to only 2% of their current range with a 3°C increase (Hilbert *et al.*, 2004).

Conclusions

The nomadic waterbirds, granivores and nectarivores of the arid inland of Australia show such dramatic changes in abundance, both locally and regionally, within and between years, that it is hard to believe that their populations ever reach equilibrium. More likely, we can consider equilibrium as an elusive target towards which they may move, but which will itself have changed before they reach it. Of course, not all birds that occur in the arid parts of Australia show such changes; about half are sedentary, showing much more subdued population changes. For these, the concept of equilibrium may still apply.

Although some islands, perhaps those distant from the mainland, may possess an equilibrium number of species, many may have been periodically connected to continents or large islands as sea levels have fallen and risen again. An example is Kangaroo Island, off the southern Australian coast. Changes in sea level have not only connected and isolated it from the mainland but also changed its area. In addition historical climatic changes and the presence, and for a time absence, of humans will also have changed the island's vegetation and avifauna. Here again, a long-term equilibrium in the number of species seems elusive. Again, the target number of species, and indeed the actual species, would be forever changing. There were many habitat islands in Australia at the time of European settlement, such as the forests of the Mount Lofty Ranges. Their bird species richness may have been near equilibrium. However, removal of 90% of the forest and fragmentation and degradation of what is left has pushed the equilibrium species richness to a much lower level. It is not inevitable that bird species will be lost, because species relaxation takes time, and extensive replanting, connecting and rehabilitation of the forest could prevent many extinctions.

Finally, we are in a period of rapid climate change. Many species are experiencing local population changes, some invading new areas and communities, others going extinct. Many populations and communities will probably be far from any equilibrium. Ongoing changes in the species mix in communities will mean that again equilibrium will be a distant goal.

References

Abbott, I. (1973). Birds of the Bass Strait. Evolution and ecology of the avifaunas of some Bass Strait islands, and comparisons with those of Tasmania and Victoria. *Proceedings of the Royal Society of Victoria*, **85**, 197–223.

Abbott, I. (1974). The avifauna of Kangaroo island and causes of its impoverishment. *Emu*, **74**, 124–134.

Abbott, I., & Grant, P. R. (1976). Nonequilibrial bird faunas on islands. *The American Naturalist*, **110**, 507–528.

Brereton, R., Bennett, S., & Mansergh, I. (1995). Enhanced Greenhouse climate change and its potential effect on selected fauna of south-eastern Australia: a trend analysis. *Biological Conservation*, **72**, 339–354.

Briggs, S. V. (1992). Movement patterns and breeding characteristics of arid zone ducks. *Corella*, **16**, 15–22.

Burbidge, A. A., & Fuller, P. J. (2007). Gibson Desert birds: responses to drought and plenty. *Emu*, **107**, 126–134.

Bye, J. A. T. (1976). Physical oceanography of Gulf St Vincent and Investigator Strait. In *Natural History of the Adelaide Region* (pp. 143–1160). Adelaide: Royal Society of South Australia.

Chambers, L. E., Hughes, L., & Weston, M. A. (2005). Climate change and its impact on Australia's avifauna. *Emu*, **105**, 1–20.

Cody, M. L. (1994). Mulga bird communities. I. Species composition and predictability across Australia. *Australian Journal of Ecology*, **19**, 206–219.

Diamond, J. M. (1972). Biogeographic kinetics: estimation of relaxation times for avifauna of southwest Pacific Islands. *Proceedings of the National Academy of Sciences of the USA*, **69**, 3199–3203.

Diamond, J. M. (1975). The island dilemma: lessons of modern biogeographic studies for the design of nature reserves. *Biological Conservation*, **7**, 129–146.

Ford, H. A. (1989). *The Ecology of Birds: an Australian Perspective*. Chipping Norton, NSW: Surrey Beatty and Sons.

Ford, H. A. (2011). The causes of decline of birds of eucalypt woodlands: advances in our knowledge over the last 10 years. *Emu*, **111**, 1–9.

Ford, H. A., & Paton, D. C. (1975). Impoverishment of the avifauna of Kangaroo Island. *Emu*, **75**, 155–156.

Ford, H. A., & Howe, R. W. (1980). The future of birds in the Mount Lofty Ranges. *South Australian Ornithologist*, **28**, 85–89.

Ford, H. A., Bell, H., Nias, R., & Noske, R. (1988). The relationship between ecology and the incidence of cooperative breeding in Australian birds. *Behavioral Ecology and Sociobiology*, **22**, 239–249.

Garnett, S. T., & Crowley, G. M. (2000). *The Action Plan for Australian Birds 2000*. Canberra: Environment Australia.

Greenslade, P. J. M. (1982). Selection processes in arid Australia. In W. R. Barker & P. J. M. Greenslade (Eds.), *Evolution of the Flora and Fauna of Arid Australia* (pp. 125–130). Adelaide: Peacock Publications.

Hilbert, D. W., Bradford, M., Parker, T., & Westcott, D. A. (2004). Golden Bowerbird (*Prionodura newtonia*) habitat in past, present and future climates: predicted extinction of a vertebrate in tropical highlands due to global warming. *Biological Conservation*, **116**, 267–277.

Keast, J. A. (1968). Seasonal movements in the Australian honeyeaters (Meliphagidae) and their ecological significance. *Emu*, **67**, 159–210.

Kingsford, R. T., & Norman, F. I. (2002). Australian waterbirds – products of the continent's ecology. *Emu*, **102**, 47–69.

Lampert, R. J. (1979). Aborigines. In *Natural History of the Adelaide Region* (pp. 81–89). Adelaide: Royal Society of South Australia.

Letnic, M., & Dickman, C. R. (2010). Resource pulses and mammalian dynamics: conceptual models for hummock grasslands and other Australian desert habitats. *Biological Reviews*, **85**, 501–521.

MacArthur, R. H., & Wilson, E. O. 1967. *Island Biogeography*. Princeton, NJ: Princeton University Press.

McAllan, I., Cooper, D., & Curtis, B. (2007). Changes in ranges: an historical perspective. In P. Olsen (Ed.), The State of Australia's Birds 2007. Birds in a changing climate. *Supplement to Wingspan* **14**, no. 4, 12–13.

Morton, S. R., Stafford Smith, D. M., Dickman, C. R., *et al.* (2011). A fresh framework for the ecology of arid Australia. *Journal of Arid Environments*, **75**, 313–329.

Morton, S. (2009). Rain and grass: lessons in how to be a zebra finch. In L. Robin, R. Heinsohn & L. Joseph (Eds.), *Boom and Bust. Bird Stories for a Dry Country* (pp. 45–74). Melbourne: CSIRO.

Olsen, P. (2007). The State of Australia's Birds 2007. Birds in a changing climate. *Supplement to Wingspan* **14**, no. 4.

Orians, G. H., & Milewski, A. V. (2007). Ecology of Australia: the effects of nutrient-poor soils and intense fires. *Biological Reviews*, **82**, 393–423.

Paltridge, R., & Southgate, R. (2001). The effect of habitat type and seasonal conditions on fauna in two areas of the Tanami Desert. *Wildlife Research*, **28**, 247–260.

Paton, D. C., Willoughby, N., Rogers, D. J., *et al.* (2010). Managing the woodlands of the Mt Lofty region, South Australia. In D. Lindenmayer, A. Bennett & R. Hobbs (Eds.), *Temperate Woodland Conservation and Management* (pp. 83–91). Melbourne: CSIRO.

Pavey, C. R., & Nano, C. E. M. (2009). Bird assemblages of arid Australia: vegetation patterns have a greater effect than disturbance and resource pulses. *Journal of Arid Environments*, **73**, 634–642.

Possingham, H. P., & Field, S. A. (2001). Regional bird extinctions and their implications for vegetation clearance policy. *Life Lines*, **7**, 15–16.

Recher, H. F., & Davis, W. E. (1997). Foraging ecology of a mulga bird community. *Wildlife Research*, **24**, 27–43.

Reid, J. (2009). Australian pelican: flexible responses to uncertainty. In L. Robin, R. Heinsohn & L. Joseph (Eds.), *Boom and Bust. Bird Stories for a Dry Country* (pp. 95–120). Melbourne: CSIRO.

Robin, L., Heinsohn, R., & Joseph, L. (Eds.) (2009). *Boom and Bust. Bird Stories for a Dry Country*. Melbourne: CSIRO.

Robin, L., & Joseph, L. (2009). The boom and bust desert world: a bird's eye view. In L. Robin, R. Heinsohn & L. Joseph (Eds.), *Boom and Bust. Bird Stories for a Dry Country* (pp. 7–34). Melbourne: CSIRO.

Rohde, K. (2005). *Nonequilibrium Ecology*. Cambridge: Cambridge University Press.

Rosenzweig, C., & 13 others. (2008). Attributing physical and biological impacts to anthropogenic climate change. *Nature*, **453**, 353–357.

Roshier, D. (2009). Grey Teal: survivors in a changing world. In L. Robin, R. Heinsohn & L. Joseph (Eds.), *Boom and Bust. Bird Stories for a Dry Country* (pp. 75–94). Melbourne: CSIRO.

Schodde, R. (1982). Origin, adaptation and evolution of birds in arid Australia. In W. R. Barker & P. J. M. Greenslade (Eds.), *Evolution of the Flora and Fauna of Arid Australia* (pp. 191–224). Adelaide: Peacock.

Schodde, R. (2006). Australia's bird fauna today – origins and evolutionary development. In J. R. Merrick, M. Archer, G. M. Hickey & M. S. Y. Lee (Eds.), *Evolution and Biogeography of Australasian Vertebrates* (pp. 413–458). Oatlands, NSW: Auscipub.

Silcocks, A., & Sanderson, C. (2007). Volunteers monitoring change: the atlas of Australian birds. In P. Olsen (Ed.), The State of Australia's Birds 2007. Birds in a changing climate. *Supplement to Wingspan* **14**, no. 4, 10–11.

Szabo, J. K., Vesk, P. A., Baxter, P. W. J., & Possingham, H. P. (2011). Paying the extinction debt: woodland birds in the Mount Lofty Ranges, South Australia. *Emu*, **111**, 59–70.

Thornton, I. W. B., Zann, R. A., & Van Balen, S. (1993). Colonization of Rakarta (Krakatau Is.) by non-migrant land birds from 1883 to 1992 and implications for the value of island equilibrium theory. *Journal of Biogeography*, **20**, 441–452.

Thornton, I. W. B., Zann, R. A., & Stephensen, D. G. (1990). Colonization of the Krakatau Islands by land birds, and the approach to an equilibrium number. *Philosophical Transactions of the Royal Society of London B*, **327**, 55–93.

Watson, D. (2011). A productivity-based explanation for woodland bird declines: poorer soils yield less food. *Emu*, **111**, 10–18.

Wilson, E. O., & Willis, E. O. (1975). Applied biogeography. In M. L. Cody & J. M. Diamond (Eds.), *Ecology and Evolution of Communities* (pp. 522–534). Cambridge, MA: Bellknap Press.

Wyndham, E. (1982). Movements and breeding seasons of the Budgerigar. *Emu*, **82**, 276–282.

Zann, R. A., Morton, S. R., Jones, K. R., & Burley, N. T. (1995). The timing of breeding by Zebra Finches in relation to rainfall in central Australia. *Emu*, **95**, 208–222.

21 Population dynamics of insects: impacts of a changing climate

Nigel R. Andrew

Climate change is recognized as one of the most serious scientific issues to understand and respond to (AAS, 2010). One of the most fundamental issues is to recognize how our biota will respond and adapt to such rapid changes at a global scale. Already global mean temperature has risen by 0.76°C this century (IPCC, 2007), and recent research indicates that current temperature increases are tracking the upper range projected by the IPCC modeled predictions and sea-level change is faster than projected (Rahmstorf *et al.*, 2007; Steffen *et al.*, 2009). Across Australia, different regions have experienced climatic changes to varying degrees, both seasonally and annually. Future predictions are for a generally warmer and drier continent by 2030 (CSIRO, 2007), but with the likely impacts of climate change being complex and highly variable across the continent, and worldwide (Walther *et al.*, 2002). Changes in the physiological tolerances and population depletion could cause major population restructure of currently common species, leading to the collapse of trophic interactions and depletion of ecosystem services.

Two of the great challenges in predicting how biological organisms will respond to a rapidly changing climate are (i) determining whether responses of organisms are idiosyncratic, or whether there are underlying generalities that can be made based on evolutionary relationships, or ecological associations, and (ii) determining whether these responses are consistent in time and space (Andrew & Terblanche, in press).

A population's capacity to respond to temperature variation is a crucial element in understanding the effects of environmental change. This response may occur via mechanistic and key physiological traits which are linked to the thermal environments (Kingsolver & Huey, 1998) and may affect fitness over evolutionary timescales (Loeschcke & Hoffman, 2007). It could also be linked to the capacity of an animal to respond to the variability and predictability of its environments, where the thermal environment can be quantified on an ecologically relevant timescale (hours, days, seasons) (Chown & Terblanche, 2007).

Temperature and water availability are of fundamental importance to animal physiology, behavior, and ecology. This is particularly true for insects (Addo-Bediako *et al.*, 2001). Insects can respond to mean temperature variation, predictability and extremes in the thermal environment. This can be quantified due to the strong mechanistic links

The Balance of Nature and Human Impact, ed. Klaus Rohde. Published by Cambridge University Press.
© Cambridge University Press 2013.

between metabolism, development and performance and the thermal environment. The body temperature of ectothermic taxa, such as insects, influences how they adapt to their environment, their capability to survive, grow, reproduce and disperse. Water loss is also crucial for insect survival, population abundance and geographic distribution (Addo-Bediako *et al.*, 2001; Chown *et al.*, 2011). Their small body size means that their ability to conserve water can differ substantially between wet and dry climates. In order to determine the influence of climate change on insects, we need an understanding of differences in key performance traits (physiological, behavioral and ecological traits) across space and time, and how flexible these trait responses may be under different climatic circumstances, and how, if at all, short- and long-term responses to climate change might be constrained or have evolved as a trade-off among traits. Insect species faced with variation in temperature and water loss could cope by (1) altering behavior, (2) variation in phenology, (3) adaptation through genetic changes across generations, (4) physiological compensation within or between generations or some combination of these factors. In the event that a species is unable to mount any of these responses extinction is highly likely.

Exemplar taxa: psyllids, ants and dung beetles

Although assessments are ideally undertaken on all taxa, model taxa and pest species are generally relied upon for experimental manipulations (Satterlie *et al.*, 2009). However, taxa such as psyllids, ants and dung beetles show great promise for assessing impacts of climate change and population dynamics. It is also important to recognize the need to assess a range of species from natural populations, rather than rely purely on laboratory-reared insects, which may show a considerable variance from the same species from natural populations, and also variance among natural populations of the same species (Hoffmann *et al.*, 2001; Terblanche *et al.*, 2006; Terblanche & Chown, 2007).

Herbivorous insects that are specific feeders on a host plant species or genus will also respond to climate change directly via changes in temperature and rainfall, but also indirectly via changes in the host plant physiology and chemistry. Dominant herbivores (such as sap-sucking psyllids) that live on dominant host plant genera, such as *Acacia* (Andrew & Hughes, 2005b), can be used as important indicator groups of climatic change due to their ubiquity across most environments in Australia. The responses of these populations to a rapidly changing climate, via the measurement of key sub-lethal and lethal physiological traits (Terblanche *et al.*, 2011), including critical thermal limits, lethal temperatures, heat/cold shock survivorship, resting and active metabolic rate, heat shock protein expression, survival, loss of activity, and recovery time and temperature, among others, enable researchers to determine what impact changes in common species responses will have on other species and on the ecosystem as a whole. For example, *Acacia* species are an integral component of the Australian ecosystem, and from the authors' research on the community structure of the assemblages, we now have a good understanding of the influence of host plant phylogeny, plant traits, and climate influences herbivore community structure (Andrew & Hughes, 2005a, 2005b, 2005c, 2007, 2008; Bairstow *et al*., 2010), yet little knowledge of their physiological and behavioral responses to temperature fluctuations and host plant quality is available. When plants are stressed by weather conditions (notably, low water supply) they have been shown to increase their concentration of nitrogen in their leaf tissue and become more palatable to insect herbivores (White, 1984). For psyllids (*Cardiospina densitexta*) feeding on *Eucalyptus fasciculosa*, increases in plant stress via changes in soil water have led to increases in juvenile survival and reproductive output (White, 1969); however, changes in other key performance traits are not known. The impact of changing temperature and moisture on plant nutritional quality is of critical importance to insect herbivores, and they play a fundamental role in regulating herbivore physiology, and population structure and abundance.

Ants are among the most ubiquitous animals in terrestrial ecosystems and play crucial roles in ecosystem functioning on all continents except Antarctica (Lach *et al*., 2010). Worldwide, there are 47 subfamilies, about 539 genera and about 21 387 species and subspecies of ants; in Australia there are 13 subfamilies, 101 genera and 1275 described species and subspecies; in Africa: 16 subfamilies, 154 genera and 3558 described species and subspecies (http://www.antweb.org/allantweb.jsp). The crucial role that ants play – including predation, seed dispersal, herbivore "farming" and as a food source for other invertebrates and vertebrates – will be modified by a changing climate. Therefore understanding how ants will respond to temperature and moisture changes is of fundamental importance in understanding, and sustaining, global biodiversity. Ants play a fundamental role in providing ecosystem services and habitat engineering, thereby making them flagship taxa to start identifying responses to changing temperature and moisture regimes. Other species such as Argentine and fire ants are serious ecosystem threats, displacing native species and modifying ecosystem structure (Lach & Hooper-Bui, 2010). Recently, Diamond *et al*. (2012) found that ants from warmer, more mesic forests at low elevations are more physiologically susceptible to the negative impacts of climate warming than in temperate areas. This is a similar projection to that made for climate change impacts for insects more broadly (Deutsch *et al*., 2008). This finding goes against other studies of vertebrate ectotherms which suggest these taxa will have widespread susceptibility to climate warming (Huey *et al*., 2009; Sinervo *et al*., 2010; Clusella-Trullas *et al*., 2011). However, such findings are likely to be biased by limited biogeographic representation of selected climatic zones (see discussions in Clusella-Trullas *et al*., 2011) and/or a small range of species relative to the diversity of a given phylogenetic group (Chown & Gaston, 2000 Andrew *et al*., 2011). Further work assessing ants along climatic gradients and across continents is required to assess whether these trends hold for larger datasets.

Dung beetles are important ecosystem service providers (Nichols *et al*., 2008; Tshikae *et al*., 2008). The importance of dung beetles is such that they are also used as a model group for comparative studies and the assessment of a variety of ecological theories relevant to all animal taxa (Peck & Forsyth, 1982; Hanski & Cambefort, 1991; Chown & Steenkamp, 1996; Horgan & Fuentes, 2005; Inward *et al*., 2011). However, their responses to climate change are unclear: they may respond physiologically to climate change directly via changes in temperature and ambient moisture availability (Terblanche *et al.*, 2010), but also indirectly via changes in dung quality and chemistry, and via symbiotic interactions with other groups (such as endosymbionts). In order to understand how dung beetle populations and communities will respond to climate change, assessments of their ecology, physiology and behavior need to be carried out to enable a more robust forecasting of biological responses to environmental change (Angilletta & Sears, 2011).

Why would insect populations respond differently to temperature and rainfall changes in different areas?

The influence that weather and climate have on insect populations and structure has long been recognized in the ecological literature (Shelford, 1911; Andrewartha & Birch, 1954). Not only do large-scale regional changes have substantial impacts on insect abundance and distribution (e.g., Jeffree & Jeffree, 1996), but microclimatic variation has a considerable impact on insect responses to their environment (Hodkinson, 2003; Suggitt *et al.*, 2011). What is also of fundamental importance is the ability of insects to respond to variability and unpredictability within their environment (Kingsolver & Huey, 1998), intensity of extreme conditions and the frequency, rate of approach to and duration of particular conditions (Sinclair, 2001; Huberty & Denno, 2004; Rako *et al.*, 2007; Jumbam *et al.*, 2008; Terblanche *et al.*, 2011).

Changes in insect phenotypes with temperature are typically assessed using thermal reaction norms. Variation in the shape of the performance curve, either across a range of environmental conditions or under a small set of conditions, is often used to quantify the kinds of changes that constitute phenotypic plasticity (DeWitt & Scheiner, 2004). Responses of the reaction norms can be used to address physiological hypotheses (e.g., Deere & Chown, 2006; Terblanche & Kleynhans, 2009) and thus likely species-specific mechanisms of coping with changes in environmental conditions.

Climate predictability

The most common ways to assess climatic variables and their associated variation include differences in seasonal means, extremes (e.g., minima & maxima), temperature ranges and coefficients of variation, and time spent above or below critical thresholds. These variables are appropriate for assessing variation in conditions, but do not enable an assessment of the predictability of the conditions into the future at various spatial scales (Chown & Terblanche, 2007). Nevertheless, climate predictability is an important component of phenotypic plasticity and evolution of climate stress responses as it is an indicator of cue reliability (Tufto, 2000; Chown & Terblanche, 2007). The simplest way to assess the predictability of weather and climatic conditions is to assess climate conditions (temperature, rainfall, humidity, etc.) at a variety of temporal scales which are ecologically relevant to an individual organism (e.g., daily, weekly, seasonally,

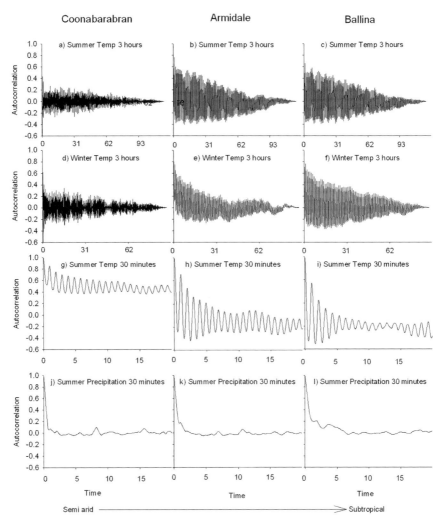

Figure 21.1. Predictability of temperature and rainfall at different temporal scales and different areas for northern NSW, 2011: Coonabarabran (31°27′S, 149°27′E, mean max temp 23.7°C, mean min temp 7.4°C, mean annual rainfall 753.1 mm), Armidale (30°29′S, 151°38′E, mean max temp 19.5°C, mean min temp 7.3°C, mean annual rainfall 796.8 mm), Ballina (28°51′S, 153°31′E, mean max temp 24.5°C, mean min temp 14.2°C, mean annual rainfall 1739.4 mm). Readings closer to 0 indicate lower predictability from one reading to the next.

yearly) (Kingsolver & Huey, 1998; Chown & Terblanche, 2007). For example, Figure 21.1 shows meteorological data collected from three climatic regions in northern NSW, Australia (Ballina, a coastal subtropical area, Armidale, a high-altitude temperate area, and Coonabarabran, a semiarid area). Predictability of temperature based on autocorrelation plots was substantially different between the regions across different temporal scales (Figure 21.1). At shorter time intervals, temperature is more predictable, particularly in the more inland sites.

Species responses to climate unpredictability

There are current shortcomings in our understanding of responses to climate unpredictability by insects for high and low temperatures. When low-temperature events are unpredictable insects may employ a range of strategies to cope, depending on the severity and the duration of the event. Rapid cold hardening (RCH) may provide a mechanism enabling activity despite sudden low-temperature periods (Kelty & Lee, 2001; Terblanche *et al.*, 2007a, 2007b). Alternatively, behavioral adjustments may be sufficient to compensate for low and high temperatures over the short and medium terms. If the plastic RCH trait evolution is costly, perhaps increasing basal resistance at the expense of plasticity can be expected (Terblanche *et al.*, 2007b). How extensive RCH is in insects, and whether any consistent environment-related pattern characterizes the expression of RCH, is, however, generally poorly understood (Chown & Nicolson, 2004). At present, approximately 30 terrestrial arthropod species are known to show RCH and only a handful of species are unable to produce a cold-induced RCH response (Lee & Denlinger, 2010). Rapid cold hardening might be common in temperate species incapable of surviving freezing events and which also dwell in unpredictable environments ("environmental predictability hypothesis"). Such environments, which seem to be especially characteristic of the temperate to high latitudes of the Southern Hemisphere (Chown *et al.*, 2004), are thought to promote strategies, including moderate freezing tolerance and RCH in individuals which are not freezing tolerant, that enable insects to cope with unexpected low-temperature periods (Sinclair *et al.*, 2003). What is further unexplored is the extent to which climate change and associated climatic predictability changes may modify geographic ranges specifically due to changes in phenotypic plasticity (Chown & Terblanche, 2007; Chown *et al.*, 2010).

Acclimation (in the laboratory) and acclimatization (in the field) can be used to assess intra-individual variability in a changing environment (Terblanche *et al.*, 2008, 2011). Moreover, acclimation responses are a form of phenotypic plasticity and are likely to play a major role in dealing with variation in climate, at least over the short term. However, over longer timescales, plastic responses may be costly, and thus short-term survival could be traded off against long-term reproductive output (e.g., Marshall & Sinclair, 2010; Basson *et al.*, 2012). Therefore, acclimation responses, which are typically assessed over relatively short timescales, need to be examined at short and long timescales to better comprehend the kinds of population responses expected under climate change scenarios.

Water loss

Insects show significant physiological and behavioral adaptations to changes in moisture availability in terrestrial environments. For example, insects from dry environments may lose water more slowly, retain more body water (possibly through increased size and resorption), or withstand desiccation for longer relative to populations or species living in wetter habitats (reviewed in Hadley, 1994; Chown *et al.*, 2011). Water loss rates are highest in areas where precipitation is also highest (Addo-Bediako *et al.*, 2001; Kleynhans & Terblanche, 2009). However, the generality of these responses is not well established, and the relative importance of behavioral as opposed to physiological modes of compensation is not understood in many insect taxa, either among species or among populations within a species. Further research needs to be conducted on insect water loss responses at sites along environmental gradients assessing both genetic (by using species or populations), environmental (by incorporating spatial variation and acclimation regimes) and gene × environment effects (species/population responses to acclimation/ seasonal variation).

Complex life cycles

The ecological and evolutionary responses to climate change by holometabolous insects (i.e., those that have distinct life stages: larvae, pupae, adults) may be highly variable as their life cycles are dependent on different climatic conditions, food sources and bio-logical interactions (Kingsolver *et al.*, 2011). For example Kingsolver *et al.* (2011) have shown that *Manduca* individuals (Lepidoptera: Sphingidae) experience highly variable conditions throughout their life – larvae more likely to experience extremely high temperatures and humidity compared to egg or pupae life stages.

   However, any insect that has a complex life cycle, which also includes hemimetabo-lous insects which require different habitats to fully complete their life cycles, is at a higher risk of climatic change variation as some life stages may be impacted on more strongly than others (Kingsolver *et al.*, 2011). This may include dragonflies (Odonata) which require aquatic ecosystems for their larval stages, and cicadas (Hemiptera) which require tree roots pre-adult dispersal, among others. It may also include animals such as locusts (Orthoptera) which require different temperatures based on their nutritional state to maximize efficiency (Coggan *et al.*, 2011) or carry out long-distance dispersal across a variety of climatic zones (Farrow, 1977; Woodman, 2012).

Understanding the responses of common species to climate change

Those species that are rare and have a restricted distribution may be highly vulnerable to human-induced extinction. These species have had a substantial amount of time and effort put into assessing their conservation value and biology (Gaston, 2008). However, the responses of common species to climate change within natural systems are still poorly understood. It is anticipated that because they are common they are resilient and have a high adaptive capability to rapid change. However, extreme ecological changes can occur when the populations of common species go through a rapid and severe fluctuation (Gaston & Fuller, 2008).

   An example of assessing common species' responses to climate change has been conducted by Andrew *et al.* (in preparation). This study focused on the thermal tolerances of meat ants *Iridomyrmex purpureus*, from samples of laboratory-acclimated ants, among seasons (summer and winter), and of winter field-fresh ants collected from a nest in Armidale, NSW, Australia. Running speeds were also assessed both in the field and in the laboratory between seasons to assess physiological plasticity of thermal traits. *Iridomyrmex* thermal tolerance assessments found that 25% of ants survived when exposed to upper temperatures of 45.8°C and lower temperatures of −8.2°C. At cooler temperatures, ant survival for 5 minutes at −18° C, −16°C, and −14°C was significantly higher for winter-acclimated ants compared to summer-acclimated samples. At −12°C exposure for 10, 15 and 30 minutes, winter field-fresh samples had lower survival compared to the two acclimated treatments. At −10°C and −8°C, summer samples had a lower survival compared to the winter treatments at 120 and 240 minutes exposure. For higher temperatures, summer ants had lower survival when exposed to lower temperatures (42°C and 43°C) for 240 minutes compared to the winter treatments. Winter field-fresh samples had a lower survival at higher temperatures (44°C, 45°C, 46°C and 48°C) compared to the two acclimated treatments for 10, 15 and 30 minutes. Winter-acclimated ants also had a higher survival at 50°C exposure for 5 minutes compared to the summer and winter field-fresh treatments. One of the most obvious and important results obtained was the substantive differences in ant survival between summer and winter, the differences of which are not directly related to laboratory-induced phenotypic plasticity. Ants taken directly from the field in winter are compromising their low-temperature survival, in comparison to the laboratory-acclimated winter ants and the summer samples. Winter field-fresh ants also do poorly at lower relative to higher temperatures. This work is being taken further with the assessment of heat shock proteins (HSPs) in ants to understand whether foraging at higher temperatures up-regulates HSPs and whether the latter may compromise ant winter survival.

Conclusion

The impact of climate change on the population dynamics of insects is still in its infancy. Environmental gradients are a useful tool for understanding the role of current climate in structuring insect assemblages and population dynamics (Harrison, 1993) and have been used as surrogates for predicting responses to future climate change (Fleishman *et al.*, 2000; Andrew & Hughes, 2004, 2005a, 2005b, 2005c, 2007; Bairstow *et al.*, 2010). Populations of species distributed across large environmental gradients may respond differently to climatic change due to differences in physiological tolerances and behavioral mechanisms when put under a varied regime of temperature and water availability, and altered trophic interactions (Davis *et al.*, 1998). This may result in significant changes in the structure and composition of present-day assemblages. Utilizing dominant, abundant and functionally important species within assemblages along clear climatic (rainfall and temperature) gradients as a basis for predicting insect responses to climate change is a clear method that can be applied internationally to enable trends to be assessed in a more strategic and generalizable fashion.

References

AAS (2010). *The Science of Climate Change: Questions and Answers*. Canberra: Australian Academy of Science.

Addo-Bediako, A., Chown, S. L., & Gaston, K. J. (2001). Revisiting water loss in insects: a large scale view. *Journal of Insect Physiology*, **47**, 1377–1388.

Andrew, N. R., & Hughes, L. (2004). Species diversity and structure of phytophagous beetle assemblages along a latitudinal gradient: predicting the potential impacts of climate change. *Ecological Entomology*, **29**, 527–542.

Andrew, N. R., & Hughes L. (2005a). Arthropod community structure along a latitudinal gradient: implications for future impacts of climate change. *Austral Ecology*, **30**, 281–297.

Andrew, N. R., & Hughes, L. (2005b). Diversity and assemblage structure of phytophagous Hemiptera along a latitudinal gradient: predicting the potential impacts of climate change. *Global Ecology and Biogeography*, **14**, 249–262.

Andrew, N. R., & Hughes, L. (2005c). Herbivore damage along a latitudinal gradient: relative impacts of different feeding guilds. *Oikos*, **108**, 176–182.

Andrew, N. R., & Hughes, L. (2007). Potential host colonization by insect herbivores in a warmer climate: a transplant experiment. *Global Change Biology*, **13**, 1539–1549.

Andrew, N. R., & Hughes, L. (2008). Abundance-body mass relationships among insects along a latitudinal gradient. *Austral Ecology*, **33**, 253–260.

Andrew, N. R., & Terblanche, J. S. (in press). Insects. In J. Salinger (Ed.), *Climate of Change: Living in a Warmer World* (Ch. 16, pp. 142–149). Auckland: David Bateman.

Andrew, N. R., Hart, R. A., & Terblanche, J. S. (2011). Limited plasticity of low temperature tolerance in an Australian cantharid beetle *Chauliognathus lugubris*. *Physiological Entomology*, **36**, 385–391.

Andrewartha, H. G., & Birch, L. C. (1954). *The Distribution and Abundance of Animals*. Chicago, IL: University of Chicago Press.

Angilletta, M. J., & Sears, M. W. (2011). Coordinating theoretical and empirical efforts to understand the linkages between organisms and environments. *Integrative and Comparative Biology*, **51**, 653–661.

Bairstow, K. A., Clarke, K. L., Mcgeoch, M. A., & Andrew, N. R. (2010). Leaf miner and plant galler species richness on Acacia: relative importance of plant traits and climate. *Oecologia*, **163**, 437–448.

Basson, C. H., Nyamukondiwa, C., & Terblanche, J. S. (2012). Fitness costs of rapid cold-hardening in Ceratitis capitata. *Evolution*, **66**, 296–304.

Chown, S. L., & Gaston, K. J. (2000). Areas, cradles and museums: the latitudinal gradient in species richness. *Trends in Ecology & Evolution*, **15**, 311–315.

Chown, S. L., & Nicolson, S. W. (2004). *Insect Physiological Ecology: Mechanisms and Patterns*. Oxford: Oxford University Press.

Chown, S. L., & Steenkamp, H. E. (1996). Body size and abundance in a dung beetle assemblage: optimal mass and the role of transients. *African Entomology*, **4**, 203–212.

Chown, S. L., & Terblanche, J. S. (2007). Physiological diversity in insects: ecological and evolutionary contexts. *Advances in Insect Physiology*, **33**, 50–152.

Chown, S. L., Sinclair, B. J., Leinaas, H. P., & Gaston, K. J. (2004). Hemispheric asymmetries in biodiversity – a serious matter for ecology. *PLoS Biology*, **2**, 1701–1707.

Chown, S. L., Gaston, K., Van Kleunen, M., & Clusella-Trullas, S. (2010). Population responses within a landscape matrix: a macrophysiological approach to understanding climate change impacts. *Evolutionary Ecology*, **24**, 601–616.

Chown, S. L., Sørensen, J. G., & Terblanche, J. S. (2011). Water loss in insects: an environmental change perspective. *Journal of Insect Physiology*, **57**, 1070–1084.

Clusella-Trullas, S., Blackburn, T. M., & Chown, S. L. (2011). Climatic predictors of temperature performance curve parameters in ectotherms imply complex responses to climate change. *The American Naturalist*, **177**, 738–751.

Coggan, N., Clissold, F. J., & Simpson, S. J. (2011). Locusts use dynamic thermoregulatory behaviour to optimize nutritional outcomes. *Proceedings of the Royal Society of London B*, **278**, 2745–2752.

CSIRO (2007). *Climate Change in Australia: Observed Changes and Projections*. www.climatechangeinaustralia.gov.au.

Davis, A. J., Lawton, J. H., Shorrocks, B., & Jenkinson, L. S. (1998). Individualistic species responses invalidate simple physiological models of community dynamics under global environmental change. *Journal of Animal Ecology*, **67**, 600–612.

Deere, J. A., & Chown, S. L. (2006). Testing the beneficial acclimation hypothesis and its alternatives for locomotor performance. *The American Naturalist*, **168**, 630–644.

Deutsch, C. A., Tewksbury, J. J., Huey, R. B., *et al.* (2008). Impacts of climate warming on terrestrial ectotherms across latitude. *Proceedings of the National Academy of Sciences of the USA*, **105**, 6668–6672.

Dewitt, T., & Scheiner, S. (Eds.) (2004). *Plasticity. Functional and Conceptual Approaches*. Oxford: Oxford University Press.

Diamond, S. E., Magdalena Sorger, D., Hulcr, J., *et al.* (2012). Who likes it hot? A global analysis of the climatic, ecological, and evolutionary determinants of warming tolerance in ants. *Global Change Biology*, **18**, 448–456.

Farrow, R. A. (1977). Maturation and fecundity of the spur-throated locust, *Austracris guttulosa* (Walker), in New South Wales during the 1974/75 plague. *Australian Journal of Entomology*, **16**, 27–39.

Fleishman, E., Fay, J. P., & Murphy, D. D. (2000). Upsides and downsides: contrasting topographic gradients in species richness and associated scenarios for climate change. *Journal of Biogeography*, **27**, 1209–1219.

Gaston, K. J. (2008). Biodiversity and extinction: the importance of being common. *Progress in Physical Geography*, **32**, 73–79.

Gaston, K. J., & Fuller, R. A. (2008). Commonness, population depletion and conservation biology. *Trends in Ecology & Evolution*, **23**, 14–19.

Hadley, N. F. (1994). *Water Relations of Terrestrial Arthropods*: New York: Academic Press.

Hanski, I., & Cambefort, Y. (Eds.) (1991). *Dung Beetle Ecology*. Princeton, NJ: Princeton University Press.

Harrison, S. (1993). Species diversity, spatial scale, and global change. In P. M. Kareiva, J. G. Kingsolver & R. B. Huey (Eds.), *Biotic Interactions and Global Change* (pp. 388–401). Sunderland, MA: Sinauer Associates.

Hodkinson, I. D. (2003). Metabolic cold adaptation in arthropods: a smaller-scale perspective. *Functional Ecology*, **17**, 562–567.

Hoffmann, A. A., Hallas, R., Sinclair, C., & Mitrovski, P. (2001). Levels of variation in stress resistance in *Drosophila* among strains, local populations, and geographic regions: patterns for desiccation, starvation, cold resistance, and associated traits. *Evolution*, **55**, 1621–1630.

Horgan, F. G., & Fuentes, R. C. (2005). Asymmetrical competition between Neotropical dung beetles and its consequences for assemblage structure. *Ecological Entomology*, **30**, 182–193.

Huberty, A. F., & Denno, R. F. (2004). Plant water stress and its consequences for herbivorous insects: a new synthesis. *Ecology*, **85**, 1383–1398.

Huey, R. B., Deutsch, C. A., Tewksbury, J. J., *et al.* (2009). Why tropical forest lizards are vulnerable to climate warming. *Proceedings of the Royal Society of London B*, **276**, 1939–1948.

Inward, D. J. G., Davies, R. G., Pergande, C., Denham, A. J., & Vogler, A. P. (2011). Local and regional ecological morphology of dung beetle assemblages across four biogeographic regions. *Journal of Biogeography*, **38**, 1668–1682.

IPCC (2007). Summary for policymakers. In S. Solomon, D. Qin, M. Manning, *et al.* (Eds.), *Climate Change 2007: The Physical Science Basis. Contribution of Working Group I to the Fourth Assessment Report of the Intergovernmental Panel on Climate Change*. Cambridge: Cambridge University Press.

Jeffree, C. E., & Jeffree, E. P. (1996). Redistribution of the potential geographical range of Mistletoe and Colorado Beetle in Europe in response to the temperature component of climate change. *Functional Ecology*, **10**, 562–577.

Jumbam, K. R., Jackson, S., Terblanche, J. S., Mcgeoch, M. A., & Chown, S. L. (2008). Acclimation effects on critical and lethal thermal limits of workers of the Argentine ant, Linepithema humile. *Journal of Insect Physiology*, **54**, 1008–1014.

Kelty, J., & Lee, R., Jr (2001). Rapid cold-hardening of *Drosophila melanogaster* (Diptera: Drosophiladae) during ecologically based thermoperiodic cycles. *Journal of Experimental Biology*, **204**, 1659–1666.

Kingsolver, J. G., & Huey, R. B. (1998). Evolutionary analyses of morphological and physiological plasticity in thermally variable environments. *American Zoologist*, **38**, 545–560.

Kingsolver, J. G., Arthur Woods, H., Buckley, L. B., *et al.* (2011). Complex life cycles and the responses of insects to climate change. *Integrative and Comparative Biology*, **51**, 719–732.

Kleynhans, E., & Terblanche, J. S. (2009). The evolution of water balance in Glossina (Diptera: Glossinidae): correlations with climate. *Biology Letters*, **5**, 93–96.

Lach, L., & Hooper-Bui, L. M. (2010). Consequences of ant invasions. In L. Lach, C. L. Parr & K. L. Abbott (Eds.), *Ant Ecology* (pp. 261–286). New York: Oxford University Press.

Lach, L., Parr, C. L., & Abbott, K. L. (2010). *Ant Ecology*. New York: Oxford University Press.

Lee, R. E., & Denlinger, D. L. (2010). Rapid cold-hardening: ecological significance and underpinning mechanisms. In D. L. Denlinger & R. E. Lee (Eds.), *Low Temperature Biology of Insects* (pp. 35–58). Cambridge: Cambridge University Press.

Loeschcke, V., & Hoffman, A. A. (2007). Consequences of heat hardening on a field fitness component in *Drosophila* depend on environmental temperature. *The American Naturalist*, **169**, 175–183.

Marshall, K. E., & Sinclair, B. J. (2010). Repeated stress exposure results in a survival–reproduction trade-off in Drosophila melanogaster. *Proceedings of the Royal Society of London B*, **277**, 963–969.

Nichols, E., Spector, S., Louzada, J., *et al.* (2008). Ecological functions and ecosystem services provided by Scarabaeinae dung beetles. *Biological Conservation*, **141**, 1461–1474.

Peck, S. B., & Forsyth, A. (1982). Composition, structure, and competitive behaviour in a guild of Ecuadorian rain forest dung beetles (Coleoptera, Scarabaeidae). *Canadian Journal of Zoology*, **60**, 1624–1634.

Rahmstorf, S., Cazenave, A., Church, J. A., *et al.* (2007). Recent climate observations compared to projections. *Science*, **316**, 709.

Rako, L., Blacket, M. J., Mckechnie, S. W., & Hoffmann, A. A. (2007). Candidate genes and thermal phenotypes: identifying ecologically important genetic variation for thermotolerance in the Australian *Drosophila melanogaster* cline. *Molecular Ecology*, **16**, 2948–2957.

Satterlie, R. A., Pearse, J. S., & Sebens, K. P. (2009). The black box, the creature from the Black Lagoon, August Krogh, and the dominant animal. *Integrative and Comparative Biology*, **49**, 89–92.

Shelford, V. E. (1911). Physiological animal geography. *Journal of Morphology*, **22**, 551–618.

Sinclair, B. J. (2001). Field ecology of freeze tolerance: interannual variation in cooling rates, freeze-thaw and thermal stress in the microhabitat of the alpine cockroach Celatoblatta quinque-maculata. *Oikos*, **93**, 286–293.

Sinclair, B. J., Vernon, P., Klok, C. J., & Chown, S. L. (2003). Insects at low temperatures: an ecological perspective. *Trends in Ecology & Evolution*, **18**, 257–262.

Sinervo, B., Méndez-De-La-Cruz, F., Miles, D. B., *et al.* (2010). Erosion of lizard diversity by climate change and altered thermal niches. *Science*, **328**, 894–899.

Steffen, W., Burbidge, A., Hughes, L., *et al.* (2009). *Australia's Biodiversity and Climate Change: Summary for Policy Makers*. Canberra: Australian Government.

Suggitt, A. J., Gillingham, P. K., Hill, J. K., *et al.* (2011). Habitat microclimates drive fine-scale variation in extreme temperatures. *Oikos*, **120**, 1–8.

Terblanche, J. S., & Chown, S. L. (2007). Factory flies are not equal to wild flies. *Science*, **317**, 1678.

Terblanche, J. S., & Kleynhans, E. (2009). Phenotypic plasticity of desiccation resistance in *Glossina* puparia: are there ecotype constraints on acclimation responses? *Journal of Evolutionary Biology*, **22**, 1636–1648.

Terblanche, J. S., Klok, C. J., Krafsur, E. S., & Chown, S. L. (2006). Phenotypic plasticity and geographic variation in thermal tolerance and water loss of the tsetse Glossina pallidipes (Diptera: Glossinidae): implications for distribution modelling. *American Journal of Tropical Medicine and Hygiene*, **74**, 786–794.

Terblanche, J. S., Deere, J. A., Clusella-Trullas, S., Janion, C., & Chown, S. L. (2007a). Critical thermal limits depend on methodological context. *Proceedings of the Royal Society of London B*, **274**, 2935–2943.

Terblanche, J. S., Marais, E., & Chown, S. L. (2007b). Stage-related variation in rapid cold hard-ening as a test of the environmental predictability hypothesis. *Journal of Insect Physiology*, **53**, 455–462.

Terblanche, J. S., Clusella-Trullas, S., Deere, J. A., & Chown, S. L. (2008). Thermal tolerance in a south-east African population of the tsetse fly Glossina pallidipes (Diptera, Glossinidae): implications for forecasting climate change impacts. *Journal of Insect Physiology*, **54**, 114–127.

Terblanche, J. S., Clusella-Trullas, S., & Chown, S. L. (2010). Phenotypic plasticity of gas exchange pattern and water loss in Scarabaeus spretus (Coleóptera: Scarabaeidae): deconstruct-ing the basis for metabolic rate variation. *Journal of Experimental Biology*, **213**, 2940–2949.

Terblanche, J. S., Hoffmann, A. A., Mitchell, K. A., *et al.* (2011). Ecologically relevant measures of tolerance to potentially lethal temperatures. *The Journal of Experimental Biology*, **214**, 3713–3725.

Tshikae, B. P., Davis, A. L. V., & Scholtz, C. H. (2008). Trophic associations of a dung beetle assemblage (Scarabaeidae: Scarabaeinae) in a woodland savanna of Botswana. *Environmental Entomology*, **37**, 431–441.

Tufto, J. (2000). The evolution of plasticity and nonplastic spatial and temporal adaptations in the presence of imperfect environmental cues. *The American Naturalist*, **156**, 121–130.

Walther, G.-R., Post, E., Convey, P., *et al*. (2002). Ecological responses to recent climate change. *Nature*, **416**, 389–395.

White, T. C. R. (1969). An index to measure weather-induced stress of trees associated with outbreaks of psyllids in Australia. *Ecology*, **50**, 905–909.

White, T. C. R. (1984). The abundance of invertebrate herbivores in relation to the availability of nitrogen in stressed food plants. *Oecologia*, **63**, 90–105.

Woodman, J. D. (2012). Cold tolerance of the Australian spur-throated locust, *Austracris guttulosa*. *Journal of Insect Physiology*, **58**, 384–390.

22 The futures of coral reefs

Peter F. Sale

Coral reefs are severely threatened and their future looks dire

Coral reefs, as we knew them in the 1970s, are likely to have disappeared entirely from the planet by 2050, if current trends in human environmental impacts continue. This claim is predicated principally on continued growth in atmospheric concentrations of greenhouse gases and continued ocean acidification (Hoegh-Guldberg *et al.*, 2007; Veron *et al.*, 2009; Sale, 2011). Indeed, late in 2011, it is difficult to see how we will mobilize quickly enough to alter our behavior sufficiently, in time to save them, although several authors now point to the spatial heterogeneity with which reefs are degrading, and suggest that in some regions reefs may survive through 2100 (Hughes *et al.*, 2010; Anthony *et al.*, 2011; Edwards *et al.*, 2011; Hoeke *et al.*, 2011). I believe the chance of such survival is vanishingly small. What will be left is eroding limestone benches, dominated by macroalgae, and with small, isolated coral colonies. In this chapter, I discuss the various possible futures for coral reefs, and the reasons for my dire prediction. Reef ecosystems suffer because their keystone species are proving particularly susceptible to certain of our environmental impacts. The fact that they could disappear within 40 years is testament to the limited capacity for homeostasis among natural ecosystems and to the capacity of humanity to alter fundamental properties of the planetary environment.

Coral reefs: fragile, transitory ecosystems of the ocean-atmosphere boundary zone

Humans have been damaging coral reef ecosystems around the world almost since humans and reefs first came into contact. There are several reasons for this (Sale, 2008). They all concern the particular dependence of the reef ecosystem on the survival and health of one group of ecologically delicate species, the corals themselves. Coral reefs are one of the few ecosystems in which major components of the physical structure of the system are produced by the organisms themselves (rainforests are approximate analogs). On reefs, the rocky structure with its complex topography is the result of calcification by a broad range of organisms but including in particular the corals themselves. Erosional forces, some also

The Balance of Nature and Human Impact, ed. Klaus Rohde. Published by Cambridge University Press.
© Cambridge University Press 2013.

biogenic, modify the structure built up, and prevailing patterns of water flow further shape the resulting reef, so that the coral reef is a dynamic equilibrium between processes of growth in which new rock is created as calcified skeletons, and processes of erosion in which the rocky structure is broken down by a variety of bioeroders, by storms, by changes in sea level, and by the routine effects of wave action and currents.

Bioeroders are numerous and diverse in coral reef systems; they range from minute sponges and worms to 1-m-long parrotfishes, and their rasping, burrowing and dissolving activities gradually reduce consolidated limestone rock derived from calcified skeletal elements of various species into sand (Glynn, 1997). Cyclonic storms break up living coral, shift rubble and sediments, scour and abrade corals and other sessile species and generally wear down coral reefs while also throwing some rock and sediment up onto reef flats to add to or become new islands up to 3–4 m elevation above high tide levels. Changes in sea level, whether due to global change in mean sea level or to subsidence or elevation of underlying rock, can lift a reef above sea level, resulting in rapid death, lower it modestly (meters) allowing it to flourish, or take it deeper at rates too quick for reef building processes to compensate, with the result that it is drowned. Normal wave action and currents have continual low-level erosional impacts on a reef. Typically, there is no rock present on a coral reef that is not biogenic. On some reefs this biogenic limestone is hundreds or thousands of meters thick, having been built up progressively as an underlying rocky substratum slowly subsided over geologically long stretches of time. Yet even such reefs function as a living ecosystem limited to waters within 100 m or so of the ocean surface. Reefs are a dynamic equilibrium played out at the sea surface (Glynn, 1997).

Hermatypic (reef building) corals have always been limited in distribution by their narrow tolerances for temperature, salinity, water clarity, pollution and depth. The modern corals (order Scleractinia) have been present since the early Mesozoic and are likely to be derived from earlier rugose corals dominant in the Paleozoic (Veron *et al.*, 1996). Most Scleractinia contain single-celled algal zooxanthellae within their tissues in a symbiosis that greatly increases their capacity to calcify. Zooxanthellate Scleractinia are necessarily restricted to shallow water by the photosynthetic requirements of the algae, but some azooxanthellate forms occur in deep water where they build very slowly growing reefs. This chapter deals with the shallow-water, zooxanthellate, reef-building corals.

As well as being restricted to shallow water by the need for light for photosynthesis, these corals possess narrow tolerances to temperature, to salinity, and to nutrients, sediments and other pollutants. They are limited to the tropics by temperature, and to clear, oceanic, low-nutrient waters. Further, because they are restricted to shallow waters, they typically occur relatively close to coastlines – coastlines that are increasingly occupied by human settlements. Not surprisingly, human activities have frequently impacted these relatively delicate species and, through impacts on them, have disrupted the functioning of coral reef ecosystems.

Human impacts on coral reefs

Coral reefs are of considerable economic value to those nations that have them off their shores. They provide direct economic value through fisheries and tourism, and indirect (and rarely measured) value through services such as the storm protection they provide to shorelines behind them. The Great Barrier Reef has been valued as providing over $6 billion annually to GDP through tourism alone (Fenton *et al.*, 2007), and its total present value was recently calculated at $51.4 billion (Oxford Economics, 2009). Most Caribbean countries receive at least 50% of their GDP from coastal activities, of which tourism and fisheries (both strongly reef-dependent) are the major components. (While many tourists seek out beaches and perhaps do not even bother to take a half-day snorkel excursion to see a reef, the reefs provide the white carbonate sand that lines the shore, and protection to the tourism infrastructure built there – the tourism is reef-dependent.)

Despite this evident economic value, human societies have frequently failed to manage coastal waters in ways that promote the persistence of coral reefs, and sometimes their management directly favors reef decline. Most prominent among the failures are impacts from inappropriate coastal development, pollution and inappropriate fishing.

Coastal development is a normal part of expanding coastal populations, and is expected to continue as coastal populations continue to grow more quickly than inland ones. At present, 14 of the 17 largest cities worldwide are coastal and many of these are tropical (UNEP, 2007). Inappropriate coastal development includes change to coastlines that removes inshore nursery habitats such as seagrass beds and mangrove forests that are vital to the lives of many reef species. Dredging, coastal "hardening" and "reclamation" that alter patterns of water flow or turbidity in the vicinity of reefs are also inappropriate and damaging (Sale, 2008). Of these, the loss of mangrove habitat may be particularly problematic because these forests also provide major protection against storms for coastal communities. The elimination of nursery habitat is also facilitated when development inshore from mangroves or seagrass habitat in backreef lagoons makes it impossible for these shallow-water systems to move back upshore as sea levels rise. This is likely to become a growing problem over the next decades as sea levels rise due to climate change.

The prevailing pattern of domestic and industrial waste disposal worldwide is by dilution in bodies of water, usually after some initial treatment in a sewage plant. All inland bodies of water ultimately flow to the coastal ocean, and their municipal waste with its nutrients, heavy metals and advanced chemicals (including pharmaceuticals, fertilizers, health and beauty products, as well as other industrial by-products) can damage nearby reefs. As well, there is the substantial, untreated runoff of agricultural wastes and products from erosion of improperly cleared land. Pollution of the coastal ocean has been problematic in specific locations for several hundred years, but with the growth in our populations and economies it is becoming truly widespread. There are now over 400 "dead zones" around the continents' shores – enormous regions where the extent of pollutant load has been sufficient to make waters anoxic for substantial periods, with the result that few fish or other species survive there (Diaz & Rosenberg, 2008). The second-largest such dead zone grows annually in the western Gulf of Mexico, from the Mississippi delta to the Texas coast, covering an area typically of $17\,000\,km^2$ (maximum $22\,000\,km^2$) during late summer; in the winter it shrinks back to as little as $5000\,km^2$ or so. Coral reefs in the vicinity of such pollution show evident impact, and reefs offshore from many ports and harbors worldwide have also declined substantially over the last 40 years.

Rectifying incidents of reef pollution can prove very difficult. Kaneohe Bay, on the Hawaiian island of Oahu is a large embayment largely closed to the open ocean by a well-developed barrier reef. Within the bay are numerous patch reefs that supported rich, coral-dominated systems until the 1970s. Coastal development surrounding the bay, including installation in the 1950s of two sewage outfalls that delivered treated effluent and run-off into the bay, resulted in a gradual increase in nutrients and in the early 1970s to a massive bloom of macroalgae (principally *Dictyosphaerium* sp.) which smothered corals (Stimson & Conklin, 2008). Extending the outfalls to deeper off-shore waters in 1978 did reduce the delivery of nutrients and slowly lowered nitrate and phosphate concentrations in the water column, but the abundant algae persisted until recently. A massive algal die-off occurred in 2006, due apparently to an unprecedented period of continuous rains, and algae have yet to recover. However, corals remain at low abundance and it is unclear whether they or the algae will ultimately reclaim the vacated space (Stimson & Conklin, 2008).

Coral reefs close to human settlements provide a diverse array of abundant fish and other edible species, many of which are relatively easy to capture in shallow clear reef waters. Reef fisheries provide both protein-rich food for coastal communities and high-value fishery products that can sustain a local economy. Perhaps because reefs are patchy in distribution and their non-sessile occupants are often strongly site-attached, local overfishing has been a long-term and frequent problem. Reef fisheries can also be depressed due to inappropriate coastal development which removes mangrove forests, sea grass beds and salt marshes. These habitats constitute important inshore nursery grounds for many reef species, which decline in numbers as access to suitable nurseries is removed. Further, while the overexploitation of reef fisheries is often treated as a "fisheries" problem requiring efforts to limit catch in order to restore populations, overfishing also has important ecosystem consequences in coral reef systems. First, some fishing methods, most notably dynamite fishing, trawl fishing, and fishing by use of cyanide and other toxic chemicals, have multiple deleterious impacts on other components of the system, leading to decline in system capacity and integrity. Second, reduction or removal of certain critical species groups, most notably the larger herbivores characteristic of reef fish faunas (particularly the parrotfishes, Scaridae), can release macroscopic algal species from predation and enable these to out-compete corals for space and light (Mumby & Steneck, 2008; Mumby, 2009). Such depression of herbivory due to overfishing, coupled with the mass die-off of another herbivore, *Diadema antillarum*, in 1983 (Lessios *et al.*, 1984), is believed to be the primary reason the reefs of northern Jamaica are now predominantly algal-covered rubble banks, while they were rich, coral-dominated slopes in the 1970s (Mumby & Steneck, 2008).

The deleterious impacts on reefs due to coastal development, pollution and fishing are all effectively local, and can be managed locally. Indeed, while there are many degraded reefs around the world, there are numerous examples of coral reefs where effective management programs have been put in place and these impacts have been minimized or eliminated. And truly remote reefs, such as Kingman Reef in the Line Islands (Sandin *et al.*, 2008), which have largely escaped human impacts can be considered pristine examples of what reefs might be. If these local impacts were our only negative impacts on reefs their future would be promising.

Unfortunately, that is not the case. Our releases of greenhouse gases during the twentieth and twenty-first centuries, and particularly our releases of CO$_2$, are now having serious global impacts on coral reefs. These global impacts, in combination with the local stressors, threaten to eliminate reefs by mid-century.

The impacts of CO$_2$ on coral reefs

Approximately one-third of the CO$_2$ released into the atmosphere by human activities dissolves into surface waters of the world's oceans. Were this not the case, climatic warming would be even more rapid than it currently is. This dissolved CO$_2$ dissociates, producing carbonate ions and a lowered pH. Average oceanic pH has been reduced approximately 0.1 pH unit since 1750 – a relatively large shift in acidity given the logarithmic pH scale. Corals are impacted by both the warming and the acidification.

Because corals live close to the top of rather narrow temperature tolerance ranges, global warming of sea surfaces interacts with seasonal temperature variation and other oscillations, most notably the ENSO – El Niño, southern oscillation – which generates marked fluctuations in SST across the Pacific and, when strong enough, the Indian and Atlantic equatorial seas. An El Niño during a warm summer can now push temperatures high enough that corals are stressed to the extent that they "bleach", expelling their symbiotic zooxanthellae. If warm conditions persist for 3 weeks or so, the stress is sufficient that many of the corals die.

While the phenomenon of bleaching has long been known, mass bleaching, in which nearly all corals over hectares of reef bleach together, was first described in 1983 and appears not to have occurred prior to 1979–1980 (Glynn, 1991). In 1983, corals bleached at a number of sites in the Galapagos Archipelago and along the Pacific coast of Panamá, and there was widespread mortality (Glynn, 1983, 1984). Since 1983, mass bleaching has been reported a number of times and has become more frequent. The most extreme case was in 1998 when corals bleached at numerous sites distributed around the world (Wilkinson, 1999).

Mortality following a mass bleaching event typically ranges from 50% to 95% of all coral present, and recovery, through the growth and reproduction of those remnant portions of colonies that survive, can be very slow. Baker and colleagues (Baker *et al.*, 2008) examined the extent of recovery at a number of sites throughout the world over periods up to a decade in length. Their results were mixed. At 46 of 58 Indian Ocean sites from East Africa to Australia there was some measurable recovery of coral cover in the four or so years between bleaching in 1998 and subsequent re-measuring, but in the Caribbean, 16 of 17 sites showed further decline in cover 4 to 5 years after particular bleaching events. Sites in the Eastern Pacific, the West Pacific and the Arabian Gulf showed no clear trend, and overall it is now recognized that the world has lost about 20% of live coral cover since 1980.

Acidification, the other impact from CO_2 release, affects corals by impeding the process of calcification they use in building their skeletons (Hoegh-Guldberg *et al.*, 2007). The chemical reaction in which calcium and carbonate ions are combined to form calcium carbonate crystals is more energetically demanding in colder water and in more acidic water. Changes in pH that have occurred in tropical surface waters are now sufficient to have caused measurable decreases in the rate of growth, and in the density of skeletons produced, in coral species that have been examined. D'eath *et al.* (2009) showed that colonies of massive *Porites* spp. on the Great Barrier Reef grew 14% more slowly during the 1990s than during the 30 immediately prior years. Their study, which used samples from a wide range of sites along the Great Barrier Reef, and samples from young and old colonies showed clear trends, and while the study was only correlative, acidification seemed an appropriate cause. About the same time, Tanzil *et al.* (2009), in a smaller study, showed very similar results for corals at Phuket, Thailand, and subsequent work elsewhere has confirmed these trends while revealing important interspecific variation (Manzello, 2010; Friedrich *et al.*, 2012).

Assuming current trends continue (and the Durban 2011 climate meetings give scant evidence that they will not), warming episodes above critical tolerance levels for corals are likely to become more frequent, and depressed growth due to continued acidification is likely to be extended. We will be killing off corals every year or so, while depressing their capacity for the growth that will be essential for recovery. Donner *et al.* (2005), in a modeling study, have suggested that warming will be so marked by 2050 that most reef regions can expect a serious mass bleaching event every other year. At that rate of bleaching events, even without depressed growth, the dynamic equilibrium between reef growth and reef decline will have been shifted and reef systems will no longer be dominated by corals.

The potential for adaptation to a changed climate

The discussion above makes no mention of acclimation or adaptation by corals to enable them to survive the changing environment that our releases of CO_2 are creating. There are several reasons why significant acclimation or adaptation is not expected. First, and most important, is the pace of the changes occurring. The world last approached temperature regimes anticipated by mid-century during the mid-Pliocene warm period three million years ago (Dowsett and Robinson, 2009; Haywood *et al.*, 2009), but the rate of increase in sea surface temperatures that the world is now undergoing appears to be substantially more rapid than during the Pliocene or at any time since, and is likely to be too rapid for many organisms to evolve adaptations to it (Veron *et al.*, 2009). Certainly, while there is spatial, specific, and inter-individual variation in susceptibility to bleaching, there is as yet no evidence that corals bleached in earlier years are now surviving warmer temperatures without bleaching. Secondly, the pace of change in pH (a reduction of 0.1 pH units in the last 250 years) is two orders of magnitude more rapid than during any period during the last 420 000 years (Hoegh-Guldberg *et al.*, 2007; Friedrich *et al.*, 2012). Thirdly, corals, with their long generation times, and their use of asexual reproduction, are species that are quite unlikely to be able to evolve quickly. Fourthly, acidification, which shifts the energetic balance of a critical chemical reaction in calcification, is changing the requirements of a fundamental process common to a broad range of invertebrates and some plants, and such fundamental processes tend to be managed by genes with little allelic variation across individuals and even across species. For all these reasons, adaptation and acclimation are expected to be of minimal importance in enabling corals to survive these impacts.

Our decisions over the next few years will decide the future they are to have

There is important evidence accumulating that not all portions of the tropics are changing at precisely the same rates. Some regions will favor coral reefs for longer into the future than others, even if our current trends in CO_2 release continue. Nevertheless, if these trends continue, coral-dominated reefs will be a thing of the past by 2050 as CO_2 concentrations in the atmosphere exceed 510 ppm.

These trends may not continue. There remains the possibility, slimmer every day, that the global community will put in place programs to aggressively phase out use of fossil fuels, and programs to encourage sequestration of atmospheric CO_2. Such programs may arise out of strong, broadly supported treaties, or through recognition of the economic self-interest to be gained by leading aggressively on CO_2 mitigation efforts in a few large countries. This shift in behavior brings many benefits beyond the retention of coral reefs – a productive, non-acidified ocean, more reliable weather and rainfall patterns, improved health and quality of life for peoples dependent on fisheries and agriculture for their food, and the prosperity that being a leader in green energy technology will provide for countries that choose this path. An aggressive effort to keep atmospheric CO_2 levels below 450 ppm, and to pull them back towards 350 ppm, will be sufficient to permit coral reefs to continue to prosper. Less than this will probably fail.

In a world of controlled CO_2, coral reefs will still be subjected to the local impacts of overfishing, pollution and inappropriate coastal development. How we deal with these issues in different places will determine the extent to which coral reefs remain, or become healthy. There are thus many futures possible for coral reefs over the next 40 years.

What loss of coral reefs might mean

Coral reefs have disappeared from the earth a number of times during the geological past. In each such case, physical and/or chemical conditions in the ocean have become incompatible with reef growth, and in each such case, corals of some species have managed to survive so that, once conditions improved, reef formation was able to begin again. In each such case, however, the duration of the reef-free period has been 10 to 20 million years in length, and has been accompanied by major losses in biodiversity. The loss of coral reefs has never been a trivial event, even though, so far, it has never been permanent (Veron, 2011).

The loss of coral reefs has four types of immediate costs; biotic, economic, esthetic, and ethical. Fully 25% of oceanic species occur on coral reefs, and while loss of reefs will not eliminate all of them, the impact on oceanic biodiversity will be substantial. Given that their loss may be one conspicuous part of a substantial change to ocean chemistry, the overall biotic effect may be very substantial. The economic costs of reef loss range from the loss of fisheries production and tourism benefits to the loss of seldom explicitly costed services, coastal protection in particular (Hoegh-Guldberg *et al.*, 2007; Veron *et al.*, 2009; Sale, 2011). At a time in which tropical storms seem likely to become more severe while coastal populations seem likely to continue to grow, this loss of coastal protection may be particularly problematic.

The esthetic and ethical impacts are perhaps even more important than these economic costs. Coral reefs have provided us with plenty of warning of the impacts of atmospheric CO_2 on their survival. The science community understands this problem, and has been able to observe, and experiment on it, over the past 20 years or so. We are better equipped to anticipate the future for coral reef ecosystems than for many other ecosystems that have only now begun to provide evidence of the changes that will occur. The solutions to this problem – reduce CO_2 emissions, and increase effective long-term sequestration – are apparent. If we fail to act, because we simply cannot muster the political will to do so, this says much about us as people. If we are willing to countenance the total loss of an ecosystem that exceeds nearly all others in the sheer complexity of its processes, and the magnificence exuberance of its existence, we lose something that is every bit as unique as the Mona Lisa, or the paintings on the caves at Lascaux. E. O. Wilson argued that losing just one species to extinction measurably diminishes the richness of the world around us. The loss of all coral reefs is surely a far bigger esthetic loss than this (Sale, 2011).

That we are causing this loss gives the prospect of loss an ethical dimension as well. If we do not at least make a concerted effort to avoid the loss of coral reefs, we countenance their destruction. We can believe that we do not have a moral obligation to protect other forms of life. After all, no other species, so far as we know, has an ethical responsibility towards other organisms. However, if we choose to view ourselves as more than other beasts, as most humanists would like to believe, we surely have an ethical responsibility to try to avoid the destruction of an entire ecosystem. The biosphere will survive without coral reefs, but what about our humanity?

References

Anthony, K. R. N., Maynard, J. A., Diaz-Pulido, G., *et al.* (2011). Ocean acidification and warming will lower coral reef resilience. *Global Change Biology*, **17**, 1798–1808. doi: 10.1111/j.1365-2486.2010.02364.x.

Baker, A. C., Glynn, P. W., & Riegl, B. (2008). Climate change and coral reef bleaching: an ecological assessment of long-term impacts, recovery trends and future outlook. *Estuarine, Coastal and Shelf Science*, **80**, 435–471.

D'eath, G., Lough, J. M., & Fabricius, K. E. (2009). Declining coral calcification on the Great Barrier Reef. *Science*, **323**, 116–119.

Diaz, R. J., & Rosenberg, R. (2008). Spreading dead zones and consequences for marine ecosystems. *Science*, **321**, 926–929.

Donner, S. D., Skirving, W. J., Little, C. M., Oppenheimer, M., & Hoegh-Guldberg, O. (2005). Global assessment of coral bleaching and required rates of adaptation under climate change. *Global Change Biology*, **11**, 2251–2265.

Dowsett, H. J., & Robinson, M. M. (2009). Mid-Pliocene equatorial Pacific sea surface temperature reconstruction: a multi-proxy perspective. *Philosophical Transactions of the Royal Society A*, **367**, 109–125.

Edwards, H. J., Elliott, J. A., Eakin, C. M., *et al.* (2011). How much time can herbivore protection buy for coral reefs under realistic regimes of hurricanes and coral bleaching? *Global Change Biology*, **17**, 2033–2048. doi: 10.1111/j.1365-2486.2010.02366.x.

Fenton, M., Kelly, G., Vella, K., & Innes, J. (2007). Climate change and the Great Barrier Reef: industries and communities. In J. E. Johnson & P. A. Marshall (Eds.), *Climate Change and the Great Barrier Reef* (pp. 745–771). Canberra: Great Barrier Reef Marine Park Authority and Australian Greenhouse Office.

Friedrich, T., Timmerman, A., Abe-Ouchi, A., *et al.* (2012). Detecting regional anthropogenic trends in ocean acidification against natural variability. *Nature Climate Change*. doi: 10.1038/NCLIMATE1372.

Glynn, P. W. (1983). Extensive "bleaching" and death of reef corals on the Pacific coast of Panamá. *Environmental Conservation*, **10**, 149–154.

Glynn, P. W. (1984). Widespread coral mortality and the 1982–83 El Niño warming event. *Environmental Conservation*, **11**, 133–146.

Glynn, P. W. (1991). Coral reef bleaching in the 1980s and possible connections with global warming. *Trends in Ecology & Evolution*, **6**, 175–179.

Glynn, P. W. (1997). Bioerosion and coral reef growth: a dynamic balance. In C. Birkeland (Ed.), *Life and Death of Coral Reefs* (pp. 69–98). New York: Chapman and Hall.

Haywood, A. M., Dowsett, H. J., Valdes, P. J., *et al.* (2009). Introduction. Pliocene climate, processes and problems. *Philosophical Transactions of the Royal Society A*, **367**, 3–17.

Hoegh-Guldberg, O., Mumby, P. J., Hooten, A. J., *et al.* (2007). The carbon crisis: coral reefs under rapid climate change and ocean acidification. *Science*, **318**, 1737–1742.

Hoeke, R. K., Jokiel, P. L., Buddemeier, R. W., & Brainard, R. E. (2011). Projected changes to growth and mortality of Hawaiian corals over the next 100 years. *PLoS One*, **6**, e18038. doi:10.1371/journal.pone.0018038.

Hughes, T. P., Graham, N. A. J., Jackson, J. B. C., Mumby, P. J., & Steneck, R. S. (2010). Rising to the challenge of sustaining coral reef resilience. *Trends in Ecology & Evolution*, **25**, 633–642.

Lessios, H. A., Robertson, D. R., & Cubit, J. D. (1984). Spread of *Diadema* mass mortality through the Caribbean. *Science*, **226**, 335–337.

Manzello, D. P. (2010). Coral growth with thermal stress and ocean acidification: lessons from the eastern tropical Pacific. *Coral Reefs*, **29**, 749–758.

Mumby, P. J. (2009). Phase shifts and the stability of macroalgal communities on Caribbean coral reefs. *Coral Reefs*, **28**, 683–690.

Mumby, P. J., & Steneck, R. S. (2008). Coral reef management and conservation in the light of rapidly-evolving ecological paradigms. *Trends in Ecology & Evolution*, **23**, 555–563.

Oxford Economics (2009). *Valuing the Effects of Great Barrier Reef Bleaching*. Report prepared for the Great Barrier Reef Foundation. Australia, 95 pp.

Sale, P. F. (2008). Management of coral reefs: where we have gone wrong and what we can do about it. *Marine Pollution Bulletin*, **56**, 805–809.

Sale, P. F. (2011). *Our Dying Planet. An Ecologist's View of the Crisis We Face*. Berkeley, CA: University of California Press.

Sandin, S. A., Smith, J. E., DeMartini, E. E., *et al.* (2008). Baselines and degradation of coral reefs in the Northern Line Islands. *PLoS ONE*, **3**, e1548. doi:10.1371/journal.pone.0001548.

Stimson, J., & Conklin, E. (2008). Potential reversal of a phase shift: the rapid decrease in the cover of the invasive green macroalga *Dictyosphaeria cavernosa* Forsskål on coral reefs in Kane'ohe Bay, Oahu, Hawai'i. *Coral Reefs*, **27**, 717–726.

Tanzil, J. T. I., Brown, B. E., Tudhope, A. W., & Dunne, R. P. (2009). Decline in skeletal growth of the coral *Porites lutea* from the Andaman Sea, South Thailand between 1984 and, 2005. *Coral Reefs*, **28**, 519–528.

UNEP (2007). *Global Environment Outlook – Environment for Development (GEO-4)*. Nairobi: UNEP. 540 pp.

Veron, J. E. N. (2011). Ocean acidification and coral reefs: an emerging big picture. *Diversity*, **3**, 262–274.

Veron, J. E. N., Hoegh-Guldberg, O., Lenton, T. M., *et al.* (2009). The coral reef crisis: the critical importance of <350 ppm CO_2. *Marine Pollution Bulletin*, **58**, 1428–1436.

Veron, J. E. N., Odorico, D. M., Chen C. A., & Miller, D. J. (1996). Reassessing evolutionary relationships of scleractinian corals. *Coral Reefs*, **15**, 1–9.

Wilkinson, C. R. (1999). Global and local threats to coral reef functioning and existence: review and predictions. *Marine and Freshwater Research*, **50**, 867–878.

Part VI

Autecological Studies

23 Autecology and the balance of nature – ecological laws and human-induced invasions

G. H. Walter

1. Introduction

Autecology and its place in ecological interpretation is only poorly understood. It has consequently been edged out of consideration as a valid ecological theory. The term "autecology" is used relatively infrequently in the ecological literature and seldom, if ever, in evolutionary biology. The understanding of autecology is confounded further because it is perceived in several fundamentally different ways. The various background perceptions of autecology therefore need to be disentangled to help establish its limits and thus determine what autecology really is. This aspect is covered in section 2. The place of autecology relative to other branches of ecology can thus be established.

The foundation statements (basic assumptions or fundamental premises) that define the basis and scope of autecology are expanded in section 3. At this stage it is sufficient to state that autecology deals with the species-specific adaptations of organisms, as they change through the various stages of the species' life cycle, and how these are involved in interactions that impact on individual organisms in nature. The consequences for interpreting the local presence and geographical distribution of species and their changing intensity of occurrence (or abundance) across space and through time can thus be determined. In this way, autecology provides a mutually exclusive alternative perspective on these issues of central concern to ecology. Clearly, other branches of ecology also deal with these central issues, including population, community and landscape ecology. Teasing apart these alternative perceptions of ecology to justify that autecology is, indeed, a true alternative begins in section 2, but is returned to periodically through the chapter.

Although autecological research is not new, and some information that is autecological in nature is collected and published, the theoretical framework structured to take such information is little known, mainly because of the common tendency to treat it dismissively (section 4), a practice that has endured for at least half a century. Autecological theory has, accordingly, had little exposure, so its insights for understanding organisms,

The Balance of Nature and Human Impact, ed. Klaus Rohde. Published by Cambridge University Press.
© Cambridge University Press 2013.

their environment and the interactions between them are not fully appreciated or are missed altogether. Testing of autecological theory and developing it fully have thus suffered and so too has our understanding of how organisms respond to environmental dynamics and anthropogenic change.

The main conclusion of this paper is that autecological theory has a little appreciated but unparalleled place in modern ecology, for it provides an alternative general theoretical framework for ecology. The discussion (section 5) justifies the position of autecology sketched above and formalizes how autecology structures investigation and interpretation. A consequence of this analysis is the light it sheds on a perennial problem in ecology – the issue of ecological laws, which remains unresolved despite the attention of ecologists (e.g., Lawton, 1999; Turchin, 2001; Colyvan & Ginzburg, 2003; Lange, 2005; O'Hara, 2005). The discussion explains how the confusing dualities in perceptions of ecological laws can be resolved. The practical consequences that stem from autecology are profound in the direction they provide for research and generalization in ecology. This is illustrated briefly with respect to issues relating to the invasiveness of species, the invasibility of environments and anthropogenic change.

2.   Background to interpreting autecology

This section begins to build an interpretation of autecology from basic principles. A few serious misconceptions about autecology are eliminated initially because they distract. Also, some of the central concepts that are used generally in ecology differ significantly in autecology from the way they are conceptualized in population and community ecology (or demographic ecology). I therefore consider the basic nature of "environment" and "ecological system" in some depth to explain how autecology has been edged out of consideration as theory in ecology and to provide a positive basis for moving to the foundation statements of autecology in section 3.

2.1.   Misconstructions and ambivalence about autecology – what autecology is not

The most common general perception of autecology is that it deals with "how a single species interacts with the environment" (Losos, 2009). Such statements are correct in principle but are developed misleadingly in at least two ways.

1. Applied ecology is often said to benefit from autecological information, for example in the biological control of pests and conservation of endangered species (e.g., Kareiva, 1994). Autecological studies are thus seen to generate useful data, but the relevant information is usually seen as factual, in the sense of being simple and descriptive, and basic to the proposed practical aims. That is, the data are unique to the species in question and not amenable to generalization.
2. Autecology is portrayed as the evolutionary arm of contemporary ecology, in the form of such aspects as niche theory, physiological ecology, foraging behavior, dispersal, adaptation and ecological speciation (Lawton, 1993; Losos, 2009). Autecology is thus

placed complementary to population and community ecology, with its theoretical basis derived implicitly from evolutionary ecology. Features considered to be autecological in this approach include competitive ability and release from enemies (e.g., Barney & Whitlow, 2008).

These views are not only contradictory but are at odds with the foundations of autecology. However, two further aspects need consideration if these differences are to be resolved so that autecology can be interpreted accurately. These relate to the very perceptions of ecological systems, for autecology differs fundamentally from demographic ecology in how the environment is defined and it does not accept that ecological systems should be treated as hierarchical. The question of species and how they are treated in ecological interpretation becomes central to these considerations, and this array of interrelated issues is dealt with next.

2.2. Demographic conceptualization of environment

Definitive statements about the foundation statements of population and community ecology are seldom made, and have to be inferred (Walter, 1995; Cooper, 2001, 2003; Walter, 2008). What can be said about them, however, is that they see populations as regulated, with the governing feedback being applied by ecological processes that have the potential to act in density-dependent fashion. The primary candidates for this role are competition, predation, parasitism and the like. The consideration of numbers (as opposed to organisms and their adaptations) thus lies at the heart of contemporary ecological theory. Its abstractions, outlook and interpretations are essentially demographic (Hengeveld & Walter, 1999; Walter & Hengeveld, 2000), a term that thus covers population and community ecology.

When the primary focus in ecology is on the influence of population density and competition, the environment must almost inevitably be treated as bipartite. It is abstracted into a biotic component (which provides a deterministic regulatory skeleton) and an abiotic component (from which external stochastic influences intrude) (e.g., Coulson et al., 2004). The consequences of such a conceptual development are profound; the two most significant of these for ecological research and interpretation are expanded below.

1. The abiotic, in the form of climate, soil chemistry and the like, is seen as singular in its influence. This explains why such factors are so often set aside from population and community ecology, sometimes in the form of "autecological factors" (e.g., Lawton & Strong, 1981); but these clearly are environmental variables that impinge on the ecology of organisms, and need to be a central part of any ecological theory. Transferring them to autecology simply removes these ecologically significant abiotic influences from the realms of ecological generalization in demographic ecology.

2. A great many interactions that have a biotic component (e.g., host or habitat searching behavior, mutualisms, biotic change to the environment) are excluded from the foundation statements of demographic ecology, where the focus is strongly on population density and competition. This approach is thus consistent with modern

perceptions of fitness and its measurement. Fitness is seen to be optimized relative to that of competitors and is measured in terms of the relative production of offspring within that competitive context (e.g., Kokko & Jennions, 2010) and such interactions as host-searching behavior are treated thus under the subdiscipline of behavioral ecology. That is, adaptive mechanisms that are central in the ecology of organisms are effectively removed from the foundation statements of ecological theory to be placed in another subdiscipline, one whose foundation statements accept that organisms are under strict selection to maximize their fitness continuously. The logical link between behavioral ecology and demographic ecology, through density-dependent competitive processes, is clear (Walter & Hengeveld, in press).

That these consequences, which follow directly from an emphasis on population density and competition, misguide ecological interpretation is justified further in section 3.

The advantage of the approach taken by demographic population and community ecology, if it were to reflect accurately what takes place in nature, would be its generality across species. Indeed, species are seen as awkward to ecological theory, and have been portrayed as idiosyncratic or individualist relative to demographic theory in ecology (beginning with the critical insights of Gleason (1926)). This helps explain why adaptations of species tend to be removed from consideration in demographic ecology (in favor of density dependence and competition) and even from behavioral ecology and evolutionary ecology (in favor of fitness benefits that are considered to enhance efficiency in various measures of performance and reproductive output within an environment that is seen as predominantly competitive). That is, the adaptive process (which is believed to be inexorable and contemporary) is given far greater significance than is given to adaptive mechanisms in modern interpretations of evolution and ecology. The idiosyncrasies of species (relative to demographic expectations) are thus removed so that organisms fit in better with the basic tenets of demographic theory and with the way in which fitness is seen to lead to optimization within a primarily competitive environment.

2.3.   Ecological systems – hierarchy and species

Demographic ecology accepts that ecological systems are hierarchical (e.g., Krebs, 2008) and builds from the population to the community and then ecosystem, but almost always omits species (Walter, 2003, p. 92). Each level is usually treated independently in the search for generalizations, and each removes any trace of the unique nature of species and thus the way in which the constituent individuals interact with the environment, for the focus is on density, resource use and competitive influences.

The search in population ecology is for common patterns in population dynamics and on the processes that lead to population regulation. The populations of all species are expected to be regulated in much the same way through density dependence, although the pattern of regulation could be obscured at times by stochastic or chaotic influences (e.g., Coulson *et al.*, 2004). Density dependence is also seen to play a key role in community ecology, although species tend to be treated typologically in relation to resource use (Walter, 1995). The population of the species of interest is seen to "fit" into the community of species as they exploit a common resource, either because an "empty niche" is available or through adaptive adjustment to divide the resource and thus promote their coexistence locally. Assembly rules for communities are therefore expected (e.g., Drake, 1990; Gaston, 2004; Leibold *et al.*, 2004; Holdaway & Sparrow, 2006). The concept of "ecological fitting" is apposite here because its primary focus is on community structure and competitive relationships (through its association with the Hutchinsonian niche) and how organisms "fit" in this regard (e.g., Agosta & Klemens (2008) and see section 3).

These perceptions about populations and communities elevate density-dependent processes as the most significant influences in ecological systems and are encapsulated in the core of demographic theory. Related theoretical developments (e.g., metapopulation theory, landscape ecology, spatial ecology, macroecology) work off the foundation statements they provide (Walter & Hengeveld, in press). The way in which organisms are conceptualized relative to ecological influences like logistic population growth and competitive exclusion is particularly significant. Organisms are believed to conform to these influences, which, being external to the organisms, represent regularities or natural laws that impose on the organisms to conform. Demographic ecological processes are thus implied (and often claimed) to represent natural laws, as discussed further in section 5.

2.4.   Conclusions

The way in which the environment, ecological systems and species are portrayed in demographic ecology helps explain why individual organisms and their specific adaptations are put aside as descriptive details with little significance for ecological theory (in population ecology) and why species are either ignored or treated typologically (as units) that fit into the demographic background of communities (Hengeveld, 1988; Walter, 1995). In this sense, modern demographic ecology seems unique among the sciences in distancing itself from the material objects whose deployment and actions it sets out to explain.

In autecology, by contrast, the individual organisms themselves are seen as central in ecological systems, in the sense that they influence ecological systems through a range of species-specific interactions with the environment. The generalizations made about them do represent law-like statements, but these are heuristic. Organisms do not conform to anything that these statements represent. Rather, the generalization encapsulates how the organisms interact with the environment. It helps observers interrogate ecological systems and interpret them through an appreciation of the organism-environment interaction and the spatio-temporal dynamics of the environment and species of interest (Walter & Hengeveld, in press), as expanded in the following section. The issue of ecological laws is pursued further in the discussion (section 5).

3.   Foundation statements in autecology and the scale of investigation

Autecology has a history as long as ecology itself and started emerging as a subdiscipline after publication of Darwin's *Origin* (Cittadino, 1990). The core of autecological theory has been lost, though, mainly through misrepresentation and conflation with other subdisciplines. Nevertheless, the autecological approach has become somewhat invisible despite its status as an alternative theory being acknowledged (Errington, 1955; Simberloff, 1989). Even autecology's most representative text, the enduring *Distribution and Abundance of Animals* (Andrewartha & Birch, 1954), hardly uses the term.

An outline of autecology is therefore developed, from its foundation statements, to demonstrate its validity as a scientific generalization. Autecology deals primarily with the spatiotemporal dynamics of organisms (Walter & Hengeveld, 2000) and has a claim to utility in applied ecology (Andrewartha & Birch, 1954; Andrewartha, 1984). This practical aspect, in particular, is suggestive of a direct connection with organisms in autecology, one that seems to be lost in the theory that underpins population and community ecology theory (Walter, 1995, 2003). The synopsis of autecology (presented below) can be compared directly with the generalizations that form the core of demographic population and community ecology because both approaches incontestably address the same issues, the numbers of organisms (abundance) and their geographical representation (whether distribution or local diversity ("community structure")). The difference between the two approaches is in their foundation statements and the way in which each approach frames its questions about ecological systems. These differences are so fundamental that the two approaches to ecology are mutually exclusive, and cannot be made complementary.

Recent analyses (Hengeveld & Walter, 1999; Walter & Hengeveld, 2000, in press) identified the following foundation statements for autecology.

1. Organisms have species-specific adaptations that mediate their interaction with the environment, and are thus the primary influences in what we call their ecology. Species, and the organisms that make up species, therefore both have special significance in autecology. That is, autecology has a duality, as expanded in subsection 3.3 (below).

2. The environment of organisms is specified in terms of physicochemical (or abiotic) aspects impinging on organisms simultaneously with the impacts of biotic influences (such as vegetation structure and host availability). The biotic is thus characterized with reference to its physical and chemical nature rather than in terms of density dependence and competition. Predation and competition are seen as secondary influences of the primary pattern set by the organism-environment interaction and are best understood in relation to the effects of those primary influences.

3. Individual organisms interact in a species-specific way with a dynamic, but structured, environment. For instance, the environment of a particular locality is structured seasonally in climatic terms, although each variable is subject to stochastic influence.

4. Each species is adapted to a particular subset of the environmental circumstances that prevail within any locality. Each ecologically significant variable may impinge on organisms in a range of different ways and we measure these as different environmental axes of differentiation. For example, temperature relations may be specified best in terms of minima for particular species, in terms of number of days without frost in early spring for others, and with respect to both for yet other species.

5. Individual organisms interact with their structured environment through their structured life cycle, with different developmental stages (or even different ages within a developmental stage) often having very different requirements and tolerances from one another. Different species thus match environmental circumstances differentially across the range of environmental axes of differentiation. The limit to their distribution may thus be imposed in different ways in different geographic areas.

These points are expanded, in the numbered points below, in preparation for a reinspection of the value and place of autecological information and theory in modern ecology, as well as for its relevance in understanding the consequences of anthropogenic ecological change (section 5). In the meantime, the significance of these points for recognizing the primary pattern in ecology, and for investigating and interpreting it, is specified.

The ways in which organisms interact with the environment, as envisaged in autecology and as outlined in the points above, mean that organisms have to match their adaptations, requirements and tolerances against the spatiotemporal dynamics of the environment. "Environmental matching" is thus an autecological perspective that differs fundamentally from the "ecological fitting" of demographic ecology, in which organisms are seen to fit into a community of species. That is, ecological fitting focuses on the avoidance or amelioration of competition whereas environmental matching focuses fully on the primary adaptations and ecological requirements of organisms through their life cycle and across the seasons. It is a spatiotemporally dynamic matching process that underpins explanation of the shifting distribution of abundance of species. Geographically, the process of environmental matching translates into the spatiotemporally dynamic distribution of organisms, and this reflects the primary pattern to be explained in ecology. The geographic distribution of a species dictates, to a large extent, the scale at which its ecology needs to be understood. Simultaneously, though, the scale at which individuals of that species interact with the environment is complementary in forming a basis for understanding the species' ecology (summarized from Walter & Hengeveld (in press)).

An understanding of autecology (and thus ecology in general) therefore encompasses the following issues and the questions that derive from them and these provide initial insight into the real utility of autecological data. Also indicated is what research has been done in regard to each question and what is still required.

3.1.   Autecology, species and adaptations

The term autecology is strongly associated with the term species, for species are seen as real and as having specific traits (or adaptations). How is a focal species selected for investigation? It may be one or more of those that make up the local diversity of a selected area, or a particular species may attract attention because of its abundance (whether low or high), recent invasion into an area, or even its erratic appearance in a locality, or for some other similar reason.

Species-specific adaptive mechanisms dictate the ecology of organisms (Walter & Hengeveld, in press). Crucial, therefore, is a clear understanding of what in principle represents a species, and what in particular demarcates the limits of the particular species under investigation. Errors in ecological interpretation and application will inevitably follow any lapse in this regard, for one of the few truisms in ecology is that no two species are exactly alike ecologically. Even closely related species differ from one another ecologically, and that is often reflected in their differential host relationships or associations, habitat requirements and geographical distributions. Mistakes are costly, as reviewed by Paterson (1991) and Walter (2003).

Even the best known of birds illustrates the subtlety and relevance of species to ecological understanding. The great tit (*Parus major*), interpreted broadly, is distributed from Britain to Japan and south to some of the larger Indonesian islands. Generalizing its habitat requirements on this basis is necessarily expansive. For instance: "The Great Tit, the single most varied and widespread parid in the world, occurs in every kind of forest, scrub or bushy habitat, and in both primary and secondary forest, and is the only member of the family specifically associated with mangroves in part of its range" (Gosler & Clement, 2007). Particular subsidiary clauses have to be incorporated, for example the one made to cover the use by great tits of mangroves on the Malay Peninsula (a vegetation type used by relatively few birds, most of which are adapted to the mangal habitat (Johnstone, 1990)). The recent clarification that the great tit comprises a complex of at least three species (Packert *et al.*, 2005), and possibly even more, should help in understanding the usual habitat and distinct distribution of each and how each is adapted to those conditions.

For organisms in general, obvious differences in habitat or host use across populations considered to be conspecific call for investigation (as in pelagic vs. pack ice killer whales in the Antarctic (Pitman & Ensor, 2003), fresh vs. brackish water breeders in mosquitoes (Paterson, 1993a) and use of alternative host species by insect parasitoids and herbivores (Fernando & Walter, 1997; Najar-Rodriguez et al., 2009)), but more subtle ecological differences are readily overlooked and such situations may be common (e.g., Condon *et al.*, 2008). Research on such cryptic species complexes has gathered momentum recently, as evidenced by the reviews and analyses available (Paterson, 1991; Clarke & Walter, 1995; Walter, 2003; Bickford *et al.*, 2007). Clearly, the variability across species is not as extensive as envisaged by the polytypic species concept (Moore, 1975; Hillis, 1988), which represents Mayr's (1942,1963) extension of Darwin's approach to species (as explained further in section 4).

The species-specific adaptive mechanisms that dictate the ecology of organisms are complex; they comprise numerous integrated components (Walter & Hengeveld, in press, for examples). This complexity implies, in turn, that adaptive change, of the extent evident in the differences between species, is likely to require very specific (and unusual) ecological circumstances. The fixation of such mechanisms across the entire species gene pool, in other words, is likely to have taken place in small populations trapped in an environment different from the one to which they were originally adapted. All individuals could thus become exposed to strong directional selection from the new environment, defined autecologically (Paterson, 1986). Autecology accepts that the features of organisms that influence their ecology are set down during this relatively brief period of adaptation.

3.2. The autecological environment is inclusive, structured and dynamic

Different species live in the same environment, but are not adapted to the environmental circumstances that prevail there in the same way as one another. Each is adapted to a particular subset of the environmental axes that prevail within a locality. These axes may be biotic (e.g., vegetation structure, host phenology) or abiotic (e.g., climatic influences). Even closely related species differ from one another in their environmental axes of differentiation, which helps explain their different relative abundances (where they exist in sympatry) and their differential geographic distributions (Walter & Hengeveld, in press).

Each axis of differentiation can be represented by a mean and a variance for any locality. Both of these statistical measures shift with time, and usually we see this as the seasonal structure of the local environment. With respect to each species, the structures of several crucial axes of differentiation need to be taken into account for a particular locality. Some areas of the globe do not have the seasons structured in so orderly a way, but structure of an equivalent nature can still be perceived, even if it spans decades, as in arid Australia (Morton *et al.*, 2011). The mean and variance of an axis of differentiation also shifts across space, so the environment (seen from the autecological perspective) has spatiotemporal structure. Environmental structure is thus dynamic in a spatiotemporal sense. Naturally, changing climate also imposes a dynamic, but at a greater scale of temporal resolution (Walter & Hengeveld, 2000). The impact of such change is profound, as is now being demonstrated, and this chapter reasons why an autecological perspective is required not only to interpret observed shifts, but to determine the limits to predictability in this regard (for published promise certainly outstrips our abilities in this respect).

3.3. Organisms, life cycles, interactions and environmental matching

The interactions of organisms with the environment are mediated by various mechanisms (or adaptations). These are species-specific and different stages in the life cycle may well have different mechanisms and requirements. This structured life cycle sequence must match the structured sequence of the environment within a locality. For the individual organisms of many species, this takes place within a single locality, but sometimes the environmental match must be met by individuals shifting across localities or habitats, if they have ecologically differentiated life stages for example. These aspects are extended and illustrated by Walter and Hengeveld (in press).

Such environmental matching is dynamic in the sense that organisms track the environmental circumstances to which they are adapted. The geographical distribution of the species shifts as a consequence. We thus have a means to understand the local presence of a species and its abundance within a locality, as well as such consequential aspects as the associations of species. This is done without primary deference to population density, competitive interactions, communities or associations, coexistence or "niche space". Conversely, the sporadic and local influences of density and competition require an autecological context for their (local) significance to be interpreted.

4.   Why has autecology been treated ambivalently?

Autecology has been treated somewhat dismissively because of two closely related conceptual developments that lie at the foundation of contemporary evolutionary and ecological interpretation. First, the reality of species has been eroded. This process, ironically enough, was initiated by Darwin and then extended by Mayr (1963). Second, and in consequence, evolutionary change is seen to be ineluctable in promoting both diversification and the achievement of full species status through competitive selection. These points are expanded below.

To prove that evolutionary change did take place, Darwin felt compelled to demonstrate that one "form" did give rise to another. He therefore focused attention on the variation inherent in species, with emphasis on information from breeders, and on a blurred distinction between species and what are referred to as local races. The idea of local race or variety was used for all manner of populations considered to differ from one another, in whatever way, and this tradition continues. Such populations have usually not even been investigated to establish their status. Some could prove, eventually, to be undetected cryptic species (sometimes erroneously called "cryptic speciation"), some subspecies (defined in population genetics terms (e.g., Popple & Walter, 2010)). The singular emphasis given to what amounts to a miscellaneous collection of phenomena is logically unacceptable in the development of theory (the fallacy of false analogy). The intricacies and ramifications of this contrivance are spelled out by Beatty (1985), for we still live with its consequences and debate swirls around it without resolution.

The common ground on species, shared by Darwin, Mayr and most modern authors (e.g., Mallet, 2010) has a significant consequence, one that seems to be singular in modern science. A process is implicitly elevated above the primary pattern we see in nature. That is, natural selection, as a competitive and inevitable directional driver of adaptation, and thus speciation, takes precedence over the observation that natural systems comprise a pattern of distinct species with unique combinations of species-specific complex adaptations. That view downplays the functional significance of fertilization mechanisms (of which the specific-mate recognition system is a part) and their species specificity whilst seeing speciation teleologically as a goal of competitive selection. This view is implicit whenever reference is made to such categorizations as populations, varieties and races as incipient species, and when populations are treated as species because they are seen as independent phylogenetic lineages (e.g., Wilson, 2011).

One simply has to reflect on the question of what gives writers such certainty that they are, indeed, dealing with incipient species to see it is underpinned by an acceptance of competitive selection as a deterministic organizational principle for efficient resource use in nature (as pointed out by Paterson (1982, 1986) with reference to the work of both Dobzhansky and Mayr). Intraspecific variation is therefore often accepted uncritically as evidence of adaptive change towards full species status (Walter, 1995, 2003). The extremely long time since the divergence of at least some cryptic species (e.g., de Vargas et al., 1999) and the fact they have not diverged morphologically in such a long time further questions such views of evolution. A significant consequence of the focus on adaptation as process (through the accumulation of fitness benefits) has removed, in evolutionary ecology and behavioral ecology for example, virtually all attention from adaptive mechanisms as informative of organisms, their ecology and their evolutionary past (see subsections 2.2 & 5.2).

Darwin (1859) returned time and again to competition between organisms (pp. 62, 63, 77, 314, 477, for example), and related it to the phrase "struggle for existence". He thus focused our attention firmly on natural selection as a competitive process (Brady, 1982; Sylvan, 1994; Gould, 2002; Paterson, 2005). The almost universally accepted precept that nature is quintessentially competitive provides the background and dynamic for modern interpretations of adaptive change and modern abstractions of ecological systems.

Darwin's emphases on the uncertain limits of species and the unquestioned strength of competitive selection have become foundation statements in modern evolution and ecology (see Cooper, 2001, 2003). Species can therefore have no place in the ecological hierarchy – generalized population processes that regulate density are seen as pre-eminent (because the expected strength and consistency of density dependence underpins competitive natural selection) and local populations of species are seen to "fit" into communities according to "assembly rules" (subsection 2.3). Competition and ongoing adaptation take center stage and species fall outside the most widely accepted vision of the organization of ecological systems. Even in ecosystem ecology, species are marginalized – they are treated typologically in food webs, whereas the process of interest, the dynamics of energy and material transformation, are mediated by individual organisms with independent species-specific spatiotemporal dynamics (see Walter, 1995).

The Recognition Concept of species (Paterson, 1993b) was developed to escape those rather vague definitions of species that are so heavily "laden with Darwin's theory" (Beatty, 1985, p. 280). The focus is firmly set on individual organisms, and emphasizes the positive aspects of male-female sexual interactions. These complex adaptations ensure fertilization takes place and they are under stabilizing selection because (i) of the necessary co-adaptation across the sexes, (ii) the whole is adapted to the usual environment of the species, and (iii) they are buffered generally by the background of a large population (Paterson, 1978, 1985, 1986, 1987, 1993b, pp. 15, 21). Species thus have an objective reality, and we define them on the basis of their sexual interactions in their usual environment (Paterson, 1981; Lambert et al., 1987). Even Mayr (1942, p. 148) was tenacious in his defense of the reality of species. Hybridization and asexuality are sometimes presented as challenges to the recognition concept and even to the reality of species, but such perceptions are misguided (as argued by Paterson (1981) and Walter (2003, pp. 133, 165ff), and demonstrated by Popple & Walter (2010) and Najar-Rodriguez et al. (2009)).

5.   Discussion

5.1.   Autecological data – description and utility

The term "autecology" is frequently used synonymously with "natural history", the latter of which is seen as largely observational and anecdotal (in the sense that it falls short of being amenable to generalization and thus not really scientific). Autecological data are thus seen as valuable, but primarily in providing the background necessary to designing tests of demographic theory or in manipulating organisms for applied purposes (e.g., Lawton, 1993; Kareiva, 1994). The approach is therefore portrayed as essentially descriptive and species-specific and thus not related directly to ecological theory.

To illustrate, consider generalizations of the host-searching behavior of parasitoid wasps. A massive diversity of these insects exists; they attack and kill mainly other insects (but also some other arthropods) and are largely specific to one (or at most a few) host species in their parasitic associations. They are thus extremely useful in agriculture, forestry and associated endeavors. They are also popular subjects in evolutionary ecology. A prominent review (Godfray, 1994) covers their host relationships with a focus on how natural selection enhances their efficiency in locating hosts, for this is seen to increase their fitness. Such a perspective has at least two significant consequences.

1. The way in which the female parasitoids actually locate their hosts is largely ignored, as this is seen as the proximate mechanism relative to their host associations and is therefore seen as simply descriptive (Hassell & Godfray, 1992). Therefore, the investigation of "the proximate cues involved is often left to those in other disciplines" (Kokko & Jennions, 2010, p. 303).
2. Because natural selection is seen as competitive, the focus of investigation is fixed on the efficiency of host discovery. More efficiency is seen to confer relatively greater fitness (even in terms of grandchildren expected (e.g., Godfray, 1994, p. 84)) and is thus selected. The emphasis on "relative" here reflects the demographic background against which selection is seen to operate, competitively. Consequently, and significantly, the way in which efficiency relates to the actual behavioral responses to environmental information is left undisclosed.

A strong contrast is evident between the approach just outlined and the theory relating to host localization in the applied science of weed and insect biological control, which is essentially autecological (Walter, 2003, p. 252ff.). Here, each aspect of the organism's behavior is seen in the context of a catenary sequence in which each component fulfills a particular requirement that helps that organism to detect and localize an appropriate host progressively. This perspective on adaptive mechanisms not only has important consequences for interpreting the adaptive process, directly in relation to the physical environment (and inclusive of both biotic and abiotic aspects (section 3.2)), but also suggests why the proposed selection on ever increasing grades of efficiency is unlikely (see Cunningham (2012)). A response to a particular chemical compound or wavelength, for example, has only limited leeway for grades of efficiency. And when the various steps in the catenary sequence are taken into account, the limits on increasing efficiency in searching become much more restricted. It is not remarkable, then, that predictions derived from optimization theory are so frequently not met (e.g., Pierce & Ollason, 1987; Rapport, 1991; Walter & Donaldson, 1994).

The sensory adaptations of organisms clearly influence their ecology, even to the point of influencing their abundance. The membracid bug *Acanophora compressa* was introduced into eastern Australia to reduce the density of the introduced weed lantana.

Although lantana is relatively widespread and forms dense thickets, *Acanophora* achieved high, damaging densities only on the ornamental fiddlewood tree (*Citharexylum spinosum*), which is not nearly as abundant as lantana. In the laboratory, *Acanophora* develops and reproduces to much the same extent across both these plant species and the adult bugs settle equally across both (Manners & Walter, 2009; Manners *et al.*, 2010). By contrast, sentinel plants of both species deployed in the field attracted adult bugs in ratios of 85% (on fiddlewood) to 15% (on lantana) (Manners, 2008). Presumably substantial numbers of bugs do not settle on lantana in the field because their sensory systems preclude or reduce this option. And because the individuals involved continue in their search mode they are presumably at much greater risk of mortality and lower reproductive output than if they had simply settled on lantana. The density of bugs across host plant species is thus dictated, to a considerable extent, by their sensory adaptations (Manners, 2008). Methods for developing mechanistic quantitative models that incorporate such data are available, to date mainly for organisms in the process of invading new areas (e.g., Hengeveld & van den Bosch, 1997).

Adaptations such as the host searching mechanism of herbivorous insects thus dictate, in part, the local ecology of a species. Other mechanisms (such as temperature-related adaptations) contribute simultaneously, and also operate sequentially with age or through the life cycle, which explains why all species are, in one sense or another, specialists (Velasco & Walter, 1992; Walter & Benfield, 1994; Walter, 2003; Rajapakse *et al.*, 2006; Rajapakse & Walter, 2007; Manners *et al.*, 2010). The sum of these interactions contributes to the spatiotemporal dynamics of a species (Walter & Hengeveld, in press).

The autecological approach to understanding the responses of organisms to anthropogenic change emphasizes the species-specificity of the interaction between organism and environment through the entire life cycle. Any correlation between invasiveness and competitive ability (Barney & Whitlow, 2008), fruit size, or flowering time (Pysek et al., 2009), for example, cannot be representative enough of the relevant interaction to be explanatory or predictive of invasions. Increasingly, autecological insights are being identified as likely to be more valuable in explaining the responses to climate change of various organisms, including small mammals (Terry *et al.*, 2011) and trees (Nunez & Medley, 2011). Autecological information thus translates into an understanding of (i) local and geographical presence, what Hengeveld and Haeck (1982) refer to as the distribution of abundance, and (ii) the local diversity of species but with the focus on a limited number of species, at most, and their environmental associations (e.g., Hengeveld, 1985). Each species, nevertheless, presents a deep ecological puzzle, soluble only on its own terms.

5.2.   Ecological laws

Demographic ecology and autecology see ecological systems in fundamentally different ways, and they thus generalize about organisms quite differently from one another.

Demographic ecology sees ecological systems as having an underlying order or organization. Indeed, order should be discernible at each level within the hierarchical structure that underpins ecological systems, namely populations, communities and ecosystems. Ecological order is expected to be manifest through regular patterns, with each type of pattern conferred by a process that operates in accordance with one or more ecological laws (even if these are portrayed as "contingency laws" (Lawton, 1999)). Density dependence is thus responsible for population equilibrium. Competitive exclusion underpins the structure (or assembly) of communities. Food webs impose order on ecosystems. Such ecological laws are expected to operate in somewhat different ways across different ecological systems, but predictably so. Demographic ecology theory focuses investigation on these processes and patterns, with considerable emphasis still on finding robust patterns (Lawton, 1999).

Such an approach would be convenient if it worked. We would have a key to understanding the ecology of all systems and species. Interpretation would transfer from one system to another with considerable certainty, even allowing for contingencies. The justification for a continued search for such a key sometimes falls on there being far too many species to study each thoroughly in its own right. Despite irregularities and exceptions to the expected order being commonly identified, the search is deemed important enough to continue. This and the failure to find robust generalizations mean that the search for ecological patterns has been driven in other dimensions and at larger scales (e.g., metacommunities and macroecology).

If the demographic approach to understanding ecology is fundamentally flawed, as argued here, the discipline denies the very nature of its own subject matter. The persistence of such an approach can be contrasted with that in other scientific disciplines. Astronomers, for example, long ago focused on what was possible at the time and worked outwards from there. Consequently, the three-body problem was acknowledged as a limitation to general prediction, and investigation was pursued within those confines and at the limits of possibility, trying to extend capability. By contrast, we see no such reservations in this respect in demographic ecology. Astronomers also showed incredible strength in discarding theory shown to be unrealistic and unhelpful, starting with the rejection of astrology, which had once been the primary motivation for stellar investigation (Margolis, 2002). Demographic ecology, by contrast, retains its failed original concept, the balance of nature, as well as the more recent derivatives from that metaphor, even if tacitly so (Cooper, 2001; Walter, 2008).

Autecology accepts the idiosyncrasy of species and the stochastic influences that undermine the establishment of the patterns expected in demographic ecology. Patterns are, nevertheless, discernible, but they are different from demographic expectations of nature. Autecology focuses more on patterns of positive assortative mating (i.e., species gene pools) and the spatiotemporal dynamics of species, for these indicate the way in which organisms interact with the environment. Ecological systems are not seen as chaotic despite the stochastic (and other) influences that disrupt the patterns expected under demographic ecology. Instead, ecological systems develop and change through the spatiotemporal dynamics of the species present and are therefore not seen to be hierarchical (Walter & Hengeveld, in press). The species that live alongside one another within a locality (the "community" of demographic ecology) do so because their adaptations, tolerances and requirements permit their existence there, and their spatial dynamics encompass that locality. It is in this sense that species are individualistic

(Walter & Paterson, 1994). (In demographic ecology, species are individualistic relative to the expectations about community-level processes.)

The primary pattern (or observation) in autecology thus holds that species are independently adapted to a subset of environmental conditions (or axes of differentiation) and they respond to the spatiotemporal dynamics of these circumstances through their adaptive mechanisms. Their response is manifest in the changing dynamics of their survival and reproductive rates relative to the spatiotemporal dynamics of the environment, defined broadly to include both biological and physicochemical factors. Theory focuses on the development of a generalization (or heuristic law) that captures this observed pattern in relation to all species and in contributing to the understanding of local diversity from this perspective. The subtleties and intricacies in the interactions of organisms with their environment (and their environment will inevitably include organisms of other species) provide the challenge for hypothesis testing and development in autecology (Walter & Hengeveld (in press) for examples). A generalized theory is also developed for each species investigated, to contribute to interpretation of the local ecology and other particular circumstances in which individuals of that species are present.

5.3.   Conclusions

Demographic ecology studies the laws it has established, for the laws themselves are seen to be operative in nature and the organisms are expected to comply with them. The laws thus represent the underlying organizational principles of ecological systems. Admittedly, hypotheses in demographic ecology are tested and supporting evidence is published, but the nature of the tests is often such as to guarantee support, for they often set out to demonstrate what is assumed. That is, the method is essentially one of verification (see Walter, 2003). In terms of theory, demographic ecology continues searching for patterns that will indicate how environments and organisms of different types will be influenced by key demographic processes (Lawton, 1999).

Autecology is fundamentally different in its approach for it studies the organisms themselves with the aid of heuristic generalizations that provide guidance on how to investigate the adaptations that underpin the organism-environment interaction. Autecology thus investigates and interprets the spatiotemporal dynamics of one or more focal species. The heuristic generalizations of autecology are unlike demographic laws because they are not seen to represent processes that are operative within ecological systems. Autecological interpretations help explain why adaptation and speciation require special ecological circumstances, with such change most likely to take place in small geographically isolated populations exposed to environmental conditions that differ from their usual ones (Paterson, 1986). That is, the adaptive process is not seen to be driving ecological systems locally, competitively and inevitably, as envisaged in evolutionary ecology, which is logically connected with demographic ecology (Walter & Hengeveld, in press). Finally, the autecological approach to understanding anthropogenic invasions by organisms should help ecologists move away from current (demographic) notions of species' invasiveness and the invasibility of environments.

Synopsis

A consideration of how autecology is generally perceived in the ecological literature is contrasted against the basic premises (or foundation statements) of autecology. This comparison reveals that autecology is frequently misrepresented and thus readily displaced as a valid alternative interpretation of ecological systems. That autecology does, indeed, represent a true alternative theory to the dominant perspective in ecology (as represented by population and community ecology (or demographic ecology)) and their offshoots (e.g., metapopulation and landscape theory) is justified further, and the statements that represent the foundations of autecology are summarized. This treatment helps resolve the controversial issue of the existence and use of laws in ecology. Demographic ecology sees organisms conforming to natural laws (as represented by logistic processes and competitive exclusion, for example), whereas autecology uses law-like generalizations about the organisms themselves, in their environment, as a heuristic to guide investigation and interpretation. The example of anthropogenic invasions illustrates the difference in these two approaches to interpreting ecological systems, and also the relevance of autecology.

Acknowledgments

Several colleagues put a lot of effort into clarifying the message in this manuscript, for which I am extremely grateful. They include Rob Hengeveld, Andrew Ridley, Raghu, Iman Lissone, Greg Daglish, Ed White, Renee Rossini and James Hereward. I am grateful, too, to Klaus Rohde in helping to steer the development of this chapter.

References

Agosta, S. J., & Klemens, J. A. (2008). Ecological fitting by phenotypically flexible genotypes: implications for species associations, community assembly and evolution. *Ecology Letters*, **11**, 1123–1134.

Andrewartha, H. G. (1984). Ecology at the crossroads. *Australian Journal of Ecology*, **9**, 1–3.

Andrewartha, H. G., & Birch, L. C. (1954). *The Distribution and Abundance of Animals*. Chicago, IL: University of Chicago Press.

Barney, J. N., & Whitlow, T. H. (2008). A unifying framework for biological invasions: the state factor model. *Biological Invasions*, **10**, 259–272.

Beatty, J. (1985). Speaking of species: Darwin's strategy. In D. Kohn (Ed.), *The Darwinian Heritage* (pp. 265–281). Princeton, NJ: Princeton University Press and Nova Pacifica.

Bickford, D., Lohman, D. J., Sodhi, N. S., *et al*. (2007). Cryptic species as a window on diversity and conservation. *Trends in Ecology & Evolution*, **22**, 148–155.

Brady, R. H. (1982). Dogma and doubt. *Biological Journal of the Linnean Society*, **17**, 79–96.

Cittadino, E. (1990). *Nature as the Laboratory. Darwinian Plant Ecology in the German Empire, 1880–1900*. Cambridge: Cambridge University Press.

Clarke, A. R., & Walter, G. H. (1995). "Strains" and the classical biological control of insect pests. *Canadian Journal of Zoology*, **73**, 1777–1790.

Colyvan, M., & Ginzburg, L. R. (2003). Laws of nature and laws of ecology. *Oikos*, **101**, 649–653.

Condon, M., Adams, D. C., Bann, D., *et al.* (2008). Uncovering tropical diversity: six sympatric cryptic species of *Blepharoneura* (Diptera: Tephritidae) in flowers of *Gurania spinulosa* (Cucurbitaceae) in eastern Ecuador. *Biological Journal of the Linnean Society*, **93**, 779–797.

Cooper, G. (2001). Must there be a balance of nature? *Biology & Philosophy*, **16**, 481–506.

Cooper, G. J. (2003). *The Science of the Struggle for Existence*. Cambridge: Cambridge University Press.

Coulson, T., Rohani, P., & Pascual, M. (2004). Skeletons, noise and population growth: the end of an old debate? *Trends in Ecology & Evolution*, **19**, 359–364.

Cunningham, J. P. (2012). Can mechanism help explain insect host choice? *Journal of Evolutionary Biology*, **25**, 244–251.

Darwin, C. (1859). *On the Origin of Species by Means of Natural Selection, or the Preservation of Favoured Races in the Struggle for Life*. Facsimile reprint, 1964, Cambridge, MA: Harvard University Press & London: John Murray.

Drake, J. A. (1990). Communities as assembled structures: do rules govern patterns? *Trends in Ecology & Evolution*, **5**, 159–164.

de Vargas, C., Norris, R., Zaninetti, L., *et al.* (1999). Molecular evidence of cryptic speciation in planktonic foraminifers and their relation to oceanic provinces. *Proceedings of the National Academy of Sciences of the USA*, **96**, 2864–2868.

Errington, P. L. (1955). Book reviews: The Distribution and Abundance of Animals. *Science*, **121**, 389–390.

Fernando, L. C. P., & Walter, G. H. (1997). Species status of two host-associated populations of *Aphytis lingnanensis* (Hymenoptera: Aphelinidae) in citrus. *Bulletin of Entomological Research*, **87**, 137–144.

Gaston, K. J. (2004). Macroecology and people. *Basic and Applied Ecology*, **5**, 303–307.

Gleason, H. A. (1926). The individualistic concept of the plant association. *Bulletin of the Torrey Botanical Club*, **53**, 7–26.

Godfray, H. C. J. (1994). *Parasitoids. Behavioral and Evolutionary Ecology*. Princeton, NJ: Princeton University Press.

Gosler, A. G., & Clement, P. (2007). Family Paridae (tits and chickadees). In J. del Hoyo, A. Elliott & D. A. Christie (Eds.), *Handbook of the Birds of the World*. Vol. 12. *Picathartes to Tits and Chickadees* (pp. 662–750). Barcelona: Lynx Edicions.

Gould, S. J. (2002). *The Structure of Evolutionary Theory*. Cambridge, MA: Belknap Press of Harvard University Press.

Hassell, M. P., & Godfray, H. C. J. (1992). The population biology of insect parasitoids. In M. J. Crawley (Ed.), *Natural Enemies. The Population Biology of Predators, Parasites and Diseases* (pp. 265–292). Oxford: Blackwell Scientific.

Hengeveld, R. (1985). Dynamics of Dutch beetle species during the twentieth century (Coleoptera, Carabidae). *Journal of Biogeography*, **12**, 389–411.

Hengeveld, R. (1988). Mayr's ecological species criterion. *Systematic Zoology*, **37**, 47–55.

Hengeveld, R., & Haeck, J. (1982). The distribution of abundance. 1. Measurements. *Journal of Biogeography*, **9**, 303–316.

Hengeveld, R., & van den Bosch, F. (1997). Invading into an ecologically non-uniform area. In B. Huntley, W. Cramer, A. V. Morgan, H. C. Prentice & J. R. M. Allen (Eds.), *Past and Future Rapid Environmental Changes* (pp. 217–225). Berlin: Springer.

Hengeveld, R., & Walter, G. H. (1999). The two coexisting ecological paradigms. *Acta Biotheoretica*, **47**, 141–170.

Hillis, D. M. (1988). Systematics of the *Rana pipiens* complex: puzzle and paradigm. *Annual Review of Ecology and Systematics*, **19**, 39–63.

Holdaway, R. J., & Sparrow, A. D. (2006). Assembly rules operating along a primary riverbed-grassland successional sequence. *Journal of Ecology*, **94**, 1092 1102.

Johnstone, R. E. (1990). Mangrove and mangrove birds of Western Australia. *Records of the Western Australian Museum, Supplement* No. 32, 1–120.

Kareiva, P. (1994). Ecological theory and endangered species. *Ecology*, **75**, 583.

Kokko, H., & Jennions, M. D. (2010). Behavioral ecology: the natural history of evolutionary theory. In M. A. Bell, D. J. Futuyma, W. F. Eanes & J. S. Levinton (Eds.), *Evolution Since Darwin: The First 150 Years* (pp. 269–290). Sunderland, MA: Sinauer Associates.

Krebs, C. J. (2008). *The Ecological World View*. Collingwood: CSIRO.

Lambert, D. M., Michaux, B., & White, C. S. (1987). Are species self-defining? *Systematic Zoology*, **36**, 196–205.

Lange, M. (2005). Ecological laws: what would they be and why would they matter? *Oikos*, **110**, 394–403.

Lawton, J. H. (1993). On the behaviour of autecologists and the crisis of extinction. *Oikos*, **67**, 3–5.

Lawton, J. H. (1999). Are there general laws in ecology? *Oikos*, **84**, 177–192.

Lawton, J. H., & Strong, D. R. (1981). Community patterns and competition in folivorous insects. *The American Naturalist*, **118**, 317–338.

Leibold, M. A., Holyoak, M., Mouquet, N., *et al.* (2004). The metacommunity concept: a framework for multi-scale community ecology. *Ecology Letters*, **7**, 601–613.

Losos, J. B. (2009). Autecology. In S. A. Levin (Ed.), *The Princeton Guide to Ecology* (pp. 1–2). Princeton, NJ: Princeton University Press.

Mallet, J. (2010). Why was Darwin's view of species rejected by twentieth century biologists? *Biology & Philosophy*, **25**, 497–527.

Manners, A. G. (2008). *The ecological principles that underpin host testing for weed biological control – a case study with the lantana sap-sucking bug, Acanophora compressa* Walker *(*Hemiptera: Membracidae*)*. Unpublished PhD thesis. Brisbane: The University of Queensland.

Manners, A. G., Palmer, W. A., Dhileepan, K., *et al.* (2010). Characterising insect plant host relationships facilitates understanding multiple host use. *Arthropod-Plant Interactions*, **4**, 7–17.

Manners, A. G., & Walter, G. H. (2009). Multiple host use by a sap-sucking membracid: population consequences of nymphal development on primary and secondary host plant species. *Arthropod-Plant Interactions*, **3**, 87–98.

Margolis, H. (2002). *It Started with Copernicus: How Turning the World Inside Out Led to the Scientific Revolution*. New York: McGraw-Hill.

Mayr, E. (1942). *Systematics and the Origin of Species from the Viewpoint of a Zoologist*. New York: Columbia University Press.

Mayr, E. (1963). *Animal Species and Evolution*. Cambridge, MA: Harvard University Press.

Moore, J. A. (1975). *Rana pipiens* – the changing paradigm. *American Zoologist*, **15**, 837–849.

Morton, S. R., Smith, D. M. S., Dickman, C. R. *et al.* (2011). A fresh framework for the ecology of arid Australia. *Journal of Arid Environments*, **75**, 313–329.

Najar-Rodriguez, A. J., McGraw, E. A., Hull, C. D., *et al.* (2009). The ecological differentiation of asexual lineages of cotton aphids: alate behaviour, sensory physiology, and differential host associations. *Biological Journal of the Linnean Society*, **97**, 503–519.

Nunez, M. A., & Medley, K. A. (2011). Pine invasions: climate predicts invasion success; something else predicts failure. *Diversity and Distributions*, **17**, 703–713.

O'Hara, R. B. (2005). The anarchist's guide to ecological theory. Or, we don't need no stinkin' laws. *Oikos*, **110**, 390–393.

Packert, M., Martens, J., Eck, S., *et al.* (2005). The great tit (*Parus major*) – a misclassified ring species. *Biological Journal of the Linnean Society*, **86**, 153–174.

Paterson, H. E. H. (1978). More evidence against speciation by reinforcement. *South African Journal of Science*, **74**, 369–371.

Paterson, H. E. H. (1981). The continuing search for the unkown and the unknowable: a critique of contemporary ideas on speciation. *South African Journal of Science*, **77**, 113–119.

Paterson, H. E. H. (1982). Darwin and the origin of species. *South African Journal of Science*, **78**, 272–275.

Paterson, H. E. H. (1985). The recognition concept of species. In E. S. Vrba (Ed.), *Species and Speciation* (pp. 21–29). Pretoria: Transvaal Museum.

Paterson, H. E. H. (1986). Environment and species. *South African Journal of Science*, **82**, 62–65.

Paterson, H. E. (1987). A view of species. *Rivista di Biologia Biology Forum*, **80**, 211–215.

Paterson, H. E. H. (1991). The recognition of cryptic species among economically important insects. In M. P. Zalucki (Ed.), *Heliothis: Research Methods and Prospects* (pp. 1–10). New York: Springer.

Paterson, H. E. (1993a). Botha de Meillon and the *Anopheles gambiae* complex. In M. Coetzee (Ed.), *Entomologist Extraordinary: A Festschrift in Honour of Botha de Meillon* (pp. 39–46). Johannesburg: South African Institute for Medical Research.

Paterson, H. E. H. (1993b) *Evolution and the Recognition Concept of Species: Collected Writings*. Baltimore, MD: Johns Hopkins University Press.

Paterson, H. E. H. (2005). The competitive Darwin. *Paleobiology*, **31**, 56–76.

Pierce, G. J., & Ollason, J. G. (1987). Eight reasons why optimal foraging is a complete waste of time. *Oikos*, **49**, 111–117.

Pitman, R. L., & Ensor, P. (2003). Three forms of killer whales (*Orcinus orca*) in Antarctic waters. *Journal of Cetacean Research and Management*, **5**, 131–139.

Popple, L. W., & Walter, G. H. (2010). A spatial analysis of the ecology and morphology of cicadas in the *Pauropsalta annulata* species complex (Hemiptera: Cicadidae). *Biological Journal of the Linnean Society*, **101**, 553–565.

Pysek, P., Krivanek, M., & Jarosik, V. (2009). Planting intensity, residence time, and species traits determine invasion success of alien woody species. *Ecology*, **90**, 2734–2744.

Rajapakse, C. N. K., & Walter, G. H. (2007). Polyphagy and primary host plants: oviposition preference versus larval performance in the lepidopteran pest *Helicoverpa armigera*. *Arthropod-Plant Interactions*, **1**, 17–26.

Rajapakse, C. N. K., Walter, G. H., Moore, C. J., *et al.* (2006). Host recognition by a polyphagous lepidopteran (*Helicoverpa armigera*): primary host plants, host produced volatiles and neurosensory stimulation. *Physiological Entomology*, **31**, 270–277.

Rapport, D. J. (1991). Myths in the foundations of economics and ecology. *Biological Journal of the Linnean Society*, **44**, 185–202.

Simberloff, D. (1989). Eminent ecologist. Herbert G. Andrewartha and L. Charles Birch. *Bulletin of the Ecological Society of America*, **70**, 28–29.

Sylvan, R. (1994). Illusion and illogic in evolution. *Rivista di Biologia – Biology Forum*, **87**, 191–221.

Terry, R. C., Li, C., & Hadly, E. A. (2011). Predicting small-mammal responses to climatic warming: autecology, geographic range, and the Holocene fossil record. *Global Change Biology*, **17**, 3019–3034.

Turchin, P. (2001). Does population ecology have general laws? *Oikos*, **94**, 17–26.

Velasco, L. R. I., & Walter, G. H. (1992). Availability of different host plant species and changing abundance of the polyphagous bug *Nezara viridula* (Hemiptera: Pentatomidae). *Environmental Entomology*, **21**, 751–759.

Walter, G. H. (1995). Species concepts and the nature of ecological generalizations about diversity. In D. M. Lambert & H. G. Spencer (Eds.), *Speciation and the Recognition Concept: Theory and Application* (pp 191–224). Baltimore, MD: Johns Hopkins University Press.

Walter, G. H. (2003). *Insect Pest Management and Ecological Research*. Cambridge: Cambridge University Press.

Walter, G. H. (2008). Individuals, populations and the balance of nature: the question of persistence in ecology. *Biology and Philosophy*, **23**, 417–438.

Walter, G. H., & Benfield, M. D. (1994). Temporal host plant use in three polyphagous Heliothinae, with special reference to *Helicoverpa punctigera* (Wallengren) (Noctuidae: Lepidoptera). *Australian Journal of Ecology*, **19**, 458–465.

Walter, G. H., & Donaldson, J. S. (1994). Heteronomous hyperparasitoids, sex ratios and adaptations. *Ecological Entomology*, **19**, 89–92.

Walter, G. H., & Hengeveld, R. (2000). The structure of the two ecological paradigms. *Acta Biotheoretica*, **48**, 15–46.

Walter, G. H., & Hengeveld, R. (In press). *Autecology: Organisms, Interactions and Environmental Dynamics*. Enfield, NH: Science Publishers.

Walter, G. H., & Paterson, H. E. H. (1994). The implications of palaeontological evidence for theories of ecological communities and species richness. *Australian Journal of Ecology*, **19**, 241–250.

Wilson, D. E. (2011). Introduction. In D. E. Wilson & R. A. Mittermeier (Eds.), *Handbook of the Mammals of the World*. Vol. 1. *Hoofed Mammals* (pp. 13–16). Barcelona: Lynx Edicions.

24 The intricacy of structural and ecological adaptations: micromorphology and ecology of some Aspidogastrea

Klaus Rohde

This chapter aims to show that morphological and ecological adaptations of some animal species are highly intricate, formed by evolutionary processes over many millions of years. Disturbances upsetting the value of such adaptations may lead to the extinction of species that cannot easily be replaced, at least in some habitats. As an example we use two species of Aspidogastrea studied over many years by light and electron microscopy, DNA analyses and autecological approaches including studies of life cycles, effects on hosts, niche selection and population dynamics.

The Aspidogastrea (= Aspidobothrea = Aspidobothria) is a small group of trematodes (flukes) with about 80 species. They represent the sister group of the digeneans. Cladistic analyses using a total evidence approach, i.e., 18S rDNA sequences, morphology, ultrastructure, life-cycle data and hosts, suggest that they separated from the digeneans over 400 million years ago (Littlewood *et al.*, 1999). They parasitize mollusks and vertebrates. Mollusks either serve as intermediate hosts containing the larvae and juveniles, or the entire life cycle can be completed in them. In all known life cycles, vertebrates become infected by eating infected mollusks (snails and/or bivalves). There are four families, and three of them, each with a single genus and one or two species, infect chondrichthyan fishes (sharks, rays and chimaeras); species of the fourth family infect teleost fishes and turtles. The greater number of host families infected suggests that chondrychthyans are the original hosts of aspidogastreans. Whereas digeneans multiply in the mollusk intermediate hosts, aspidogastreans do not: one egg never gives rise to more than one larva.

Lobatostoma manteri

The larva has strongly developed, muscular anterior and posterior suckers, a pharynx and short intestine (Figure 24.1A). It lacks locomotory cilia. Electron-microscopic studies have revealed the presence of nine types of sensory receptors, differing in the absence or

The Balance of Nature and Human Impact, ed. Klaus Rohde. Published by Cambridge University Press.
© Cambridge University Press 2013.

presence of a cilium, the length of the cilium, and the presence or absence of a ciliary rootlet and its shape (Rohde & Watson, 1992) (Figure 24.1E). Juveniles from snails have a remarkable variety of receptors described by electron microscopy. There are at least eight and possibly 14 types, some ciliate, others without a free cilium, some located deep below the surface (Rohde & Watson, 1989). Based on light and scanning electron microscopy, Rohde (1973) estimated the total number of receptors at around 20 000–40 000, with the greatest concentration on the anterior head lobes.

Rohde (1973, 1975) described the structure and development of the species at Heron Island, Great Barrier Reef (Figure 24.2). Maximum length of worms from the carangid teleost fish *Trachinotus blochii* is 4 mm (7 mm compressed). The very long uterus, extending through much of the anterior and posterior parts of the body, is filled with large numbers of eggs which develop during their passage through the uterus. Eggs containing fully developed larvae are laid. Intermediate hosts at Heron Island are three species of prosobranch snails belonging to three families, *Cerithium* (*Clypeomorus*) *moniliferum*, *Planaxis sulcatus* and *Peristernia australiensis*. In various experiments, eggs containing fully developed larvae were exposed to light, mechanical and temperature stimuli. Hatching could not be induced. It is therefore safe to assume that larvae normally do not hatch in the free environment. However, snails kept together with eggs containing fully developed larvae became infected. In snails dissected soon after exposure to eggs, empty eggshells and free larvae were found in their stomach. Eggshells were also present in the snails' fecal balls. Hence, we must conclude that snails become infected by eating eggs. Larvae hatch in the stomach and migrate immediately along the ducts of the digestive gland into the digestive follicles, where they feed on the secretion and probably epithelial cells of the follicles, using the acetabulum (posterior sucker) for adhesion to the epithelium. In heavily infected snails follicles disappear. Larger developing juvenile worms were found in the stomach and digestive gland of snails, in a cavity formed by enlargement of the main duct and one or more (?) side ducts of the digestive gland near the stomach in *Cerithium*, and in the stomach and main ducts of the digestive gland in the much larger *Peristernia*. They may creep between stomach and digestive gland. Snails could not be infected experimentally with juvenile *Lobatostoma* from other snails. Experiments to infect vertebrates (baby green sea turtles, various fish species) were not very successful: only a few worms were recovered after the first hours and 1 day after infection. However, there can be no doubt that final hosts, juvenile snub-nosed dart *Trachinotus blochii*, become infected by eating snails, crushing the thick shell between their pharyngeal plates (Figure 24.1B,C,D). Of 11 *Trachinotus blochii* dissected immediately after capture, six contained large numbers of shell fragments (often hundreds) in the intestinal tract, two contained only a few fragments, and three none. In the only habitat at Heron Island where *Lobatostoma* was found in significant numbers, large numbers of shell fragments of *Cerithium* were found on the sea floor. *Cerithium* fits exactly between the anterior vomer and the pharyngeal plates of the fish, apparently used to crush the snails.

Pre-adult juveniles from snails could be kept alive in sea water, sea water diluted with rainwater (1:5), Tyrode and frog Ringer solution (maximum 52 days in dilute sea water). Worms from fish died in sea water during the first 1 or 2 days, but stayed alive for a long

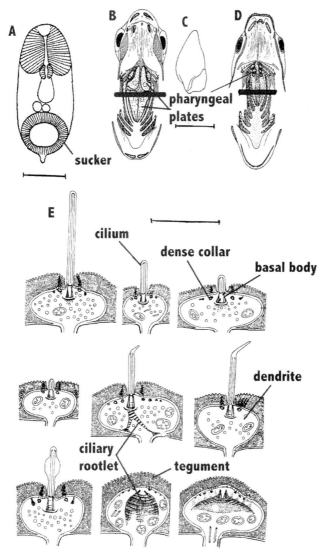

Figure 24.1.   *Lobatostoma manteri*. (A) Whole mount of larva. (B) Juvenile *Trachinotus blochii* (Carangidae), head opened along thick line, note very large pharyngeal plates used to crush thick-shelled snails. (C) *Cerithium* (*Clypeomorus*) *moniliferum*, the main intermediate host of *Lobatostoma* at Heron Island. Note that the snail fits exactly into the mouth of *Trachinotus blochii*. (D) Juvenile carangid, note that the pharyngeal plates are not enlarged; the fish is a suitable experimental host for *Lobatostoma* when force-fed with juveniles, but does not become naturally infected. (E) Sensory receptors of larva. Note that receptors differ in the absence or presence of the cilium, its length, the number of dense collars, and the presence, absence and shape of the ciliary rootlet. Scale bars 0.5 mm (A), 10 mm (B,C,D), 0.002 mm (E). Modified from Rohde, 1973 (A,B,C,D), with the permission of CUP, and Rohde and Watson, 1992 (E), with the permission of Elsevier.

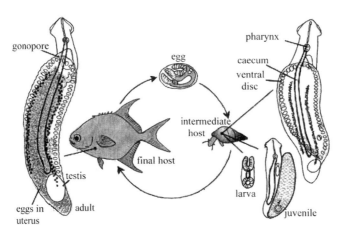

Figure 24.2.   Life cycle of *Lobatostoma manteri*. Adult worms live in the digestive tract of teleost fish (*Trachinotus blochii*). Eggs containing infective larvae are shed in the feces and are eaten by snails in which the larvae hatch and develop to juveniles. Snails are eaten by fish and the juveniles mature to adult worms in them.

time in frog Ringer solution and dilute sea water, in which they produced eggs infective to snails (maximum 13 days in frog Ringer). This indicates that worms from snails are more tolerant to osmotic pressure than worms from fish, likely to be an adaptation to their mode of infection: there is a greater variability in the osmotic pressure in the stomach of at least some species of fish, which juvenile worms enter first, than in posterior parts of the digestive tract, as shown for the flounder *Platichthys flesus* (stomach: sea water to 40% sea water; rectum 28–31% sea water) (Mackenzie & Gibson, 1970), a result – however – not obtained for some other fish (references in Rohde, 1973).

Of many fish of 27 species in 15 families examined, only juvenile *Trachinotus blochii* (Carangidae) (65–120 mm long) was found to be infected. Intensity of infection was 1–25 in 17 fish, of which only three were not infected. Great numbers of mollusks (five species of snails, one species of bivalves) were examined, but only *Cerithium moniliferum*, *Planaxis sulcatus* and *Peristernia australiensis* were found to be infected.

Rohde and Sandland (1973) and Rohde (1975) examined host-parasite relations of *Lobatostoma manteri* in snails. Whereas *Cerithium moniliferum* harbors usually a single large juvenile in the digestive gland, with metaplasia of the duct epithelium, hyperplasia of the connective tissue between the glandular follicles and amebocytes, and necrosis of some follicles, the much larger *Peristernia australiensis* harbors up to six parasites in the stomach and ducts of the digestive gland, with a thickening of the subepithelial connective tissue layer. In other words, reactions were much stronger in the former snail species. Between January 1971 and April 1972 the relative number of snails of both species infected with Digenea as well as *Lobatostoma* decreased strongly: in fact, no *Lobatostoma* at all were found in *Cerithium* in April 1972, and the population had not recovered by mid-1975 (Rohde, 1981) (Figure 24.3). Also, during the period of high frequencies of infection, *Cerithium* containing larval Digenea contained *Lobatostoma* more frequently than *Cerithium* without larval Digenea; snails infected with both

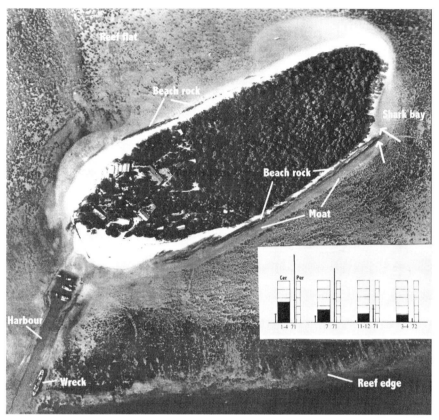

Figure 24.3. Vertical photograph of Heron Island. Inset: infections of snails (*Cerithium moniliferum* and *Peristernia australiensis*) in Shark Bay between January-February 1971 and March-April 1972. Black areas in columns indicate frequency of infection with Digenea, *Peristernia* not infected; narrow columns beside wide and medium-wide columns indicate infection with *Lobatostoma*; their total height gives density of infection = number of parasites/number of host specimens examined, height up to cross-bar gives frequency of infection: height of wide columns = 100%. The coral cay is about 900 × 300 m in size. *Cerithium moniliferum* occurs on beachrock all around the island and along the harbor, but infections with *Lobatostoma manteri* occured in significant numbers only in Shark Bay (arrows). Photograph courtesy of RAAF No. 2 squadron. Inset modified from Rohde and Sandland (1973) with the permission of Springer-Verlag.

disappeared first. During the period of high frequency of infection, the relative number of egg-producing *Cerithium* was not affected by infection with *Lobatostoma*, in spite of the severe tissue reaction caused by it.

Multicotyle purvisi

Worms reach a length of 10 mm or more (uncompressed), and their uterus is relatively short, restricted to the anterior part of the body. Many eggs are produced but shed before they can develop to advanced cleavage stages or larvae. Development to larvae occurs in eggs that have been shed into water. Larvae are covered with a layer of minute microfila, unique among the Platyhelminthes, and possess ten ciliary tufts and two ocelli (eye spots) (Figure 24.4A). Rohde (1971a) describes the development and hatching of *Multicotyle* under experimental conditions as follows. Mature specimens could be kept alive in 0.7% NaCl at 19–20°C for up to 8 days. Eggs containing 1, 2 or 3 cells were produced during the first days. They developed in rainwater. Hatching occurred at 25–31°C after 31–32 days, at 28–31°C after 26 days. Optimal temperature for development was 28–29°C, and even a minor increase in temperature led to death. Temperatures correspond to those found in Malaya, where hosts and parasites are naturally found. Cilia of ciliary tufts began to beat before hatching, and all larvae observed swam immediately after hatching. If kept at normal 24 hour light and temperature fluctuations, hatching occurred with few exceptions in the early morning hours. If cultures were kept in the dark for days, larvae hatched without the normal stimulus (light). Larvae swim with the anterior end extended, slowly rotating around the longitudinal axis in a screw-like manner, either along the bottom, straight to the surface, or irregularly in the water. They often remain attached to the surface, slowly rotating, and then sink to the bottom with the posterior end directed downwards, or they swim fast and actively to the bottom, with the anterior end directed downwards. Frequently, the larva also floats in the middle of the water column, carried sidewards by currents while its anterior end is directed upwards. At the bottom it may lie in a contracted fashion, exploring the substratum with its anterior end. Life span of larvae in water varied from 5 to over 33 hours. Experimental intermediate hosts were freshwater snails of three families, *Pila scutata*, *Taia polyzonata* and *Bithynia* (*Digoniostoma*) *siamensis*. Young larvae were found in the anterior kidney, older larvae in the posterior kidney; the route of infection within the snail host was not examined. Speed of growth depends on temperature and host species. Freshwater turtles, *Siebenrockiella crassicollis*, could successfully be infected with juveniles from snails, but feeding with eggs containing fully developed larvae or with free-swimming larvae did not result in infection. In extensive studies conducted in the early and mid 1960s, adult worms were found in the stomach and occasionally the upper part of the duodenum of *S. crassicollis* slaughtered for food at a Chinese shop in Kuala Lumpur (300 worms in 11 of 753 animals), *Cyclemys* (= *Cuora*) *amboinensis* (15 worms in 8 of 1383 animals), *Dogania subplana* (occasionally infected) and other not identified swamp and freshwater turtles. Thus, the distribution of the parasite in host populations is highly aggregated: only a few host individuals are infected, but these are usually heavily infected. Adult worms collected from naturally infected hosts had never less than 17 or 18 transverse rows of alveoli on the adhesive disk, suggesting that very young juveniles are not infective to turtles. During later visits to Malaya in the 1970s, 1980s and 1990s, turtles were much more difficult to obtain and among the few animals examined very few were infected, although quantitative records were not kept.

   The observations about the remarkable complexity of larval behavior correspond to the remarkable complexity of the nervous system and sensory receptors, studied in great detail over many years using light and electron microscopy. Thus, Rohde (1971b) demonstrated that larvae of *Multicotyle purvisi* already possess the basic pattern of the nervous system of adults. As in other platyhelminths, the nervous system is a "ladder" system, consisting of longitudinal nerve cords (connectives) connected by a large number of ring-commissures. But whereas digenean trematodes and various turbellarians among the platyhelminths have a single set of ring-commissures, in *Multicotyle* there are two, an external and an internal ring. One of the anterior rings is very large, representing the brain or cerebral ganglion. Furthermore, the number of connectives in the anterior part of the body is far greater than in any other platyhelminth examined. The alveoli of the ventral adhesive disk are innervated by a complex pattern of nerves, and there are nerve plexuses around the intestine and in a connective tissue septum separating the adhesive disk from the main body. Ultrastructural examination revealed that a part of the main ventral connective is surrounded by a sheath, not known in any other platyelminth (Rohde, 1970). Clusters of neurosecretory cells are found around the junctions of connectives and commissures dorsal to the adhesive disk. Early light-microscopic investigations which revealed a remarkable array of receptor types (Rohde, 1966) were later confirmed by electron-microscopic studies (Rohde & Watson, 1990a, 1990b, 1991). Figure 24.4 shows diagrams of the various types. Two ocelli (eye spots), located dorsally at the anterior end of the pharynx (Figure 24.4A), consist of one pigment cell and two rhabdomere cells each (Figure 24.4B). The rhabdomeres, possessing an array of microvilli, are light sensitive. Light reaches them only from certain directions, facilitating a directed response to light, i.e., ocelli very likely contribute to the positive and negative phototaxis responsible for swimming upwards or downwards. All non-ciliate (Figure 24.4D) and ciliate receptors (Figure 24.4E, F) are terminal swellings of dendrites (nerve fibers), most if not all of them with sometimes strongly modified cilia. Four receptors (D) lack free cilia at the surface, but at least two of them contain basal bodies of cilia; the other two contain dense collars and/or neurosecretory vesicles, which indicate their sensory nature. Ciliate receptors differ in the length and shape of the free cilia, the number of dense collars, and the presence or absence and shape of the ciliary rootlet. Physiological studies of the function of the various receptor types have not been done, but they are likely to contribute to host finding, migration in the intermediate host to the kidneys, and prevention of damage to the host (kidneys of snails are very delicate and parasites cannot be interested in killing the host before their development is completed).

  Comprehensive review papers on the findings about *Lobatostoma* and *Multicotyle* can be found in Rohde (1972, 1994).

Comparison of the two species

A comparative evaluation of the findings on the two species shows that they differ remarkably in their life cycles and structural complexity. *Lobatostoma* does not possess a single free-living stage: eggs are eaten by snails, and snails are eaten by fish. Correspondingly, sensory receptors of larval *Lobatostoma manteri* are relatively uniform, apparently used for finding the way from the stomach into the digestive gland. The marine environment at Heron Island with its strong currents and wave action would make a life cycle involving eggs that have to mature on the ground over a long time and very small free-swimming larvae costly. The species has responded by producing numerous fully developed eggs in a very long uterus. *Multicotyle*, on the other hand,

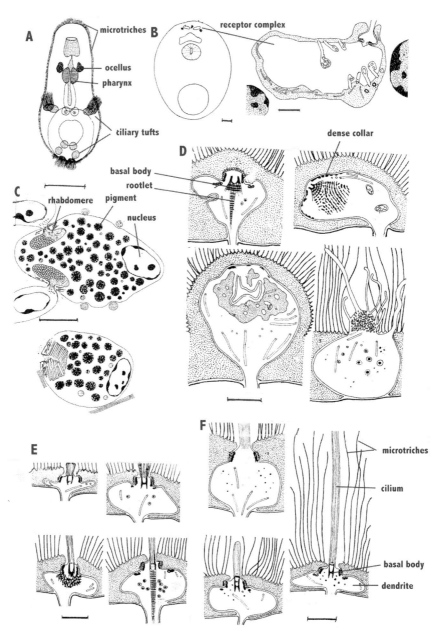

Figure 24.4. *Multicotyle purvisi.* (A) Whole mount of larva, note the layer of microtriches covering most of the animal, the 10 ciliary tufts, and the two ocelli (pigmented eyes). (B) Anterior receptor complex, two sections. note the cavities with one receptor ending in a modified cilium protruding into it, and the clusters of pendulum-like modified cilia. (C) Ocellus (eye spot); note the pigmented cell with one nucleus, and two light-sensitive rhabdomeres, each with one nucleus. (D) Four types of nonciliated receptors. (E and F) Seven types of ciliated receptors. Scale bars 0.05 mm (A), 0.01 mm (left) and 0.001 mm (right) (B), 0.001 mm (D, E and F). From Rohde, 1968 (A), Rohde and Watson, 1990c (B), Rohde and Watson, 1991 (C), Rohde 1990 (D), and Rohde, 1990 (E and F). With the permission of Springer-Verlag.

has a free-swimming larva that must reach a snail host; it achieves this by hatching in the morning when hosts apparently become active, swimming to the surface and lying suspended in the water until being inhaled by a snail. It can afford this because the freshwater habitats of the snails are calm and lack the strong currents and waves prevalent in the marine habitat of *Lobatostoma*. The great variety of receptors is necessary to secure inhalation by a snail and for finding the way to the snail's kidneys.

The variety and number of receptors of the juveniles and adults is very large in both aspidogastrean species. Possible functions of the receptors are prevention of excessive damage to the snail host, finding the way to the proper niche within the digestive tract of the final host, mate-finding and feeding. In this context, it is important that – although Aspidogastrea are hermaphroditic – mating appears to be essential. Eggs of *Lobatostoma* from single fish infections always had the haploid and not the diploid number of chromosomes, and cleavage did not proceed beyond the few-cell stage (for a discussion of mating in parasites see Rohde, 2002). For completion of the life cycle of *Lobatostoma*, the presence of final hosts able to crush the very strong shells of the snail hosts is essential. Only fish with pharyngeal plates strong enough to do this are suitable final hosts, although numerous species can be infected experimentally by force-feeding. Also essential is the availability of habitats in which eggs have a chance to be eaten by snails, and snails by fish. Although suitable snails occur all around Heron Island, significant infections occurred only in Shark Bay (Figure 24.3). Even there survival of the species is precarious. *Lobatostoma* occurred in reasonable numbers only in 1971/72, and in later years it disappeared almost completely, the result of changing migration habits of the fish, or of changing environmental conditions? In the context of the prevailing topic in this book, equilibrium or nonequilibrium, it is clear that the population of *Lobatostoma* at Heron Island is not even close to an equilibrium. Intermediate hosts are also infected with larval digenean trematodes, but there is no evidence for interspecific competition: *Lobatostoma* was more frequent where and when Digenea were also most frequent. And there is no evidence whatsoever that interspecific competition was in any way involved in the evolution of the very intricate morphological and behavioral adaptations of the two aspidogastrean species. In other words, there is no evidence for co-evolution with other, competing species. This is also the case for other aspidogastreans. *Rugogaster* infects the cecal glands of chimaeras, *Multicalyx* the gall bladder and bile ducts of chimaeras, sharks and rays, and *Stichocotyle* the bile ducts of rays. Each species has probably adapted to its host and microhabitat within the host over many millions of years. There are many more potential host species than aspidogastrean species infecting them in particular micro-habitats, although a few additional species may well be found in the future. And there are no potentially competing species of other taxa in these microhabitats. In other words, there are many vacant niches. It seems, then, that at least for the aspidogas-treans, the autecological paradigm formulated by Hengeveld and Walter (1999), and discussed in the previous chapter, is more useful than the demographic one. Demographic ecologists have not studied aspidogastreans because they are rare and unlikely to compete with each other or, to any significant extent, with species of other taxa (see Chapter 23 on "Autecology and the balance of nature").

Acknowledgment

Some aspects of this chapter are based, with permission of Cambridge University Press, on Chapter 10 in my book *Nonequilibrium Ecology*, Cambridge University Press, 2005.

References

Hengeveld, R., & Walter, G. H. (1999). The two coexisting ecological paradigms. *Acta Biotheoretica*, **47**, 141–170.

Littlewood, D. T. J., Rohde, K., Bray, R. A., & Herniou, E. A. (1999). Phylogeny of the Platyhelminthes and the evolution of parasitism. *Biological Journal of the Linnean Society*, **68**, 257–287.

Mackenzie, K., & Gibson, D. I. (1970). Ecological studies of some parasites of plaice *Pleuronectes platessa* L. and flounder *Platichthys flesus* (L.). In A. E. R. Taylor and R. R. Muller (Eds.), *Aspects of Fish Parasitology* (pp. 1–42). Symposium of the British Society of Parasitology, no. 8.

Rohde, K. (1966). Sense receptors *of Multicotyle purvisi* Dawes (Trematoda, Aspidobothria). *Nature*, **211**, 820–822.

Rohde, K. (1968). Die Entwicklung von *Multicotyle purvisi* Dawes, 1941 (Trematoda, Aspidogastrea). *Zeitschrift für Parasitenkunde*, **30**, 278–280.

Rohde, K. (1970). Nerve sheath in *Multicotyle purvisi* Dawes. *Naturwissenschaften*, **57**, 502–503.

Rohde, K. (1971a). Untersuchungen an *Multicotyle purvisi* Dawes, 1941 (Trematoda, Aspidogastrea). I. Entwicklung und Morphologie. *Zoologische Jahrbücher, Abteilung für Anatomie*, **88**, 138–187.

Rohde, K. (1971b). Untersuchungen an *Multicotyle purvisi* Dawes, 1941 (Trematoda, Aspidogastrea). III. Licht- und elecktronenmikroskopischer Bau des Nervensystems. *Zoologische Jahrbücher, Abteilung für Anatomie*, **88**, 320–363.

Rohde, K. (1972). The Aspidogastrea, especially *Multicotyle purvisi* Dawes, 1941. *Advances in Parasitology*, **10**, 77–151.

Rohde, K. (1973). Structure and development of *Lobatostoma manteri* sp. nov. (Trematoda, Aspidogastrea) from the Great Barrier Reef, Australia. *Parasitology*, **66**, 63–83.

Rohde, K. (1975). Early development and pathogenesis of *Lobatostoma manteri* Rohde (Trematoda: Aspidogastrea). *International Journal for Parasitology*, **5**, 597–607.

Rohde, K. (1981). Population dynamics of two snail species, *Planaxis sulcatus* and *Cerithium moniliferum*, and their trematode species at Heron Island, Great Barrier Reef. *Oecologia*, **49**, 344–352.

Rohde, K. (1990). Ultrastructure of the sense receptors of adult *Multicotyle purvisi* (Trematoda, Aspidogastrea). *Zoologica Scripta*, **19**, 233–241.

Rohde, K. (1994). The minor groups of parasitic Platyhelminthes. *Advances in Parasitology*, **33**, 145–234.

Rohde, K. (2002). Niche restriction and mate finding in vertebrate hosts. In E. E. Lewis, J. F. Campbell & M. V. K. Sukhdeo (Eds.), *The Behavioral Ecology of Parasites* (pp. 171–197). Wallingford: CAB International.

Rohde, K., & Sandland, R. (1973). Host-parasite relations in *Lobatostoma manteri* Rohde (Trematoda, Aspidogastrea). *Zeitschrift für Parasitenkunde*, **41**, 115–136.

Rohde, K., & Watson, N. (1989). Sense receptors in *Lobatostoma manteri* (Trematoda, Aspidogastrea). *International Journal for Parasitology*, **19**, 847–858.

Rohde, K., & Watson, N. (1990a). Non-ciliate sensory receptors of larval *Multicotyle purvisi* (Trematoda, Aspidogastrea). *Parasitology Research*, **76**, 585–590.

Rohde, K., & Watson, N. (1990b). Uniciliate sensory receptors of larval *Multicotyle purvisi* (Trematoda, Aspidogastrea). *Parasitology Research*, **76**, 591–596.

Rohde, K., & Watson, N. (1990c). Paired multiciliate receptor complexes in larval *Multicotyle purvisi* (Trematoda, Aspidogastrea). *Parasitology Research*, **76**, 597–601.

Rohde, K., & Watson, N. (1991). Ultrastructure of pigmented photoreceptor of larval *Multicotyle purvisi* (Trematoda, Aspidogastrea). *Parasitology Research*, **77**, 485–490.

Rohde, K., & Watson, N. A. (1992). Sense receptors of larval *Lobatostoma manteri* (Trematoda, Aspidogastrea). *International Journal for Parasitology*, **22**, 35–42.

Part VII

An Overall View

.

25 The importance of interspecific competition in regulating communities, equilibrium vs. nonequilibrium

Klaus Rohde

The view that competition is an important "regulatory" factor in nature is widespread among ecologists. A discussion of the evolutionary significance of interspecific competition is therefore crucial in the context of this book.

Definitions, kinds of competition, historical considerations, factors that bring about competition, and examples of the effects of competition on species and populations were discussed in detail in Rohde (2005). Here we restrict ourselves to a brief outline of the points made in that book, supplemented by evidence presented in the various chapters of the present book.

Definition, limiting factors responsible for competition, and occurrence of competition

Interspecific competition can be defined as an interaction between individuals of different species that arises because of shared requirements for a limiting resource, leading to reduced survival, growth and/or reproduction of at least some of the individuals (adapted from Begon *et al.*, 1996), although Levin (1970) has shown that resource limitation is not the only factor involved. Thus, even two species that feed on different food resources that are not in limited supply cannot indefinitely coexist if they are limited by the same predator. An important criterion for species coexistence is that limiting factors (whether food resource, predation, etc.) differ and are independent.

According to a model proposed by Edmunds *et al.* (2003), as interspecific competition becomes more intense, a scenario with two stable competitive exclusion equilibria is replaced by one with two competitive exclusion equilibria and a stable coexistence cycle. In other words, two species may well coexist on one limiting resource, even if (or because) there is increased competition.

The significance of competition in plant and animal communities varies considerably and is often doubtful. Lawton (1984a) mentions examples which suggest that certain ecological patterns may be entirely or partly determined by interspecific competition, but he points out that each of these patterns can also be explained by other hypotheses. According to Lawton, attempts to demonstrate interspecific resource limitation have

A

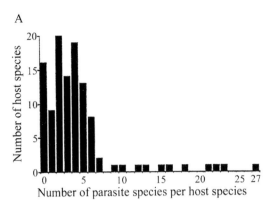

B

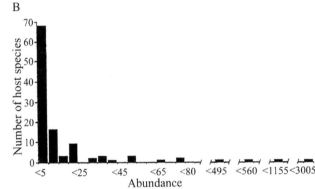

Figure 25.1 (A) Number of species of metazoan ectoparasites on the heads and gills per species of marine teleost (5666 fish of 112 species). Note: maximum number 27 parasite species, most species with less than five. If 27 is considered to be the maximum possible on all fish species, the percentage of empty niches would be 84.1%. (B) Abundance (= mean number of metazoan ectoparasites of all species per host species). Note: maximum abundance more than 3000, but most species with less than five. From Rohde, K. (1998b). Reprinted by permission of the editor of *Oikos*.

failed for insects inhabiting bracken (*Pteridium aquilinum*), because these insects are rare and therefore unlikely to compete for resources. The same applies to ectoparasites of fish, which usually occur at low population densities in habitats in which resources are not limited (blood, mucus: both in unlimited supply as long as hosts are alive: Rohde, 1991). Furthermore and importantly, hosts of parasites usually have few parasite species at low abundances, with hardly any opportunity to compete with each other (Figure 25.1) (see Chapter 6). Species of the fruitfly genus *Drosophila* are among the best-studied organisms. Barker (1983), reviewing work on competition in that genus, concluded that the idea of interspecific competition being responsible for niche segregation is attractive, but that evidence for this is not conclusive. Concerning attempts to assume past competition as explanations for present patterns, Wiens (1984) emphasized, with regard to the "ghost of competition past", that hypotheses explaining present patterns by past competition are not testable. In many communities of birds, insects, parasites and fish, among others, nonequilibrium conditions were shown to reduce the effects of interspecific competition, or competition may not exist at all (e.g., Dayton, 1971; Wiens 1974;

Sale 1977; Connell, 1979; Grime, 1979; Price 1980). Andrewartha and Birch (1984) concluded that the exaggerated emphasis on competition is fallacious, because most natural populations are rare and substantial proportions of the resources are never used up. Density-dependent factors and competition therefore do not become operative. They list three possible ways by which numbers of animals in natural populations can become limited. They are (1) shortage of resources, (2) inaccessibility of resources, and (3) shortage of time during which an increase in reproductive rates is positive. Based on the study of various animal populations, they conclude that the first is probably least and the last most important in nature.

Nevertheless, it is beyond doubt that resources are sometimes limited and that species compete for them. Brown *et al.* (1979) have shown experimentally that competition for seeds is important in determining the community structure of desert granivorous rodents, and Pimm (1978), also experimentally, has shown that, in hummingbirds, competition occurs when resources are predictable, but decreasingly so when resources are less predictable.

Character displacement, limiting similarity, competitive exclusion, niche restriction and segregation

Many authors (references in Rohde, 2005) have postulated that interspecific competition, in the course of evolution, has led to some displacement, i.e., "competitive character displacement", between species, resulting in less competition between them. The question was asked: how different must competing species be in order to avoid competitive exclusion, i.e., extinction of one or some of the competitors? In other words, under which conditions does Gause's principle apply? And, is divergence between species really necessary to permit coexistence?

May and MacArthur (1972) already realized that the limiting similarity in a deterministic system is zero, i.e., even very similar competitors will not go extinct. Hubbell and Foster (1986, cit. Rosenzweig, 1995), have shown that extinction even in identical species may take very long and, in populations of a few thousand, extinction time may be as long as speciation time; hence, Gause's principle is wrong. But nature is not strictly deterministic, natural processes often occur in a stochastic manner, and in such systems extinction still may take a long time, but where two species are too similar, one will finally be pushed over the rim by some "accident" (in Rosenzweig's words; see also Crawley, 1986).

Furthermore, character displacement can have causes other than competition. It may be fortuitous; for example, Rohde (1979b, 1991) has demonstrated that the feeding organs (pharynx, oral suckers) of monogeneans infecting the gills of the same host species and using the same food, blood, differ in size and shape, although food resources (but sometimes possibly space for attachment) are not in limited supply. More importantly, character displacement may be the result of reinforcement of reproductive barriers (e.g., Miller, 1967 and Kawano, 2002). Andrewartha and Birch (1984) have pointed out that emphasis on competition is wrong; one should expect to find similar species in similar habitats, and the

important question really should be: why can species with different ecological requirements live together? Importantly, random selection of microhabitats in largely empty niche
space may also lead to niche segregation, even if interspecific effects have never occurred
in evolutionary time and are not occurring now (Rohde, 1977b). Restriction of niches is not
necessarily the result of competition, it is to be expected in empty niche space, in which
niches are randomly selected by species (ibid.).

Rohde (1979a) gave a detailed discussion of evidence on niche restriction in parasites
and concluded, with respect to microhabitat restriction, that

(1) interspecific competition in parasites occurs and may lead to competitive exclusion
    or changes in microhabitat width in some or all co-occurring species (interactive site
    segregation), but there is no evidence that such effects lead to evolutionary changes
    and avoidance of competition, i.e., to selective site segregation;
(2) parasites with coinciding or overlapping microhabitats often show no interactions;
(3) related species commonly have widely overlapping microhabitats;
(4) the effect of species-intrinsic factors (facilitation of mating) on niche restriction is
    indicated by the finding that competing species often do not exist and cannot have
    existed in the past;
(5) circumstantial evidence suggests that niche restriction may be due to selection for
    increasing intraspecific contact and thus mating ("mating hypothesis of niche
    restriction");
(6) the probability that two species have completely coinciding niches is infinitesimally
    small and niche differences should not be used as evidence for competition.

Evolutionary divergence

Concerning tests for divergences due to competition, Connell (1980) has shown that tests
for competition are inadequate. He used strict criteria for demonstrating divergence of
competitors: (1) there actually has been divergence in resource use between competitors;
(2) competition and not some other mechanism was responsible for the divergence; and
(3) divergence has a genetic basis and is not simply phenotypic. He concludes that there is
little support for the coevolutionary divergence of competitors, and that it is probable
only in low-diversity communities, because in rich communities the likelihood of two
species being neighbors and therefore competitors over long periods is small.

Oscillations and chaos, unpredictability in the outcome of competition

Nonlinear dynamics (chaos) in populations may lead to unpredictability in the outcome
of interspecific competition (e.g., Hawkins, 1993; Neubert, 1997; examples in Rohde,
2005). The thorough studies of plankton have provided convincing evidence that under
certain conditions competition can lead to competitive exclusion (see Chapter 4, "The
paradox of the plankton"). Thus, a species with phenotypic plasticity (adaptability to

changing light regimes) always out-competed two other species without plasticity. However, in multi-species systems competition may lead to oscillations and chaos even when environmental conditions are constant, as shown for example in experiments lasting over 2300 days. There may be an intricate interplay of facilitation and competition, and it is often impossible to predict the outcome of competition.

Interactive and non-interactive communities

Wiens (1984) distinguished interactive communities (structured by interactive processes, mainly competition), and non-interactive communities (communities largely "structured" by individualistic responses of species). Although evidence in many cases is poor, it nevertheless seems that competition is of some importance in many communities. Schoener (1983) found evidence for some degree of interspecific competition in 76% of the species studied in papers reviewed by him, and Connell (1983) in not more than 43% of species. It is, however, important to note that species are not selected randomly for such studies. Very often, negative results are not reported and species are selected in which competition is likely. There is evidence for differences in the importance of competition between groups: according to Lawton and MacGarvin (1986), there are important differences between insects and small herbivorous mammals. According to them, interspecific competition is not of great importance in insects, because numbers in populations are kept low by enemies (parasites, predators, disease), whereas in small herbivorous mammals it is important (Schoener 1983); Rathcke (1976a, 1976b), Strong *et al.* (1979), Strong (1981), and Price (1980) give examples of communities in which there is no evidence for competition (see also some contributions in Esch *et al.*, 1990). For insects of *Pteridium aquilinum*, which are among the best-studied communities (see Lawton, 1999), and for other phytophagous insects (see Strong *et al.*, 1984) there is little or no evidence for interspecific competition (see also Kennedy, 1990, for helminth communities of freshwater fish). Krasnov, in Chapter 7, "Ectoparasites of small mammals: interactive saturated and unsaturated communities", has shown that, although ectoparasites of small mammals are not random assemblages, interactions are predominantly positive, apparently the result of facilitation by host behavior, immunology and ecology, or due to various parasite characteristics. Nevertheless, the only experimental study of interspecific interactions between flea larvae demonstrated strong and asymmetric interspecific competition.

Type I and type II communities

Many authors have discussed how species found in communities are acquired from the regional species pool through filters, i.e., large-scale biogeographic processes (distance, isolation), landscape filters (patch size, density, configuration) and habitat availability (Lawton, 1999, references therein). Concerning the relationship between regional and local species richness, Cornell and Lawton (1992; also Lawton, 1999, further references therein), following an approach introduced by Terborgh and Faaborg (1980), distinguish type I and type II systems, although in the real world systems may be intermediate. In type I systems there is proportional sampling, i.e., local richness rises proportional to regional richness. In type II systems interspecific competition prevents the local richness ever reaching the richness of the regional species pool.

Lawton (1999) concludes that type I communities (or communities close to them) are more common than type II communities (see also Cornell & Karlson, 1997; Srivastava, 1999 and Lawton, 2000, further references therein). And even some of the examples for type II communities are probably not correct (Lawton, 1999), and mechanisms other than interspecific competition may be responsible for them . Computer simulations by Rohde (1998a) have shown that an asymptotic relationship between local and regional species richness may simply arise from different likelihoods of species occurring in a community because of different life spans and colonization probabilities; and Rosenzweig and Ziv (1999) have shown that a linear relationship between local and regional richness cannot always be distinguished from a power curve, "echoing" species-area curves.

On the other hand, Caswell and Cohen (1993) have shown that type I patterns can also arise in communities with strong competition, if patches of species are knocked out by environmental disturbances, and Godfray and Lawton (2001) presented a model in which type I patterns may exist even if competition limits species numbers (see also Shurin, 2000). Altogether, linear or curvilinear relationships between local and regional richness do not say anything about the mechanisms responsible for the patterns.

Effects on infra- and component communities

Infracommunities of parasites are defined as all populations of all species infecting a single host individual, while component communities are all the populations of all parasite species infecting a host population. Effects of competition on infra- and component communities may be quite different, even if competition occurs. The studies by Sousa (1990, 1992, 1993) and Kuris (1990) on the community structure of larval trematodes of the snail *Cerithidea californica* at a number of sites on the Californian coast demonstrated that trematode species interact at the infracommunity level. This is indicated by the findings that mixed (multispecies) infections were fewer than expected under the null hypothesis (trematodes are randomly and independently distributed), that in the few multiple infections rediae of one species were observed to feed on rediae, sporocysts and cercariae of another in a hierarchical order, and that parasite species at the bottom of the hierarchy were replaced over time by species higher in the hierarchy. It has to be noted, however, that some of the observed interactions were predatory rather than competitive. Evidence from observations at the component community level did not show any significant effects of interspecific competition: trematode diversity increased with snail size and over time, and neither number of uninfected hosts nor variation in host size was correlated with parasite diversity; a greater proportion of older than younger snails were infected. It could not be excluded that there were some competitive interactions, but any reductions due to competition were more than compensated for by increases in both the number and equitability of other parasite species in older host populations (Sousa, 1990). Interestingly, trematodes are large in relation to their snail hosts, and interspecific effects might therefore have been expected.

Saturated and unsaturated communities, vacant niches

It seems likely that competition is greater in communities saturated with species and individuals than in unsaturated communities. As pointed out above, according to Andrewartha and Birch (1984) most animal species are rare, i.e., live at low population densities and hardly ever exhaust resources. Rohde (1980a) found the same for ectoparasites of marine fish and suggested that most free-living species are rare as well, living under conditions where not interspecific competition but selection to enhance mating opportunities and reinforcement of reproductive barriers are important. He suggested further that in large animals with great vagility, particularly birds and mammals, and in free-living insects with great vagility and large populations that can rapidly spread into "vacant niches", competition is probably important. He suggested further that competition does not reduce sizes of niches in the course of evolution, but there will be a gradual filling of niche space, partly by sympatric speciation, and by separation of overlapping niches in order to avoid hybridization (see Chapter 6, "Community stability and instability in ectoparasites of marine and freshwater fish"); niches do not expand into empty niche space, because the reproductive capacity of most species is too low to guarantee mating in suboptimal niches. The hypothesis that vagility and size and population sizes of species determine the importance of interspecific competiton leading to "structuring" of communities was tested by Gotelli and Rohde (2002) using null-model analyses. They found indeed that communities of large animals (birds, mammals) are highly structured, whereas communities of herbs and fish parasites are not (Figure 25.2). That niche size does not decrease with high diversity, in advanced tropical ecosystems, is for instance shown by the meta-analysis by Vázquez and Stevens (2004), who found no evidence for the latitude–niche width hypothesis (see also chapters on latitudinal diversity gradients).

The autecological vs. demographic paradigm

In view of the fact that conditions very frequently do not remain constant over extended periods, and because the outcome of interspecific competition is often unpredictable for various reasons, as shown in several chapters in this book, Hengeveld and Walter (1999) and Walter and Hengeveld (2000, further references therein) suggested that the autecological paradigm is more realistic than the demographic one. According to the demographic paradigm, intra- and interspecific competition are important, leading to coevolution of species by optimization processes. This is possible because abiotic environmental factors are believed to be on average constant. In the autecological paradigm, in contrast, optimization is not thought to be possible because of spatial and temporal variability of the environment. The demographic paradigm emphasizes the question of why so many species share the same resources, the autecological paradigm

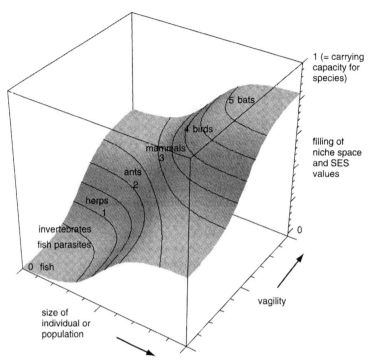

Figure 25.2.   Diagram representing filling of extant niche space as indicated by SES values (standardized effect sizes, see Gotelli & Rohde, 2002) for various vertebrate and invertebrate groups. Note that taxa comprising large (relative to their habitat) species and/or species with great vagility (bats, birds, mammals, ants) have greater SES values than taxa containing small species and/or species with little vagility. Data from Gotelli and Rohde (2002) and Gotelli and McCabe (2002). From Rohde (2005).

emphasizes the question of how species arose and how they persist in the variable and heterogeneous environment. The former assumes that nature is balanced by the establishment of population equilibria due to demographically maintained biotic processes, communities are structured and saturated with species, and emphasis is on quantification and comparison of differences in reproductive outputs of species. The latter stresses survival and reproduction, attention is paid to behavioral characters of species, and populations are considered to be temporary and dynamic aggregations of individuals in fluctuating environments. For a detailed discussion see Chapter 23 ("Autecology and the balance of nature – ecological laws and human-induced invasions").

Conclusions

White (1993) pointed to the growing evidence that interspecific competition, although it occurs, and in contrast to intraspecific competition, is relatively uncommon, and of debatable significance for evolution. Rohde (2005) concluded that much of the evidence for interspecific competition is faulty, largely because of the difficulty of formulating valid null hypotheses. Even where it occurs, its outcome may be largely unpredictable due to chaos, environmental stochasticity, etc., which reduces the likelihood that it has significant evolutionary effects, i.e., that it will lead to evolutionarily stable strategies (see Chapter 26 "Evolutionarily stable strategies: how common are they?").

Concerning equilibrium or nonequilibrium, various chapters in this book have provided evidence that both may occur, depending on environmental conditions and the groups studied. Morrison, in Chapter 9, "Island flora and fauna: equilibrium and nonequilibrium", discusses conditions that favor equilibrium or nonequilibrium. Few disturbances of low intensity, taxa with high dispersal rates, with high disturbance resistance or on close islands are among factors which tend to favor equilibrium. Forrester and Steele (Chapter 1) provide evidence that density dependence, i.e., regulation by competition, is important for reef fishes when sites ("refuges") are in short supply, whereas density-independent behavior prevails when sites are not fully utilized, i.e., when larval settlement is insufficient to colonize all the sites. Ford ("Equilibrium and nonequilibrium in Australian bird communities – the impact of natural and anthropogenic effects", Chapter 20) shows that nomadic granivorous and nectarivorous water birds in arid inland Australia undergo drastic changes in abundance both locally and regionally that vary within years and between years. On the other hand, population changes in sedentary birds are more "subdued". He postulates that climate change may lead to greater disequilibrium. According to McGill (Chapter 8), birds are a group for which equilibrial behavior could be expected to be more likely than in most other groups, because they are homeothermic and long-lived, with high parental investment in few offspring, and should therefore be well buffered against climatic variability. However, analysis of large datasets of North American birds found more support for a nonequilibrium approach, with populations showing great variability between years. There was some support for negative density-dependent regulation in populations, but less than half of the between-years variation was explained by it. Communities also showed strong variability with no trend towards a constant level, with no sign of strong species interactions. Aggregate community richness showed some indication of equilibrium; however, with long-term up or down trends. Clarke and Lawes (Chapter 5) demonstrate rapid changes between alternative stable states, and as an example gives the "complete floristic transformation of semi-arid grasslands over 26 years". Krasnov and Rizzoli ("Population dynamics of ectoparasites of terrestrial hosts", Chapter 2) conclude that strong effects of host-related and environmental factors and those related to co-occurring ectoparasites lead to nonequilibrium. Experimental and field studies by Scheffer, Huisman, Weissing and others (see Chapter 4) have shown that even in homogeneous and constant environments plankton may never reach equilibrium, because multi-species competition may lead to oscillations and chaos, contributing to the maintenance of a great biodiversity, and, importantly, in contrast to many communities in which nonequilibrium conditions occur in largely non-saturated niche space with little interspecific competition, nonequilibrium and chaos in plankton may be caused by such competition. The detailed description of the life cycles and population dynamics of two species of Aspidogastrea (Chapter 24) provide clear evidence for nonequilibrium behavior: for these species at least the autecological paradigm is more suited than the demographic one; species are far from equilibrium.

All over, nonequilibrium behavior appears to prevail, and it is likely that climate change will significantly increase it.

Acknowledgment

Parts of this chapter are based, with permission of Cambridge University Press, on the relevant chapters in my book *Nonequilibrium Ecology*, Cambridge University Press, 2005.

References

Andrewartha, H. G., & Birch, L. C. (1984). *The Ecological Web*. Chicago, IL: University of Chicago Press.

Barker, J. S. F. (1983). Interspecific competition. In M. Ashburner, H. L. Carson & J. N. Thompson Jr. (Eds.), *The Genetics and Biology of Drosophila* (pp. 285–341). London: Academic Press.

Begon, M. J., Harper, J. L., & Townsend, C. R. (1996). *Ecology*. Oxford: Blackwell Scientific.

Brown, J. H., Reichman, O. J., & Davidson, D. W. (1979). Granivory in desert ecosystems. *Annual Review of Ecology and Systematics*, **10**, 201–227.

Caswell, H., & Cohen, J. E. (1993). Local and regional regulation of species-area relations: a patch-occupancy model. In R. E. Ricklefs & D. Schluter (Eds.), *Species Diversity in Ecological Communities. Historical and Geographical Perspectives* (pp. 99–107). Chicago, IL: University of Chicago Press.

Connell J. H. (1979). Tropical rainforests and coral reefs as open non-equilibrium systems. *Symposium of the British Ecological Society*, **20**, 141–163.

Connell, J. H. (1980). Diversity and the coevolution of competitors, or the ghost of competition past. *Oikos* **35**, 131–138.

Connell, J. H. (1983). On the prevalence and relative importance of interspecific competition: evidence from field experiments. *The American Naturalist*, **122**. 661–696.

Cornell, H., & Karlson, R. H. (1997). Local and regional processes as controls of species richness. In D. Tilman & P. Kareiva (Eds.), *Spatial Ecology. The Role of Space in Population Dynamics and Interspecific Interactions* (pp. 250–268). Princeton, NJ: Princeton University Press.

Cornell, H. V., & Lawton, J. H. (1992). Species interactions, local and regional processes, and limits to to the richness of ecological communities: a theoretical perspective. *Journal of Animal Ecology*, **61**, 1–12.

Crawley, M. J. (Ed.) (1986). *Plant Ecology*. Oxford: Blackwell Scientific.

Dayton, P. K. (1971). Competition, disturbance, and community organization: the provision and subsequent utilization of space in a rocky intertidal community. *Ecological Monographs*, **41**, 351–389.

Edmunds, J., Cushing, J. M., Constantino, R. F., *et al.* (2003). Park's *Trilobium* competition experiments: a non-equilibrium species coexistence hypothesis. *Journal of Animal Ecology*, **72**, 703–712.

Esch, G. E., Bush, A., & Aho, J. (1990). *Parasite Communities: Patterns and Processes*. London: Chapman and Hall.

Godfray, H. C. J., & Lawton, J. H. (2001). Scale and species numbers. *Trends in Ecology & Evolution*, **16**, 400–404.

Gotelli, N. J., & McCabe D. J. (2002). Species co-occurrence: a meta-analysis of J. M. Diamond's assembly rule model. *Ecology*, **83**, 2091–2096.

Gotelli, N. J., & Rohde, K. (2002). Co-occurrence of ectoparasites of marine fishes: null model analysis. *Ecology Letters*, **5**, 86–94.

Grime, J. P. (1979). *Plant Strategies and Vegetation Processes*. London: Wiley.

Hawkins, A. (1993). Complex interactions between dispersal and dynamics: lessons from coupled logistic equations. *Ecology*, **74**, 1362–1372.

Hengeveld, R., & Walter, G. H. (1999). The two coexisting ecological paradigms. *Acta Biotheoretica*, **47**, 141–170.

Hubbell, S. P., & Foster, R. B. (1986). Biology, chance, and history and the structure of tropical rain forest tree communities. In J. Diamond & T. J. Case (Eds.), *Community Ecology* (pp. 314–329). New York: Harper and Row.

Kawano, K. (2002). Character displacement in giant rhinoceros beetles. *The American Naturalist*, **159**, 255–271.

Kennedy, C. R. (1990). Helminth communities in freshwater fish: structured communities or stochastic assemblages. In G. Esch, A. O. Bush & J. M. Aho (Eds.), *Parasite Communities: Patterns and Processes* (pp. 131–156). London: Chapman and Hall.

Kuris, A. M. (1990). Guild structure of larval trematodes in molluscan hosts: prevalence, dominance and significance of competition. In G. Esch, A. O. Bush & J. M. Aho (Eds.), *Parasite Communities: Patterns and Processes* (pp. 69–100). London: Chapman and Hall.

Lawton, J. H. (1984a). Herbivore community organization: general models and specific tests with phytophagous insects. In P. W. Price, C. N. Slobodchikoff & W. S. Gaud (Eds.), *A New Ecology. Novel Approaches to Interactive Systems* (pp. 329–352). New York: John Wiley & Sons.

Lawton, J. H. (1999). Are there general laws in ecology? *Oikos* **84**, 177–192.

Lawton, J. H. (2000). *Community Ecology in a Changing World*. Norbünte, Oldendorf: Ecology Institute.

Lawton, J. H., & MacGarvin, M. (1986). The organization of herbivore communities. In J. Kikkawa & D. J. Anderson (Eds.), *Community Ecology: Pattern and Process* (pp. 163–186). Melbourne: Blackwell.

Levin, S. A. (1970). Community equilibria and stability, and an extension of the competitive exclusion principle. *The American Naturalist*, **104**, 413–423.

Miller, R. S. (1967). Pattern and process in competition. *Advances in Ecological Research, 4* (pp. 1–74). London: Academic Press.

May, R. M., & MacArthur, R. H. (1972). Niche overlap as a function of environmental variability. *Proceedings of the National Academy of Sciences of the USA*, **69**, 1109–1113.

Neubert, M. G. (1997). A simple population model with qualitatively uncertain dynamics. *Journal of Theoretical Biology*, **189**, 399–411.

Pimm, S. L. (1978). An experimental approach to the effects of predictability on community structure. *The American Zoologist*, **18**, 797–808.

Price, P. W. (1980). *Evolutionary Biology of Parasites*. Princeton, NJ: Princeton University Press.

Rathcke, B. J. (1976a). Insect plant patterns and relationships in the stem-boring guild. *The American Midland Naturalist*, **99**, 98–117.

Rathcke, B. J. (1976b). Competition and coexistence within a guild of herbivorous insects. *Ecology*, **57**, 76–87.

Rohde, K. (1977b). A non-competitive mechanism responsible for restricting niches. *Zoologischer Anzeiger*, **199**, 164–172.

Rohde, K. (1979a). A critical evaluation of intrinsic and extrinsic factors responsible for niche restriction in parasites. *The American Naturalist*, **114**, 648–671.

Rohde, K. (1979b). The buccal organ of some Monogenea Popyopisthocotylea. *Zoologica Scripta*, **8**, 161–170.

Rohde, K. (1980a). Warum sind ökologische Nischen begrenzt? Zwischenartlicher Antagonismus oder innerartlicher Zusammenhalt? *Naturwissenschaftliche Rundschau*, **33**, 98–102.

Rohde, K. (1991). Intra- and interspecific interactions in low density populations in resource-rich habitats. *Oikos*, **60**, 91–104.

Rohde, K. (1998a). Is there a fixed number of niches for endoparasites of fish? *International Journal for Parasitology*, **28**, 1861–1865.

Rohde, K. (1998b). Latitudinal gradients in species diversity. Area matters, but how much? *Oikos* **82**, 184–190.

Rohde, K. (2005). *Nonequilibrium Ecology*. Cambridge: Cambridge University Press.

Rosenzweig, M. L. (1995). *Species Diversity in Space and Time*. Cambridge: Cambridge University Press.

Rosenzweig, M. L., & Ziv, Y. (1999). The echo pattern of species diversity: pattern and processes. *Oikos*, **22**, 614–628.

Sale, P. F. (1977). Maintenance of high diversity in coral reef fish communities. *The American Naturalist*, **111**, 337–359.

Schoener, T. W. (1983). Field experiments on interspecific competition. *The American Naturalist*, **122**, 240–285.

Shurin, J. B. (2000). Dispersal limitation, invasion resistance, and the structure of pond zooplankton communities. *Ecology*, **81**, 3074–3086.

Sousa, W. P. (1990). Spatial scale and the processes structuring a guild of larval trematode parasites. In G. Esch, A. O. Bush & J. M. Aho (Eds.), *Parasite Communities: Patterns and Processes* (pp. 41–67). London: Chapman and Hall.

Sousa, W. P. (1992). Interspecific interactions among larval trematode parasites of freshwater and marine snails. *The American Zoologist*, **32**, 583–592.

Sousa, W. P. (1993). Interspecific antagonism and species coexistence in a diverse guild of larval trematode parasites. *Ecological Monographs*, **63**, 103–128.

Srivastava, D. S. (1999). Using local-regional richness plots to test for species saturation: pitfalls and potentials. *Journal of Animal Ecology*, **68**, 1–16.

Strong, D. R., Jr. (1981). The possibility of insect communities without competition: Hispine beetles on *Heliconia*. In R. Denno & H. Dingle (Eds.), *Insect Life History Patterns: Habitat and Geographic Variation* (pp. 183–194). New York: Springer.

Strong, D. R., Jr., Szyska, L. A., & Simberloff, D. S. (1979). Tests of community-wide character displacement against null hypotheses. *Evolution*, **33**, 897–913.

Strong, D. R., Lawton, J. H., & Southwood, T. R. E. (1984). *Insects on Plants.Community Patterns and Mechanisms*. Oxford: Blackwell Scientific.

Terborgh, J. W., & Faaborg, J. (1980). Saturation of bird communities in the West Indies. *The American Naturalist*, **116**, 178–195.

Vázquez, D. P., & Stevens, R. D. (2004). The latitudinal gradient in niche breadth: concepts and evidence. *The American Nauralist*, **164**: E1–E19.

Walter, G. H., & Hengeveld, R. (2000). The structure of the two ecological paradigms. *Acta Biotheoretica*, **48**, 15–46.

White, T. C. R. (1993). *The Inedaquate Environment: Nitrogen and the Abundance of Animals*. Berlin: Springer Verlag.

Wiens, J. A. (1974). Habitat heterogeneity and avian community structure in North American grasslands. *The American Midland Naturalist*, **91**, 195–213.

Wiens, J. A. (1984). Resource systems, populations and communities. In P. W. Price, C. N. Slobodnikoff & W. S. Gaud (Eds.), *A New Ecology. Novel Approaches to Interactive Systems* (pp. 397–436). New York: John Wiley & Sons.

26 Evolutionarily stable strategies: how common are they?

Klaus Rohde

Here we return to the question asked in the Introduction to this book: how common are evolutionarily stable strategies and states? These two concepts were developed in the context of games theory.

Background

Games theory was developed by von Neumann and Morgenstern (1944), although the French mathematician Cournot (1838) studied some aspects, further developed by Nash (1950). Its most important contribution to evolutionary biology is the concept of the evolutionarily stable strategy (ESS). It is central to modern evolutionary ecology, and Dawkins (1976) suggests that it may be "one of the most important advances in evolutionary theory since Darwin". It was introduced into ecology by Maynard Smith and Price (1973), and can be derived from the concept of the Nash Equilibrium (Nash, 1950), according to which none of a number of players in a game can gain by changing her/his strategy unilaterally. Maynard Smith (1982) gave a detailed account of applications of game theory to evolutionary theory, including ESS. However, parts of his book rely heavily on mathematics. Dawkins's (1976) *The Selfish Gene* contains a discussion of ESS and many examples, clearly explained without any mathematics. A recent detailed review of applications of game theory and ESS to social behavior was given by McNamara and Weissing (2010).

According to Maynard Smith (1982), "An 'ESS' or 'evolutionarily stable strategy' is a strategy such that, if all the members of a population adopt it, no mutant strategy can invade". A strategy is a genetically determined behavioral "policy" ("course of action"). There may be more than one ESS for a population, and their number may even be "huge", even when interactions are relatively simple (McNamara & Weissing, 2010). The type(s) of ESS depend on many characteristics of the members of a population, such as their genetic relatedness, population size, whether members of a population can learn from previous experience, whether they reproduce asexually or sexually, whether contests are symmetric or asymmetric, etc. A symmetric game is one in which the adversaries start in

The Balance of Nature and Human Impact, ed. Klaus Rohde. Published by Cambridge University Press.
© Cambridge University Press 2013.

similar situations and can choose the same strategies with the same potential payoffs (the changes in reproductive success (fitness) due to the strategy). The game using dove-hawk strategies discussed below is an example of a symmetric game.

It is important to realize that an ESS is not necessarily a strategy that is "best" for all the members of the population, i.e., guarantees the greatest fitness (reproductive success) for them in the long term. The reason is that genes (any proportion of genetic material potentially lasting long enough for natural selection to act on it as a unit) have no "foresight". They are selected on the basis of the present conditions in their environment. Equally important is that evolution does not always lead to an ESS (McNamara & Weissing, 2010), as discussed further below, and that over-emphasis on fitness may be misleading, if genetic details are neglected (Weissing, 1996).

The following very simple theoretical example, discussed by Maynard Smith (1982) and Dawkins (1976), demonstrates the principle of an ESS. We assume that only two strategies are possible, i.e., hawk (fight as hard as possible, retreat only when badly hurt) and dove (threaten only in a mild way, never hurt the adversary). We further assume that individuals have not learned from previous experience, i.e., they cannot tell in advance who is a dove and who is a hawk. In an initial population entirely consisting of doves, all doves do well; however, the introduction of a single mutant hawk puts all the others at a disadvantage, it always beats them and the hawk gene spreads through the population. But a hawk now, with increasing frequency, encounters other hawks and gets hurt when fighting. Consequently, even a single dove is at an advantage, because it will never get hurt, and the dove gene will now become more common. A stable hawks/doves ratio is reached when the average payoff for hawks equals that for doves. There may be oscillations around the stable point, but they are not necessarily large.

Actual examples for animals are much more complex. Usually there are not two, but – in addition – other possible strategies. Important strategies are the so-called conditional strategies (strategy depends on the behavior of the adversary). Examples are that of a retaliator (behaves like a hawk when he meets another hawk, like a dove when he meets another dove), bully (treats everybody in a hawk-like spirit, but retreats when attacked by a hawk himself), prober-retaliator (usually behaves like a retaliator, but occasionally escalates the fight). For a mixture of these five possible strategies, reasoning like that used for the hawk/dove strategy leads to the conclusion that a mixture of retaliators and prober-retaliators tends to become dominant, i.e., is an evolutionarily stable strategy. And Dawkins suggests that this is indeed "often" (approximately at least) the case in nature. It is important to re-emphasize that an ESS is not necessarily the "best" strategy for all. For example, a pure dove strategy might be better, because it involves less "cost" invested in fighting, but genes, which are the units on which selection acts, cannot conspire to set up the best of all possible worlds, they have no foresight, and they are therefore exposed to the possibility of invasion by mutants which behave badly, that is, not as doves.

Other possible strategies are found, for example, in species that never fight seriously but posture only. The individual with the greatest patience wins ("war of attrition"), and the ESS is that each of the adversaries goes on posturing for an unpredictable time, the time spent on posturing depending on the value of the resource.

Most contests in nature are asymmetric; the adversaries differ for example in size, strength, fighting spirit, sex, or secondary sexual characteristics, such as antlers and plumage. Also, territory owners almost invariably win over non-owners.

The concept of an evolutionarily stable strategy is analogous to those of a developmentally stable strategy (DSS) and a culturally stable strategy (CSS) (Dawkins, 1976). In a CSS, transmission of information is not via genes, but by cultural inheritance from one generation to the next. In a DSS, transmission of information is by learning within a generation. For illustrating a DSS, Maynard Smith uses the example of a pair of pigs in an experimental "Skinner box". In the box, pressing down a lever at one end released food from a dispenser at the other. The only stable behavior pattern that established itself was as follows: the dominant pig pressed the bar and then rushed to the other end to feed, while the subordinate pig fed at the dispenser until the dominant pig arrived, and then moved "politely" away. The strategy was stable, even when the amount of food was such that the subordinate pig got more of it. The only other strategy, i.e., the subordinate pig pressing the bar, was not stable because the dominant pig would prevent it from eating.

How common are evolutionarily stable strategies in nature?

We have to address the question of how common evolutionarily stable strategies really are in nature. As pointed out by Dawkins (1976), we know little about the actual costs and benefits involved in calculating the payoffs for an ESS. Therefore, it is likely that many empirical examples that suggest an ESS are at best just that, a suggestion. Which factors determine that an ESS can be established? We consider four important aspects, (1) stability of the environment, (2) nonlinear dynamics, (3) speed of evolution, and (4) genetic constraints. (5) We also consider briefly the possible role of ESSs (or perhaps better, strategies akin to EESs), in interspecific competition.

(1) An evolutionarily stable state of a population is defined by Maynard Smith as a genetic composition which is "restored by selection after a disturbance, provided the disturbance is not too large". This shows that establishment of evolutionarily stable states and strategies critically depends, in addition to the degree of interaction between contestants and the time necessary for establishment of an ESS, on the stability of environmental conditions. In a stable world in which equilibrium is the rule, ESSs can be expected to be much more common than in an unstable world, in which nonequilibrium prevails. In other words, it is to be expected that frequent and drastic abiotic and biotic changes in the environment which affect the fitness (reproductive success) of potential contestants in evolutionary "games" will make it more difficult to establish evolutionarily stable strategies, because the establishment of an ESS cannot keep up with the changes (see also Rohde, 2005, pp.10, 13 and 34). If the establishment of an ESS takes a long time, even a single strong environmental disruption with long-lasting effects may make it impossible for an ESS to become established. An established ESS may be affected by environmental instability, for example, by reducing population size so much that encounters with possible contestants are radically reduced, facilitating the invasion by mutants which are less fit, on a more or less random basis. Alternatively, a strategy stable under

certain conditions may become less stable, permitting establishment of a different ESS. For example, it may well be that in the "war of attrition" mentioned above, the strategy of posturing in an unpredictable way and for a duration depending on the value of the resource is replaced by one which demands greater aggressiveness, because resources are severely depleted.

(2) Even long-term constant environmental conditions do not necessarily guarantee establishment of ESSs. For instance, much theoretical work has gone into investigating the evolution of sexually selected traits, of which male ornaments and elaborate behavioral displays are the best known examples (discussion in one of the most recent papers on the topic, Sander van Doorn & Weissing, 2006, further references therein). According to one hypothesis, male ornaments indicate to the female the good quality of his genes (review in Maynard Smith, 1991). However, whereas females' interests are served best by reliable ornaments, males may cheat, i.e., they may attract females even in the absence of good genes. This discrepancy may lead to complicated evolutionary dynamics, many aspects of which are not fully understood. Indeed, predictions depend to a large degree on the model used and the assumptions made. In the model used by Sanders van Doorn and Weissing, sexual conflict prevents establishment of an ESS. Equilibrium assumptions are therefore not necessarily a reliable basis for predicting the outcome of sexual selection. More generally, as shown by Abrams *et al.* (1993), evolution may even cause divergence from an ESS, that is, result in characters that minimize fitness and prevent evolution towards characters that maximize fitness. According to McNamara and Weissing (2010), evolution modeled as walks on "fluctuating fitness landscapes", i.e., landscapes that fluctuate as the result of selection, shows that evolutionary results may not be "obvious" or may even be "counterintuitive". Also, the extensive computer simulations of Wolfram (2002) have shown that in all the systems (programs) used, relatively short "instructions" led to complex patterns. The same may apply to genetic programs, i.e., the possibility must be considered that single or few mutations will frequently lead to complex characters affecting fitness, making the establishment of long-lasting stable equilibria more difficult (Rohde, 2005, pp.186–188).

(3) Rohde (1992) has proposed that the greater species richness of animals and plants in tropical habitats can be explained by an accelerated evolution there, due to direct temperature effects on generation times, mutation rates and speed of selection. Much subsequent work has provided support for this hypothesis (Rohde, 2005, pp. 159–165; e.g., Wright *et al.* 2006, 2010, 2011; Gilman *et al.* 2010; and Chapters 11 and 12 in this book on "Latitudinal diversity gradients: equilibrium and nonequilibrium explanations" and "Effective evolutionary time and the latitudinal diversity gradient"). Applied to the establishment of ESSs, it means that such strategies will develop faster in the tropics than elsewhere, provided conditions are otherwise the same, although this does not necessarily imply that the relative number of ESSs is greater (see also Rohde, 2005, pp. 186–188). It may even mean that ESSs are relatively rarer at low than at high latitudes, if origination of species is faster than establishment of ESSs.

(4) Weissing (1996, further references therein) and others have pointed out that genetic constraints may prevent maximization of fitness and hence the establishment of an ESS. However, although such constraints may be important in the short term, they

may often disappear in the long term as the result of continuing influx of mutations. Weissing concludes that (a) long-term stability (Nash strategies) can be expected only in populations at local fitness optima, (b) that "in monomorphic populations, evolutionary stability is necessary and sufficient to ensure long-term dynamic stability", and (c) that long-term stability can also be expected in non-linear frequency-dependent selection, for multiple loci, and for quite general mating systems. Nevertheless, game theoretical characterization of stability in polymorphic populations is probably not possible; many systems do not attain long-term stable equilibria; and even if such equilibria exist, it is not understood whether and how a series of gene substitution events can explain them.

(5) The original definition of ESS by Maynard Smith refers to intraspecific contests. Can one apply the concept of ESS to contests between populations of different species? Dawkins (1976) briefly discusses interspecific contests in the context of ESS, assuming that individuals belonging to the same species compete more strongly for resources than individuals belonging to different species. He gives as an example European birds: robins defend territories against other robins, but not against great tits.

Rohde (2005) has shown that interspecific competition for resources occurs, but in many cases has little evolutionary significance. For example, many parasite species, sometimes with great infection intensities, may be found on the gills of one species and even on one individual of freshwater or marine fish. Different species live intermingled in the same microhabitat on the gills and use the same food, they segregate only when they co-occur with individuals of a species belonging to the same genus (congenerics), not because they compete for resources, but because selection has led to the avoidance of contact with congenerics in order to prevent hybridization (for details see Chapter 6, "Community stability and instability in ectoparasites of marine and freshwater fish"). More generally, the outcome of "contests" between species is very often unpredictable, as shown by theoretical and experimental studies which demonstrated occurrence of chaos and unpredictability in multispecies contests of plankton species (e.g., Huisman & Weissing, 2001; Rohde, 2005, pp. 65–67; Chapter 4 on "The paradox of the plankton" in this book). Furthermore, the prevalence of vacant niches in many ecosystems (Rohde, 2005, pp. 39–48) reduces the significance of interspecific competition and with it the likelihood that stable communities can evolve. On the other hand, competition may lead to nonequilibrium dynamics, also preventing establishment of a "stable" community structure (see Chapter 4, "The paradox of the plankton"). Our conclusion, then, is that ESSs (or better, strategies akin to EESs) which "regulate" interactions between populations of different species are not likely. "Structure" of communities is often unstable in an evolutionary sense and largely determined by evolutionary "accidents". Various statistical tests applied to parasite communities have demonstrated that this is indeed the case (see Chapter 6, "Community stability and instability in ectoparasites of marine and freshwater fish"). The same is likely for communities of many other animals (examples in Rohde, 2005,). For a detailed discussion of interspecific competition see Chapter 25, "The importance of interspecific competition in regulating communities, equilibrium vs. nonequilibrium".

In summary, evolutionarily stable strategies are most likely when conditions remain relatively unchanged, and they develop faster in the tropics. They are less common when conditions tend to change significantly over relatively short time spans, and they develop more slowly at high latitudes. Even under long-term environmental stability, nonlinear evolutionary dynamics may prevent establishment of ESSs and may even lead away from them. The establishment of strategies akin to ESSs between populations of different species that structure multi-species communities by interspecific competition is unlikely.

Acknowledgments

This is an updated and revised version of an article originally published as an appendix on the website of my book *Nonequilibrium Ecology*. Included here with permission of Cambridge University Press. I wish to thank Dietrich Stauffer for critical comments on the earlier version.

References

Abrams, P. A., Matsuda, H., & Harada, Y. (1993). Evolutionarily unstable fitness maxima and stable fitness minima of continuous traits. *Evolutionary Ecology*, 7, 465–487.

Cournot, A. (1838). *Recherches sur les principes mathématiques de la théorie des richesses.*

Dawkins, R. (1976). *The Selfish Gene*. Oxford: Oxford University Press.

Gillman, L. N., Keeling, D. J., Gardner, R. C., & Wright, S. D. (2010). Faster evolution of highly conserved DNA in tropical plants. *Journal of Evolutionary Biology*, 23, 1327–1330.

Gillman, L. N., McCowan, Luke. S. C., & Wright, S. D. (2012). The tempo of genetic evolution in birds: body mass and climate effects. *Journal of Biogeography*, 21 May. doi: 10.1111/j.1365-2699.2012.02730.x.

Huisman J., & Weissing F. J. (2001). Fundamental unpredictability in multispecies competition. *The American Naturalist*, 157, 488–494.

Maynard Smith, J., & Price, G. R. (1973). The logic of animal conflict. *Nature*, 246, 15–18.

Maynard Smith, J. (1982). *Evolution and the Theory of Games*. Cambridge: Cambridge University Press.

Maynard Smith, J. (1991). Theories of sexual selection. *Trends in Ecology & Evolution*, 6, 146–151.

McNamara, J. M., & Weissing, F. J. (2010). Evolutionary game theory. In T. Székely, A. J. Moore & J. Komdeur (Eds.), *Social Behaviour: Genes, Ecology and Evolution*. Cambridge: Cambridge University Press.

Nash, J. (1950). Equilibrium points in n-person games. *Proceedings of the National Academy of Sciences of the USA*, 36, 48–49.

Rohde, K. (1992). Latitudinal gradients in species diversity: the search for the primary cause. *Oikos* 65, 514–527.

Rohde, K. (2005). *Nonequilibrium Ecology*. Cambridge: Cambridge University Press.

Sander van Doorn, G., & Weissing, F. J. (2006). Sexual conflict and the evolution of female preferences for indicators of male quality. *The American Naturalist* 168, 742–757.

von Neumann, J., & Morgenstern, O. (1944). *Theory of Games and Economic Behaviour*. Princeton, NJ: Princeton University Press.

Weissing, F. J. (1996). Genetic versus phenotypic models of selection: can genetics be neglected in a long-term perspective? *Journal of Mathematical Biology*, **34**, 533–555.

Wolfram, S. (2002). *A New Kind of Science*. Champaign, IL: Princeton, NJ: Wolfram Media.

Wright, S., Keeling, J., & Gillman, L. (2006). The road from Santa Rosalia: a faster tempo of evolution in tropical climates. *Proceedings of the National Academy of Sciences of the USA*, **103**, 7718–7722.

Wright, S. D., Gillman, L. N., Ross, H. A., & Keeling, D. J. (2010). Energy and tempo of evolution in amphibians. *Global Ecology and Biogeography*, **19**, 733–740.

Wright S. D., Ross H. A., Keeling D. J., McBride P., & Gillman, L. N. (2011). Thermal energy and the rate of genetic evolution in marine fishes. *Evolutionary Ecology*, **25**, 525–530.

27 How to conserve biodiversity in a nonequilibrium world

Klaus Rohde, Hugh Ford, Nigel R. Andrew and Harold Heatwole

Why should we conserve biodiversity?

Peter Sale, in his important book *Our Dying Planet* (2011), presented and critically examined economic and ethical/esthetic arguments for conserving biodiversity. We refer the reader to that book for an overview. An ethical responsibility of humans towards nature is usually denied: man has responsibility towards other humans but not towards animals. There are exceptions, for example in Schopenhauer's philosophy compassion with fellow humans and animals is the foundation of ethical behavior (Rohde, 2010). In other words, man has the responsibility not to harm any animal needlessly but to safeguard its survival, which implies protection of its habitat, and this may well be an attitude held by many.

The esthetic value of protecting biodiversity is even more controversial. It is almost impossible to define such a value. One is left with pointing out that many of the most important works of art were and are inspired by nature, by a forest, a plant, an animal, and that people enjoy forests and other undisturbed habitats. In the Italian Renaissance, the period when Western modern culture really took off, the development of science and the artistic appreciation of nature's beauty went hand in hand, sometimes in the same person (e.g., Leonardo da Vinci). Some great physicists (Einstein for example) have used the esthetic beauty of mathematical equations as evidence for their truth. One can argue that esthetics is as defining for humanity as is scientific exploration. It should not be forgotten that humans evolved in environments with rich floras and faunas, and that change to a life surrounded by concrete and in an environment vastly impoverished from its previous condition could have unforeseen consequences for mental and physical health.

However, many if not most people will not be convinced by ethical and esthetic arguments. What counts most in the world in which we live is the economic value of things, and there evidence for the enormous value of many species and ecosystems is overwhelming. A few examples may suffice. The most diverse ecosystems in the oceans, coral reefs, support a wide variety of fish and other marine organisms used by man. Coral reefs, as we know them today, may be extinct in a few decades (see Chapter 22, "The futures of coral reefs"). Mangroves serve as breeding grounds for fish, crabs and prawns, which form the major source of protein for many people, and they buffer the coastline against erosion and wave

The Balance of Nature and Human Impact, ed. Klaus Rohde. Published by Cambridge University Press.
© Cambridge University Press 2013.

action. According to Sale (2011), during the large Indonesian earthquake/tsunami coasts without mangroves suffered much more damage than did those protected by mangroves.

A majority of agricultural plant species (providing one-third of human nutrition) depend on pollinators for reproduction. The European honeybee (*Apis mellifera*) is one of the most important pollinators, and it is threatened by human impact, such as the use of insecticides including neonicotinoids (Henry *et al*. 2012; and Whitehorn *et al*., 2012, for the bumblebee *Bombus terrestris*), as well as by mites and insect diseases caused by viruses and fungi. For example, up to 50% mortality has been reported from parts of Northern Ireland due to CCD (colony collapse disorder). More recently, the Asian honeybee *Apis cerana* has been introduced into northern Australia; it is more difficult to manage than is the European species, produces less honey and is the natural host of parasites which – if introduced into Australia – could spread to the European honeybee.

Even species that do not strike us as particularly useful, such as parasites and harmful microbes, may contribute significantly to the functioning of ecosystems, although we know little about those functions. Many plant and animal species may be the source of yet unknown substances useful to humans; thus, marine algae are used already for "alginate" bandages, and the search for useful compounds continues. Frogs have a multitude of cutaneous secretions that have provided a pharmacopeia of medicines, second only to plants (Erspamer, 1995); many amphibians are rapidly declining toward extinction (see Chapter 18) and will probably take important potential pharmaceuticals with them into extinction before humans have examined them for potentially life-saving products. Marine plankton is at the base of the marine food chain, and preserving its diversity and abundance is of importance to humans. Marine phytoplankton has been declining over the past 100 years (Boyce *et al*., 2010). Introductions of plant or animal species by man have had destructive effects on crops and native faunas and floras; the number of such species is large (see Global invasive species database) and all efforts should be made to avoid a further increase in numbers. Unnecessary pollution by chemicals including some pharmaceuticals represents an increasing danger to ecosystems, crops, fisheries and our health (SixWise.com), and can be expected to increase further with human population growth.

The impact of humans on biodiversity

Human activity is responsible for much environmental degradation and loss of biodiversity. According to Mike Sandiford, Professor of Geology and director of the Melbourne Energy Institute, University of Melbourne, the attitude among many that human impact on the environment cannot possibly be so significant as to cause serious climate change is demonstrably wrong (Sandiford, 2011). He gives some examples to stress this point. In geological history rivers and glaciers have moved approximately 10 billion tonnes of sediment from the mountains to the sea each year. Now, each year 7 billion tonnes of coal and 2.3 billion tonnes of iron ore are mined, and in addition about the same amount is shifted to create access to the mines. Natural erosion in Australia is responsible for the removal of about 50–100 million tonnes each year, but in a single planned mining development 14 billion tonnes of rock will be extracted over 40 years. Even worse are the effects that become apparent when we consider human use of energy. One thousand billion tonnes of carbon dioxide were emitted during the twentieth century by burning of fossil fuels and by cement production. Another 30 billion tonnes are added now each year and the rate of this emission doubles about every 30 years. The rate of energy consumption now is 16 000 billion watts annually. All this adds up to the result that the consumption rate of energy by humans exceeds one-third of the Earth's natural rate of heat-loss. We will release more energy annually by 2060 than is generated by the processes due to plate tectonics (earthquakes, volcanoes). Before the end of the century, we will consume every second as much energy as was generated by the Hiroshima atomic bomb. Ocean acidity has increased by about a quarter since before the industrial revolution. Measurements of temperature in boreholes in Australia give results consistent with measurements made by climate scientists. Human activities have altered the surface of more than half of all ice-free land that is habitable. Considering all this, some geologists have proposed the term "Anthropocene" for our present geological epoch, which is an era dominated by human-induced environmental degradation. This is leading to a mass extinction resembling previous mass extinctions in geological history, when it is estimated more than 90% of species became extinct (Sandiford, 2011) (for further details see Chapter 13 "The physics of climate: equilibrium, disequilibrium and chaos").

Many benefits to humans from ecosystems only become obvious when humans have destroyed or damaged them, as shown by the following example of forests. A high diversity of tree species in forests may provide stability in the face of environmental impact. The conservation of forests is critical to prevent landslides and preserve the capacity of the soil to absorb water. Thus, deforestation in the mountains probably contributed to large-scale flooding and landslides at the foot of the Himalayas and in other parts of the world. The huge wildfires that occurred in Russia in 2010, never before recorded in Russian history, were very likely caused by human activities (Wikipedia, 2010). Extreme weather conditions over the past decade (flooding, droughts, temperature extremes, hurricanes, etc.) in various countries are not normal, and according to recent evaluations could be linked to climate change (*Science Daily*, 2012).

Large predatory animals at the top of the food chain are important members of ecological communities, and their removal by hunting or fishing has important consequences for the functioning of ecosystems, and may perhaps even lead to the collapse of these systems. For instance, the removal of top predators from fragmented landscapes may lead to an increase in the number of smaller predators, which then have a disproportionate impact on their prey. In some parts of the world predation by medium-sized mammals and birds on birds' nests near the edge of a patch results in high failure rates. Subsequently, loss of insectivorous birds may in turn result in outbreaks of herbivorous insects, which have an impact on trees and other plants (Mäntylä *et al.*, 2011). This is just one example of a trophic cascade, with potentially many species going extinct following the loss of a few species.

Overriding all other human impacts is the specter of human-induced climate change. There is abundant evidence of extension or contraction of ranges, leading effectively to whole communities shifting poleward (see Chapter 10, "The dynamic past and future of arctic vascular plants: climate change, spatial dynamics, and genetic diversity"). In Europe between 1990 and 2008 butterfly and bird assemblages effectively moved 37 and 114 km northward respectively (Devictor *et al.*, 2012). However, both lagged far behind the mean temperature, which moved 249 km northwards over the same time. This indicates that communities are indeed currently in nonequilibrium. The lag between birds and butterflies, taken as a surrogate for all insects, means that birds may be lagging behind changes in their prey. There is also evidence that within localities, the peak in abundance of major foods may precede the peak in breeding of insectivorous birds. Different levels of response to climate change have the potential to lead to species experiencing new predators, parasites and competitors (see Chapter 15).

Arctic ecosystems are likely to experience large temperature increases and this will have a major impact on plant and animal communities. However, because Arctic species have survived repeated warming episodes in the past, effects on many plant species will be largely on their geographical distribution, although restriction of their ranges will lead to reduction in gene diversity, in the worst-case scenario in all 27 species of vascular plants, of which one-third will lose more than 50% of gene diversity (see Chapter 10, "The dynamic past and future of arctic vascular plants: climate change, spatial dynamics, and genetic diversity").

The consequences of inaction could wipe out many species and ecosystems on which human civilization as we know it depends. Global plant species richness will be profoundly affected by climate change, and predictions suggest that, "while in most temperate and arctic regions, an increase of . . . plant species richness is expected, the projections indicate a strong decline in most tropical and subtropical regions. Countries least responsible for past and present greenhouse gas emissions are likely to incur disproportionately large future losses . . . whereas industrialized countries have projected moderate increases" (Sommer, *et al.* 2010). All changes in regional species richness of plants, whether at high or low latitudes, will probably threaten native floras. In consequence, numerous animals and micro-organisms will also be threatened.

Given the huge impact of humans on nature, how can we preserve the diversity of life on Earth? Each species of animal or plant, each habitat and each ecosystem may require a different approach for maintaining diversity, which would be very difficult to implement. Some authors in this book have addressed the problem for their group or ecosystem, such as amphibians, birds and coral reefs. Here we emphasize two points common to all systems, namely (1) the necessity to make allowance for environmental fluctuations (nonequilibrial conditions) in conservation programs, and (2) the necessity to provide straightforward and correct information to the public.

Concerning (1), the first question to be answered is:

What do we know about the factors responsible for the loss of biodiversity?

In spite of continual progress, we still do not fully understand how most ecosystems function, for instance what consequences the removal of certain plant and animal species will have on the structure and composition of ecosystems. Much ecological theory and its application to conservation and management is built around the concept of equilibrium in populations and communities. A common approach is to set aside fragments of natural but disturbed land, call them nature reserves or national parks, and manage them by benign neglect on the assumption that they will return to their original state. Such an approach is unrealistic: it does not consider how nonequilibrial fluctuations affect the survival of species and the structure of ecosystems (see for example Chapter 21, "Population dynamics of insects impacts of a changing climate").

Chapter 23, "Autecology and the balance of nature – ecological laws and human-induced invasions", shows that each species is differently adapted to its environment and for each species different measures are required to safeguard its survival in a changing world. Chapter 24, "The intricacy of structural and ecological adaptations: micromorphology and ecology of some Aspidogastrea", describes a detailed example of how intricate morphological and ecological adaptations can be, making replacement after extinction difficult or impossible. The chapters on "The paradox of the plankton" (Chapter 4), "The importance of interspecific competition in regulating communities, equilibrium vs. nonequilibrium" (Chapter 25) and "Evolutionarily stable strategies: how common are they?" (Chapter 26) show that a view of nature as a neatly regulated mosaic of populations that return to earlier equilibria after significant disturbance is false. This makes it very difficult to predict how communities will respond both to human impacts and to conditions when such impacts are removed. We do know, however, that unexpected ecological changes may happen very rapidly (see for example Chapter 5, "A burning issue: community stability and alternative stable states in relation to fire"). In view of what is at stake, we cannot wait until a collapse with disastrous consequences has occurred: we need to plan for it and take precautionary measures.

Probably the greatest impact of humans to date on land has been the destruction, fragmentation and degradation of natural ecosystems, principally for agriculture, urbanization and timber; in aquatic systems fisheries have had a major impact. Extinctions will be frequent in patches of remnant vegetation, which may include nature reserves, because many populations are small and at risk from random fluctuations. If the total area of lost habitat is large, then there is likely to be ongoing extinction into the future. There has been much debate about the size of habitats that must be preserved to protect threatened species, about the effectiveness of corridors connecting such habitats, whether conservation measures should concentrate on particular species or entire ecosystems. Such discussions are usually based on present conditions, often ignoring the impact of more severe environmental fluctuations, which might occur in the future, or poleward movement of ecosystems, which would lead to a shrinking or shifting of habitats and disruption of corridors, making conservation measures ineffective.

The example of amphibians, discussed in Chapter 18, "Worldwide decline and extinction of amphibians", makes the difficulties facing effective conservation programs particularly clear. Cutaneous respiration of amphibians makes them highly vulnerable to desiccation, effects of toxic substances, endocrine disruptors, and changes in the abiotic environment. They undergo seasonal migrations between aquatic and terrestrial habitats, exposing them to different sets of predators and parasites.

Decline and extinction, in some cases, could be traced to pollution, habitat destruction and fragmentation, diseases such as chytridiomycosis, UV radiation as the result of the depletion of the ozone layer, and invasive species. Causes of decline and extinction may vary among localities and species, and may be synergistic, but we know little about such synergistic effects and urgently need baseline studies before effective conservation programs can begin. Climate change will very likely become increasingly important and lead to extinctions, suggested by studies in Monteverde, Chile, where 40% of amphibian species declined rapidly following particularly warm and dry weather. Importantly, loss of a single taxon may have cascading effects, affecting the entire ecosystem, particularly important for amphibians which play dual roles (i.e., are composed of "econes", in the terminology of Heatwole), being members of aquatic and terrestrial ecosystems at different stages of their life cycle. Conservation measures mentioned by Heatwole include reduction in pollution, safe passages under roads, elimination of introduced species and connections between fragmented habitats. However, all these measures proceed from the situation as it exists now. It is vastly more difficult to incorporate future environmental fluctuations (nonequilibrium scenarios) in planning programs, because we do not know how strong such fluctuations will be or how frequently they will occur. Therefore it is essential that the root causes of mass extinctions and decline, that is human over-population and over-consumption, are addressed.

In Chapter 19, "Climatic change and reptiles", Lillywhite states that "every major taxon of reptile is threatened with ongoing diminishment of abundance and diversity, largely a result of anthropogenic causes". "Because of the scale of changes to habitat, and the rapidity with which new threats to communities of organisms appear, it is difficult to separate causation related to climate from a myriad other factors. Indeed, the various threats to wildlife are often interrelated and synergistic." Studies of 48 lizard species in Mexico have shown that 12% of local populations have disappeared since 1975, very likely as the result of climate change. On the other hand, species at higher latitudes may benefit from warmer conditions. "Despite a growing body of research attempting to demonstrate or predict the influence of climatic change on reptiles, the base of evidence required to predict the impacts of climatic change remains relatively poor."

Insect population dynamics will be profoundly affected by climate change, but, as in the case of amphibians and reptiles, studies attempting to elucidate mechanisms are in their infancy (see Chapter 21, "Population dynamics of insects: impacts of a changing climate").

All chapters point to the urgent need for more studies. Such studies must incorporate an assessment of the potential for the development of new, "emerging" diseases, which can affect humans and animals, but about which we know next to nothing (see Chapter 15). Increasing globalization opens the door for the introduction of new agents of disease, and elevated temperatures raise the likelihood of the rapid spreading of disease even further. Again: we know next to nothing about the effects nonequilibrium conditions might have on the spread of diseases, although it is certain that they make forecasts more difficult.

How to deal with nonequilibrium in conservation programs

Here we discuss the views of a number of authors, including those who have contributed chapters to this book.

Zimmerer (2000) and Wallington *et al.* (2005) reviewed applications of nonequilibrium concepts to the conservation of biodiversity. Zimmerer pointed out that the balance of nature is increasingly questioned and that "the emphasis on flux is a bold contrast to conservation principles that are rooted in the belief of nature-tending-toward-equilibrium". The latter are still at the basis of most conservation programs, although "let-burn policies on natural fire disturbances, provisions for unexpected fluctuations of wildlife populations" are guided by nonequilibrium assumptions. As pointed out by Zimmerer, however, such management measures are aimed at temporal properties, whereas spatial parameters "are almost entirely based on equilibrium assumptions". Wallington *et al.* (2005), citing various authors, also stress that "many conservation policies and plans are still based on equilibrium assumptions". According to them, nonequilibrium ecology recognizes that disturbances such as fire, insect plagues, etc., are intrinsic properties of communities and not external events that must be eliminated. Furthermore, history is important in determining the structure of ecosystems, which cannot be fully explained by factors such as soil and temperature. Disturbances may "flip" a system from one into another "stability domain" (see Chapter 5, "A burning issue: community stability and alternative stable states in relation to fire"). Since humans are integral parts of ecosystems, research should not be entirely on "pristine" areas, but include systems altered by humans as they are today. Landscapes are mosaics changing all the time, either "in different stages of succession" or in "shifting mosaic steady states" (see Chapter 14, "Episodic processes, invasion and faunal mosaics in evolutionary and ecological time").

An example of a flip, as mentioned by Wallington *et al.*, is when ecosystems have alternative equilibrium states (see Chapter 17, "Anthropogenic footprints on biodiversity", and Chapter 5, "A burning issue: community stability and alternative stable states in relation to fire"). Small environmental changes, either natural or human-induced, may cause the replacement of one ecosystem by another. There is a dynamic balance between grass-dominated savannas, with scattered trees, and dense scrub or forest. The level of grazing and dominant grazing animals, the frequency of fire and disturbance by humans may all contribute to whether the vegetation is savanna or forest. In southern Africa, Wigley *et al.* (2010) showed that the area of forest had increased on land subject to a wide range of uses. These included nature reserves with the full suite of native grazers, commercial land used for grazing cattle and communal land being subjected to intensive use, including browsing by goats and collection of firewood. They proposed that the increase in forest trees and loss of grasses was due to an increase in the CO_2 level. The change is not simply an increase in the number of trees in savannas, but a complete "biome shift" to forest, which differed from savanna in structure, composition and properties, such as flammability (Parr *et al.*, 2012). This shift has "cascading functional consequences", with the potential loss of many savanna species (Parr *et al.*, 2012).

Current human management may be unable to maintain savanna without addressing the issue of climate change.

Wallington *et al.* (2005) summarized the implications of modern views on ecology as follows: conservation reserves are not static entities, hence simply protecting them does not guarantee that they will remain in their present state or return to their previously undisturbed state. Natural disturbance regimes must be maintained. Because of the importance of historical factors such as land use, which make it impossible to "explain" the composition of ecosystems exclusively by climate, soil and geomorphology, it "will likely be difficult to predict impacts of climate change" in detail. Conservation managers should keep a "watching brief" over ecosystems, since variable rates of change have to be addressed continually. Single components of an ecosystem should be targeted, but conservation of particular species must take the entire system into consideration. Small reserves cannot "be left to their own devices or be managed in isolation". Altogether, conservation management must not be passive but active: "Conservation efforts that attempt to wall off nature and safeguard it from humans will ultimately fail" (Wallington *et al.*, 2005). An understanding of alternative stable states is important (see Chapter 5 "A burning issue: community stability and alternative stable states in relation to fire"). It is very important to include "societal values in decisions about biodiversity management", that is, to involve the public.

Coral reefs, discussed in detail by Peter Sale in Chapter 22, "The futures of coral reefs", provide a good example for demonstrating the devastating impact that humans have on the environment. He writes "Coral reefs, as we knew them in the 1970s, are likely to have disappeared entirely from the planet by 2050, if current trends in human environmental impacts continue". The main reasons, according to him, are increasing atmospheric concentrations of greenhouse gases and ocean acidification. He is pessimistic that humanity can change its habits sufficiently to reverse the process or even stop it. It is likely that changes will not be gradual but occur in spurts, because of the essentially nonequilibrium nature of the biotic and abiotic environment. This will make effective conservation measures even more difficult, in particular because various vested interests are conspiring against them (see the last section of this chapter).

In view of our limited knowledge of possible developments (see Chapter 22 on "The futures of coral reefs") "exact" forward planning is hardly possible. This also applies to actions regarding consequences of climate change. As already pointed out, we cannot expect a gradual warming; increases in temperature will very likely occur in spurts, leading to extreme events such as heavy rainfall, droughts, cyclones, etc., and will differ in different parts of the world. Even rises in sea levels will not be the same in all seas (Han *et al.*, 2010). Reduction in the emission of greenhouse gases is essential, but political obstacles (as discussed below) will make effective programs difficult, and this includes efforts to reduce human population growth, opposed by many religious groups. Chapter 15 illustrates the effects nonequilibrial conditions and warming will have on human and animal health, and it should serve as a warning to those who oppose effective measures to prevent catastrophic rises in temperature.

Considering the complex interactions of various anthropogenic stressors, such as loss and degradation of habitats, overexploitation, species invasions and climate change,

Mora and Zapata (Chapter 17) conclude that "threatened species should be managed as if all stressors at play were responsible for their decline. Arguably, a less contentious strategy could focus on what drives such stressors in the first place, which will probably reveal the role of our patterns of consumption and ongoing population growth". Considering the likely increase of conditions leading to nonequilibrium induced by man, one should add that all measures should assume a bad-case scenario, that is, make allowance for more extreme weather events like droughts, flooding, storms, etc., than are common today.

The societal role in combating overexploitation and climate change: information policy

Concerning the issue of providing correct information to the public, it is essential that media publish scientifically correct and up-to-date information to assist the wider community in addressing the impacts of human-induced climate change and other human-induced adverse effects on the environment, such as over-exploitation in fisheries and hunting, agriculture, deforestation, habitat fragmentation and species invasions (see Chapter 17 by Mora and Zapata). People must be convinced that costly measures are necessary, because, as stated by Lackey (2004, cit. Wallington *et al.*, 2005), "the choice of restoration and management goals should ultimately be a societal one". In other words, an important component of any planning is political. Scientists cannot and should not avoid involvement in the politics of climate change and environmental degradation in general.

The Internet now permits effective involvement in NGO campaigns, for example against deforestation in the Amazon, Borneo, Sumatra or Tasmania, or in favor of effective water management policies in southeastern Australia. It also provides easy access to a great range of sources; however, an oversupply of data may drown useful and important information, and studies have shown that the vast majority of people still receive most information on news from TV and the printed press, i.e., newspapers and magazines. Thus, research by the Australian Broadcasting Authority several years ago indicated that 88% of the population relied on free-to-air television, 76% on radio and newspapers, 10% on pay television and 11% on the Internet as their sources of news and of information on current affairs. It is likely that the relative importance of the Internet has somewhat increased since the survey, but overall television and the printed press are still very important.

How objective is the information supplied by the media? We give just some examples concerning climate change to show that it is often far from being objective. In the USA, the Murdoch-run *Wall Street Journal* refused to publish a letter from 255 scientists from the National Academy of Sciences supporting the mainstream view on climate change (held by 97% of scientists involved in climate research), defending "the rigor and objectivity of climate science" and calling for an end to "McCarthy-like threats" of criminal persecution of researchers (Gleick, 2012). However, it published a letter supporting an opposing view (*The Wall Street Journal*, 2012; Trenberth, 2012) signed by 16 scientists, few of them active in climate research, and at least half in some way involved with the giant oil company Exxon.

Some states in the USA have introduced legislation to include "skepticism" about climate change in school curricula.

A more detailed examination of Australian media, with which we are most familiar and which we therefore discuss in greater detail, displays the obvious bias in favor of climate skeptics over mainstream scientists. Lord Monckton, the climate change denier who has no background in climate research, is an effective speaker touring the world and sponsored by mining magnates such as the mining boss and now (May 2012) richest woman in the world, Gina Rinehart. According to a press report he appeared in the media 455 times during his recent visit to Australia. In contrast, the climate scientist James Hansen, head of the Goddard Institute for Space Studies and adjunct professor of Earth and Environmental Sciences at Columbia University, who visited Australia at about the same time, appeared just 61 times. Lord Monckton had earlier in the USA referred to the distinguished Australian economist Professor Ross Garnaut, the senior advisor on climate change to the Australian government, as a Nazi because of his active involvement in policy on climate change (the charge was later withdrawn). Leading climate scientists in Australia are routinely subjected to offensive emails (Smith, 2011).

Australian media are extremely concentrated. The two largest Australian media corporations are run by Murdoch and Fairfax. Murdoch-run titles account for nearly two-thirds (64.2%) of Australian metropolitan circulation and Fairfax-owned papers account for a further quarter (26.4%). In Queensland, for example, one of the largest Australian states, the only widely available newspapers are controlled by Murdoch. Information about climate change in the media is rather biased. This is particularly obvious in tabloids, such as the *Daily* and *Sunday Telegraph* (Murdoch), which are heavily biased in favor of the so-called climate change deniers. A check of the *Sunday Telegraph* over several months in early 2012 revealed that there was practically no or very little reporting on the views of recognized climate scientists, but that there were repeatedly lengthy articles by climate change "sceptics", including the Catholic Archbishop of Sydney, George Pell. Even in non-tabloid Fairfax newspapers like the *Sydney Morning Herald*, as much emphasis or more is given to the views of climate skeptics as to those of climate scientists. For example, the *SMH* on May 19–20, 2012, published a full-page report on present views on climate change, in which (beside two articles by a local councillor and a climate commissioner) as much space was given to a climate scientist as to the skeptic Des Moore, director of the Institute for Private Enterprise and former deputy secretary, Treasury, who had the following to say: "while sea levels have been steadily increasing, the rate of increase is clearly not threatening ... models ... have been wide of the mark …. Examples of scepticism include the signatures of 30 000 US scientists for a petition that specifically rejects the theory of dangerous warming ...". Des Moore has not provided any evidence for his views, including the source for the ridiculously high number of "scientists" to whom he refers.

At present there are attempts by groups close to the mining industry to gain a greater influence over private media. Gina Rinehart has just bought 10% of an important TV channel, which gave her a seat on the board; she has (mid-2012) increased her stakes in Fairfax to over 13% and is now the largest shareholder. According to press reports (*SMH*, May 29, 2012) she is "believed to want two board positions" of Fairfax. Another Australian mining magnate, Clive Palmer, has publicly announced plans to invest heavily in Fairfax.

Mining magnates and industry lobby groups can only succeed, however, if they have the support of some scientists. One should not dismiss out of hand the opinion of scientists who hold dissenting views. For example, Professor Murry Salby, who holds the climate chair at Macquarie University Sydney, in a recent (early 2012) talk to the Sydney Institute, an Australian right-wing "think tank", claimed that rises in temperature precede rises in CO_2, and not vice versa, and that human emissions of CO_2 are quite insignificant compared with emissions due to natural causes, which represent 96% of all emissions; recent rises in temperature and with it rises in CO_2 are due to the "recovery" from the Little Ice Age, which was not man-induced. His view, however, is contradicted by evaluations of most climate scientists and by recent findings of an international team published in *Nature* (Shakun *et al.*, 2012). The Australian Professor of Mining Geology Ian Plimer (http://en.wikipedia.org/wiki/Ian_Plimer; retrieved May 23, 2012) is a prominent spokesman for the climate change skeptics with great public exposure, opposing effective government action on climate change, and a mining tax. For him, human-induced climate change is a "myth".

According to Malcolm Turnbull, a prominent Liberal, the leader of the Australian Liberal party (one of the two large parties), Tony Abbott, has publicly declared that man-induced climate change is "crap" (ABC, 2009). Also according to Turnbull, Nick Minchin, the former leader of the Liberals in the Senate, has declared that all the fuss about global warming really is nothing but a left-wing conspiracy (ABC, 2009). Lefties who lost their cherished Communist cause now need a new one, and they found it: global warming. However, some prominent Liberals including Malcolm Turnbull oppose this view. In the United States, the atmospheric physicist Fred Singer, who had earlier called evidence for the cancer risks of passive smoking "junk science", is an outspoken critic of the view that global warming is man-made. He founded the Science & Environmental Policy Project, which works against measures to prevent global warming, financed by private sources, and has been reported to have cooperated with Exxon and other industry groups in discrediting research on climate change, accusing researchers of suppressing data, smearing opponents and misusing the peer-review process (see http://en.wikipedia. org/wiki/Fred_Singer; retrieved May 20, 2012).

The opinions of religious leaders on combating the effects of climate change and other harmful effects of humans should be influential to those with religious convictions. Indeed, Pope Benedict XVI, during his recent visit to Germany (late 2011), in a speech to the German parliament, the Bundestag, singled out the "ecological" movement ("Ökobewegung") for particular praise, although he did not refer to climate change. However, the Catholic Archbishop of Sydney and Primus of Australia Cardinal George Pell had the following to say in one of Rupert Murdoch's tabloids (*The Sunday Telegraph*): "some of the hysteric and extreme claims about global warming are also a symptom of pagan emptiness . . . belief in a benign God who is master of the universe has a steadying psychological effect …. In the past pagans sacrificed animals and even humans to placate capricious and cruel gods. Today they demand reduction in carbon dioxide emissions". In a later edition of the same tabloid, he was quoted as having said that climate change skeptics had won hands down.

A lecture tour by the climate change skeptic Lord Monckton, sponsored by Gina Rinehart, began at the catholic Notre Dame University of Western Australia.

Among other denominations, views on climate change also differ widely. Some object to any actions on climate change ("Man cannot destroy the Earth, only God can") whereas others strongly support it (God has given us the responsibility to save the Earth) (*Sydney Morning Herald* 2012 on the ARRCC: "Australian Religious Response to Climate Change").

Numerous reports in the Australian press dealt with the so-called "climate-gate" scandal, i.e., the leaked emails from the University of East Anglia just before the Copenhagen conference on climate change. Several thorough investigations have now found that there was no scandal, the reputation of the scientists involved is intact (see http://en.wikipedia.org/wiki/Climatic_Research_Unit_email_controversy; retrieved May 17, 2012). However, reports in the Australian press on these investigations and their outcome, clearing the scientists, have been minimal.

It is important that scientists take a stand against misinformation appearing in the media, which – however – does not mean that dissenting views by climate scientists should be dismissed just because they are held by a minority of scientists: such views should be discussed at a scientific level and not be politicized. The stand against misinformation should be taken at all levels, in the widely available printed press, on television and radio, as well as at the community level such as in local newspapers. The public has a right to be truthfully informed. Cutting-edge research discussed in the chapters of this book is meant to provide such information.

As pointed out above, climate change is just one of the dangers facing nature and humankind; others include over-exploitation of resources caused by population growth and over-consumption, pollution and invasive species, and habitat destruction such as deforestation. With regard to climate change, we conclude with a quote from an Amazon "description" of the important book by the American physicist and climate expert Joseph Romm (2006), which suggests that everything is not yet lost. There is hope if we do something now. Because of the prominent role of the USA as an economic and scientific power, a leading role will have to be played by that country, but countries such as Australia with its enormous coal and gas reserves must play their role as well.

"Global warming is the story of the twenty-first century. It is the most serious issue facing the future of humankind, but American energy and environmental policy is driving the whole world down a path toward global catastrophe. According to Joseph Romm, we have ten years, at most, to start making sharp cuts to our greenhouse gas emissions, or we will face disastrous consequences. The good news, he writes, is that there is something we can do – but only if the leadership of the U.S. government acts immediately and asserts its influence on the rest of the world."

References

ABC (2009). Turnbull ups the white-ante–ABC News (Australian Broadcasting Corporation). Available at: www.abc.net.au. Retrieved July 12, 2009.

Boyce, D. G., Lewis, M. R., & Worm, B. (2010). Global phytoplankton decline over the past century. *Nature*, **466**, 591–595.

Devictor, V., van Swaay, C., Brereton, T., *et al.* (2012). Differences in the climatic debts of birds and butterflies at a continental scale. *Nature Climate Change*, **2**, 121–124

Erspamer, V. (1995). Bioactive secretions of the amphibian integument. In H. Heatwole, G. T. Barthalmus & A. Y. Heatwole (Eds.), *Amphibian Biology*. Vol. 1. *The Integument* (Ch. 9, pp. 178–350). Chipping Norton: Surrey Beatty & Sons.

Gleick, P. (2012). *Remarkable Editorial Bias on Climate Science at the Wall Street Journal.* Available at: http://www.forbes.com/sites/petergleick/2012/01/27/remarkable-editorial-bias-on-climate-science-at-the-wall-street-journal/. Retrieved February 27, 2012.

Global invasive species database. *100 of the World's Worst Invasive Alien Species*. Available at: http://www.issg.org/database/species/search.asp?st=100ss. Retrieved April 15, 2012.

Han, W., Meehl, G. A., Rajagopalan, B., *et al.* (2010). Patterns of Indian Ocean sea-level change in a warming climate. *Nature Geoscience*, **3**, 546–550.

Henry, M., Béguin, M., Requier, F., *et al.* (2012). Common pesticide decreases foraging success and survival in honey bees. *Science*. doi: 10.1126/science.1215039.

Lackey, R. T. (2004). Restoration ecology: the challenge of social values and expectations. *Frontiers in Ecology*, **2**, 45–46.

Mäntylä, E., Klemola, T., & Laaksonen, T. (2011). Birds help plants: a meta-analysis of top-down trophic cascades caused by avian predators. *Oecologia*, **165**, 143–151.

Parr, C. L., Gray, E. F., & Bond, W. J. (2012). Cascading biodiversity and functional consequences of a global change-induced biome switch. *Diversity and Distributions*, **18**, 493–503.

Rohde, K. (2010). *Arthur Schopenhauer: Ethics and Theory of Justice*. http://krohde.wordpress.com/article/arthur-schopenhauer-ethics-and-theory-xk923bc3gp4-106/. Retrieved January 1, 2012.

Romm, J. N. (2006). *Hell and High Water: Global Warming–the Solution and the Politics–and What We Should Do*. New York: William Morrow.

Sale, P. (2011). *Our Dying Planet. An Ecologist's View of the Crisis We Face*. Berkeley, CA: University of California Press.

Sandiford, M. (2011). The scale of the effect we have on the planet is yet to sink in. *Sydney Morning Herald*, 23 May. Available at: http://www.smh.com.au/opinion/society-and-culture/the-scale-of-the-effect-we-have-on-the-planet-is-yet-to-sink-in-20110522-1eyqk.html . Retrieved March 5, 2012.

Science Daily (2012). Extreme Weather of Last Decade Part of Larger Pattern Linked to Global Warming. Available at: http://www.sciencedaily.com/releases/2012/03/120325173206.htm. Retrieved March 25, 2012.

Shakun, J. D., Clark, P. U., He, F., *et al.* (2012). Global warming preceded by increasing carbon dioxide concentrations during the last deglaciation. *Nature*, **484**, 49–54.

SixWise.com. *Pharmaceutical Pollution: What it is, and How Pharmaceutical Pollution Threatens Your Health*. Available at: http://www.sixwise.com/newsletters/06/02/16/pharmaceutical_pollution_what_it_is_and_how_pharmaceutical_pollution_threatens_your_health.htm. Retrieved April 15, 2012.

Smith, D. (2011). Academics fear climate change hate mail might deter future researchers. *The Sydney Morning Herald*, June 11–12.

Sommer, J. H., Kreft, H., Kierl, H. J., *et al.* (2010). Projected impacts of climate change on regional capacities for global plant species richness. *Proceedings of the Royal Society of London B*. doi: 10.1098/rspb.2010.0120. Available at: http://rspb.royalsocietypublishing.org/content/early/2010/03/20/rspb.2010.0120.abstract. Retrieved April 15, 2012.

Sydney Morning Herald (2012). Religious leaders back mining tax. Available at: http://www.smh.com.au/environment/climate-change/religious-leaders-back-carbon-tax-20110602-1fie4.html. Retrieved April 19, 2012.

The Wall Street Journal. (2012). No need to panic about global warming. January 26. Available at: http://online.wsj.com/article/SB10001424052970204301404577171531838421366.html. Retrieved March 3, 2012.

Trenberth, K. (2012). Check with climate scientists for views on climate. *The Guardian*, February 1. Available at: http://online.wsj.com/article/SB10001424052970204740904577193270727472662.html. Retrieved February 19, 2012.

Wallington, T. J.,. Hobbs, R. J., & Moore, S. A. (2005). Implications of current ecological thinking for biodiversity conservation: a review of salient issues. *Ecology and Society*, **10**, 15. Available at: http://www.google.com/url?sa=t&rct=j&q=&esrc=s&source=web&cd=5&ved=0CEwQFjAE&url=http%3A%2F%2Fresearchrepository.murdoch.edu.au%2F1743%2F1%2FEcological_Thinking_for_Biodiversity.pdf&ei=YkeST9J8x7mJB__c8fED&usg=AFQjCNE8_QWNEBEK76_WUESSuAV4wuPvCg. Retrieved April 21, 2012.

Whitehorn, P. R., O'Connor, S., Wackers, F. L., & Goulson, D. (2012). Neonicotinoid pesticide reduces bumble bee colony growth and queen production. *Science*, **336**, 351–352. doi:10.1126/science.1215025.

Wigley, B. J., Bond, W. J., & Hoffman, M. T. (2010). Thicket expansion in a South African savanna under divergent land use: local vs. global drivers. *Global Change Biology*, **16**, 964–976.

Wikipedia (2010). Russian wildfires. Available at: http://en.wikipedia.org/wiki/2010_Russian_wildfires. Retrieved March 3, 2012.

Zimmerer, K. S. (2000). The reworking of conservation geographies: nonequilibrium landscapes and nature-society hybrids. *Annals of the Association of American Geographers*, **90**, 356–369.

Index